Pathways to a Smarter Power System

Pathways to a Smarter Power System

Edited by

Akın Taşcıkaraoğlu

Ozan Erdinç

ELSEVIER

ACADEMIC PRESS

An imprint of Elsevier

Academic Press is an imprint of Elsevier
125 London Wall, London EC2Y 5AS, United Kingdom
525 B Street, Suite 1650, San Diego, CA 92101, United States
50 Hampshire Street, 5th Floor, Cambridge, MA 02139, United States
The Boulevard, Langford Lane, Kidlington, Oxford OX5 1GB, United Kingdom

Library of Congress Cataloging-in-Publication Data
A catalog record for this book is available from the Library of Congress

British Library Cataloguing-in-Publication Data
A catalogue record for this book is available from the British Library

ISBN 978-0-08-102592-5

For information on all Academic Press publications
visit our website at https://www.elsevier.com/books-and-journals

Publisher: Matthew Deans
Acquisition Editor: Lisa Reading
Editorial Project Manager: Hilary Carr
Production Project Manager: Kamesh Ramajogi
Cover Designer: Victoria Pearson

Typeset by SPi Global, India

Dedication

I would like to dedicate this book to my lovely wife, Fatma Yıldız Taşcıkaraoğlu, our children, Selim and Melisa, and also my parents.

Akın Taşcıkaraoğlu

I dedicate this book to my beloved wife F. Gülşen Erdinç, and my little angels Mahmut Pamir (my son) and Pera (my daughter). I also convey my best wishes and my special thanks as always to my family, who have raised me with full love and support throughout my life.

Ozan Erdinç

Contents

Contributors

T. Adefarati (29), Department of Electrical, Electronic and Computer Engineering, University of Pretoria, Pretoria, South Africa; Department of Electrical and Computer Engineering, University of Sharjah, Sharjah, United Arab Emirates

M. Hadi Amini (289), Carnegie Mellon University, Pittsburgh, United States

R.C. Bansal (29), Department of Electrical, Electronic and Computer Engineering, University of Pretoria, Pretoria, South Africa; Department of Electrical and Computer Engineering, University of Sharjah, Sharjah, United Arab Emirates

Stefan Böcker (343), TU Dortmund University, Dortmund, Germany

João P.S. Catalão (183), C-MAST, University of Beira Interior; INESC TEC and the Faculty of Engineering of the University of Porto, Porto; INESC-ID, Instituto Superior Técnico, University of Lisbon, Lisbon, Portugal

Ambrish Chandra (67,225), École de Technologie Supérieure (ETS), Montréal, QC, Canada

Archie C. Chapman (115), The University of Sydney, School of Electrical and Information Engineering, Sydney, NSW, Australia

Wai-Hei Choi (247), University of Macau, Macau, China

Peter Connor (397), University of Exeter, Penryn, United Kingdom

Juan Manuel Corchado (183), BISITE Research Group, University of Salamanca, Salamanca, Spain; Osaka Institute of Technology, Osaka, Japan

Ashok Kumar Das (371), Center for Security, Theory and Algorithmic Research, International Institute of Information Technology, Hyderabad, India

Payman Dehghanian (317), The George Washington University, Washington, DC, United States

Vasco Delgado-Gomes (89), Faculty of Sciences and Technology, Universidade NOVA de Lisboa, CTS-UNINOVA, Lisbon, Portugal

Nils Dorsch (343), TU Dortmund University, Dortmund, Germany

Ozan Erdinç (1), Yildiz Technical University, Istanbul, Turkey

Ayşe Kübra Erenoğlu (1), Yildiz Technical University, Istanbul, Turkey

Oscar Fitch-Roy (397), University of Exeter, Penryn, United Kingdom

Amin Shokri Gazafroudi (183), BISITE Research Group, University of Salamanca, Salamanca, Spain

Mazen Ghandour (67,225), Faculty of Engineering, Lebanese University, Beirut, Lebanon

Abdelhamid Hamadi (225), École de Technologie Supérieure (ETS), Montréal, QC, Canada

Guannan He (289), Carnegie Mellon University, Pittsburgh, United States

Hussein Ibrahim (67,225), Institut Technologique de Maintenance Industrielle (ITMI), Cégep de Sept-Îles, Sept-Îles, QC, Canada

Soummya Kar (289), Carnegie Mellon University, Pittsburgh, United States

Irfan Ahmad Khan (289), Carnegie Mellon University, Pittsburgh, United States

Chi-Seng Lam (247), University of Macau, Macau, China

Sara Lumbreras (205), Universidad Pontificia Comillas, Madrid, Spain

João F. Martins (89), Faculty of Sciences and Technology, Universidade NOVA de Lisboa, CTS-UNINOVA, Lisbon, Portugal

Sleiman Mhanna (115), The University of Sydney, School of Electrical and Information Engineering, Sydney, NSW, Australia

Javad Mohammadi (289), Carnegie Mellon University, Pittsburgh, United States

Panayiotis Moutis (289), Carnegie Mellon University, Pittsburgh, United States

Francisco Prieto-Castrillo (183), BISITE Research Group, University of Salamanca, Salamanca, Spain; Media Laboratory, Massachusetts Institute of Technology, Cambridge; Harvard T.H. Chan School of Public Health, Harvard University, Boston, MA, United States

Anabela Gonçalves Pronto (89), Faculty of Sciences and Technology, Universidade NOVA de Lisboa, CTS-UNINOVA, Lisbon, Portugal

Andres Ramos (205), Universidad Pontificia Comillas, Madrid, Spain

Miloud Rezkallah (67,225), Institut Technologique de Maintenance Industrielle (ITMI), Cégep de Sept-Îles, Sept-Îles; École de Technologie Supérieure (ETS), Montréal, QC, Canada

Mihai Sanduleac (89), Faculty of Power Engineering, Polytechnic University of Bucharest, Bucharest, Romania

Miadreza Shafie-khah (183), C-MAST, University of Beira Interior, Portugal

Saber Talari (183), C-MAST, University of Beira Interior, Portugal

Akın Taşcıkaraoğlu (1), Mugla Sitki Kocman University, Mugla, Turkey

Mohammad Tasdighi (317), S&C Electric, Chicago, IL, United States

Gregor Verbič (115), The University of Sydney, School of Electrical and Information Engineering, Sydney, NSW, Australia

Lei Wang (247), University of Macau, Macau, China

Bo Wang (317), The George Washington University, Washington, DC, United States

Jay Whitacre (289), Carnegie Mellon University, Pittsburgh, United States

Christian Wietfeld (343), TU Dortmund University, Dortmund, Germany

Man-Chung Wong (247), University of Macau, Macau, China

Sherali Zeadally (371), College of Communication and Information, University of Kentucky, Lexington, KY, United States

Preface

The smart grid is an area of research drawing increasing interest, especially in the last 10 years. The word "smart" has been used for variety of reasons in different areas, but the most vital reason is the "advertisement" of something (such as a product). However, how can a product, technique, or concept be smart? On the other hand, What is the level of this smartness? As there is no index definitely defining the level of smartness, we generally check how much the decision can be realized by "itself."

From an electric power systems point of view, the term "smart" refers to the modernization of electric power system planning and operation with the aid of improved information and communication technologies (ICT) penetration. However, as the change in ICT industry is rapid, adopting the possibly newest solutions to the electric power systems will always have a time delay. Besides, in order to describe an electric power system as "smart," ICT developments should end somewhere and the implementations in electric power system should then reach the possibly highest point in ICT evolvement. Thus, instead of making the electric power system smart, we indeed make it smarter at each step compared to the mature concept. Thus, this book considers the pathways to such a "smarter" power system and includes the recent developments in different areas of research conveyed by leading and well-known authors living and working in different regions of the world.

We hope that this book will be useful for existing and new researchers in the area, as well as students and industrial partners, to make the process of changing to a smarter power system more rapid and effective.

The Editors

Chapter 1

History of Electricity

Ayşe Kübra Erenoğlu*, Ozan Erdinç* and Akın Taşcıkaraoğlu†
*Yildiz Technical University, Istanbul, Turkey, †Mugla Sitki Kocman University, Mugla, Turkey

Chapter Outline

1.1 Introduction

Electricity is one of the most awesome phenomena and nature's greatest force that has undoubtedly changed everything completely in the world. It has attracted mankind's attention for centuries, and a high number of spectacular scientific experiments has been conducted to reveal its mysterious and secret nature. However, in order to understand the basic principles of electricity, it is necessary to figure out the character/construction of atoms as the smallest building blocks of matter. An atom consists of a central nucleus, which is made up of positively charged and uncharged particles, namely protons and neutrons, respectively. Additionally, electrons as negatively charged particles orbit around the nucleus as shown in Fig. 1.1, and the cumulative flow of these free electrons results in electric energy.

Although many people may believe that electricity was discovered by a scientist, there is no indication of this in the rich history of electricity. In other words, it is not possible to point out when and by whom electricity was discovered, or rather "invented." Instead, we know that numerous scientists have conducted a considerable amount of experiments to control electricity. The

Pathways to a Smarter Power System. https://doi.org/10.1016/B978-0-08-102592-5.00001-6

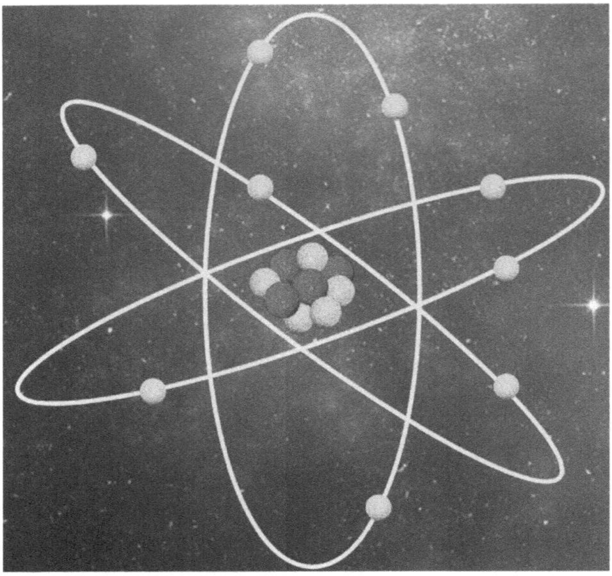

FIG. 1.1 Illustration of an atom molecule.

greatest challenge for scientists was to find a way of transforming electricity, which is like a natural resource, and to make it usable for mankind.

1.2 The Chronological Development of Electricity

This chronological investigation is intended to provide the highlights of electricity and magnetism history, considering scientific developments, experiments, and breakthroughs. It can be deduced from texts and reports dating from 2750 BCE that the origins of electricity extend back to antiquity (prehistory), through the ancient Egyptians observing shocks coming from electric fishes (electric catfish, electric eels, etc.). They called these fish the "Thunderers of the Nile," and described them as the "protectors" of all other fish. In addition, these shockwaves were used to treat patients' illnesses such as headache and gout, which was reported in later times by ancient Greek, Roman and Arabic physicians and ancient writers [1]. Electrical phenomena were also studied by Thales of Miletus, the first scientist to recognize the electrical properties of amber, in around 600 BCE. He observed the static electricity in the nature with his experiment. According to this experiment, when amber was rubbed with clothes, this enabled it to attract light objects such as feathers and straw, as illustrated in Fig. 1.2. Moreover, Thales carried out some experiments on lodestone and was aware of its magnetic properties that could attract iron, as shown in Fig. 1.3. His findings provided a significant contribution in the understanding of electricity for many hundreds of years together with his series of observations [1].

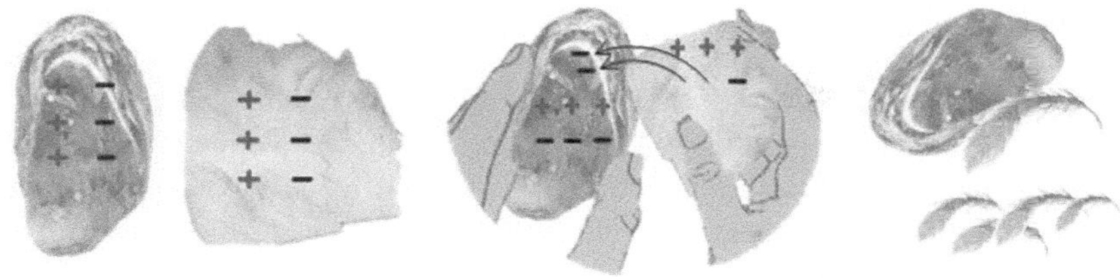

FIG. 1.2 Static electricity experiment with amber and cloth.

FIG. 1.3 Naturally magnetized mineral of loadstone.

1.2.1 Scientific Advancement on Electricity in the 17th Century

The 17th century was the most crucial era in the history of electricity. William Gilbert, often credited as the father of electricity, published in 1600 his famous book *De Magnete, Magneticisque Corporibus, et de Magno Magnete Tellure*, which comprised his first scientific and systematic research and previous literature about electricity and magnetism and their natural manifestations [2]. He was the first scientist to use the terms "electric attraction," "electric force," and "magnetic pole" [3]. Thus, this book marked a crucial turning point and the end of the ideas and feelings era, and the beginning of the experiments, theories, and inventions era. Therefore, Gilbert was apparently accepted as the pioneer and founder of the experimental method. In addition, he coined the Latin word *electricus* from the Greek term for amber: *electron* (λεκτρον) [4]. Otto von Guericke and Robert Boyle, who were educated and inspired by William Gilbert, contributed to expanding knowledge about electricity with their investigations. The presence of a vacuum was proved by Guericke and he created a partial vacuum in 1650, which was a significant phenomenon particularly for further electronic research. In 1660, he went on to invent the electrostatic generator, which soon began to be used broadly and became a fundamental instrument of scientific experiments [5]. Robert Boyle, as a contemporary of Guericke, performed experiments and showed that electric forces could be transmitted through a vacuum after the second half of the 17th century [6].

1.2.2 Scientific Advancement on Electricity in the 18th Century

In the 18th century, which covered part of the Age of Enlightenment, it can be seen that studies on this subject were limited to static electricity. Some of the most prominent scientists of that age were natural philosopher Francis Hauksbee, physicist Pieter Van Musschenbroek, and scientist Benjamin Franklin, who worked to expand knowledge of the electric technology by increasing the number of experiments in the scientifically productive environment.

In the first decade of the 18th century, there was a significant occasion in the Royal Society and one of the well-known demonstrations of proving existence of electricity, shown with a striking experiment by Francis Hauksbee. He built a static electricity generator, which includes a glass globe that could be accelerated and spun rapidly, allowing electric charges to be produced and as a result sparkling light inside the globe, by rubbing the glass globe with a hand. It had a spectacular effect on other scientists and they were stunned after this experimentation. Subsequently, Hauksbee's papers were published in the *Society's Philosophical Transactions* and also *Physico-Mechanical Experiments on Various Subjects* consisting of his findings published in 1709, which was widely read in the 18th century [7]. His studies paved the way for other scientists to conduct widely known inventions at that time, namely Charles Francois Du Fay, Pieter Van Musschenbroek, and Benjamin Franklin [8]. In 1733, Charles Francois Du Fay carried out experiments to clarify inexplicable phenomena related to electricity, and he observed two distinct types of electric charges, which he called resinous (−) and vitreous (+), known as positive and negative today [9]. After a series of experiments and observations, there was a great achievement so that electricity could be produced with a generator, and scientists began trying to understand its nature and properties. There was a great number of engrossing questions—for example, if electricity (thought to be a kind of mysterious fluid) flows like water, is it possible to store it like water in the insulator glass jar? Soon after, in 1746, an earth-shaking invention came from Dutch mathematician and physicist Pieter Van Musschenbroek, when he developed the first energy storage device: the Leyden Jar. This is one of the most ubiquitous electronic components today, known now as a capacitor, though it was discovered accidentally. It was presented as a solution for the fundamental challenge of storing electric charges. In Musschenbroek's experimentation, the Leyden Jar was filled up with water and there were connections between Hauksbee's generator with a conductive (metal) rod inside.

Musschenbroek always placed the jar on an insulator, and 1 day he forgot the insulator and held the jar in his hand. When he touched the top point, all electricity discharged on his body, since these charges were conserved inside the jar. Thus, he established the fundamental principle of the Leyden Jar with his assistant and managed to accumulate electricity in the glass bottle for hours, and even days. This had a profound impact on the worldwide scientific community and the jar became an extraordinarily globalized scientific item, gaining interest from his contemporary researchers [10]. However, no one understood the underlying philosophy behind the Leyden Jar and could not answer the questions of why or how it worked.

One scientist, who was politically and philosophically at war with the Royal Society, reasoned out all possibilities in trying to come up with an answer. His name was Benjamin Franklin. He dedicated himself to conducting his theories practically and he carried out various analyses systematically and successfully in order to aid future science research as an enlightened thinker. On June 15, 1752, Franklin conducted his most significant kite experiment, using one of the simplest yet most brilliant and dangerous methods for supporting his treatise that lightning was an electrical phenomenon. The kite's string was wet hemp, which was capable of conducting electricity from the lightning in the thunderstorm down to the iron key that hung on the end of the string. Franklin also used dry silk ribbon instead of metal wire to prevent the lightning from discharging onto his body or even touching his hand. He witnessed a spark jump from the key to his knuckle, and he thus identified that lightning was certainly electrical in nature. After he represented a rational explanation of his theory, this accomplishment opened a different door and he discovered that lightning rods could be placed on the tops of buildings for transferring charges to the ground by using metal wire without any harm, as illustrated in Fig. 1.4. Moreover, he coined new English electrical terms in the literature: *conductor*, condenser, charge, discharge, electrify, etc. [11]. Over 200 years after his discoveries, Franklin's insight is still valid and his contributions should not be ignored in the history of electricity.

Another scientist of the 18th century, the chemist and natural philosopher Henry Cavendish, was curious about how electric fishes could emit powerful shocks to their environment without a spark, and he carried out his experiments based on this idea. Firstly, he created an artificial "Torpedo Fish" by using leather, wood soaked in salt water, and tin (for imitating the organs of fish) in order to carry out his investigations and examine the source of emitting power intensity. After his experiments were successful, he concluded that the organs of the torpedo were similar to Leyden Jars and they store electricity as a battery, but in a small quantity. When the number of organs increased, it was possible to obtain a greater potential (he called this the degree of electrification) which is similar to connect batteries serially in circuit and, as a result, increase the voltage. In 1776, his findings and observations about electric potential and the relationship between potential and current were published [12]. There was evidence

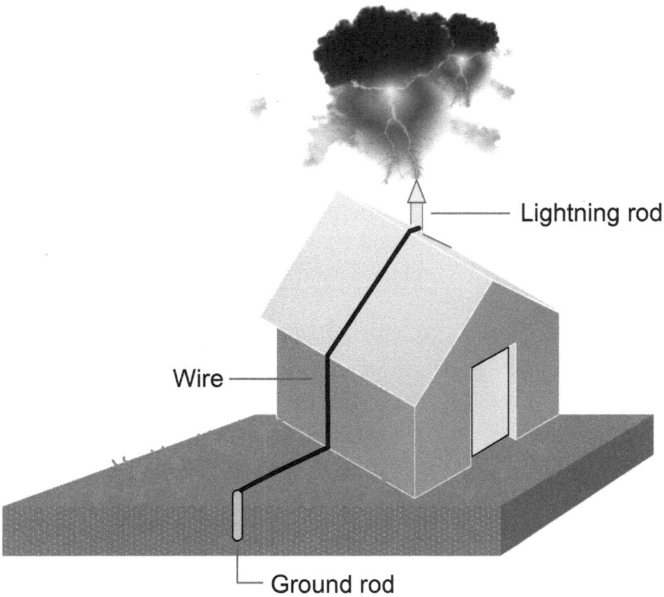

FIG. 1.4 Lightning rod.

that Cavendish's experiments played a crucial role in addressing the issue of "animal electricity" and paved the way for one of the notable scientists, Luigi Eliseo Galvani.

In the second half of the 18th century, Italian physicist and physician Luigi Eliseo Galvani investigated the nature of electricity in animal tissue and made extraordinary experiments on dead frog bodies that laid the foundations of the modern field of bioelectricity. He used either Hauksbee's electricity generator or lightning as a source of power, and attached the copper wire to the nerves of the frog and iron wire to the muscle of the frog. It was an astonishing moment when the frog's leg responded and twitched as if it were alive when electricity was applied. After his experiments, Galvani concluded that there was a third form of electricity in the animal tissue coming from its own innate vital force, called "animal electricity," for the first time in the 18th century, and he published his book *De Animali Electricitate* in October 1786 [13]. He also added that this electricity produced by the contact of dissimilar metals in a moist environment was not originated only by animal innate force. Thus, it is evident that he was a pioneer of the electrophysiology and neuroscience, and provided outstanding contributions to today's electrical basis of nerve impulses. His findings not only became the inspiration of novelist Mary Shelley when writing *Frankenstein*, especially its eponymous character, but also opened the way for the electric battery's invention by physicist Alessandro Volta, who stood exactly on the opposite side of Galvani. That is, he stated that it was not possible to produce electricity from the frog's leg, but that this was caused by using different types of metals. In the history of electricity, Volta's experiments have been thought of as a key factor in opening a new era for the scientific community.

1.2.3 Scientific Advancement on Electricity in the 19th Century

The 19th century saw the Industrial Revolution; major changes and outstanding developments took place in the fields of science, technology, communication and information when never came true before in history. Accelerated technological innovations resulted in a scientific revolution that affected society from top to bottom, and earth-shattering inventions were demonstrated by scientists. Notable experimentalists of this era such as Michael Faraday, Thomas Alva Edison, and Nikola Tesla quietly pursued their passion in order to achieve advances in power technology which is still developing in our modern world. During this time, one of the most relatively sophisticated breakthroughs of its time occurred.

At the beginning of the 19th century, Italian physicist Alessandro Volta was interested in electricity like many scholars. He was already a prominent scientist of the time, but he contributed arguably the greatest invention to the world, which effectively made him immortal. It is well known that Volta and his scientific colleague Galvani had debated on their research/ideas and thus this could be accepted as one of the most interesting episode in the history of electricity. Volta was not convinced by the concept of animal electricity; as already noted, he asserted that it was not possible for the frog's leg to produce electricity, and this was only caused by connecting dissimilar metals. The twitching frog's leg was just an indicator of the existence of the electricity. This disagreement opened doors to a broad exploration and he carried out numerous observations and experiments in order to prove his theory. The voltaic pile, the predecessor of the modern battery, was invented and this could produce continuous and sustainable electricity after this extensive research. The first electric pile included of a series of zinc and silver discs, divided by a piece of cardboard which was soaked in salt water or acid, and it ended bottom to top with wires (see Fig. 1.5). This pile was capable of creating a small amount of electric current, but Volta expanded the number of connected piles in order to increase the intensity of electric shock that would be used in his experiments [14]. The voltaic pile's fame spread rapidly and became a widely known scientific component. The London Royal Society published Volta's findings and demonstrations for the first time in 1800. Today, the potential difference unit of "volt" has been designated with his name for his remarkable scientific discovery and his honor. Further work was conducted by British electrochemist and electrician Humphry Davy, who built a huge battery consisting of 800 voltaic piles underneath the Royal Institution, for utilizing its power in electrolysis experiments. This battery, the largest in the world, was used in electrochemical decompositions and enabled him to obtain potassium, sodium, barium, strontium, calcium, and

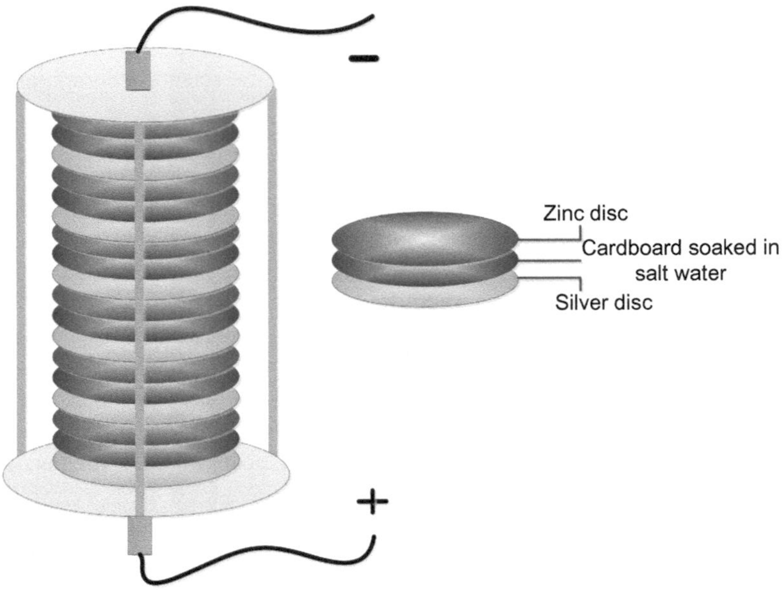

FIG. 1.5 The first voltaic pile.

magnesium for the first time from 1807 to 1808. In addition, Davy obtained blindingly bright light using this high voltage battery's output between two carbon filaments and developed one of the most important inventions of arc light in 1809, as illustrated in Fig. 1.6. Therefore, it should be noted that electric lightning was known almost 8 years before Thomas Alva Edison made a particular scientific discovery, but it was not in a usable form for home illumination. Furthermore, Davy developed a system for protecting miners from methane gas in the coal industry; he presented his device, the Davy lamp, in 1815 [15].

The discovery of electromagnetism by chemist Hans Cristian Øersted triggered a huge amount of innovative scientific inquiries and unleashed a revolution in modern science and industrial process. In fact, until Øersted's first

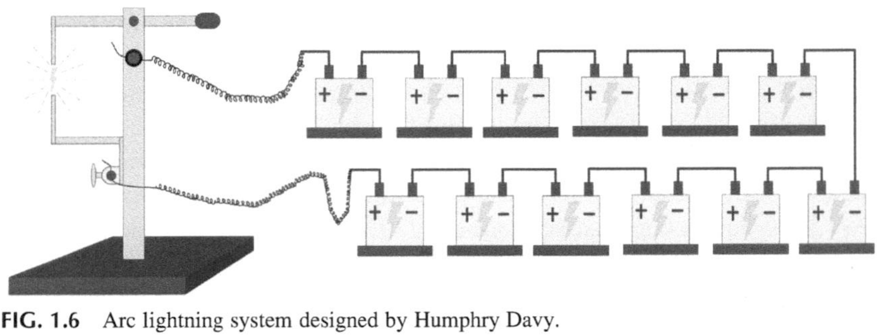

FIG. 1.6 Arc lightning system designed by Humphry Davy.

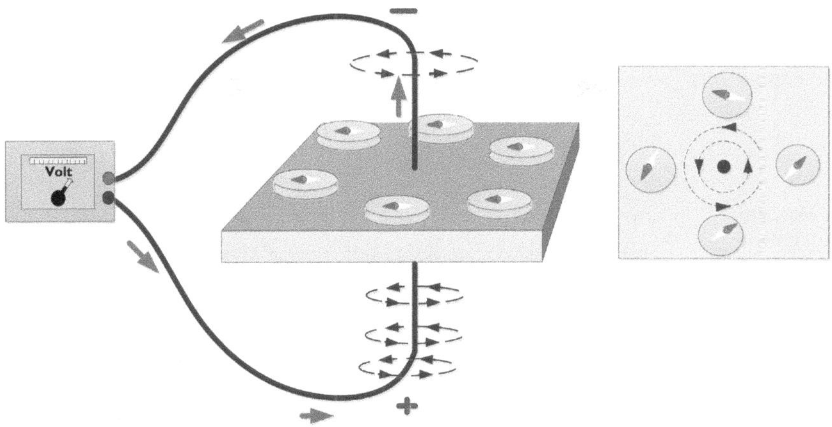

FIG. 1.7 Øersted's experiment.

scientific explanations, what we know about electromagnetism was largely based on observations by the ancient Greeks. The first serious and scientific analyses of electromagnetism emerged in 1820 when Øersted was in lecture. When he conducted the experiment by using a voltaic pile for his university students, he noticed that changing circuit conditions such as switching on/off caused the nearby compass needle to move as illustrated in Fig. 1.7. Therefore, it was evident that current-carrying wire created a magnetic field around it, and there was a strong and consistent association between electricity and magnetism according to the reported results [16]. However, the major contributions to the electromagnetism came from self-educated Michael Faraday (1791–1867), the assistant of Humphry Davy; Faraday discovered and proved the electromagnetic-induction which is the fundamental working principle of the instruments currently used in modern technology. In fact, electricity had been thought of as a kind of mysterious fluid by some scientists and a great deal of previous research into electricity had focused on two different fluids, such as positive and negative, for many years. Unlike other electricity-focused scientists, Faraday introduced electricity as a force for the first time in its history and provided an exciting opportunity to advance our knowledge of relationships between magnetism, electricity, and motion. In his first experiment on electromagnetism in 1821, he used a voltaic pile, mercury bath, and copper wire (see Fig. 1.8). This straightforward apparatus was actually the first electric motor and predecessor of today's sophisticated electrical machines. One filament of the battery was connected to the mercury bath, which is an excellent conductor, and the other one was attached to the stiff copper wire. This basic device was capable of converting electrical energy into mechanical energy thanks to the magnetic fields created from both the bar magnet and the current-carrying wire around it (Fig. 1.8). Thus, the interacting magnetic fields enabled the copper wire to complete its circuit and rotate clockwise. Faraday performed a similar

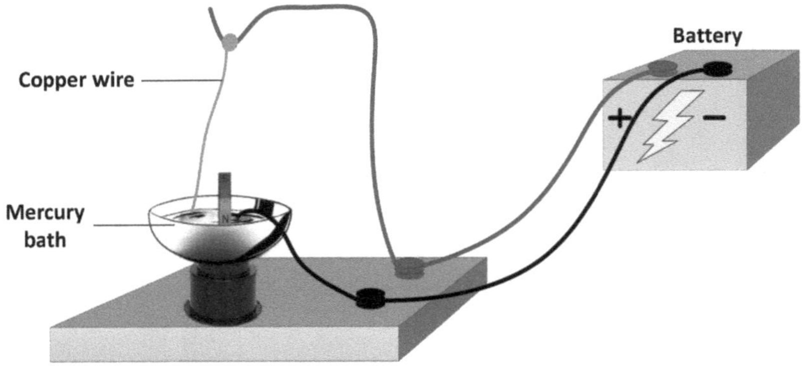

FIG. 1.8 Michael Faraday's first electric motor.

series of experiments and tried to find answers to his questions dealing with the possibility of reversing energy transformation, so both magnetic field and motion could produce an electric current in the wire. The second section of this study took place in the Royal Institute and an earth-shaking invention took shape in 1831 when Faraday discovered electro-magnetic induction, enabling him to obtain steady and continuous electric current. He moved a magnetic rod in and out of a cylindrical coil of copper wire, and a small amount of current was observed at the end of the wire (see Fig. 1.9). He then indicated that changing magnetic field enables the induction of electric current, which is one of the basic working principles of machines; he subsequently presented the first electric generator thanks to his dedicated research [17].

These findings became a worldwide phenomenon and could be evaluated as tremendously important scientific discoveries that also have profound implications

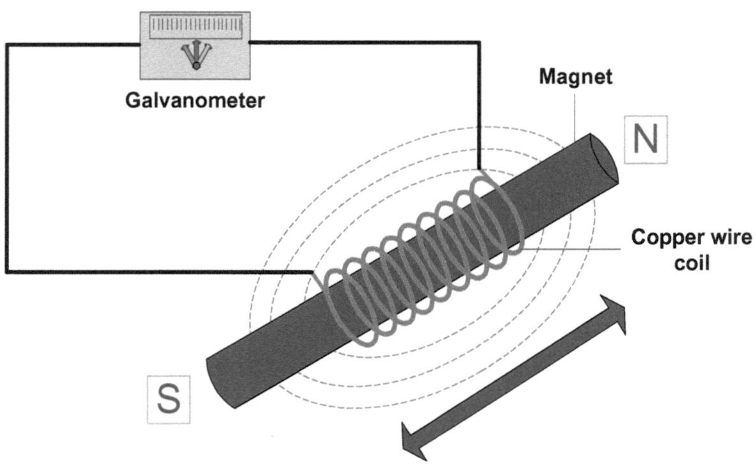

FIG. 1.9 Faraday's first electric generator.

in our modern social life. The discovery of electromagnetic induction was a crucial turning point that paved the way for utilizing electricity in a wide range of applications thanks to scientists' endeavors. Afterwards, the world began to change extraordinarily and one of the most major changes that first occurred in the field of communication was the breakthrough of the telegraph. In the 1830s, Samuel Morse (1791–1872) developed the electric telegraph with the aid of other scientists' contributions based on the principle of electromagnetism. The telegraph transferred the current pulses or ripples over a wire between two stations, in which one of them is the receiver and the other one is the transmitter. Long-distance communication became a difficult task due to sending complex messages from location to location with electric wires. To address this issue, Morse developed the alphabet and matched the dashes and dots with every English letter in order to provide practical transmission throughout a wire. The first telegraph message was conveyed from Washington, District of Columbia to Baltimore, Maryland in 1844 and it was expanded in the year 1866 from Atlantic Ocean to Europe [18]. Consequently, Samuel Morse launched a new epoch in communication with the invention of the telegraph, which was the forerunner of radio, telephone, internet, fax machines, etc.

The era of wireless telecommunication was based on James Clerck Maxwell's theories and equations; he in turn was inspired by Hans Cristian Ørsted, André-Marie Ampère, and especially Michael Faraday. Maxwell was one of the greatest scholars in the field of physics, and his groundbreaking theories, also known as Maxwell's equations, influenced his scientific colleagues widely. The mathematical form of the Faraday's physical ideas about electricity and magnetism was described by him in order to aid understanding of electromagnetic phenomena. In addition, he established a mechanical model aiming at illustrating Faraday's law of induction, and this experimentation enabled him to develop a new theory of electromagnetic movement, which was his greatest contribution to the knowledge of electrical science. He proposed for the first time that light is the form of electromagnetic wave, and there was a strong association between electricity, magnetism, and light. This unified model is also called Maxwell's synthesis. His two significant articles, *On Faraday's Lines of Force* and *On Physical Lines of Force*, were published in 1856 and 1861, respectively [19]. Eight years after Maxwell's death, Heinrich Hertz experimentally observed the presence of electromagnetic waves in the laboratory, which behaved as Maxwell had anticipated in 1887. In addition, Maxwell's revolutionary ideas laid the establishment for Einstein's special theory of relativity and quantum mechanics, thus having a great influence on 20th-century physicists. In 1901, there was a great achievement on wireless communication when Guglielmo Marconi successfully broadcast the first transatlantic radio signal from Britain to Canada with electromagnetic waves thanks to Maxwell's studies [20]. This type of wireless telegraphy was essential and could be regarded as a predecessor of today's long-distance radios.

Toward the end of the 19th century, the world witnessed the genuinely innovative scientific revolution in electrical science and social life. There was also an extraordinarily quickening rate in industrialization thanks to scientists and businesspeople. One of the most well-known electrical engineers and entrepreneurs, Thomas Alva Edison (1847–1931), made significant contributions to improving technology and addressed the biggest problem of that time: lighting. Although today Edison is known as an inventor of the light bulb, his first systematic researches, which was based on the phonograph and patented in 1877, led him to be a globally famous man at that time. He was also the founder of the "Wizard of Menlo Park" in New Jersey, which was the first industrial research laboratory, and he organized a team consisting of skilled engineers from different fields of science [21]. One of the greatest achievements of this team was researching incandescence and testing a large amount of organic fibers (6000) to decide which one was the best to use as a light bulb filament.

However, Edison was neither the first nor the only scientist aiming to bring electricity from the laboratory and make this system available for ordinary people instead of gaslight. Studies over the past eight decades have provided important information on lighting systems, which started with Sir Humphry Davy's arc lightning for street and railway stations, and progressed with contributions from other prominent scientists, such as Warren de la Rue and Joseph Wilson Swan. However, Edison made the first commercial, practical, and affordable incandescent lamp for home illumination in 1879, which was the major milestone of harnessing electricity, and he patented it the following year [22]. After a series of researches and explorations, carbonized bamboo was chosen as an alternative to platinum filament and one of the most ubiquitous components of light bulbs was used, being reliable, inexpensive, and long-lasting. By the time Edison was 84, he had an incredible 1093 patents for phonograph, electric light and power, telegraph, and the earliest motion picture cameras [23]. Moreover, he was also a competitive manufacturer and businessman who focused on marketing his patents and obtaining usable form for them in industrial situations. Therefore, America's first low-voltage central power plant, Pearl Street, was constructed for representing his invention to the public with the aim of meeting customers' energy demands on September 4, 1882 in New York City. At that time, the power required for electric lights or motor was normally supplied by on-site generating systems since it was not possible to mention presence of electric grid. Therefore, Pearl Street power station marked a turning point for shifting from small-scale producing units to centralized large industrial-scale power plants, which were the forerunner of our modern electric generation and distribution systems [24]. However, some important challenges remained to be handled, of which one was to produce enough power from the dynamometer to match demand. The dynamometer can transform mechanical energy to electrical energy, but there was no sufficiently powerful dynamometer at that time. Thus, Edison used a 27-ton machine in order to provide 100 kW power output and light up the street. According to his insight, it was the best way to deliver electricity to customers with underground cables considering telegraph lines and buildings. However, it would be difficult to convince

New York City's politicians about digging the ground and burying the cables. Furthermore, the other important point was to deliver electricity to consumers within secure/safe ranges requiring low voltage systems, and there was no way to change DC voltage levels from high to low at that time. Therefore, losses were increased in the lines and DC systems could not represent an effective solution to transmit electricity economically especially over long distances.

To address the mentioned issue, the solution came from one of the other giants of electrical engineering, the brilliant and eccentric genius Nikola Tesla, who changed the world's perspective completely and made outstanding contributions to the development of the alternating-current (AC) system, which constitutes the basis of our modern-day power system. However, this invention apparently caused one of the most exciting debates and skirmishes, widely known as the "War of Currents" between AC and DC, between Thomas Edison and Nikola Tesla, who was his former employee. In a DC system, electric charges can only flow in one direction by the driven force of voltage. On the other hand, the flow of electric charges reverses periodically in the AC concept proposed by Tesla, and it is possible to change the voltage level by using a transformer consisting of iron cores and copper windings. This was one of the most useful things ever discovered, allowing bulk power stations to be established out of town and enabling the output voltage levels of the generator to be boosted in order to transmit power with high voltages, and accordingly low losses. At the same time, it enabled delivery of energy to consumers with low voltages by taking into account safe operational limits. However, Edison spread misinformation about the AC system and claimed that this system was dangerous, and even electrocuted animals at public gatherings. Meanwhile, a hydroelectric power plant was installed in the United States using Tesla's significant invention of the polyphase AC induction motor in order to harness Niagara Falls' power production capability in 1896. Afterward, the city of Buffalo in New York was electrified by this power station, and not only did this this issue affect Buffalo, but electrification of the whole world started with this attempt [25].

As a consequence, it should be noted that the world has been driven by technological developments and revolutionary ideas, but the modern power grid is still predominantly powered by using AC and AC equipment. In addition, the distribution system is based entirely on the fundamental principles proposed by Tesla. Large power stations have become widespread all over the world, and output voltage is boosted while sending the transmission lines. At the end-point, the voltage is stepped down in order to protect consumers from dangerous situations. However, it is too early to claim that the "War of Currents" has finished and Tesla has won. Today, there are a great number of components powered by DC such as computers, light emitting diodes (LEDs), electric vehicle batteries, etc. In addition, high voltage DC (HVDC) systems represent a more economical solution at longer distances and have a profound impact on reducing total installation costs [26].

The aforementioned scientific discoveries and innovations have caused electricity to be available for everyone. The spectacular journey of electrification began with the arc lighting systems and proceeded with incandescent lighting systems revitalizing science, technology, and social life profoundly. Needless to say, power systems were mature, small-scale, and localized in the first stage of the development era. In addition, two distinct types of power generation plants based on AC and DC were used to serve energy to end-users [27]. Subsequently, a great number of larger power plants was constructed to meet the increased demand of consumers and enabled utility companies to make a profit by trading electricity. Therefore, electricity became widespread and propelled us into the modern electrified era. The timeline of multiple discoveries, studies, and developments from the 1900s to the current time is summarized in Fig. 1.10.

In Pearl Street Station, powered by DC, there were some restrictions caused by high losses due to low voltage level and the short delivery distance of produced energy. However, the invention of the transformer and induction motor brought different insights to the company owners, leading them to activate an action plan of an industrial-scale power grid.

Until the early 1990s, electric utilities were supervised by a regulated monopoly system that was vertically integrated, that is, an individual utility could provide all three main parts of electric power systems: power plants, transmission lines, and distribution systems [28]. At the end of the 20th century, the complex structure of energy market began to change rapidly thanks to developed countries' attempts and legislation, which paved the way for wholesale competition in 1996. Governments supported the idea of opening access to utility grid and encouraged the power facilities to get involved in this brand-new structure, allowing them to manage transmission lines as independent system operators (ISOs) or regional transmission organizations (RTOs) [29].

1.3 Interconnected Electric Power System

The electrical grid is one of the most sophisticated and complex structures consisting of countless cluster of transformers, substations, generating units, and long-distance transmission lines aiming to deliver power to consumers. It was expanded from the small-size local designs and became a great interconnected structure by meeting hundreds of millions of electricity consumers' daily needs with extraordinarily long high-voltage and low-voltage lines. The fundamental components of this structure are the power plants for power generation, the transmission lines, and the distribution system, as shown in Fig. 1.11, each of which is detailed below.

1.3.1 Power Plant

The utility grid starts exactly where the electricity is generated. The power plants are generally established near the energy sources, which are generally

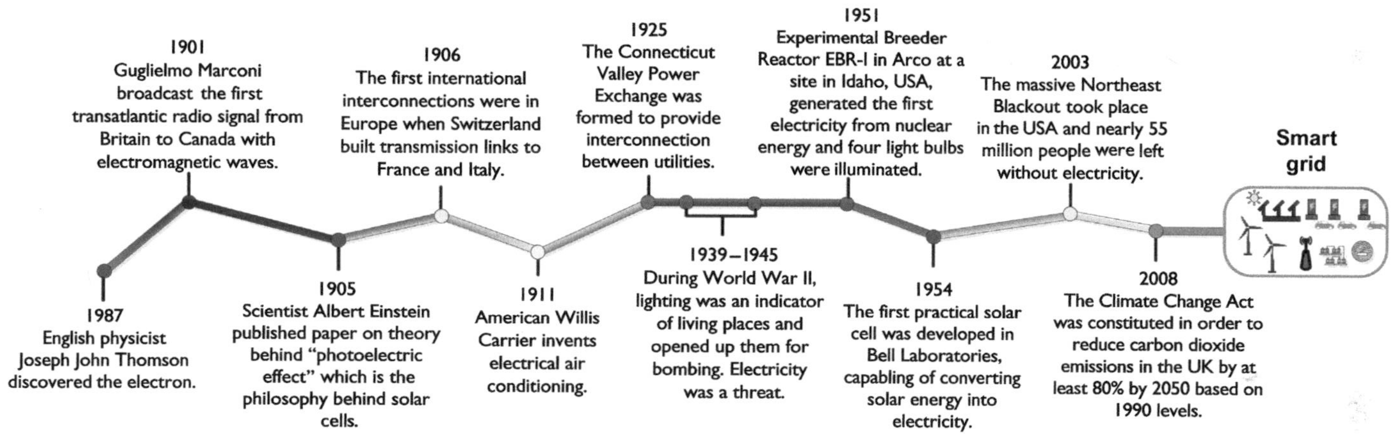

FIG. 1.10 The timeline of changes/discoveries from the 1900s to now.

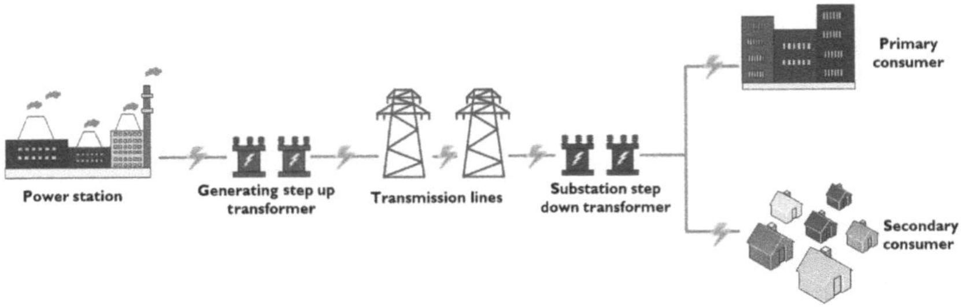

FIG. 1.11 Traditional electricity delivery system.

built up as coal or natural gas thermal power stations, hydroelectric dams, and nuclear reactors far away from the consumers. Coal and nuclear power plants present less flexibility in terms of regulating output power when needed, as ramp-up/down duration takes longer time compared to natural-gas-fired plants. Therefore, in order to prevent energy shortages and detrimental effects, it is a common method to use back-up generating systems. The grid operator monitors reserves, grid conditions, and the consumer and supply side continuously to develop control axioms even in unexpected situations such as blackouts [30].

In the history of power generation, steam turbines converting thermal energy of steam into mechanical energy and into electrical energy invented by Sir Charles Parsons in 1884 were a major milestone for energy production. This invention enabled consumers to obtain affordable low-cost electricity and led to electricity becoming a broadly utilized source even in rural areas. The first coal-fired steam generators provided low-pressure steam, which was used to drive DC dynamometers. However, improvements in efficiency (from 1.6% to 15% between 1884 and 1910) allowed the driving of AC generators. Moreover, it is evident in the history of electrification that vastly improved pulverized coal steam generators were used widely throughout the 19th century [31].

1.3.2 Transmission Lines

High voltage transmission lines provide the transfer of electricity from power plants to consumers with higher efficiency, lower expenses, and low losses, especially over long distances. However, there is a crucial problem that electricity generators produce power in low voltages. Thus, the output voltage must be increased with step-up transformers. Overhead power lines, which are not insulated and are less expensive compared to the underground power cables, can be used in transferring electricity. Today, the highest level of transmission voltage is changing in the range of 700–800kV, but normally 110kV or above voltages are accepted as widely used [32]. It is evident that these values are high, considering what we need in our homes, so these voltages are converted to low voltage levels with step-down transformers and sent to the distribution network within the safe constraints. The management of transmission systems over long distances is one of the most impressive engineering feats supported by developing control actions to provide an efficient, safe, robust, and low-cost power grid all the time. High voltage AC, high voltage DC, super conducting technology, and wireless power transmission are the most common methods to transfer electrical power for long distances.

When the 20th century's electrical grid system is examined, it can be seen that there were more than 4000 individual electric utilities which were operated separately for their own profits. Due to the fact that electricity demand increased especially after World War II, the utilities began to establish connected transmission systems for the purpose of increasing their profits while reducing energy cost, and managed the system together in the context of commercial targets [33].

1.3.3 Distribution Network

These kind of electrical networks begin after step-down transformer and the point at which the transmission lines end. Electricity is procured by primary and secondary consumers at the final destination. The distribution system voltage level generally changes between the ranges of 0.4 and 33kV. With appropriate management and planning strategies, it is possible to obtain an coordinated, reliable, and cost-effective power grid supplying both present and future probable loads.

There are two types of electric power distribution systems, among which the radially connected one was preferred in the early development age of electrification. However, this kind of distribution system causes some important obstacles in terms of supplying adequate electric service to consumer in case of any feeder or transformer fault. Hence, the main ring electrical power distribution system is presented as a solution to handle the mentioned drawbacks. It is ensured that consumer demand is still accounted for by the other feeder where no failure occurs.

1.4 Modernization of the Electric Power System With Smart Grid

The world had witnessed two industrial revolutions occurring in the 18th and 19th centuries, as a first revolution and a second revolution which utilized coal and petroleum's energy production capability, respectively. Although they paved the way for rapid industrialization and economic development, undesirable side effects such as natural pollution occurred due to harnessing fossil fuel-based energy resources excessively since the last industrialization.

In the last few decades, it has been seen that the energy consumption of the emerging modern world has grown extraordinarily due to increasing population and rapid urbanization. This demand is expected to increase by 48% between 2012 and 2040 based on the Energy Information Agency's statistical data [34]. All this causes greenhouse gas emission (GHG) into the climate to accelerate at an unprecedented rate, which has led to growing concerns about environmental issues such as global warming, carbon footprint, and acid rain. Today, it is not possible to ignore the profound impact of GHGs on the environment especially after humankind witnessed its highest value during the years from 1983 to 2012, considering the last 1400 years [35]. Unfortunately, this issue induced an increase in average temperature of the earth by nearly 0.85°C and as a result oceans warmed, polar ice melted, and sea levels rose. To address these issues, there have been significant attempts to reduce GHGs in the power and energy industries which are in charge of 40% of the global CO_2 production. For example, the European Union (EU) has aimed at providing significant reduction in GHGs as a proportion of 20% by 2020, which is one of the most popular commitments, also known as 20-20-20 target [36]. Similarly, the United Kingdom (UK) Climate Change Act 2008 announced that it has the main goal of reducing emissions by 80% by the year 2050 compared to a 1990 rate and revitalizing the awareness of environmental issues [37].

On the power supply side, harnessing the renewable energy sources (RESs) is considered to be one of the most promising approaches to deal with the aforementioned problems by making the power system more efficient and less polluting [38]. In the last few decades, there has been a strong trend toward integrating RESs (solar, wind, hydro, geothermal, biomass, etc.) into the power grid, and advanced management strategies have been developed in order to provide a sustainable, economic, robust, and resilient electrical grid. In this sense, there are two important targets for increasing installed RESs capacity as a proportion of 10% and 49% for the EU countries Malta and Sweden, respectively [39]. Additionally, according to the National Renewable Energy Laboratory report, the United States has also committed to expand RES capacity from 30% to 90% by 2050 [40].

Wind and solar power farms were first penetrated as complementing sources in the hybrid systems for the purpose of rural applications [41]. However, increased demand, pollution, and further technological achievements led distributed energy systems to gain importance worldwide and become the primary choice to address electricity generation. Modular and distributed on-site generation units present a feasible solution in terms of reducing grid dependence and losses. Therefore, it is widely accepted that all of these are forerunners of the modern power grid which is actually based on noncentralized large power plants but small-scale low voltage energy systems, namely microgrids (MGs). MGs are autonomous systems consisting of local loads, generation units, and energy storage systems with distinct electrical boundaries [42] (see Fig. 1.12). These systems currently rely on different kind of RESs such as photovoltaic (PV) farms, wind turbines, and biomass. It can thus be concluded that the 21st century is a time of significant change and upheavals in the power system considering revolution in the energy supply side within the smart grid concept [43].

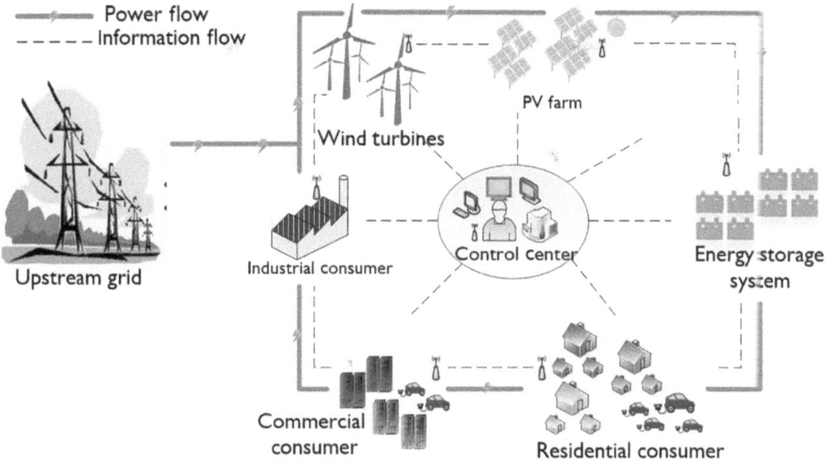

FIG. 1.12 A small-scale microgrid structure.

From another perspective, it is evident that transportation industry also causes to increase GHG emission at alarming rate in our modern power grid. According to the relevant reports, it is seen that this system is responsible for a major portion of global GHG emission by 33% due to its massive reliance on fossil fuels [44]. In 1990, about 24.9% of total US GHG emission was accounted in traditional transportation system. In 2012, it accounted for about 28.2%, and the largest share of growth was announced as 103% in total emissions between 1990 and 2012 [45]. In order to reduce emissions and fuel consumption, electric vehicles (EVs) are presented as a promising green solution of the automotive industry due to stringent legislations, environmental awareness, and rising gasoline prices. EVs have a vast range of benefits consisting of improving air quality, decreasing noise emissions, and some economic savings.

Currently, there are some important partnerships and protocols to obtain efficient, zero-emission transportation systems. In this respect, the United Kingdom constituted the National Energy Efficiency Action Plan in 2007 for the purpose of decreasing GHG emission by 18% for the year 2016, which regarded as 272.7 TWh energy saving [46]. The EU has also made an effort to succeed significant reduction in CO_2 emissions globally. Therefore, it is clear that electrification of the powertrain became one of the main goals of governments, policy makers, and industry based on abovementioned efforts. As a result, the world has seen a stunning revolution in transportation systems. In other words, penetrating environmentally friendly low-emission EVs has been rigorously suggested instead of petroleum-based conventional vehicles. To expand our knowledge, a detailed research by the International Energy Agency revealed that there were more than 1.26 million EVs on the road worldwide in 2015 [47], and uptake of EVs is rising day by day. Today, different companies present various types of EVs, as shown in Table 1.1. In addition to the environmental

TABLE 1.1 Electrical Characteristics of Electrical Vehicles

EV Types	Battery Capacity (kWh)	Charging/ Discharging Rate (kW)	EPA Range (miles)
BMW i3	33	7.7	114
Chevrolet Bolt	60	7.2	238
Fiat 500e	24	6.6	84
Ford Focus	23	6.6	115
Hyundai Ioniq	28	7	124
Kia Soul	27	6.6	111
Mercedes B-Class	28	10	87
Nissan LEAF	40	6.6	151
Tesla Model X	100	17	237
Volkswagen *E*-Golf	36	7.2	125

benefits, EVs can also be used in smart grid applications as a flexible load such as energy sources, energy storage units, or load. There is a possibility to provide bidirectional power flow through grid-to-vehicle (G2V) and vehicle-to-grid (V2G) modes when needed. From the system operator side, this flexibility can aid in maintaining power quality and regulating voltage/frequency fluctuations.

Apart from all aforementioned advantageous, issues such as high penetration of both RESs and EVs into the power system can cause severe operational challenges from the system operator's point of view. The main drawback for RESs is that they are completely weather-dependent, that is, they provide highly volatile and intermittent power output that need some serious managements and improvements for smooth penetration. Otherwise, these may cause notable consequences in the electrical grid such as unbalanced supply-demand, voltage flicker, frequency fluctuations, and other power quality issues.

On the other hand, it is widely known that distribution system is not robust to withstand extra unpredictable new loads such as EVs demanding high charging power. Therefore, integration of large amount of EVs may lead to undesired operational consequences such as higher losses, transformer or feeder overloading, and phase imbalances [48]. All of these are significant technical problems that must be avoided in order to ensure a sustainable, reliable, and economic distribution network operation. Firstly, charging stations are represented as a solution that can provide regulated/controlled charging power for EVs and aid power system operator to understand EVs' impact on the network.

Moreover, they enable the development of management strategies and reliable forecasting models for responding to system requirements. However, the stochastic nature of arrival/departure times, state of energy, and expected duration times of EVs are one of the most common operational problems that may be faced by system operators continuously.

The traditional power grid structure had limited construction design, which could only transfer electricity from the large-scale central energy plants established far away from the consumers, enabling only one-way communication and energy exchange infrastructure, restricted energy storage capability, and passive loads. However, it is obvious that today's power system has become a sophisticated energy architecture considering the aforementioned technological revolutions which emerged in both supply and demand side (see Fig 1.13). Over the past decade, there have been increasingly rapid advances on the supply side that RES-based distributed generation units and lighter-weight generators (e.g., gas-fired turbines) have become widely used generating plants instead of the large synchronous generators. A great amount of variable speed drives such as EVs have penetrated into power grids, which paved the way for low-emission electrified transportation system. Moreover, different types of energy storage systems such as ultracapasitors, flywheels, and even EV batteries have begun to be used as ancillary services in order to mitigate supply-demand imbalances [49]. On the demand side, prosumers (consumers also with on-site production facilities) together with the heterogeneity of the appliances also cause major uncertainties in electrical system. However, it is possible to take advantage of flexible load-based energy reduction capabilities by demand response

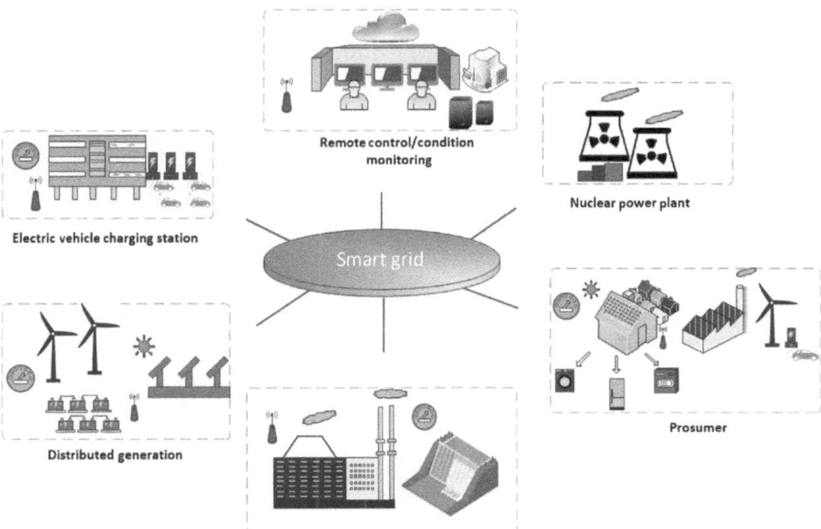

FIG. 1.13 Smart grid conceptual diagram.

(DR) programs. Additionally, a supervisory control and data acquisition (SCADA) system has been developed as an industrial automation system to increase energy efficiency, but decrease extra costs by monitoring system conditions every time. Therefore, it would not be wrong to note that since the beginning of the 21st century, this emerging modern power grid has been witnessing spectacular revolutions in all main system components, which may be described as a third industrial revolution. However, these rapid changes/transformations are seriously impacting the grid by triggering a huge amount of discussions owing to the increased concerns about safety problems; questions have also been raised about the capability of our current power networks. Development of new approaches plays a critical role in long-term power system planning while considering the need for "smarter" electrical grids. In this sense, the paradigm of the smart grid was represented as an attractive solution in order to coordinate the above-mentioned frameworks optimally and to provide more flexible, fast responding, resilient, and reliable power systems. A conceptual definition of the smart grid by EU's viewpoint is as follows: "A smart grid is an electricity network that can intelligently integrate the actions of all users connected to it—generators, consumers and those that do both—in order to efficiently deliver sustainable, economic and secure electricity supplies" [50].

Unlike the main thoughts such as smart grid is a new grid infrastructure, it is actually the state-of-the-art combination integrating communication and information technologies into the conventional grid to transmit electricity in an economic and efficient way. One of the most tremendous benefits of a smart grid is that system operators can monitor the main components, and improve control axioms as well as management strategies to prevent undesired conditions (such as disturbances and outages) thanks to the digital technology (phasor measurement units, automated capacitor banks, and feeder switching). From the consumer's perspective, advanced metering infrastructure provides information about power consumption and electricity prices in real-time to allow consumers to manage and alter their load profiles with the DR programs. The last decade has seen a growing trend toward DR strategies with the aim of compensating uncontrolled, intermittent power generation systems output and providing supply-demand balance every single time. It empowers consumers to participate in electricity markets actively and accomplish particular objectives such as load leveling, reliability requirements, and voltage regulation within the context of a smart grid. Around the world, many different DR programs are currently applied in real-world electric power system operations. Fundamentally, these applications can be classified as follows:

- programs aiming the control of a specific load (air conditioner, pool pump, irrigation pump, water heaters, etc.);
- programs aiming to manipulate the total power consumption of an end-user premise (through contracted reduction, building energy management system, stand-by generators, etc.);

- programs focusing on changing the load pattern of regions composed of multiple end-users; and
- programs enabling direct or indirect participation of end-users to the electricity market.

There are many different approaches under each DR strategy type together with breakdowns, but only the first type of DR practical evidence is shown in Table 1.2. For more detailed information about DR strategies, reference [51] can be also examined.

Smart grid applications (see Fig. 1.14) can be recognized as a feasible solution to the new challenges, which improves significant operational strategies to deal with uncertainties and increases in RESs penetration. In addition, it leads to the modernization of the electrical grid while ensuring a robust, sustainable, and flexible grid structure thanks to all the aforementioned developments. Today, this concept has been comprehensively and systematically examined with a growing body of literature and it is possible to identify some real implementation fields in many countries such as the United States, Germany, Canada, Japan, China, and India [52].

TABLE 1.2 Practical Evidence for DR Programs

Controllable Load	Practical Evidence
Air conditioners	Pacific Gas and Electric Company—PG&E (California — Commercial and Residential—"*SmartAC program*"
	CPS Energy (Texas)—Commercial and Residential—"*Smart Thermostat Program*"
	Endeavor Energy (Australia)—Residential—"*CoolSaver Program*"
	Western Power Company (Australia)—"*Air Conditioned Trial*"
Pool pumps	Endeavor Energy (Australia)—Residential—"*PoolSaver Program*"
	Energex Company (Australia)—Residential—"*Pool Rewards Program*"
Water heaters	Energex Company (Australia) —Residential—"*Hot Water Rewards Program*"
Irrigation pumps	Southern California Edison (SCE) Company—"*Agricultural and Pumping Interruptible Program*"
	American Electric Power (AEP)&EnerNOC—Texas—"*Irrigation Load Management Program*"

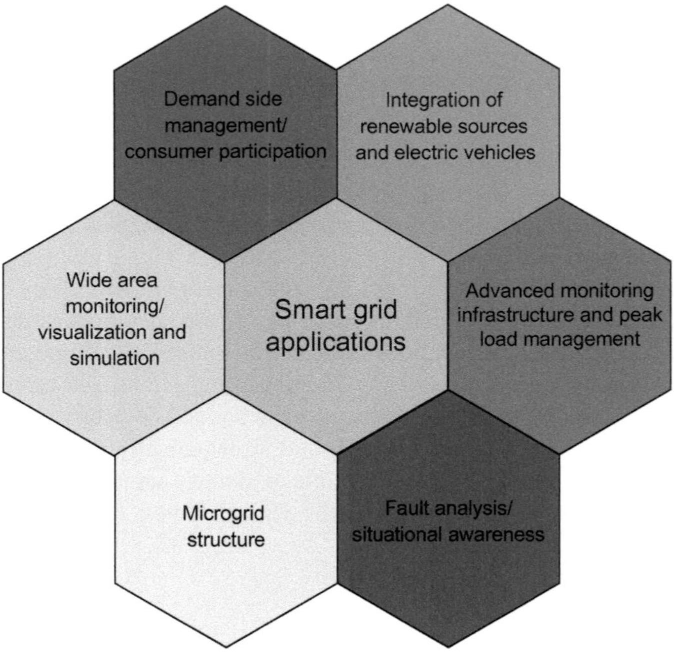

FIG. 1.14 Smart grid applications.

Therefore, we are at the edge of "Pathways to a Smarter Power System" where we can foresee a future power system structure with adapted advanced technologies enabling a far different world than today.

References

[1] Electricity History, https://www.ideallighting.co.uk/2016/09/09/electricity-history, 2018.

[2] A. Van Helden, William Gilbert. From The Galileo Project, 2018, 1995. http://galileo.rice.edu/sci/gilbert.html.

[3] Gilbert, William (1544–1603). In The Hutchinson Dictionary of Scientific Biography, 2018. From http://www.credoreference.com/topic/gilbert_william_1544_1603.

[4] H.W. Meyer, A History of Electricity and Magnetism, The Massachusetts Institute of Technology, Norwalk, CT, 1971 Burndy Library Publication No.27.

[5] Otto von Guericke, Prussian Physicist, Engineer, and Philosopher. Written by: The Editors of Encyclopaedia Britannica, 2018. https://www.britannica.com/biography/Otto-von-Guericke.

[6] J.L. Heilbron, Electricity in the 17th and 18th Centuries: A Study of Early Modern Physics, University of California Press, 1979.

[7] F. Hauksbee, Physico-Mechanical Experiments on Various Subjects, University of Cornell Press, 2013. Digitally Available.

[8] Francis Hauksbee—Inventor, https://www.awesomestories.com/asset/view/Francis-Hauksbee-Inventor, 2018.

[9] Charles-François De Cisternay Du Fay, Science and its Times: Understanding the Social Significance of Scientific Discovery, 2018. http://www.encyclopedia.com/science/encyclopedias-almanacs-transcripts-and-maps/charles-francois-de-cisternay-du-fay.
[10] Discovering Electricity—The Leyden Jar, 2018. https://www.awesomestories.com/asset/view/THE-LEYDEN-JAR.
[11] Benjamin Franklin, American Author, Scientist, and Statesman. Written by: Theodore Hornberger, Gordon S. Wood, 2018. https://www.britannica.com/biography/Benjamin-Franklin.
[12] C. Jungnickel, Cavendish: The Experimental Life, Bucknell University Press, Lewisburg, PA, 1999.
[13] R.B. Campenot, Animal Electricity: How We Learned That the Body and Brain Are Electric Machines, Harvard University Press, 2016.
[14] Battery Facts, Alessandro Volta (1745–1827), 2018. http://www.batteryfacts.co.uk/BatteryHistory/Volta.html.
[15] Sir Humphry Davy (1778–1829), History, BBC, 2018. http://www.bbc.co.uk/history/historic_figures/davy_humphrey.shtml.
[16] Hans Christian Ørsted, Danish Physicist and Chemist, Written by: The Editors of Encyclopaedia Britannica, 2018. https://www.britannica.com/biography/Hans-Christian-Orsted.
[17] N. Forbes, B. Mahon, Faraday, Maxwell, and the Electromagnetic Field: How Two Men Revolutionized Physics, Prometheus Books, New York, 2014.
[18] Morse Code & The Telegraph, Morse's First Telegram Marked the Beginning of the Telecommunications Revolution, http://www.history.com/topics/inventions/telegraph, 2018.
[19] J.C. Maxwell, T.F. Torrance, A Dynamical Theory of the Electromagnetic Field, Wipf and Stock Publishers, 1996.
[20] Guglielmo Marconi, Italian Physicist, Written by: Reginald Leslie Smith-Rose, 2018. https://www.britannica.com/biography/Guglielmo-Marconi.
[21] Thomas Alva Edison, Encyclopedia of World Biography, 2018. https://www.encyclopedia.com/people/science-and-technology/electrical-engineering-biographies/thomas-alva-edison.
[22] National Archives and Records Administration, Thomas Edison's Patent Drawing for an Improvement in Electric Lamps, 2018. http://www.archives.gov/exhibits/american_originals_iv/images/thomas_edison/patent_drawing.html.
[23] History of Electricity, 2018. http://www.need.org/Files/curriculum/infobook/Elec3I.pdf.
[24] R.L. Bradley Jr., Edison to Enron: Energy Markets and Political Strategies, Scrivener Publishing and John Wiley & Sons, Hoboken, NJ, 2011.
[25] N. Tesla, D.H. Childress, The Fantastic Inventions of Nikola Tesla, Adventures Unlimited Stelle, Illinois, 1993.
[26] R. Rudervall, J.P. Charpentier, R. Sharma, High voltage direct current (HVDC) transmission systems technology review paper, in: Presented at Energy Week 2000; March 7–8, 2000.
[27] W. Galli, Electric Transmission 101: Operational Characteristics, Clean Line Energy Partners LLC, 2013.
[28] Capitalizing on the Evolving Power Sector: Policies for a Modern and Reliable U.S. Electric Grid, Energy & Infrastructure Program Energy Project, 2018. http://bipartisanpolicy.org/wp-content/uploads/sites/default/files/Energy_Grid_Report[1].pdf.
[29] The History Chanel. The Energy Crisis (1970s), 2018. https://www.history.com/topics/energy-crisis.
[30] How the Electricity Grid Works. Union Concerned Scientists Science for a Healthy Planet and Safer World, 2018. https://www.ucsusa.org/clean-energy/how-electricity-grid-works#.WrNk84huaM9.

[31] History of Power: The Evolution of the Electric Generation Indusrty, Abby Harvey, Aaron Larson and Sonal Patel, 2018. http://www.powermag.com/history-of-power-the-evolution-of-the-electric-generation-industry.
[32] Electric Power Transmission. Wikipedia, 2018. https://en.wikipedia.org/wiki/Electric_power_transmission.
[33] Electricity Explained. How Electricity Is Delivered to Consumers. U.S. Energy Information Administration, 2018. https://www.eia.gov/energyexplained/index.cfm?page=electricity_delivery.
[34] International Energy Outlook 2016, US Energy Information Administration, Washington, DC, May 2016.
[35] R.K. Pachauri, L. Meyer, Climate Change 2014 Synthesis Report, Intergovernmental Panel on Climate Change Technical Report, https://www.climate-service-center.de/imperia/md/video/csc/syr_ar5_spmcorr1.pdf, 2018.
[36] The 2020 Climate and Energy Package. Directorate-General for Climate Action, 2018. https://ec.europa.eu/clima/policies/strategies/2020_en.
[37] Department of Energy & Climate Change, The Carbon Plan—Reducing Greenhouse Gas Emissions, 2018. https://www.ukgbc.org/sites/default/files/3702-the-carbon-plan-delivering-our-low-carbon-future.pdf.
[38] Department of Energy & Climate Change. Energy Consumption in the UK, 2018.
[39] European Commission: Renewable energy, 2018. http://ec.europa.eu/energy/en/topics/renewable-energy.
[40] M. Hand, S. Baldwin, E. DeMeo, J.M. Reilly, T. Mai, D. Arent, G. Porro, M. Meshek, D. Sandor, Renewable Electricity Futures Study—National Renewable Energy Laboratory, https://www.nrel.gov/docs/fy12osti/52409-1.pdf, 2018.
[41] L.J. Ricalde, E. Ordonez, M. Gamez, E.N. Sanchez, Design of a smart grid management system with renewable energy generation, in: IEEE Symposium on Computational Intelligence Applications in Smart Grid, 2011.
[42] A.T. Eseye, D. Zheng, H. Li, J. Zhang, Grid-Price dependent optimal energy storage management strategy for grid-connected industrial microgrids, in: IEEE Green Technologies Conference, 2017, , pp. 124–131.
[43] K. Meng, D. Wang, Z.Y. Dong, X. Gao, Y. Zheng, K.P. Wong, Distributed control of thermostatically controlled loads in distribution network with high penetration of solar PV, CSEE J. Power Energy Syst. 3 (2017) 53–62.
[44] M. Rahmani-andebili, Optimal power factor for optimally located and sized solar parking lots applying quantum annealing, IET Gener. Transm. Distrib. 10 (2016) 2538–2547.
[45] Y. Wang, W. Ding, L. Huang, Z. Wei, H. Liu, J.A. Stankovic, Toward urban electric taxi systems in smart cities: the battery swapping challenge, IEEE Trans. Veh. Technol. 67 (2018).
[46] D. Liyuan, Q. Yue, W. Wenjing, Q. Guoliang, Current situation of energy saving and emission reduction of urban public transportation and green development suggestion, in: IEEE International Conference on Intelligent Transportation, Big Data & Smart City, 2016.
[47] Global EV Outlook 2016. Beyond One Million Electric Cars. IEA, 2018 https://www.iea.org/publications/freepublications/publication/Global_EV_Outlook_2016.pdf.
[48] M.F. Shaaban, A.H. Osman, M.S. Hassan, Optimal coordination for electric vehicles in smart grids with high penetration of PV generation, in: IEEE Modelling Symposium, European, 2016.
[49] An assessment of energy technologies and research opportunities, 2018. https://www.energy.gov/sites/prod/files/2015/09/f26/Quadrennial-Technology-Review-2015_0.pdf.

[50] European Technology Platform Smart Grids: Vision and Strategy for Europe's Electricity Networks of the Future. Directorate-General for research sustainable energy systems, 2018. https://ec.europa.eu/research/energy/pdf/smartgrids_en.pdf.
[51] O. Erdinç, N.G. Paterakis, J.P. Catalão, An overview of demand response: key-elements and international experience, Renew. Sustain. Energy Rev. 69 (2017) 871–891.
[52] M.A. Mohamed, A.M. Eltamaly, H.M. Farh, A.I. Alolah, Energy management and renewable energy integration in smart grid system, in: IEEE International Conference on Smart Energy Grid Engineering, 2015.

Chapter 2

Energizing Renewable Energy Systems and Distribution Generation

T. Adefarati[*,†] and R.C. Bansal[*,†]
[*]*Department of Electrical, Electronic and Computer Engineering, University of Pretoria, Pretoria, South Africa,* [†]*Department of Electrical and Computer Engineering, University of Sharjah, Sharjah, United Arab Emirates*

Chapter Outline

2.1 Introduction

The assessment of a power system has attracted public attention due to the sudden increase in environmental impacts such as GHG emissions, thermal releases, ozone layer depletion, climate change, and noise, as well as depletion of fossil fuels, sudden increase of power demand, and fluctuation of fuel prices. The global energy requirement is increasing on a daily basis at a rate that is higher than the human population growth. As a result, the high demand of a reliable power supply is caused by introduction of complex electronic gadgets that require uninterrupted power supply. On the other hand, it has been reported by many international organizations that more than 70% of the global blackout

Pathways to a Smarter Power System. https://doi.org/10.1016/B978-0-08-102592-5.00002-8

occurs in the developing countries where approximately 1.1 billion people do not have access to electricity [1]. Owing to this, distributed energy resources (DERs) have become the potential alternative to reduce the negative impacts of power outages that can affect the economic activities of the commercial, industrial, and residential customers. Moreover, DERs are power generation technologies connected directly or near the customer load points to produce electricity and at the same time to reduce the energy consumption through the market mechanism put in place by the utilities. DER technologies comprise of the diesel generator, microturbine, gas turbine, steam turbine, bioenergy technologies, photovoltaic (PV), fuel cell, wind turbine generator (WTG), inverter, electric vehicle, small hydro, smart grid features battery storage system (BSS), and controllable loads [2]. These components work together or independently to form distributed generation technologies that operate as grid-connected systems or standalone power systems.

The renewable energy technologies utilize natural resources that are naturally replenishing to generate energy for various applications. Most of the renewable DERs have low operation and maintenance (O&M) costs, low GHG emissions, and are easy to install [3]. The DER technologies can be seen as energy resources that are globally accepted by the utilities for load balancing, peak curtailment, demand response, peak load management, and energy efficiency in real time power systems. In addition, they can be used in the planning process to improve the reliability of the power system by ensuring that loads will not exceed supply during the system peak. In addition to this, DERs can provide ancillary services such as voltage support and frequency control at low operating costs and higher efficiency than traditional power systems. The combination of numerous DERs that have base and peak load, and economic and operational characteristics, will maximize the utilization of the transmission and distribution lines while running on the full load. For instance, gas turbines that operate with higher fuel costs and low capital costs when compared with other baseload generating units can be brought online quickly. This indicates that gas turbines operate within a limited number of hours since that they are used by the utilities for peak load demand. Similarly, small hydro is a sustainable energy source that operates at better efficiency [4]. It has some economic features such as no fuel costs, provision of backup during electricity interruptions, reduction of the overall cost of electric energy, and no GHG emissions, but its operation is limited owing to nonavailability of water during the dry season [5]. As a result, small hydro works at low capacity factors with the limited water supply. However, hydropower systems can be quickly brought online and ramped up and down easily.

The traditional power system with a large number of thermal power plants, hydro power plants, gas turbine, and reciprocating engines was initially designed with a unidirectional power flow. This shows that power only flows from the power stations to the consumers in one direction. However, with the advent of DERs and the smart grid system, coupled with the state of the art of the technologies in the power system, the operation of the power network has been changed from unidirectional to bidirectional power flow. Due to the application of the smart grid features in the power system, DERs are now being utilized by various households to generate their power requirements at the commercial level with the application of net metering and feed in tariff. The integration of the smart grid features such as demand response, smart meters, and peak load demand management allows power consumption to be reduced during peak periods. This assists the utilities to reduce the transmission line congestion and balance the power in the grid. The reliability of a power supply can be guaranteed with the integration of a number of DERs into a power system during peak periods when the power generated exceeds the power demand. At this particular period, a backup generation is required to balance the shortfall between the power generation and demand. Moreover, in a situation where the source of power supply is not predictable owing to the intermittent characteristics of the wind and solar resources, the BSS can be deployed to improve the reliability and balance the grid power supply. The BSS is utilized in such conditions to smooth out the effect of local renewable energy resources [6]. Furthermore, the penetration of DER technologies into the power system is growing on a daily basis due to their affordability and economic and technical benefits. The aforementioned benefits have encouraged many countries to allocate more resources that will augment DERs' investments and improve the performance of their power systems.

This chapter is intended to help the power utilities, designers, government agencies, and retailers to understand the concept of DER technologies and their economic and environmental importance. The technical challenges and benefits of a high penetration of DER in a microgrid system are presented in this chapter. This chapter can be used as a guide to invest heavily in the new technology and techniques to study the effects of DERs penetration in the power systems. It also sheds more light on the methods that can be used to increase the capacity and efficiency of DERs, as well as to reduce costs or risks that are associated with their applications in the power system.

2.2 Distributed Energy Resources

The DERs are small scale power sources that combine various array of generation, storage technologies and energy monitoring and control facilities to provide a potential alternative for improvement of the conventional power system [7]. The DERs are installed at or near the location where electricity power is being utilized and can either be connected to the grid or operate as the standalone systems and operate independently of the grid. The DERs are the generation and storage technologies that are classified based on the sizes, types, and capacity for residential, commercial, and industrial power and heat solutions. Similarly, the DER technologies are encouraged by the policy makers to maximize their benefits such as production of electricity, heating, and cooling.

Owing to this, some benchmarks, market mechanisms, and regulatory structures have been introduced by government agencies to effectively harness the benefits of the DERs. These frameworks are initiated to guarantee the security of power supply at all times.

2.2.1 Importance of Distributed Energy Resources

The DER technologies have been widely accepted to provide power, heat, and cooling solutions for numerous customers across the globe, owing to their important technical, economic and environmental benefits with better power system resilience and efficiency. DERs can also improve grid balancing and affordable extension of the grid to unconnected communities. The other aspects of importance of DER in the power system are as follows:

- Reduce the COE and improve the security of power supply.
- Enhance the reliability of power system and reduce the cost that related to power outages.
- Reduce GHG emissions and improve fuel utilization on site.
- Defer the need to install additional generating units.
- Reduce fuel cost risk and fuel supply risk.
- Provide ancillary support in some cases.
- Improve energy efficiency and local economic development.

2.2.2 Application of Distributed Energy Resources

The sudden increased in the power demands coupled with the shortage of electric power supply, power quality problems, and COE spikes have encouraged utilities to seek better sources of power supply on the basis of technical, economic, and environmental benefits [8]. The acceptance of DERs is centered on the lower operating costs and being easy and faster to install when compared with the fossil fuel-based generating units. DER technologies can be used for various applications based on commercial readiness, economics, availability, and environmental considerations. These factors are used to select the appropriate DER technologies to provide fast solutions to numerous power problems. Table 2.1 can assist the utilities to make the best decisions when selecting DER technologies based on the benefits that are attached with their utilizations.

2.2.2.1 Standalone

A standalone power system is an off-grid power system for remote areas that are not tied to the utility grid owing to some economic and technical constraints [9]. The standalone system is independent of the grid and can operate effectively in conjunction with nonrenewable and renewable DERs, as well as electric storage systems. The application of the standalone system in the rural areas is more cost-effective than extending the transmission lines to such locations.

TABLE 2.1 Application of DER Technologies

DER Technologies		Power Quality	Standby Power	Standalone System	CHP	Peak Shaving	Low Cost of Energy
Nonrenewable DER technologies	Diesel generator	×	✓	✓	✓	✓	✓
	Gas turbine	×	✓	✓	✓	✓	✓
	Fuel cell	×		✓	✓		
	Steam turbine	×	✓	✓	✓	✓	✓
	Microturbine	×	✓	✓	✓	✓	
Renewable DER technologies	PV	×	×	✓	×	✓	✓
	Wind turbine generator	×	×	✓	×	✓	✓
Battery storage technologies	SMES	✓	×	×	×	×	×
	Flywheel	✓	✓	×	×	×	×
	Battery system	✓	✓	×	×	×	×

In addition to this, the standalone system is also utilized by individuals who live close to the utility grid and wish to acquire autonomy from the utilities to show their commitment to sustainability of renewable energy. The standalone systems are generally designed to combine various DER technologies to generate a reliable power, reduce O&M costs, and reduce the inconvenience associated with power interruptions. These approaches include application of DERs to reduce the value of electricity required to meet consumer power demands from the utility grid. The application of DERs for standalone operation depends on some paramount factors such as fuel cost, emissions, O&M costs, and capital cost. The aforementioned factors have made renewable DERs more cost-effective than utilizing nonrenewable DERs alone.

2.2.2.2 Peak Shaving

DER technologies are utilized to generate electric power during the periods when the power purchase from the utility is costly. This includes peak demand periods when the power demand is very high and the utilities cannot meet such power demand owing to limited power generation capacity. DERs are used for the peak shaving application as a measure to reduce peak demand rather than to subject some customers to unwarranted power outages. Some DER units are designed to supply power during the peak load periods, in so doing providing numerous economic benefits to the consumers and utilities. The purpose of utilizing DERs for peak shaving applications is to increase the generation capability of the network and at the same time to reduce the economic effect of load shedding on some highly variable loads. Similarly, DER provides a quick response to power demand, making it the optimal solution for the peak shaving application. The application of DER technologies for a peak shaving solution limits the costs that are associated with load shedding.

2.2.2.3 Standby Power

The DERs can be used as a power solution in circumstances where the power supply from the utilities is not reliable owing to a large number occurrence of power outages that takes a long duration to restore power supply. The application of DERs in a microgrid system can provide the optimal solution to meet the load demand when the power supply is interrupted. Similarly, owing to some benefits such as a reliable operation and reduced operational down time, they can supply some critical loads during the power outages pending the time when power service from the utilities will be restored back [10]. The optimal selection of nonrenewable DERs due to unstable power supply from the utilities depends on how frequently the units should start up and how many hours the units will work per year. The power system with the combination of a variety of DERs for standby applications has many benefits, and it is one of the best power solutions based on operating costs and power outages reduction.

2.2.2.4 Emergency Power

DERs are also designed for emergency power solutions in off-grid and grid-connected systems. DERs for emergency power solutions encompass the full range of enabling technologies such as renewable and nonrenewable DERs, energy storage systems, and intelligent control technology. DERs' power solutions also ensure the power supply sustainability and cost-effectiveness in an extensive scope of utilizations. DERs are used for emergency power solutions in a variety of settings such as residential, commercial, industrial, data centers, communication facilities, and modern security monitoring centers. The advantages of DERs in emergency power applications include quick start-up, rapid raping to full load, low duration of power interruptions, no interruption of the security system, and support of continuous operation in various sectors of the economy. However, the operating hours of the diesel generators, gas turbines, and microturbines are limited owing to high emissions of air pollutants.

2.2.2.5 Load Curtailment

DERs are utilized in the power system to meet global energy and environmental objectives based on the reliable, affordable, and clean energy benefits. DER is a platform to optimize energy systems across multiple pathways such as electrical, thermal, fuels, water, communication, and physical scales. Load curtailment is a program designed by the utilities potentially to reduce the power demand or electrical energy usage during the peak load periods. The application of numerous DERs reduces congestion of the grid, which is common during the peak hours when the sum of power demand from the load centers is greater than the power generated by the utilities [11]. This will reduce the congestion of the system and the crisis that is associated with such an event. In this context, many consumers have been encouraged to install DERs on their premises as a measure to prevent the cost associated with load curtailment. The number of DERs installed at different places with various capacities can be used by the utilities and consumers to meet the required power demand at peak periods. This will reduce the economic impact of the load curtailment on the consumers.

2.2.2.6 Combined Heat and Power

Combined heat and power (CHP) is a DER technology comprised of the diesel generator, microturbine, gas turbine, and fuel cell for production of both power and heat solutions. The CHP plants are designed by their manufacturers to recover wasted thermal energy from the conventional generating units for residential, industrial, municipal, and commercial heating applications. DER technologies that produce waste heat can be utilized in a CHP technology due to the higher fuel utilization efficiency and better financial prospects [10]. The CHP is an energy-efficient innovation that produces power. It is designed to capture the heat that would otherwise have been wasted to provide valuable thermal energy for industrial and domestic applications. It tends to be situated at an individual office and facilities that need electrical and thermal energy simultaneously.

2.2.2.7 Demand Response

Demand response is a reduction in the amount of power required from the utilities by customers in a response to an escalation in the COE or motivation payments that are strategically designed to bring down utilization of electric energy at a particular time [12]. The application of DER technologies for a demand response solution is based on the reduction of power supply from the distribution network operators with the local injection of power at the load centers. The supply from the utilities is reduced and the overall system stability is maintained (if not enhanced) as the reserve supply capacity is increased. In some situations, load shedding is employed to help augment the utility control or cost, but the benefits are limited to monetary compensation to sites that can afford to reduce nonessential electrical loads at a requested time.

2.3 Smart Grid System

A smart grid is defined as the electrical network that incorporates numerous generating units, smart grid features, and loads, and efficiently convey power that is adequate in capacity, increases grid efficiency, sustainability, accessibility, self-healing, safety, and reliability [8]. Smart grid advancement has a tendency to be driven by the objective of enhancing the performance of electric power systems based on the collaborations between utilities and customers. The rapid demand for installations of DERs in the traditional power system requires a computer-aided coordination system that can be used for power generation, transmission, distribution, and consumption [9]. Owing to the rapid growth in the global power demand, fluctuation of the prices of fossil fuels and the reduction in DERs costs over the last few years, the opportunities for DER technologies in the smart grid-connected power system seem to be increasing. A smart energy system is able to integrate and optimize both electrical and mechanical plants in order to lower the COE, reduce greenhouse gas (GHG) emissions, and enhance the efficiency of power systems. The combination of various DER technologies can also be used to proliferate the values of the smart grid.

2.3.1 Benefits of DERs in the Smart Grid System

The benefits of DERs in the smart grid system can be summarized as follows [10]:

- The smart grid allows higher penetration of DERs with minimal operational cost while enhancing the reliability of power supply and power quality.
- DER technologies play a proactive role in the cost of energy savings, accomplished by enhancing the energy efficiency and reducing the peaks demand.

- The application of DER technologies in the smart grid reduces the operating costs.
- The application of smart grid technology provides numerous electricity benefits to consumers by efficiently utilizing the power system components.
- The utilization of smart grid features can safely fulfill consumer power demands at lower operational and environmental costs.

2.3.2 Methods of Connecting DERs into the Grid

A smart grid allows a number of consumers to power their appliances with a number of DERs based on the availability of local renewable energy resources. The excess power generated by each consumer can be fed into the grid. The system is configured in a way that whenever local renewable resources are unavailable, the power demand of the consumers is met from the utility grid. However, the utilities face numerous problems in order to connect small DERs to the smart grid, based on the reliability, power quality, safety, rate, and metering. The utilities allow net metering and feed in tariff in some countries.

2.3.2.1 Net Metering

Smart meters are used to measure the difference between the electricity generated by the individual consumers and that taken from the grid. This indicates that utilities bill their consumers based on the net energy on a monthly or yearly basis [10]. The platform provided by the utilities with the application of the net metering allows the consumers to supply any excess power generated into the grid and credit for future energy utilization. Net metering programs have been introduced in many countries for residential and commercial customers based on their specific requirements. This allows a single bi-directional smart meter to be utilized for recording the energy that consumers draw from the utility grid and the excess energy that consumers feed back into the utility grid.

2.3.2.2 Feed-in Tariffs

Feed-in tariffs is a method designed to accelerate investments in DER technologies by allowing energy produced by the consumer to be compensated for the energy they feed back into the grid. Feed-in tariffs require two unidirectional meters in order to measure outflow and inflow of electricity from each customer independently. The first meter records energy drawn from the utility grid, while the second meter records excess energy generated by each consumer that is fed back into the utility grid. This enables electricity consumption and electricity generation to be priced separately.

2.4 Classification of Distributed Energy Resources

DERs consist of energy generation components, storage systems, energy management technologies, power electronic interfaces, and communication and control devices for efficient operation of a power system. DER technologies are modular units that are universally accepted as means of providing a potential alternative to a traditional power system in order to reduce economic effects of power interruption, operating costs, quantity of emission, and cost of energy. There are three types of DERs based on the capacity and technology points of view. The first classification is based on the capacity of the generating units, as presented in Table 2.2, and depends on the type of utilities and customers as well as on the applications. The capacity of DERs can be used by the utilities as a prerequisite to make a managerial decision with regards to which technology is suitable for a particular application. The second classification of DERs is based on the type of technology (fuels) designed and adopted by the original equipment manufacturers (OEMs) for its effective operation. The general characteristics of renewable DERs technologies, nonrenewable DG technologies, and electric storage systems as well as their benefits are presented in Tables 2.2 and 2.3, respectively.

2.4.1 Nonrenewable Distributed Energy Generation Technologies

Nonrenewable DER technologies are energy sources such as gas turbine, microturbine, and diesel generators that use fossil fuel to generate electrical power. The applications of nonrenewable DERs have played proactive roles in the availability of electricity to the populace. These generating units are installed at the load centers to increase the reliability of the power system, especially during the peak and emergency periods, to meet the power demand and avoid exposing customers to unnecessary power interruptions. The development of nonrenewable energy DG technologies that make use of a variety of fossil fuels for their operations and distribution system automation have become one of the attractive options for distribution system reinforcement. However, high GHG emissions have defeated the primary aim of using nonrenewable DER

TABLE 2.2 Classification of DG Technologies Based on the Capacity of the Unit [13]

Type of DG Technologies	Capacity
Micro DG	1 W–5 kW
Small DG	5 kW–5 MW
Medium DG	5–50 MW
Large DG	50–300 MW

TABLE 2.3 Comparison of Different DG Technologies [14–17]

	Nonrenewable DG Technologies			Renewable DG Technologies			
DG Technologies	*Reciprocating Engines*	*Gas Turbines*	*Micro Turbines*	*Fuel cells*	*PV*	*Wind*	*Hydro*
Noise level	High	Moderate	Moderate	Low	No	Low	No
Availability	90%–95%	90%–98%	90%–98%	More than 95%	Subject to solar resource	Subject to wind speed resource	Subject to water level
Capacity range	Diesel: 20 kW to10 MW Gas: 50 kW to 5 MW	1–20 MW	30–250 kW	50 kW to 1 MW	1–20 kW	0.2–3 MW	Small: 100 kW to 100 MW Micro: 25 kW to 1 MW
Efficiency (%)	Diesel: 36–43 Gas: 28–42	21–40	25–30	35–60	8–35	35–45	60–90
Fuel Pressure (psi)	<5	120–500 (compressor)	40–100 (compressor)	0.5–4.5	N/A	N/A	N/A
Start up	10 s	10 min to 1 h	60 s	3 h to 2 days	N/A	N/A	N/A
CO_2 emission (g/kWh)	Diesel: 650 Gas: 500–620	580–680	720	430–490	No direct emission	No direct emission	Small: 10–12 Micro: 16–20
CO emission (g/kWh)	Diesel: 2.8 Gas: 1.8	0.42	0.47	0	No direct emission	No direct emission	Negligible
SO_2emission (g/kWh)	Diesel: 0.032 Gas: 1.25	0.032	0.037	0.024	No direct emission	No direct emission	Small: 0.024–0.029 Micro: 0.038–0.046

Continued

TABLE 2.3 Comparison of Different DG Technologies [14–17]—cont'd

	Nonrenewable DG Technologies			Renewable DG Technologies			
DG Technologies	*Reciprocating Engines*	*Gas Turbines*	*Micro Turbines*	*Fuel cells*	*PV*	*Wind*	*Hydro*
NO_x emission (g/kWh)	Diesel: 10 Gas: 0.2–1	0.3–0.5	0.1	0.005–0.001	No direct emission	No direct emission	Small: 0.046–0.056 Micro: 0.071–0.086
Fuel	Diesel, heavy oil, natural gas, bio-diesel, biogas and landfill gas	Natural gas	Natural gas, landfill and biogas	Hydrogen and methanol	Solar insolation	Wind speed	Water
Power to heat ratio	0.5–1	0.5–2	0.4–0.7	1–2	N/A	N/A	N/A
Installation cost/kW (US$/kW)	Diesel: 125–300 Gas: 250–600	300–600	500–750	1500–3000	1550–3830	900–1400	30–250
Hours to overhauls	25,000–30,000	30,000–50,000	5000–40,000	10,000–40,000	N/A	N/A	N/A
Part load	Good	Poor	OK	Good	OK	OK	OK
O&M cost (US$/MWh)	Diesel: 5–10 Gas: 7–15	3–8	5–10	5–10	1–4	10	0.045–0.09

technologies due to some technical and environmental limitations. The nonrenewable DER technologies are not sustainable and naturally not replenish. They will soon run out of use because of the high rate of power demand for nonrenewable DERs is faster than the rate of restoring them inside the Earth [9]. The nonrenewable DERs have been widely deployed in the power system for CHP, peak load, demand response, load curtailment, and standby power applications with or without having significant interaction with the utility distribution networks.

2.4.2 Gas Turbine

The gas turbine is an internal combustion engine that employs gas as the working fluid for conversion of the kinetic energy of the moving fluid to the mechanical energy by using the reaction or impulse. The converted mechanical energy can be used for power generation as well as the prime mover for several devices such as industrial pumps, marine propulsion, locomotive propulsion, and automotive propulsion [18]. It can be classified into different categories based on the type of duty, type of the combustor, shaft configuration, and degree of packaging [19]. The applications of gas turbines for power generation and mechanical drive technologies have increased tremendously for over a decade due to advances in technology. It has an advantage of flexible fuel capability, which means that it can utilize a number of fuels such as synthetic fuels and natural fuels [20]. The gas turbine can be utilized for prime, continuous, and standalone power solutions in remote communities that have no access to the utility grid as well as grid-connected microgrid systems. In addition to this, the gas turbine can be used for baseload, peak shaving, and ancillary applications. The gas turbines can be classified into two categories: open cycle and closed cycle. The operations of the closed and open cycle gas turbines can be compared based on the cost of installation, type of fuels used for the operation, efficiency, cycle of operation, and quality of heat input.

2.4.2.1 Open Cycle Gas Turbine

The open cycle gas turbine operates on the principle of drawing the fresh atmospheric air at ambient conditions into the compressor where the air is compressed by a centrifugal or an axial flow compressor [21]. The compressor will raise the temperature and pressure of the atmospheric air to the levels specified by the original equipment manufacturer (OEM). The compressed air at the higher pressure is fed into a combustion chamber, where the fuel will be burned at an essentially constant pressure [22]. The high-pressure hot gases drive the turbine and developed power owing to the rotation of the turbine shaft can be used for a number of applications such as power generation and prime movers for industrial devices [23]. The exhaust gases that leave the turbine are released into the atmosphere, as shown in Fig. 2.1 [22]. The cycle is classified as an open cycle system since the exhaust gases are not recirculated but released into the atmosphere.

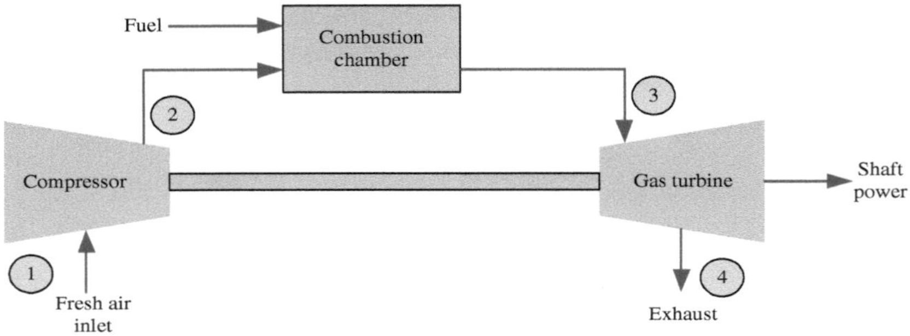

FIG. 2.1 An open cycle gas turbine engine [22].

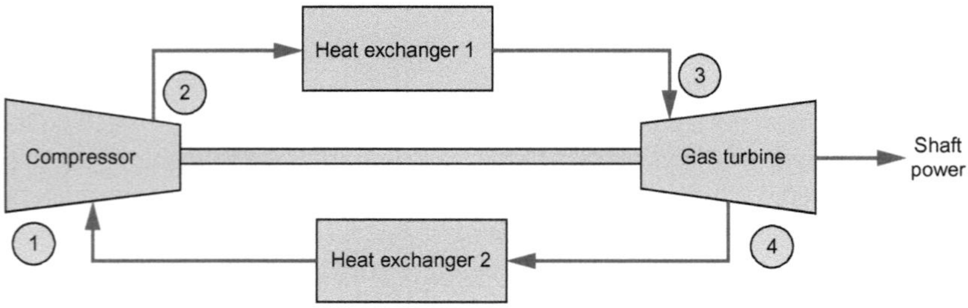

FIG. 2.2 A closed cycle gas turbine engine [22].

2.4.2.2 Closed Cycle Gas Turbine

A closed cycle gas turbine is a turbine in which the temperature and pressure of the atmospheric air that enter the compressor are increased [21]. The compressed air at high pressure and temperature enters the heat exchanger, where it is heated by an external source. The high pressure and temperature air are fed into the turbine for expansion to take place [22]. The power is developed in the closed cycle gas turbine owing to the high-pressure working fluid that increases over the turbine. The exhaust working fluid is not rejected into the atmosphere, but cooled by the cooling chamber and recirculated for a continuous operation of the system [23]. The cycle is termed as a closed gas turbine engine since the same working fluid is returned back to the compressor before the process is completeed, as shown in Fig. 2.2 [22].

2.4.2.3 Advantages and Disadvantages of Gas Turbines

The benefits derived from the application of gas turbines to provide power and thermal solutions have led to many research works to be focused on how to improve the performance, fuel flexibility, emissions, life cycle cost, and thermal utilization of gas turbines. The benefits of gas turbines have offered fast power solutions that can deliver at a site within a short period. The advantages and disadvantages of gas turbines are presented in Table 2.4.

TABLE 2.4 Advantages and Disadvantages of Gas Turbines

Advantages	Disadvantages
• Higher reliability and availability.	• Limited ability to use solid fuels when compared with steam turbines.
• Lower operating costs and GHG emissions.	• Any slight deviation of the fuels from the specifications can reduce the efficiency of gas turbine and increase maintenance factor and early overhaul.
• The large quantity of the waste heat stream from the exhaust can be utilized in other processes.	• Gas turbines require special and expensive equipment owing to high operating temperatures.
• High power density and fuel flexibility.	• The power generated by the gas turbine reduces as the ambient temperature increases.
• Gas turbines have a very low initial cost.	• External source of power such as black start generator needed to start self-sustained operation.
• Gas turbines can be built faster at reduced construction costs.	• The power required to drive the pumps and compressors reduces the net power outputs of the gas turbines.
• Gas turbines are capable of reaching full load operation in a matter of minutes.	• Consume more fuel to drive the compressors and pumps.
• Quick start-up and easy to transport to the sites.	• The high operating temperature of the gas turbines can reduce the lifespan of some components.
• High power-to-weight ratio and high heat grade fuels are available.	• The efficiency levels of an open cycle gas turbine engines are lower than open cycle gas turbine engines, since thermal energy is wasted when the exhaust is released.
• Smaller size and weight compared to steam turbines, and less space is needed for installation.	• Gas turbines operate with a low part load efficiency.

2.4.3 Diesel Generator

Diesel generators can be utilized as standalone, emergency, standby, and peak shaving units because of the following characteristics: availability, quick start up, reliability, fast ramp up, durability, etc. [8]. The diesel generators have high operating cost and generate electricity on demand when compared with renewable DERs. The lifetime of diesel generators is within the range of 15,000–30,000 h based on the proper O&M practices and atmospheric conditions of the environment [2, 23, 24]. The diesel generator is designed in such a way that it can operate at 30%–80% of its nominal rating specified by the by the OEMs. The diesel generator that utilizes the electronic control unit (ECU) is mostly used by the utilities. The ECU is a device that is incorporated into the diesel generator to sense the changes in the generator load demand and compensate by regulating the fuel delivery to the diesel engine. However, there are several factors to consider during the conceptual and design stages of a power system that is tailored to utilize a diesel generator [23]. These factors include the following: cooling system, starting equipment, maintenance, altitude, control system, abnormal environmental conditions, fuel quality, speed governing, drive type, and ambient temperature. Efficiency is another factor that must be considered when choosing a diesel engine for a given application. However, the efficiency of a diesel generator depends on the load factor, capacity, engine design, ambient conditions, fuel control, cooling system, and operating speed [25]. The successful operation of a diesel generator depends on the performance of the components that constitute the system, as presented in Fig. 2.3.

2.4.3.1 Fuel Cost of the Diesel Generator

The fuel cost of diesel generator is a function of capacity of the load connected, type of the diesel generator, rating of the diesel generator, the quality of the fuel, ambient temperature, and humidity [26]. The relationship between the consumption and the power produced by the diesel generator is shown in Fig. 2.4 [27]. The fuel cost of the diesel generator can be expressed as follows [28]:

$$FC_i = a_i P_{gen}^2(i,t) + b_i P_{gen}(i,t) + c_i \ (\$/\text{h}) \tag{2.1}$$

where a_i, b_i, and c_i are cost coefficients of the diesel generator, FC_i is the fuel cost of diesel generator, and P_{gen} is the power generated by the diesel generator.

The power produced by the diesel generator can be expressed in Eq. (2.2) as follows [3]:

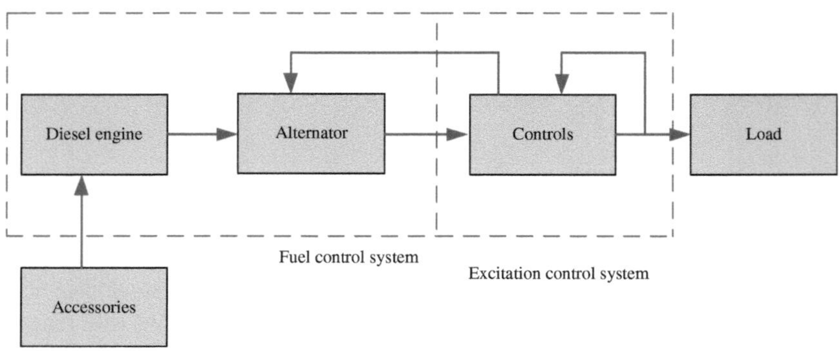

FIG. 2.3 Block diagram of diesel generator [25].

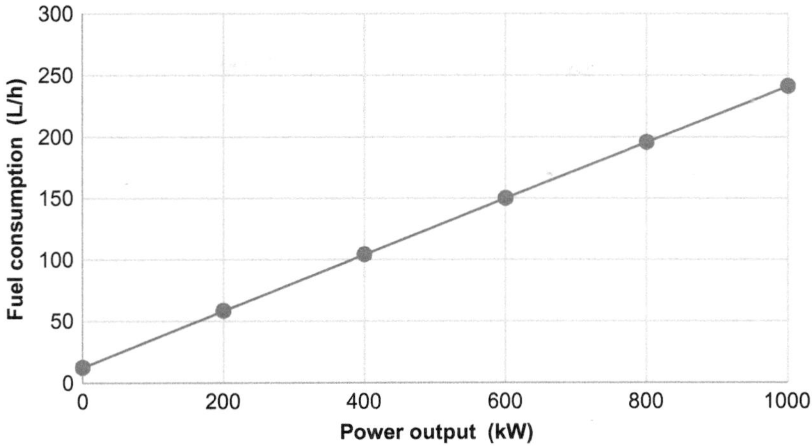

FIG. 2.4 Diesel generator fuel curve.

$$P_{gen} = P_n \times N_{gen} \times \eta_{gen} \tag{2.2}$$

where N_{gen} is number of the diesel generators, P_{gen} is the power generated (kW), P_n is the nominal power generated by the diesel generator (kW), and η_{gen} is efficiency of the diesel generator.

The power generated by the diesel generator is designed to be within the maximum and minimum rated capacity specified by the manufacturers as presented in Eq. (2.3) [28]:

$$P_{gen}^{min} \leq P_{gen}(t) \leq P_{gen}^{max} \tag{2.3}$$

2.4.3.2 Diesel Generator Efficiency

The efficiency of a diesel generator is a combination of the efficiency of the diesel engine and the alternator. The efficiency of a diesel generator based on the combination of the major components ranges between 30% and 55%, while the standalone efficiency of the diesel engine and alternator varies between 35% and 60% and 85% and 95%, respectively. The efficiency of a diesel generator depends on the design, size or capacity, mechanism for fuel control, operating speed, type of cooling mechanism, etc. However, the efficiency of a diesel generator during operation deviates from the design value because of load conditions, ambient conditions, and O&M practices [29]. The efficiency patterns based on the size of the diesel generator indicates that efficiency increases with the size of generator, as shown in Fig. 2.5 [27].

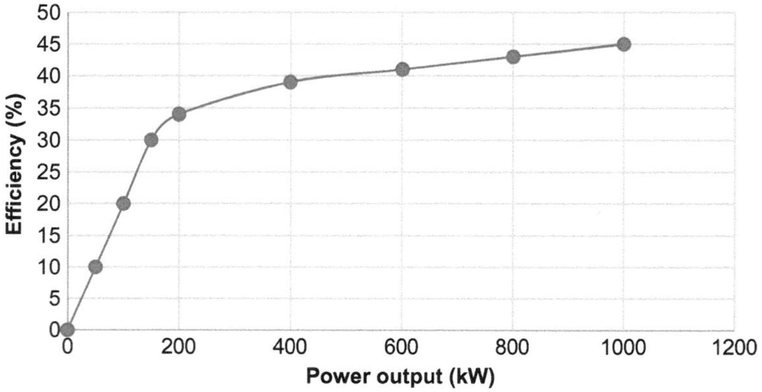

FIG. 2.5 Efficiency curve of diesel generator.

TABLE 2.5 Advantages and Disadvantages of the Diesel Generator

Advantages	Disadvantages
• More efficient with part load operation flexibility and commercially available.	• Generally more expensive.
• No environmental hazards to store the fuel.	• Noise pollution.
• High efficiency, quick start-up and operation with less risk of ignition.	• Diesel generators have numerous parts that can lead to a higher cost.
• The cost to maintain the generator is a lot lower than for gas generators.	• Diesel generators are not environmentally friendly.
• Diesel generators are cheaper than other generators and can withstand heavy loads.	• Servicing and repairs are more expensive when compared with other DERs.
• Diesel generators can operate for a long duration and durable with a longer lifetime.	• The time taken for the installation process is longer than other DERs.
• Diesel generators are more fuel-efficient that petrol generators.	• The initial cost of diesel generators is expensive.
• Diesel generators are cheaper to run than petrol generators.	• Starting diesel generators can be a serious problem in the winter and rainy season.

2.4.3.3 Advantages and Disadvantages of Diesel Generators

The diesel generator is widely utilized in a number of fields such as power sector, mines, railways, factories, enterprises, and hospitals for power and heat solutions. The advantages and disadvantages are presented in Table 2.5.

2.4.4 Microturbine

Microturbines are new technologies that burn gaseous and liquid fuels such as air, steam, and hot gases to produce a high-speed rotation that turns an alternator attached to the shaft of the engine for power generation applications. They can be utilized for small and medium-sized commercial and industrial energy applications because of their operating characteristics such as modular size, low emission, and low O&M cost [23]. Microturbines are the simplest form of gas turbines and provide a high electrical efficiency compared with gas turbines of the same size. A microturbine is a small combustion turbine with a power output that ranges from 25 to 500kW with the following major parts: compressor, combustor, turbine, alternator, and recuperator (see Fig. 2.6). Some of the applications of microturbines include standalone, standby power units, peak shaving units, and combined heat and power generation applications [23]. They can also be utilized to boost a standalone and grid-connected power generation capacity, improve power system reliability, and supply power to remote areas where extension of the transmission and distribution (T&D) lines are difficult due to some financial barriers. In addition to this, microturbines can be used as an efficient prime mover for many mechanical drives such as compressor and air-conditioning. The simple design of a microturbine system improves the reliability and reduces O&M costs of a power system. The efficiency of a microturbine can be expressed as follows [30]:

$$\eta_{mt} = \frac{P_{mt}(elect) + P_{mt}(therm)}{mf_r \times FH_r} \tag{2.4}$$

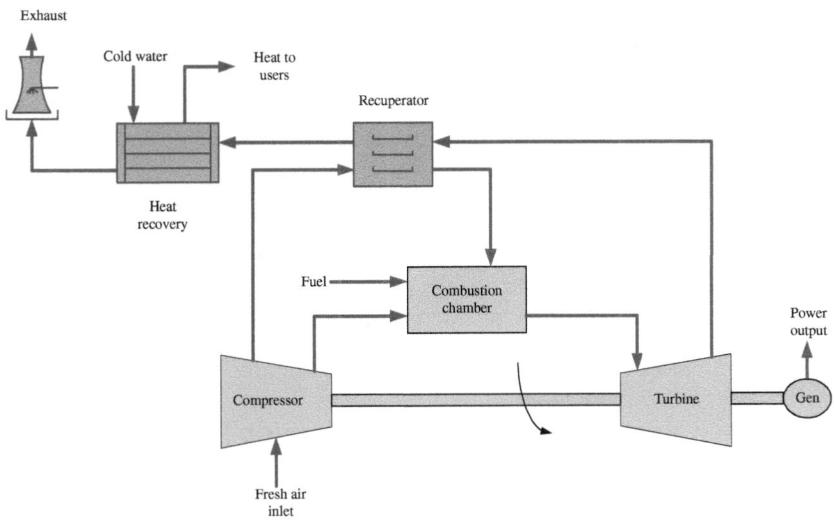

FIG. 2.6 Operation of a microturbine.

TABLE 2.6 Advantages and Disadvantages of Microturbines

Advantages	Disadvantages
• Defer T&D lines upgrading and can utilize waste fuels.	• Low mechanical efficiency.
• Good efficiency in cogeneration, low GHG emissions and no vibrations.	• Reduce power output and efficiency with increase in the ambient temperatures and elevation.
• Strengthens energy security and small numbers of moving parts.	• Restricted operations owing to reduced temperature cogeneration applications.
• Can be utilized for standalone power system.	• Operate with low fuel to electricity efficiency.
• Fuel flexibility, long maintenance intervals, lightweight and compact size.	• High cost.
• Respond quickly to change in power demand and have low O&M costs.	• Inadequate thermal energy for industrial applications.

where η_{mt} is the efficiency of a microturbine, $P_{mt}(elect)$ is the net electrical power output of a microturbine, $P_{mt}(therm)$ is the thermal power recovered from a microturbine, mf_r is the mass flow of gas (kg/s), and FH_r is the lower heating rate of the fuel (kJ/kgf).

The fuel cost of a microturbine can be expressed as follows [30]:

$$FC_{mt} = PNG\sum_{k}^{n}\frac{P_k}{\eta_k} \tag{2.5}$$

where FC_{mt} is the fuel cost of a microturbine, PNG is the price of natural gas, P_k is the electrical power generated at interval k, and η_k is the cell efficiency at interval k.

2.4.4.1 Advantages and Disadvantages of Microturbines

Microturbines offer many benefits for grid and off grid power systems. The advantages and disadvantages of microturbine technologies are presented in Table 2.6.

2.5 Renewable Energy Distributed Generator Technologies

Renewable DER technologies such as wind turbine generator (WTG), photovoltaic (PV) and mini/small hydro power base their mode of operations on the accessibility of the local renewable energy resources such as water, wind, and solar radiation that can be naturally replenished [31]. The availability of local renewable energy resources is an important factor for selecting the potential sites of renewable DER technologies based on the specifications of the OEMs. The renewable DERs are viewed as environmentally friendly technologies because of no direct GHG emissions during their operations. Renewable DERs can be connected to the grid or operate as a standalone system to serve residential, industrial, and commercial loads. The utilization of renewable DERs will reduce the number of people who are currently living without electricity around the world. Recent surveys carried out by the World Energy Council have projected that the world power output from renewable DERs will proliferate from 23% as it was in 2010 to approximately 34% in 2030 [32]. Similarly, according to a renewables 2018 global status report, the global power generation is classified as follows: nonrenewable energy (73.5%), wind (5.6%), hydropower (16.4%), bio-power (2.2%), PV (1.9%), and geothermal, concentrated solar power and ocean (0.4%) [33]. This statistical record has shown that about 24.5% of the energy utilized globally comes from renewable DERs. Moreover, DERs are projected to contribute 47.7% of the global power supply by 2040 [34].

2.5.1 PV System

The PV system comprises of numerous panels that are connected in series or parallel to generate the required power demanded in view of the voltage and current ratings specified by their respective manufacturers [24]. The PV system is a potential alternative to the conventional power system because of the environmentally friendly nature of solar resources [35]. The configurations of the PV system depend on the pattern of the usage, operational requirements, component topologies, and the load requirements. The PV system is one of the most rapidly growing DERs for power generation because it has emerged as a potential alternative that addresses some of the public concerns over the operations of conventional generating units. It has several advantages over nonrenewable DER technologies. The rapid increase in the PV system capacity in the world is due to the high competitiveness of the system with other renewable and nonrenewable DER technologies, government policies to encourage utilization of the PV systems for power generation, high demand of electric energy, a limited supply of fossil fuels, rising prices of fossil fuels, and long life span [9, 11, 36, 37].

The application of the PV system in the microgrid system is universally accepted as a potential power source that can reduce over dependence on the conventional generating units. The security of the power supply can be improved since the fluctuation in the price of crude oil does not affect the operation of a microgrid system that is integrated with a number of PV system. The economic benefits of the PV system can be used to stabilize the optimal operation of the traditional power system that has the PV units in the long run.

The statistical data presented by a renewables 2018 global status report (REN21) showed that the global installed capacity of the PV system at the end of 2017 was approximately 402 GW, and it has been projected to increase to about 872 GW by 2030 [33, 38].

2.5.1.1 Classifications of the PV System

The PV systems are classified by the utilities and customers based on their functions, configuration of the components that constitute the system, operational requirements, type of electrical load, and the mode of connecting the PV systems to other power sources. The PV systems are classified as standalone or grid-connected power systems, as shown in Fig. 2.7. They are designed by the utilities to operate independently or in parallel with the utility grid with the application of the BSS and other DERs technologies. The PV systems can produce both DC and AC power output based on the configuration of the utilities and consumers.

2.5.1.2 Grid-Connected PV Systems

The grid-connected PV systems are designed by the utilities and consumers to operate in conjunction with other DERs and the utility grid based on some benefits. The grid-connected PV systems consist of the bi-directional inverter, PV panels, battery system, smart meter, direct current (DC) bus system, and alternating current (AC) bus, as shown in Fig. 2.8 with their respective functions. A bi-directional interface is made between the PV system and the utility power output with the application of an inverter. The bi-directional interface or features introduced into the system allows the power generated by the PV panels

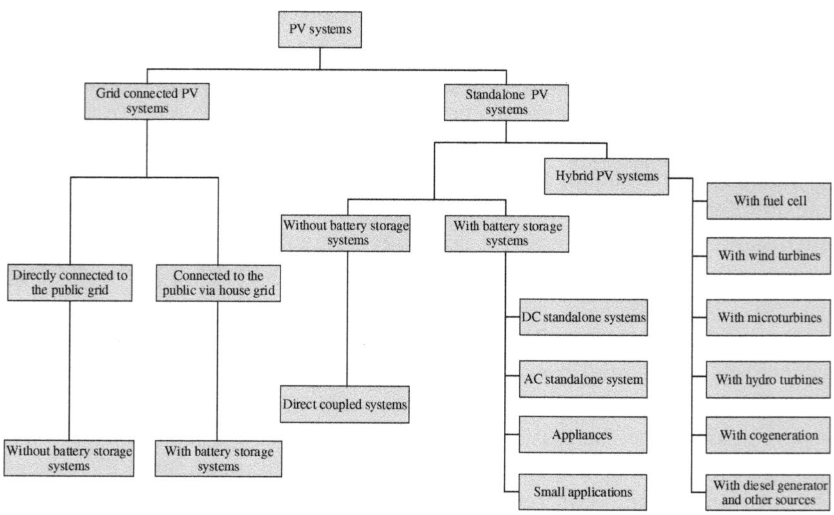

FIG. 2.7 Classification of PV systems.

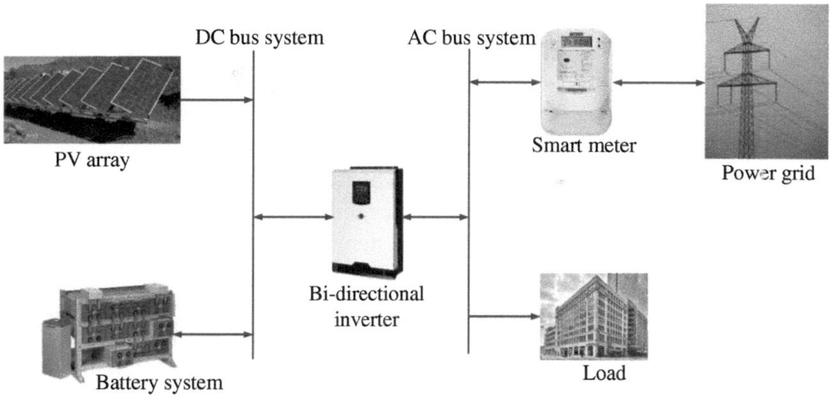

FIG. 2.8 Grid-connected PV system.

to supply the loads directly connected to the AC bus system and sends the excess power to the grid whenever the power output of the PV system is more than the power required by the consumers. The opposite occurs whenever the power demand is more than the output of the PV system. Some safety features are introduced into the grid-connected PV systems to prevent the system from feeding the grid whenever the utility grid is down due to faults or under maintenance [20].

2.5.1.3 Standalone PV Systems

The PV systems for standalone power application are designed by the customers to supply both DC and AC electrical loads. They can be powered by the PV panels and operate independently of the utility grid based on the specifications of the utilities and customers [20]. The simplest type of the standalone PV system that can power both AC and DC loads is presented in Fig. 2.9.

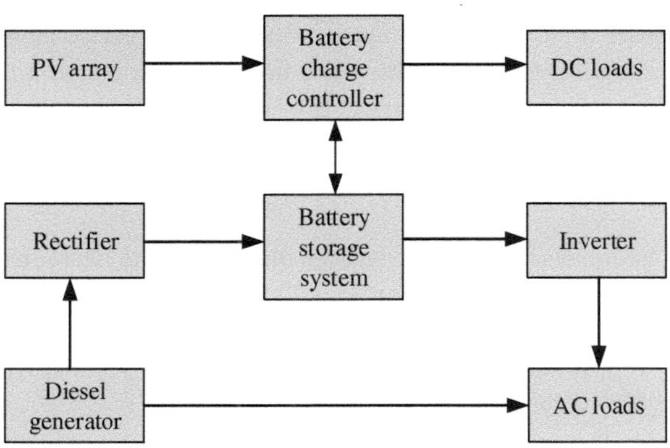

FIG. 2.9 Standalone PV system for AC and DC loads.

2.5.1.4 Modeling of PV Systems

The power generated by the PV systems is a function of the solar irradiance, ambient temperature, location of the sun in the sky, the conversion efficiency of the PV panels, sizes of the PV panels, and other input parameters from a manufacturer's data sheet [39]. Calculation procedures about the operations of the PV systems are discussed in this section. The rate at which the radiation energy from the sun incidents on the Earth's surface must be estimated as a prerequisite to evaluate the performance of the PV system. This is crucial for the optimal design and operation of a power system that utilizes the PV system for power generation applications. The operation of the PV system can be analyzed by monitoring the global radiation that is available on the Earth's surface. This is subject to the location of the site and the capacity of the PV system.

The global solar radiation incidents on a tilted plane is a combination of beam radiation, reflected radiation, and diffuse radiation [40]. The actual percentage of each component of global radiation varies with respect to weather conditions and location. Diffuse radiation is the solar radiation that is absorbed, scattered, or reflected by water vapor, dust particles or pollution when passing through the atmosphere. The amount of diffuse radiation is directly proportional to the quantity of the pollution in the atmosphere. The percentage of diffuse radiation also increases during the winter period. Beam solar radiation is the radiation that is obtained from the sun without being scattered by the climate. It is the radiation that is received from the sunrays that travel in a straight line from the sun to the Earth. It accounts for 70%–80% of the global radiation during the summer period and in the sunny regions of the world. The reflected solar radiation is the component of radiation that is reflected from surfaces, hills, trees, houses, and water bodies other than air particles. The reflected radiation generally accounts for a small percentage in the global radiation, but can contribute as much as 15% in snowy regions.

The amount of solar radiation incidents on the PV panel is strongly affected by the installation angle, hour angle, tilt of the PV panel, PV azimuth angle, location of the PV panel based on the latitude, hourly global solar radiation, and the diffusion fraction of the solar irradiation. The performance of the PV system can be assessed by estimating the hourly radiation on a tilted surface or horizontal surface. The solar irradiation incident (I_{pv}) on the PV array on an hourly basis is represented in Eq. (2.6) as follows [41]:

$$I_{pv} = (I + I_d)R_b + I_d \tag{2.6}$$

where R_b is a geometric factor that denotes the ratio of the beam irradiance incident on a tilted plane to that on a horizontal plane at time t, I_g is the global solar radiation incidents on a tilted plane (kWh/m^2), I_b is a mixture of beam radiation (kWh/m^2), I_r is a reflected radiation (kWh/m^2), and I_d is a diffuse radiation (kWh/m^2) [42].

The efficiency of the PV module can be expressed as follows [43]:

$$\eta_{pv} = \eta_R \left[1 - 0.9\beta\left(\frac{I_{pv}}{I_{pv,NT}}\right)(T_{C,NT} - T_{A,NT}) - \beta(T_a - T_r)\right] \tag{2.7}$$

where η_R is the efficiency of the PV module that is measured at a reference temperature (T_r), β is the temperature coefficient for cell efficiency (typically 0.0004–0.0005/°C), $I_{pv,\ NT}$ is the average solar irradiation incident on PV array at NT test conditions per hour, $T_{C,\ NT}$ (typically 45°C) is the cell temperature at NT test conditions, $T_{A,\ NT}$ (20°C) is the ambient temperature at NT test conditions, T_r is typically 20°C under standard test conditions, T_a is the ambient temperature, and NT is the nominal operating cell temperature.

The power output of PV system can be estimated using the following expression [44]:

$$P_{pv} = \eta_{pv} \times A_{pv} \times I_{pv} \tag{2.8}$$

where P_{pv} is the output power of the PV system (kW), A_{pv} is the surface area of the PV system (m^2), η_{pv} is the efficiency of the PV module, and I_{pv} is the solar irradiation incident on the PV array (kWh/m^2).

2.5.1.5 Advantages and Disadvantages of the PV Systems

The PV system is an electricity production resource that is directly connected to a local distribution system or connected to a host facility within the local distribution system [45]. The technological advancements, climate change policies, and increasing consumers' loads have increased the prospect of the PV system in the grid-connected and standalone power system. The PV system presents the opportunities to optimize overall system investments and provide a range of grid services. The benefits derived from the provision of a reliable power supply and explore opportunities to incorporate the PV system into electricity markets are numerous. The advantages and disadvantages of the PV system are presented in Table 2.7.

2.5.2 Wind System

The wind turbine generator (WTG) converts mechanical energy to electrical energy for various applications. A large number of the WTG units are used for production of domestic power supply and the excess power is fed into the utility grid. The WTG units have become significant sources of renewable energy that are used on the global note as a strategy to reduce overreliance on fossil fuels and GHG emissions. The arrays of the WTG units can be connected to the grid or operate autonomously as a standalone system. These have encouraged the incorporation of the WTG units into the utility electric grid [46].

TABLE 2.7 Advantages and Disadvantages of the PV Systems

Advantages	Disadvantages
• The electricity generated is clean without noise pollution.	• The power output of PV system is solar resources dependent.
• No fuel and reduced the cost of energy.	• The PV systems are fragile and can easily damage.
• No direct GHG emissions and environmentally friendly.	• The efficiency of the PV system is relatively low.
• The unused space on rooftops of existing buildings can be utilized for installation of PV system.	• Not reliable owing to intermittency and unpredictability of solar resources.
• Low O&M costs and no mechanically moving parts.	• Needs additional components to convert DC voltage to AC voltage.
• The PV system can be constructed to any size based on the power requirements of the consumers.	• Requires a lot of space.
• Provide an effective solution to energy demand during peak periods.	• The ESS that is used to smooth out the effect of solar resources intermittency is expensive.
• Economically viable and environmental sustainability.	• Utilization of toxic chemicals during the production of the PV panels.

The global installed capacity of the wind system was estimated to be 539 GW at the end of 2017 [33]. The global installed capacity of the wind power capacity has been anticipated to increase to approximately 2000 GW by 2030. This indicates that the wind system has been projected to supply between 16.7% and 18.8% of the global electric power by 2030 [47]. In addition to this, the greenhouse gas emissions are expected to be reduced by over 3 billion tons by 2030 with the application of the wind system [47].

2.5.2.1 *Modeling of the WTG*

The output power of the wind system is a function of the swept area of the rotor, efficiency of the wind turbine, characteristics of the wind system and aero-turbine performance [48]. Similarly, the intermittent changes in the wind speed with height affect the design of WTGs and evaluation of wind resources [27]. Consequently, a model of the wind speed based on the change with elevation is essential for the wind system assessment. The power generated by the WTG depends on the wind speed at the hub height and the operating characteristic of the wind speed. The output power of the WTG can be calculated by changing the measured wind speeds obtained at the site to corresponding values at the hub height, as presented in Eq. (2.9). The wind speed at a hub height can be estimated as follows [49, 50]:

$$v = v_r \left(\frac{H_{hub}}{H_r} \right)^{\alpha} \tag{2.9}$$

where v is the wind speed at the hub height (H_{hub}), v_r is the wind speed at reference height (H_r), and α is the power law exponent that signifies the ground surface friction co-efficient. The value of α varies with respect to the elevation, characteristics of the topography, wind speed, time, and season. The value of α is usually between 0.14 and 0.25, and it is normally 1/7 in a situation when there is no specific wind data for the site.

The relationship between the power output of the WTG and wind speed can be expressed as follows [51, 52]:

$$P_{wtg}(v(t)) = \begin{cases} 0 & v(t) < v_{ci} \\ a \times v^k - b \times P_r & v_{ci} \leq v(t) \leq v_r \\ P_r & v_r \leq v(t) \leq v_{co} \\ 0 & v(t) > v_{co} \end{cases} \tag{2.10}$$

where $a = \frac{P_r}{\left(v_r^k - v_{vci}^k\right)}$, $b = \frac{v_{ci}^k}{\left(v_r^k - v_{ci}^k\right)}$ and k is the Weibull shape parameter.

The formula for estimation of the power output of the WTG is expressed as follows [53]:

$$P_{wind} = \frac{1}{2} \rho A v^3 C_p \eta_g \eta_b \tag{2.11}$$

where ρ is the air density, v is the wind speed, η_g is the generator efficiency, η_b is the gear/bearing efficiency, A is the turbine rotor swept area, and C_p is the coefficient performance.

2.5.2.2 *Advantages and Disadvantages of the WTG System*

The WTG system supports a select number of rural electrification projects that are strategically located and potentially paired with other distributed energy resources and smart-grid technologies. The wind system also supports grid modernization and net metering policies and support grid interoperability that is aligned with market renewal. The operation of the WTG system is subject to the availability and intensity of the wind resource. The advantages and disadvantages of the WTG system are presented in Table 2.8.

2.5.2.3 *Types of the Wind Turbines*

The wind turbines consist of horizontal axis wind turbine (HAWT) and vertical axis wind turbine (VAWT) that are universally utilized for various applications.

TABLE 2.8 Advantages and Disadvantages of the WTG System

Advantages	Disadvantages
• Robust and simple in nature.	• Intermittency of speed affects its operation.
• Durable with a lifetime of 20 years.	• Requires ESS to smooth out the effects of stochastic characteristic of wind speed.
• Low cost of energy and O&M.	• Wind speed is unpredictable.
• No GHG emissions. • Increases energy security and job creation.	• Wind energy is not a reliable source owing to wind fluctuation.
• Renewable and sustainable and no fuel cost.	
• Environmentally friendly and provides a clean energy with no waste product.	
• WTG has an unlimited and free renewable resource to operate effectively.	

A commonly use HAWT is designed by the manufacturers in such a way that its blades is rotating on an axis parallel to the ground. On the other hand, a VAWT has its blades rotating on the axis that is perpendicular to the ground. The HAWTs and VAWTs can be used to meet the universal power demand since wind resources are readily available with the capability of generating enough power for the global population.

2.5.2.4 Horizontal Axis Wind Turbines

The generally utilized type of wind turbine is HAWT that consists of a rotor, tail, generator, and a frame upon which the rotor stands to produce power. The components of the HAWT are mounted on top of a tall tower since the blades of HAWT require a high-speed air to rotate effectively. The HAWT can be utilized in the areas where the wind factor is high since it is able to withstand strong forces. The shaft has a gear that rotates the generator when the wind hits the blade of the turbine. Due to this, the HAWT generates electrical power that will be sent to the grid or consumes locally with the application of a standalone power system. The HAWT is also designed with the integration of some important components that improve its efficiency. The nacelle, which has an anemometer, controller, and wind vane, is purposely incorporated into a HAWT to monitor the wind speed and wind direction. The HAWT is also designed with the incorporation of a yaw motor that turns the nacelle as the wind changes direction. The HAWT has a safety device that can slow down the shaft speed in a situation where there is a turbulent wind that can damage the turbine. This prevents the damage and associated cost that is related to such extreme conditions.

2.5.2.5 Advantages and Disadvantages of Horizontal Axis Wind Turbines

The operation of the HAWTs is a function of the availability and intensity of the wind resource. Some basic advantages and disadvantages of HAWTs are presented in Table 2.9.

2.5.2.6 Vertical Axis Wind Turbines

The VAWT is designed to proliferate swept area and enhance power generation capacity and as well as to maintain the intrinsic beauty of the original design. It is designed with the incorporation of main motor shaft that is set to transverse with the wind speed. The advancement in the design of the VAWT permits the main components of the turbine to be located at its base. With this arrangement, it is very easy to carry out maintenance on the turbine since the main components such as generator and gearbox are located very close to the ground. This will reduce the maintenance cost. In addition to this, the VAWT is designed in such a way that it does not need to be pointed into the direction of the wind; as a

TABLE 2.9 Advantages and Disadvantages of HAWTs

Advantages	Disadvantages
• The blades are mounted on to the side of a HAWT to stabilize the wind turbine.	• Difficult to transport and local opposition to aesthetics.
• Capable of reducing damage in a situation where wind turbine is subjected to extreme wind conditions.	• Difficulty in operating near ground winds.
• The tall tower base permits access to stronger wind in sites with wind shear.	• Difficult to install, it requires tall cranes and skilled operators.
• The pitch of the blades is designed to be changed based on the direction of the wind.	• Difficult to carry out operation and maintenance and required heavy construction.
• The towers of the wind turbine can be utilized to produce a higher power.	• It requires a massive tower base to support the components of a HAWT.
• Most are self-starting and high efficiency.	• The operation of a HAWT acts as an interference to signals of various telecom companies.
• Tall towers can be placed on rough parcels of land or in offshore territories.	• It requires very weighty and costly generator, gearbox, and blades.

result of this, it does not require the wind-sensing and orientation mechanisms. The wind turbines that have vertical axes are starting to become more popular as a way for generating localized electricity particularly for new constructions. The benefit of vertical axis turbines is that they can be placed much closer to the ground and are ideal for rooftop arrays.

2.5.2.7 Advantages and Disadvantages of Vertical Axis Wind Turbines

The operation of the VAWTs is subject to the availability and intensity of the wind resource. Some basic advantages and disadvantages of VAWTs are presented in Table 2.10.

2.6 Battery Storage System

The intermittent natures of the local renewable energy resources coupled with the need to have uninterrupted power supply at all times with minimal fuel and emission costs has necessitated the incorporation of battery system into a standalone or grid-connected power system. The battery system is used in combination with the renewable DERs to reduce the effects of stochastic natures of the local renewable energy resources in a microgrid system [54]. This will reduce the effects of power fluctuation and improve the reliability of power supply. The selection of appropriate battery storage technologies is based on the superior

TABLE 2.10 Advantages and Disadvantages of VAWTs

Advantages	Disadvantages
• Easy to maintain and install as compared to other wind turbine.	• The blades of VAWTs are always spinning back into the wind and cause dragging.
• Not directional and lower construction cost.	• Operate in lower, more turbulent wind.
• VAWTs are designed to generate electric power in any wind direction.	• VAWTs operate with a low starting torque that require energy to start turning.
• The supporting tower produces less noise and low maintenance downtime.	• The efficiency of VAWTs is very low compared to HAWTS that has three blades.
• Low production and transportation costs as compared to HAWTs.	• The VAWTs are not efficient since they need an initial push to operate.
• VAWTs do not need to be pointed to the wind direction, and the operations of VAWTs do not require yaw drive and pitch mechanism.	• The VAWTs operate with a turbulent vibration since the air that flows close the ground creates turbulent flow.

TABLE 2.10 Advantages and Disadvantages of VAWTs—cont'd

Advantages	Disadvantages
• Easy to transport from one place to another and more efficient in gusty winds.	• The continuous vibration of the mechanical part of VAWTs increases wearing and tearing of the bearing. This increases of maintenance costs.
• The operations of VAWTs do not pose any risk to human and animals since blades rotate at low speeds.	• They can create noise pollution.
• The VAWTs are suitable for locations or sites with extreme or turbulent wind speed.	
• The VAWTs have greater surface areas for energy capture.	

performance of one battery storage technology over another. The performance of the battery system is a function of some parameters such as cost, ambient temperature, environmental impacts, state of charge, duty cycle, voltage effects, flexibility, rate of charging, energy density, and rate of discharging. The listed factors can be used to determine the lifetime and the best choice of the battery system. The battery system can be classified into several categories in light of the response times, capacities, functions, technologies, and form of energy stored in the system [55]. The battery system is classified into different categories, as presented in Fig. 2.10 and Table 2.11.

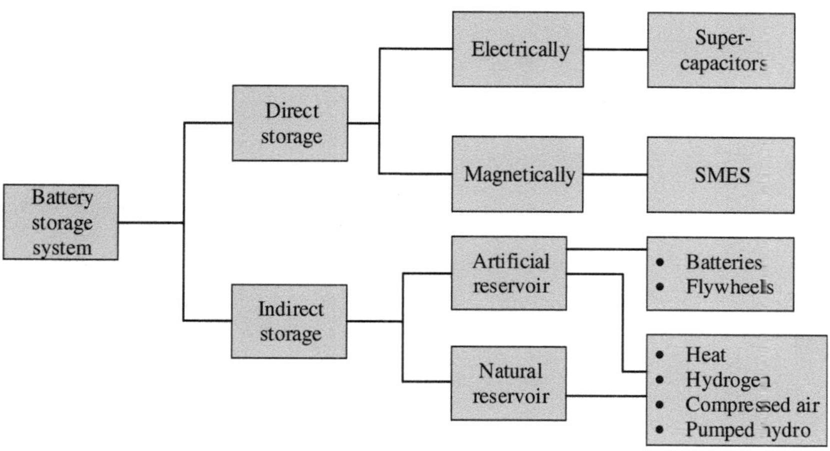

FIG. 2.10 Classifications of electric storage systems.

TABLE 2.11 Classification of Electric Storage Systems [9, 56–60]

Classes	Technology	Available Power Rating	Power Capital Cost ($/kW)	Energy Capital Cost ($/kWh)	Lifetime (years)	Efficiency %	Energy Density (Wh/kg)	Environmental Impacts
Mechanical	Pumped hydroelectric storage (PHS)	10–5000 MW	600–2000	5–100	40–60	70–82	0.5–1.5	Negative
	Compressed air energy storage (CAES) underground	Up to 400 MW	400–800	2–50	20–40	70–89	30–60	Negative
	Flywheel energy storage (FES)	Up to 250 kW	250–350	350	15	93–95	10–30	Almost
Electrical	Capacitors	Up to 50 kW	200–400	500–1000	5	60–65	0.05–5	Small
	Super capacitors	Up to 300 kW	100–300	300–2000	10–30	90–97	2.5–15	Small
	Superconducting magnetic energy storage (SMES)	10–100 MW	200–300	1000–10,000	20+	95–98	0.5–5	Benign

Thermochemical	Solar fuels	Up to 10 MW	3–24	-	20–25	20–30	800–100,000	Small
Chemical	Hydrogen storage with fuel cells	0.3–50 MW	1500–3000	3–23	5–15	33–42	800–10,000	Small
Thermal energy storage (TES)	Aquiferous low temperature TES (AL-TES)	Up to 5 MW	100–400	20–50	10–20	30–62	80–250	Small
	Cryogenic energy storage (CES)	100–300 kW	200–300	3–30	25+	85–95	150–250	Small
Electrochemical: conventional rechargeable batteries	Lead-acid	Up to 20 MW	300–600	200–302	5–15	70–90	30–50	Negative
	NiCd	Up to 40 MW	500–1500	800–1500	10–20	60–73	50–75	Negative
	NaS	5 kW–8 MW	1000–3000	300–500	10–15	75–90	150–240	Negative
	ZEBRA	Up to 300 kW	150–300	100–200	10–20	86–88	100–120	Negative
	Li-ion	1–100 kW	1200–4000	600–2500	5–15	85–95	75–200	Negative
Electrochemical: flow batteries, energy storage (FBES)	VRB	30 kW–3 MW	600–1500	150–1000	5–10	65–85	10–30	Negative
	ZnBr	50 kW–2 MW	700–2500	150–1000	5–10	60–70	10–50	Negative
	PSB	1–15 MW	700–2500	150–1000	10–15	65–85	10–15	Negative

2.7 Conclusion

The technology development and concern about environmental effects have made DERs emerge as the potential alternative to improve the efficiency of power system. The potential of distributed energy technologies can be achieved with the application of multiple resources thorough microgrid or hybrid systems. The microgrid system has some benefits such as improved reliability, reduced transmission and distribution line losses, ancillary voltage support, improved power system efficiency by utilizing waste heat from the exhaust to generate thermal energy for many applications through combined heat and power system, and reduced rate of power interruption and associated costs. In this chapter, an overview of DERs has been introduced as a measure to examine the economic and environmental effects of fossil fuel-based conventional generating units. The operation of conventional generating units can be limited with the incorporation of renewable DERs into a power system in order to reduce the amount of GHG emissions, improve the energy efficiency, reduce the adverse effects of pollution on human health, and conserve natural resources. Owing to the environmental effects of the conventional generating units as clarified in this chapter, it is not ideal to depend on traditional power systems that operate only with the fossil fuel-based generating units. In view of this, the best option is to find the most suitable alternative that will reduce the adverse environmental effects that are associated with the operation of traditional methods but also meet the power demand at reduced operating costs. Due to the economic, technical, and environmental benefits of renewable energy resources, the utilization of renewable energy distributed energy technologies has been proposed in this chapter as one of the promising solutions to solve the global power crisis and environmental challenges as a result of GHG emissions. The benefits of DERs can be maximized with their optimal integration into distribution systems. This will not only guarantee the best utilization of the resource, but also minimize investment and operation costs. This chapter provided an assessment of the state of the art of DER technologies and presented the challenges and up-to-date developments of their applications in the power system.

References

[1] International Energy Agency. World energy outlook 2017—executive summary, 2017. Available online: https://www.iea.org/access2017/, Accessed September 2018.

[2] A.F. Zobaa, M.M. Canteli, R.C. Bansal, Power Quality: Monitoring, Analysis and Enhancement, InTech, 2011.

[3] T. Adefarati, R.C. Bansal, Reliability and economic assessment of a microgrid power system with the integration of renewable energy resources, Appl. Energy 206 (2017) 911–933.

[4] T. Bhatti, R.C. Bansal, D. Kothari, Small Hydro Power Systems, Dhanpat Rai & Co., New Delhi, 2004.

[5] S. Doolla, T. Bhatti, R. Bansal, Load frequency control of an isolated small hydro power plant using multi-pipe scheme, Electr. Power Compon. Syst. 39 (1) (2011) 46–63.

[6] T. Adefarati, R.C. Bansal, The impacts of PV-wind-diesel-electric storage hybrid system on the reliability of a power system, Energy Procedia 105 (2017) 616–621.
[7] T. Adefarati, R.C. Bansal, J.J. Justo, Techno-economic analysis of a PV–wind–battery–diesel standalone power system in a remote area, J. Eng. 2017 (13) (2017) 740–744.
[8] T. Adefarati, R.C. Bansal, J.J. Justo, Reliability and economic evaluation of a microgrid power system, Energy Procedia 142 (2017) 43–48.
[9] T. Adefarati, R.C. Bansal, Integration of renewable distributed generators into the distribution system: a review, IET Renew. Power Gener. 10 (7) (2016) 873–884.
[10] FEMP US Department of Energy, Using distributed energy resources—a how-to guide for federal energy managers, Cogener. Compet. Power J. 17 (4) (2002) 37–68.
[11] M. Ahmad, Operation and Control of Renewable Energy Systems, John Wiley & Sons, 2017.
[12] J. Zhu, Optimization of Power System Operation, John Wiley & Sons, 2015.
[13] W. El-Khattam, M.M. Salama, Distributed generation technologies, definitions and benefits, Electr. Pow. Syst. Res. 71 (2) (2004) 119–128.
[14] D.Q. Hung, M. Nadarajah, DG allocation in primary distribution systems considering loss reduction, in: A.F. Zobaa, R.C. Bansal (Eds.), Handbook of Renewable Energy Technology, World Scientific Publishing Co, 2011, pp. 587–635.
[15] N. Strachan, A. Farrell, Emissions from distributed vs. centralized generation: the importance of system performance, Energy Policy 34 (17) (2006) 2677–2689.
[16] A.M. Borbely, J.F. Kreider, Distributed Generation: The Power Paradigm for the New Millennium, CRC press, 2001.
[17] International Renewable Energy Agency: Renewable power generation costs in 2014, Available online: http://www.irena.org/menu/index.aspx?mnu=Subcat&PriMenuID=36&CatID=141&SubcatID=494, Accessed September 2018.
[18] Gas turbine for power generation: introduction, Available online: http://www.wartsila.com/energy/learning-center/technical-comparisons/gas-turbine-for-power-generation-introduction, Accessed September 2018.
[19] Gas turbine engines, Available online: http://petrowiki.org/gas_turbine_engines, Accessed September 2018.
[20] Florida solar energy center creating energy independence, Available online: http://www.fsec.ucf.edu/en/consumer/solar_electricity/basics/types_of_pv.htm, Accessed September 2018.
[21] A. Polyzakis, C. Koroneos, G. Xydis, Optimum gas turbine cycle for combined cycle power plant, Energ. Conver. Manage. 49 (4) (2008) 551–563.
[22] Gas turbine power plant with regeneration, reheating and intercooling, Available online: https://basicmechanicalengineering.com/gas-turbine-power-plant-with-regeneration-reheat-intercooling/, Accessed September 2018.
[23] T. Adefarati, N.B. Papy, M. Thopil, H. Tazvinga, Non-renewable distributed generation technologies: a review, in: Handbook of Distributed Generation, Springer, 2017, pp. 69–105.
[24] T. Adefarati, R.C. Bansal, Reliability assessment of distribution system with the integration of renewable distributed generation, Appl. Energy 185 (2017) 158–171.
[25] Diesel generator, Available: http://www.eec-fncci.org/content-learn-dieselgenerator, Accessed September 2018.
[26] B.K. Das, Y.M. Al-Abdeli, G. Kothapalli, Optimisation of stand-alone hybrid energy systems supplemented by combustion-based prime movers, Appl. Energy 196 (2017) 18–33.
[27] S. Kong, R. Bansal, Z. Dong, Comparative small-signal stability analyses of PMSG-, DFIG- and SCIG-based wind farms, Int. J. Ambient Energy 33 (2) (2012) 87–97.
[28] K. Kusakana, Operation cost minimization of photovoltaic–diesel–battery hybrid systems, Energy 85 (2015) 645–653.

[29] Diesel generators: Improving efficiency and emission performance in India, Available online: http://shaktifoundation.in/wp-content/uploads/2017/06/ICF-2014-Diesel-Generators-Improving-Efficiency-and-Emission-Performance-in-India.pdf, Accessed September 2018.

[30] F.A. Mohamed, H.N. Koivo, System modelling and online optimal management of microgrid using mesh adaptive direct search, Int. J. Electr. Power Energy Syst. 32 (5) (2010) 398–407.

[31] A. Arulampalam, N. Mithulananthan, R. C. Bansal, and T. Saha, Micro-grid control of PV-wind-diesel hybrid system with islanded and grid connected operations, IEEE International Conference on Sustainable Energy Technologies (ICSET), pp. 1–5, 2010.

[32] J. Salvatore, World Energy Perspective: Cost of Energy Technologies, World Energy Council, London, UK, 2013.

[33] Renewables 2018 global status report, Available online: http://www.ren21.net/wp-content/uploads/2018/06/17-8652_GSR2018_FullReport_web_final_.pdf, Accessed on September 2018.

[34] A. Demirbas, Global renewable energy projections, Energy Sources B 4 (2) (2009) 212–224.

[35] H. Tazvinga, M. Thopil, P.B. Numbi, T. Adefarati, Distributed renewable energy technologies, in: Handbook of Distributed Generation, Springer, 2017, pp. 3–67.

[36] L. Ma, et al., Multi-party energy management for smart building cluster with PV systems using automatic demand response, Energ. Buildings 121 (2016) 11–21.

[37] H. Tazvinga, T. Hove, Photovoltaic/Diesel/Battery Hybrid Power Supply System, VDM Publishers, 2010.

[38] M. Taylor, K. Daniel, A. Ilas, E.Y. So, Renewable Power Generation Costs in 2014, International Renewable Energy Agency, Masdar City, Abu Dhabi, 2015.

[39] A.F. Zobaa, R.C. Bansal, Handbook of Renewable Energy Technology, World Scientific, 2011.

[40] L. Olatomiwa, S. Mekhilef, O.S. Ohunakin, Hybrid renewable power supply for rural health clinics (RHC) in six geo-political zones of Nigeria, Sustain. Energy Technol. Assess. 13 (2016) 1–12.

[41] M. Collares-Pereira, A. Rabl, The average distribution of solar radiation-correlations between diffuse and hemispherical and between daily and hourly insolation values, Sol. Energy 22 (2) (1979) 155–164.

[42] G. Tina, S. Gagliano, S. Raiti, Hybrid solar/wind power system probabilistic modelling for long-term performance assessment, Sol. Energy 80 (5) (2006) 578–588.

[43] T. Hove, A method for predicting long-term average performance of photovoltaic systems, Renew. Energy 21 (2) (2000) 207–229.

[44] F. Blaabjerg, D.M. Ionel, Renewable energy devices and systems–research, Electr. Power Compon. Syst. 43 (8–10) (2015) 837–838.

[45] R. C. Bansal and T. Bhatti, Small Signal Analysis of Isolated Hybrid Power Systems: Reactive Power and Frequency Control Analysis. Alpha Science, 2008.

[46] L. Gidwani, H. Tiwari, R. Bansal, Improving power quality of wind energy conversion system with unconventional power electronic interface, Int. J. Electr. Power Energy Syst. 44 (1) (2013) 445–453.

[47] Global Wind Energy Council (GWEC), Available online: http://www.gwec.net/global-figures/graph, Accessed September 2018.

[48] W. Chong, M. Naghavi, S. Poh, T. Mahlia, K. Pan, Techno-economic analysis of a wind–solar hybrid renewable energy system with rainwater collection feature for urban high-rise application, Appl. Energy 88 (11) (2011) 4067–4077.

[49] M.R. Patel, Wind and Solar Power Systems: Design, Analysis, and Operation, CRC press, 2005.

[50] R. Belfkira, L. Zhang, G. Barakat, Optimal sizing study of hybrid wind/PV/diesel power generation unit, Sol. Energy 85 (1) (2011) 100–110.

[51] P.C. Del Granado, Z. Pang, S.W. Wallace, Synergy of smart grids and hybrid distributed generation on the value of energy storage, Appl. Energy 170 (2016) 476–488.

[52] R. Rocchetta, Y. Li, E. Zio, Risk assessment and risk-cost optimization of distributed power generation systems considering extreme weather conditions, Reliab. Eng. Syst. Saf. 136 (2015) 47–61.

[53] S. Ashok, Optimised model for community-based hybrid energy system, Renew. Energy 32 (7) (2007) 1155–1164.

[54] B. Battke, T.S. Schmidt, Cost-efficient demand-pull policies for multi-purpose technologies–The case of stationary electricity storage, Appl. Energy 155 (2015) 334–348.

[55] A. Evans, V. Strezov, T.J. Evans, Assessment of utility energy storage options for increased renewable energy penetration, Renew. Sustain. Energy Rev. 16 (6) (2012) 4141–4147.

[56] X. Luo, J. Wang, M. Dooner, J. Clarke, Overview of current development in electrical energy storage technologies and the application potential in power system operation, Appl Energy 137 (2015) 511–536.

[57] B. Zakeri, S. Syri, Electrical energy storage systems: a comparative life cycle cost analysis, Renew. Sustain. Energy Rev. 42 (2015) 569–596.

[58] S. Mekhilef, R. Saidur, A. Safari, Comparative study of different fuel cell technologies, Renew. Sustain. Energy Rev. 16 (1) (2012) 981–989.

[59] F. Blaabjerg, D.M. Ionel, Renewable energy devices and systems–state-of-the-art technology, research and development, challenges and future trends, Electr. Power Compon. Syst. 43 (12) (2015) 1319–1328.

[60] S. Ahmad Hamidi, D.M. Ionel, A. Nasiri, Modeling and management of batteries and ultracapacitors for renewable energy support in electric power systems–an overview, Electr. Power Compon. Syst. 43 (12) (2015) 1434–1452.

Chapter 3

Energy Storage in Smart Grids

Hussein Ibrahim*, Miloud Rezkallah†,*, Mazen Ghandour‡ and Ambrish Chandra†

Institut Technologique de Maintenance Industrielle (ITMI), Cégep de Sept-Îles, Sept-Îles, QC, Canada, †École de Technologie supérieure (ETS), Montréal, QC, Canada, ‡Faculty of Engineering, Lebanese University, Beirut, Lebanon

Chapter Outline

3.1 Introduction

Until now, the conventional power grid (CPG), which consists of various power subsystems elements, such as energy sources, power transformers, transmissions line, loads, etc., has continued to operate based on the old concept of supply follows demand. Before, the electricity was generated at a central plant, but now more energy provided by renewable energy sources, such as

Pathways to a Smarter Power System. https://doi.org/10.1016/B978-0-08-102592-5.00003-X

wind, solar, hydro, biomass, etc., is injected into the CPG. Although these energy sources are intermittent and stochastic, it is possible to predict the generated power from these sources and manipulate other generation and demand accordingly [1]. Therefore, a bigger storage system that can store the predicted renewable power and restore it for demand to follow supply. In other words, it is possible to switch from supply-follows-demand to demand-follows-supply using the stored energy. However, effective management of power in CPG infrastructures requires advanced sensing and communication technologies. This requires the deployment of information and communication technology (ICT) in CPG and informing the consumers in real-time about consumption through their smart meters. According to [2], in the future SPG, millions of customers will produce their own power in their houses, factories, schools, electric vehicles, etc., and share it with the grid. Therefore, ICT will transform the power grid into an energy-sharing intergrid. This new philosophy is still in the concept stage. It requires flexible operating strategies and innovative components and equipment. To realize this concept, some countries such as Germany [3] and the United Kingdom [4] are focused on smart metering, because smart meters are the most relevant elements in SPG. According to [3], by 2032 more than 50 million meters in Germany are supposed to be electronic and many of them smart. Currently, 90.5% of installed meters right now are electromechanically [4]. Many research works are reported in the literature, where authors have proposed new methodologies and advanced equipment that are required to switch to the new concept of power grid, where smart metering and ICT represent the major contribution in smart grid investigations [5–11]. This is also confirmed in [12], where smart meters and management of energy storage (ES) at the customer level are suggested as first investigations in SPG.

3.2 Convectional Power Grid Versus Smart Power Grid

Generally, the CPG (as shown in Fig. 3.1) is characterized by four operations: power generation, transmission, distribution, and control [7–13]. Power generation is centralized and most installed meters are electromechanical. Grid monitoring, as well as its restoration, is manually operated. Communication occurs in one direction, and customers do not have many choices. As shown in Fig. 3.2 [15], new functionalities and smart infrastructure are integrated to manage the power smartly, and restore it automatically. Compared to the CPG, in the SPG, power and information flows are two-way. Furthermore, integration of the storage system at the distribution level helps to stabilize the grid by instantly compensating the power in demand, as well as the fluctuating behavior of renewable energy sources.

Integration of the smart metering, or as it is also called, the next generation of the power measurement in the power grid, plays a key role in SPG. A different electrical energy metering standard, as well the effect of harmonics on the measured data, is given in [14]. The smart meter security issue and privacy as well as all solutions are given in detail in [14, 16–18].

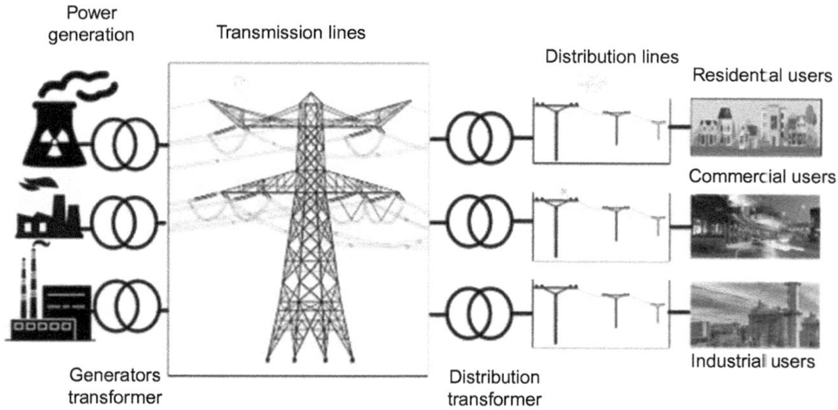

FIG. 3.1 Simplified scheme of the conventional power grid.

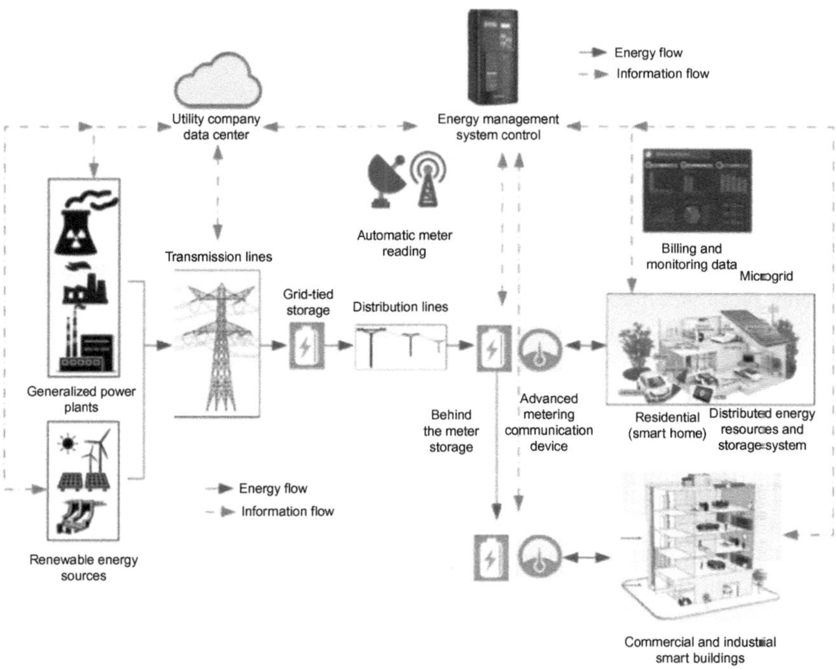

FIG. 3.2 Simplified scheme of future smart grid [14].

3.3 Role and Application of Energy Storage in Smart Grid

As already indicated, the ES system plays a key role in the stability of the grid by compensating for the intermittency and variability of RESs. According to [19–21], ES can provide many services at different levels in a power grid. The services and benefits are classified into five categories, as presented in

TABLE 3.1 ESS Services in Power Grid [19]

Category	Description
Bulk energy service	• Arbitrage • Reserve capacity • Compensate the fluctuation power provided from RESs
Ancillary service	• Maintain the system frequency constant • Operating reserve
Transmission service	• Upgrade delay • Congestion relief
Distribution service	• Upgrade delay • Maintain the voltage constant • Outage mitigation
User-end service	• Power quality improvement • Balance the instantaneously power demand • Peak shaving • Regulate the voltage and frequency in islanding mode operation

Table 3.1 [22]: (1) bulk energy; (2) ancillary; (3) transmission; (4) distribution; and (5) energy management services. As we know that ES requires information from an SPG to achieve its complete potential, so collecting data in real time using smart meters such as generating power from RESs, power demand, price of energy, power quality, etc. are encouraged to integrate the ESS technologies at different levels of the power grid as is detailed in Fig. 3.2. Regarding the bulk energy services, ESS stores bulk energy and uses it during peak demand to arbitrage the production price. It can satisfy the peak demand and compensate the intermittency of the generated power from RESs. For ancillary services, the ESSs regulate the system by balance the power between the generation and load during transition and peak demand, and for transmission service, ESS acts on the system stability by providing power to prevent discharging of devices during peak hours [23]. With regard to distribution services, ESS can regulate the voltage and frequency during the islanding mode and balance the power between the local generation and demand. Furthermore, ESS installed at customers level can easily use for peak shaving.

In this chapter, a detailed study of energy storage technology in microgrid based on solar photovoltaic systems for smart grid application is detailed. Advanced control strategies to control the battery ESS (BESS), achieve maximum power point tracking, regulate constant the frequency as well as the voltage, and improve the power quality at the point common of coupling are presented. The proposed configuration operates in grid connected and islanding modes.

3.4 Application

Fig. 3.1 shows the proposed microgrid configuration based on solar photovoltaic systems and BESS dedicated to smart house applications in smart grids. Using this configuration, wind turbine and diesel engine-based variable speed generators, as well as battery chargers for electrical vehicles, can easily be connected to the DC bus. The same three-phase inverter with the proposed control strategies can be used. The proposed configuration operates in connected and islanding modes. In connected mode, the solar photovoltaic system provides a clean and stable power to the connected load. It charges the BESS and injects a clean power into grid during peak hour demand. In islanding mode, the solar photovoltaic system supplies the load. In both modes, BESS balance the power in the system, compensate for the fluctuation of generated power from the solar photovoltaic system, and support it in islanding mode by supplying the load together.

3.5 Control System

In this section, the developed control strategies for DC-DC buck boost power converter and three-phase inverter are presented.

3.5.1 Control of the DC-DC Buck Boost Converter

The ESS, which represents the lead acid battery pack as shown in Fig. 3.3, is connected to the common DC bus through a DC-DC buck boost converter. It is controlled using sliding mode control with boundary layer to achieve high performance from the solar photovoltaic system, compensate for the fluctuation of the generated power, balance the power in the system during transition, and when system operate in islanding mode [24].

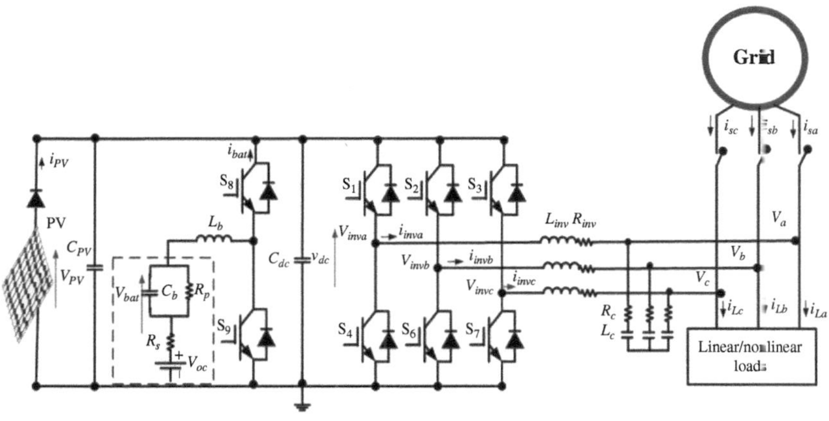

FIG. 3.3 Microgrid configuration based on solar photovoltaic system and BESS.

3.5.2 Mathematical Model of the DC-DC Power Converter

As presented in Fig. 3.1, the DC-DC power converter consists of battery voltage (V_{bat}) controlled switches (S_8 and S_9), inductor (L_b), output DC voltage (V_{dc}), and a capacitor (C_{dc}). The state-space equations of the DC converter when the switch is ON are described as follows:

$$\begin{cases} \dfrac{di_{bat}}{dt} = \dfrac{V_{bat}}{L_b} \\ \dfrac{dV_{dc}}{dt} = -\dfrac{V_{dc}}{RC_{dc}} \end{cases} \tag{3.1}$$

where R is load resistance.

If the switch is OFF, the state-space equations are written as follows:

$$\begin{cases} \dfrac{di_{bat}}{dt} = \dfrac{V_{dc}}{L_b} \\ \dfrac{dV_{dc}}{dt} = -\dfrac{i_{bat}}{C_{dc}} - \dfrac{V_{dc}}{RC_{dc}} \end{cases} \tag{3.2}$$

Replacing, x_1 by i_{bat}, and x_2 by V_{dc}, then; $\dot{x}_1 = \frac{di_{bat}}{dt}$, $\dot{x}_2 = \frac{dV_{dc}}{dt}$. The state-space equations, which are expressed in Eqs. (3.1) and (3.2), are expressed as follows:

$$\begin{cases} \dot{x}_1 = A_1 x + B_1 u \\ V_{dc} = C_1 x \end{cases} \tag{3.3}$$

where $A_1 = \begin{bmatrix} 0 & 0 \\ 0 & -\dfrac{1}{RC_{dc}} \end{bmatrix}, B_1 = \begin{bmatrix} \dfrac{1}{L_b} \\ 0 \end{bmatrix}, C_1 = [0 \;\; 1]$, and $u = v_{bat}$.

$$\begin{cases} \dot{x}_2 = A_2 x + B_2 u \\ V_{dc} = C_2 x \end{cases} \tag{3.4}$$

where $A_2 = \begin{bmatrix} 0 & \dfrac{1}{L_b} \\ -\dfrac{1}{C_{dc}} & -\dfrac{1}{RC_{dc}} \end{bmatrix}, B_1 = \begin{bmatrix} 0 \\ 0 \end{bmatrix}, C_1 = [0 \;\; 1]$, and $u = v_{bat}$.

Using Eqs. (3.3) and (3.4), one can obtain the average state-space equation of the DC-DC power converter as follows:

$$\begin{cases} \dot{x}(t) = Ax(t) + Bu(t) \\ V_{dc} = Cx(t) \end{cases} \tag{3.5}$$

where $A = \begin{bmatrix} 0 & \dfrac{1-d}{L_b} \\ -\dfrac{1-d}{C_{dc}} & -\dfrac{1}{RC_{dc}} \end{bmatrix}, B = \begin{bmatrix} \dfrac{d}{L_b} \\ 0 \end{bmatrix} v_{bat}$, and $C = [0 \;\; 1]$.

3.5.3 Sliding Mode Control for the DC-DC Power Converter

The scheme of the developed control strategy of the DC-DC power converter is presented in Fig. 3.4. The model of the DC-DC power converter as given in Eq. (3.6) is nonlinear. Therefore, to realize high dynamic performance during transition, and when the system operates in islanding mode with optimal accuracy regulation and robust trajectory tracking in presence of uncertain parameters, a sliding mode control approach is proposed. The control design of the general control (d) for the DC-DC power converter is obtained as follows:

3.5.4 Selection of the Sliding Surface

The sliding surface (σ) as expressed in Eq. (3.6) is selected to ensure reaching of the surface with the desired dynamics of the corresponding sliding motion [25, 26]:

$$\sigma = \beta_1 \left(x_1 - x_1^*\right) + \beta_2 \left(x_2 - x_2^*\right) \tag{3.6}$$

where β_1 and β_2 denote the sliding gains, and they should be positive.

x_1^* and x_2^* are the battery current and DC-link voltage references, respectively. The value of the DC-link voltage reference is equal to output PV voltage (V_{PV}), which is selected equal to 350 V. As shown in Fig. 3.5, V_{PV} corresponds

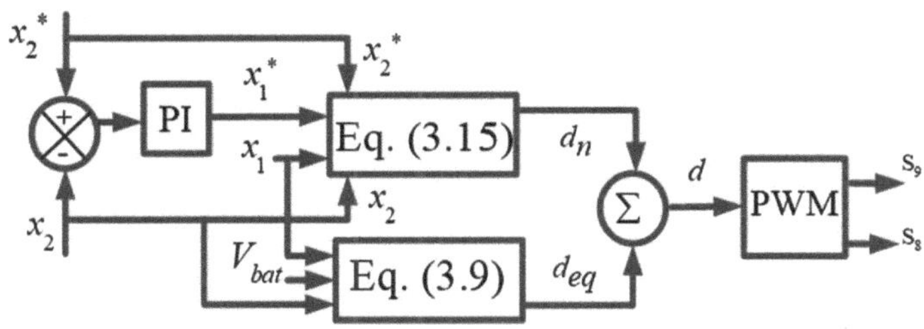

FIG. 3.4 Control of the DC-DC power converter.

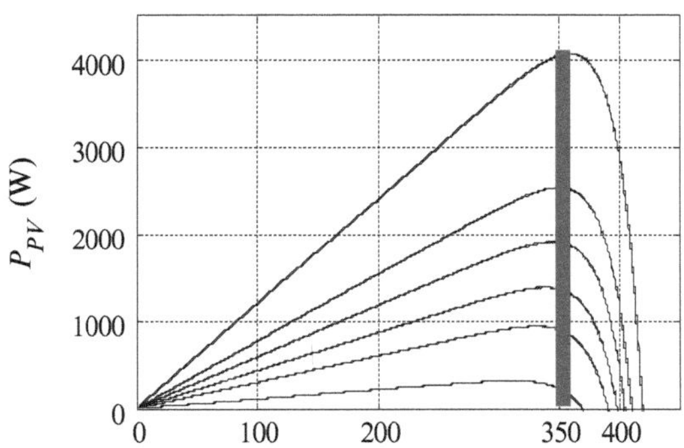

FIG. 3.5 $P_{PV}=f(V_{PV})$ with fixed temperature and solar irradiation change.

to the maximum extracted power from the solar photovoltaic system for different solar irradiations. Therefore, by controlling the DC-link voltage, one can obtain the maximum of power without using any MPPT technique.

3.5.5 Equivalent Control

The structure of the desired control (d) is written as follows [24]:

$$d = d_{eq} + d_n \tag{3.7}$$

where d_n denotes the nonlinear control and d_{eq} is the equivalent control, which is obtained using Eqs. (3.5) and (3.6), and by setting the derivative of Eq. (3.6) to zero as follows:

$$\dot{\sigma} = 0 \tag{3.8}$$

Replacing $\dot{x}_1$ and $\dot{x}_2$ with their equality, the following expression is obtained:

$$d_{eq} = \frac{x_2(\beta_1 RC_{dc} - \beta_2 L_b) - \beta_2 RL_b x_1}{\beta_2 RL_b x_1 + \beta_1 RC_{dc}(x_2 - V_{bat})} \tag{3.9}$$

3.5.6 Stability Analysis

The objective of the approach using sliding mode control is to guarantee the convergence of the operating points. To ensure the stability of the control, the Lyapunov stability function is used. According to Lyapunov's theorem, a nonlinear time variant system is globally and uniformly stable if it satisfies the following conditions [27, 28]:

$$\begin{cases} V(0) = 0 \\ V(X) > 0 \\ V(\dot{X}) < 0 \\ Then\, X = 0\, is asymptotically stable \end{cases} \tag{3.10}$$

For the DC-DC power converter, the Lyapunov function is defined as follows [24]:

$$V = \frac{1}{2}\sigma^2 \tag{3.11}$$

The system is considered asymptotically stable as detailed in Eq. (3.10); the derivative of Eq. (3.11) should be negative:

$$\dot{V} = \sigma\dot{\sigma} < 0 \tag{3.12}$$

Replacing Eqs. (3.6) and (3.8) in Eq. (3.12), one obtains the following expression:

$$\dot{V} = \sigma\dot{\sigma} = \left(\beta_1\left(x_1 - x_1^*\right) + \beta_2\left(x_2 - x_2^*\right)\right)\left(\beta_1\dot{x}_1 + \beta_2\dot{x}_2\right) < 0 \tag{3.13}$$

where x_1^* and x_2^* denote the references of the battery current, which represents the output of the PI controller with antiwindup for DC link voltage regulator and DC link voltage, respectively. The DC link voltage is equal to 350 V.

Replacing terms $\dot{x}_1, \dot{x}_2, x_1, x_2$, and d by their equalities in Eq. (3.12) gives:

$$\dot{V}=\sigma\dot{\sigma}=-\left[\overbrace{\left(\frac{V_{dc}^2\beta_2^2}{RC_{dc}}\right)+\left(\frac{i_{bat}^2V_{bat}\beta_2\beta_1}{V_{dc}}\right)+\left(\frac{V_{bat}^2\beta_2\beta_1}{L_b}\right)+\left(\frac{V_{bat}\beta_2\beta_1}{L_b}x_2^*\right)+\left(\frac{V_{dc}i_{bat}\beta_2\beta_1}{RC_{dc}}\right)+\left(V_{dc}i_{bat}\beta_2^2x_1^*\right)}^{\text{Term1}}\right]$$
$$+\left[\overbrace{\left(\frac{V_{dc}^2\beta_1}{L_b}\right)+\left(V_{bat}\beta_1^2i_{bat}\right)+\left(\frac{V_{dc}\beta_2}{RC_{dc}}\right)+\left(\frac{V_{dc}\beta_2\beta_1}{RC_{dc}}x_1^*\right)+\left(\frac{V_{bat}V_{dc}\beta_2\beta_1}{L_b}\right)+\left(\frac{V_{bat}\beta_1^2i_{bat}}{L_b}\right)+\left(\frac{V_{bat}\beta_1^2i_{bat}}{C_{dc}V_{dc}}x_2^*\right)+\left(\frac{i_{bat}V_{bat}\beta_2\beta_1}{C_{dc}V_{dc}}x_1^*\right)}^{\text{Term2}}\right]<0 \quad (3.14)$$

Replacing the parameters L_b, C_{dc}, V_{bat}, and V_{dc}, as well as the optimal gains of the sliding mode control β_1 and β_2 by their value given in [24], the first term with negative sign in Eq. (3.14) is larger than the second term. This leads to the condition given in Eq. (3.13) being satisfied, which confirms that the system is asymptotically stable.

To ensure the robustness, the second of the desired control d_n, which represents the discontinuous control, is written as follows [35–37]:

$$d_n=\beta_3 sat(\sigma,\varnothing) \quad (3.15)$$

where β_3 is positive control gain.

To reduce the chattering of the control signal, the saturation function with boundary conditions is defined:

$$sat(\sigma,\varnothing)\begin{cases} 1 & \sigma>\varnothing \\ \dfrac{\sigma}{\varnothing} & |\sigma|\le\varnothing \\ -1 & \sigma<-\varnothing \end{cases} \quad (3.16)$$

where Φ denotes the sliding layer, which is selected between 0.5 and −0.5 [24].

The optimal gains (β_1, β_2, and β_3) of the sliding mode control are selected equal to $\beta_1=0.001$, $\beta_2=0.8$, and $\beta_3=5$, respectively. It can be observed in Fig. 3.6C that based on the optimal values of β_1, β_2, and β_3, the signal reaches the origin rapidly.

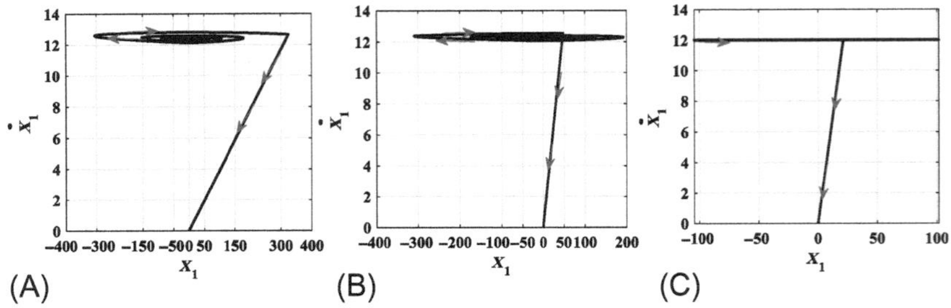

FIG. 3.6 Diagram of $\dot{x}_1 = f(x_1)$ with: (A) $\beta_1 = 0.001$, $\beta_2 = 0.8$, and $\beta_3 = 0.05$; (B) $\beta_1 = 0.01$, $\beta_2 = 0.8$; and $\beta_3 = 0.09$; and (C) $\beta_1 = 0.001$, $\beta_2 = 0.8$, and $\beta_3 = 5$.

3.5.7 Control of the Three-Phase Inverter

A three-phase inverter (VSI) is operated to control the voltage and its frequency, balancing and leveling of loads, and harmonics mitigation at PCC. To maintain constant frequency, the VSI is forced to operate at desired frequency. For the voltage and frequency control at PCC in the presence of perturbations and non-linear load, proportional resonant (AWPRC) and proportional integral (AWPI) controllers with antiwindup are suggested to avoid the saturation phenomenon. As shown in Fig. 3.7, two different control approaches are combined to achieve all desired tasks simultaneously when the system is operating in grid and islanding modes. A remote-controlled switch selector is used to select the right control based on the information received from the data center using smart meters. To simplify the study, one supposes that condition to select the right control is based on the source current. Therefore, if the source current is equal to zero, the selector selects d_{1abc}, otherwise d_{2abc} stays selected.

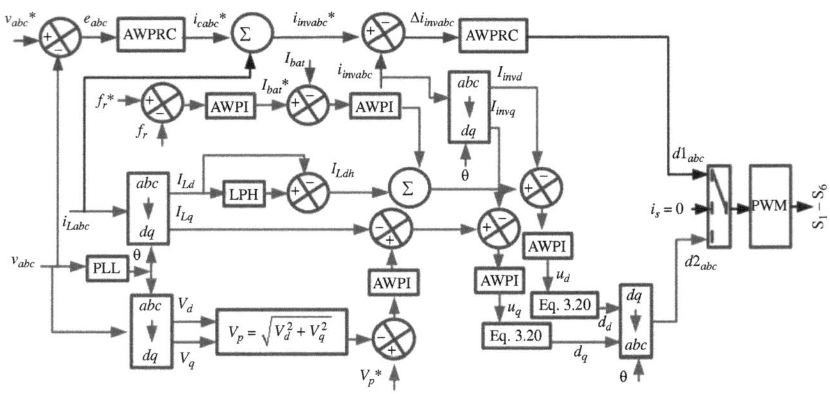

FIG. 3.7 Control strategy for three-phase VSI.

Regarding the duty cycles (d_{1abc}) of the first level of developed control strategy shown in Fig. 3.7, these are obtained using PCC voltages as follows:

$$\begin{cases} v_a^* = V_p \sin(\omega_s t) \\ v_b^* = V_p \sin(\omega_s t - 2\pi/3) \\ v_c^* = V_p \sin(\omega_s t + 2\pi/3) \end{cases} \tag{3.17}$$

where V_p denotes peak of phase voltage and ω_s represents the angular frequency as follows:

$$\omega_s = 2\pi f_s \tag{3.18}$$

where f_s is the system frequency.

The inverter references currents ($i_{inv(abc)}{}^*$) are obtained by adding the outputs of outer loops, which represent the RC passive filter references currents ($i_{L(abc)}$), which the measured load currents ($i_{L(abc)}$) as:

$$i_{inv(abc)}{}^* = i^*_{C(abc)} + i_{L(abc)} \tag{3.19}$$

The obtained inverter references currents are compared with sensed measured inverter currents ($i_{inv(abc)}$). The errors are fed to AWPRC and the outputs, which represent the control signals (d_{1a}, d_{1b} and d_{1c}), are fed to PWM generator to obtain pulses for the three-phase VSI.

Regarding the duty cycles (d_{2abc}) of the second level of the developed control strategy shown in Fig. 3.7, a phased locked loop (PLL) is required to estimate the frequency and to synchronize the three-phase VSI with PCC.

The reactive component of the load current is subtracted from the output of the AWPI voltage regulator. The active and reactive components of source currents are estimated using Park's transformation. Sensed three-phase load and VSC currents are converted to d-q-equivalent. The errors between load and d-q axis VSI currents (i_{qinv} and i_{dinv}) are fed to AWPI current controllers (d_d and d_q) as detailed in Eq. (3.8). Inverse Park transformation is used to obtain the duty cycle (d_{2abc}), which is fed to PWM to generate the switching gate pulses for VSI.

$$\begin{cases} d_d = \dfrac{-u_d + v_d + L_f \omega_s i_{qinv}}{V_{dc}} \\ d_d = \dfrac{-u_q + v_q - L_f \omega_s i_{dinv}}{V_{dc}} \end{cases} \tag{3.20}$$

where u_{dq}, v_{dq}, i_{invdq}, L_f, and V_{dc} denote the AWPI output current controllers in d-q axis, PCC voltages, output VSI currents, output VSI inductor, and DC link voltage, respectively.

3.6 Results and Discussion

Tests of the proposed microgrid configuration based on the solar photovoltaic system and BESS for smart houses in SPG shown in Fig. 3.3 and their developed control strategies presented in Figs. 3.4 and 3.7 are presented in this session. Three different seniors are tested using Matlab/Simulink. Furthermore, the behavior of the BESS during transition is experimentally validated. In the simulation tests, grid-connected operation mode as well as islanding mode are tested. In addition, the transition between the both modes is also tested. For all tests, systems supplying a balanced nonlinear load and the solar radiations vary. In order to simplify the tests, the state of the source current (i_s) is used as information to select the right control.

3.6.1 Performance Under Presence of Nonlinear Load and Solar Irradiation Change When System Is Operating in Grid-Connected Mode

In Fig. 3.8, the waveforms of the grid voltage (v_{sa}) and current (i_{sa}), load current (i_{La}), inverter current (i_{inva}) of phase "a," DC-link voltage and its reference (V_{dc} and V_{dcref}), battery current and it reference (i_b and i_{bref}), and the output solar photovoltaic current (i_{pv}) are presented. It can be observed that all end-user services, which are indicated in Table 3.1, are satisfied.

It may be observed that the source current is balanced and sinusoidal in the presence of nonlinear load. The DC-link voltage is well regulated; it follows the out solar photovoltaic voltage when solar irradiation changes. The BESS balance the power in the system, and one may observe that BESS is charging in this mode of operation, the battery current follows its reference, and the solar photovoltaic current varies with variation of the solar photovoltaic system, which confirms the robustness of the proposed control strategies for both power converters.

In Fig. 3.9, the zoomed waveforms of Fig. 3.8 between $t=0.4$ s and $t=0.6$ s are presented. It is observed that the output filter current contains the fundamental; this is because of the source current, which is used to charge the BESS. The source current is sinusoidally balanced.

3.6.2 Performance Under Presence of Nonlinear Load and Solar Irradiation Change When System Is Operating in Islanding Mode

In Fig. 3.10, the waveforms of the load voltage (v_{La}) and source current (i_{sa}), load current (i_{La}), inverter current (i_{inva}) of phase "a," DC-link voltage and its reference (V_{dc} and V_{dcref}), battery current and it reference (i_b and i_{bref}), and the output solar photovoltaic current (i_{pv}) are presented.

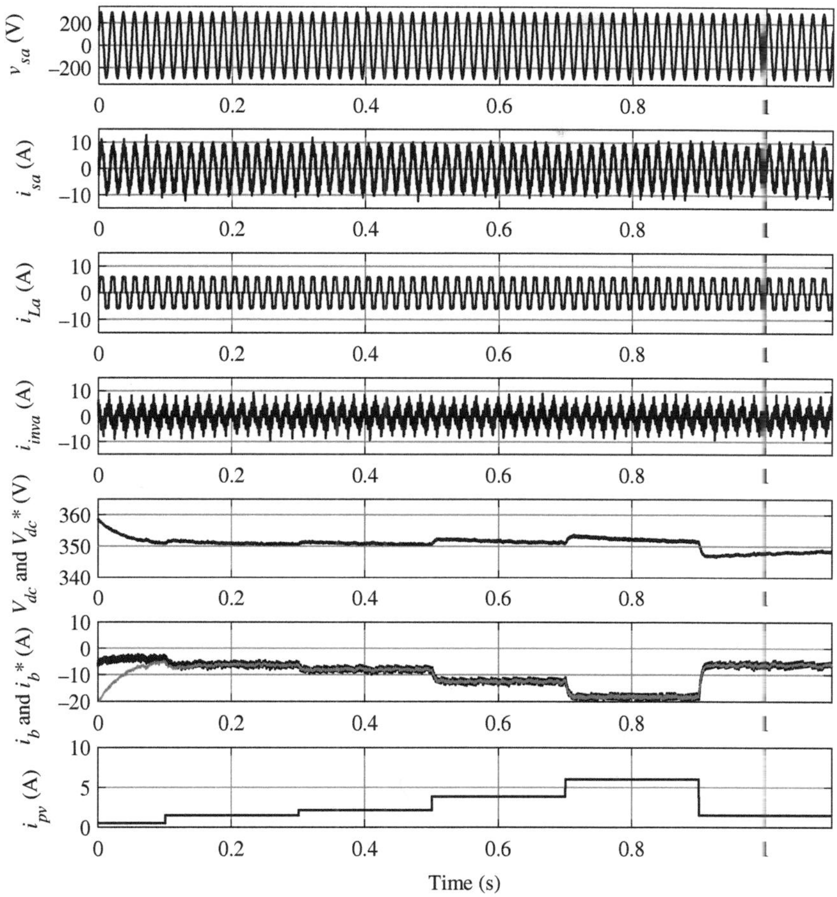

FIG. 3.8 Dynamic performance under presence of nonlinear load and solar irradiation change when system is operating in grid connected mode.

Seeing that the source current is equal to zero, the second level of control is selected. One observes that the source current is equal to zero, and the load is supplied locally using the generated power from the solar photovoltaic system and from BESS. It is observed in this operation mode that the battery current is with positive sign that is mean that BESS. As was detailed earlier, the proposed control strategy for VSI in this mode of operation regulates the load voltage and provides power to the connected load. It can be seen that the PCC voltage is regulated and sinusoidal in the presence of nonlinear load. In addition, the load is supplied continuously.

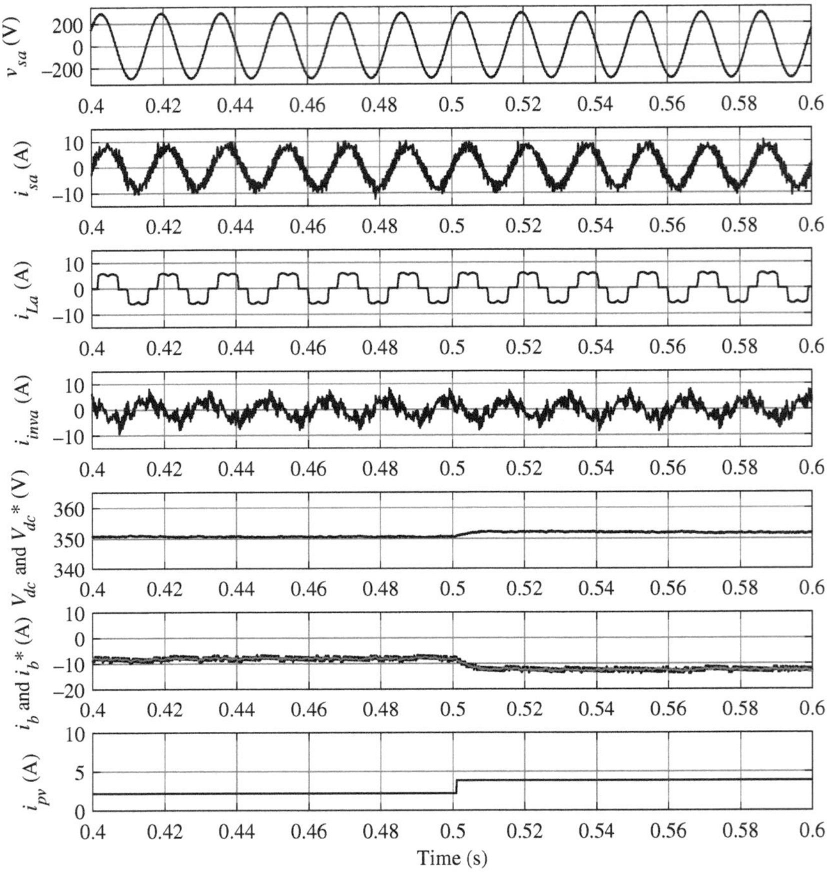

FIG. 3.9 Zoomed waveforms of Fig. 3.8 between $t=0.4$ s and $t=0.6$ s.

In Fig. 3.11, the zoomed waveforms of Fig. 3.10 between $t=0.4$ s and $t=0.6$ s are presented. One may observe that the PCC voltage is well regulated at its rated value and is sinusoidal. It can clearly be seen that the BESS balance power in the system by helping the solar photovoltaic system to supply the connected load.

3.6.3 Performance Under Presence of Nonlinear Load and Solar Irradiation Change When System Switch From Grid-Connected Mode to Islanding Mode

In Fig. 3.12, the waveforms of the load voltage (v_{La}) and source current (i_{sa}), load current (i_{La}), inverter current (i_{inva}) of phase "a," DC-link voltage and its reference (V_{dc} and V_{dcref}), battery current and it reference (i_b and i_{bref}), and the output solar photovoltaic current (i_{pv}) are presented.

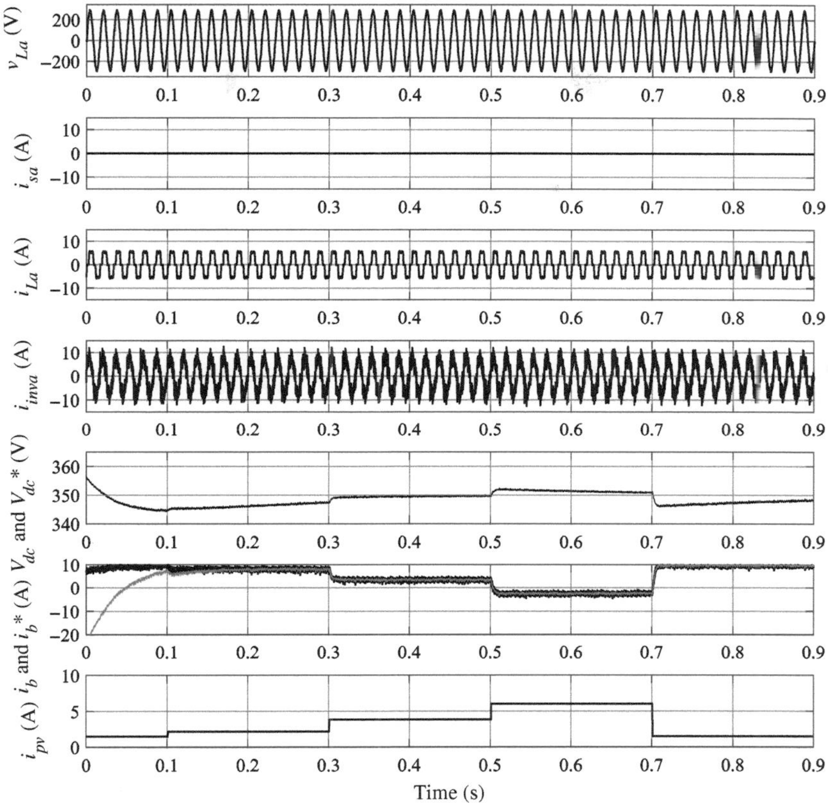

FIG. 3.10 Dynamic performance under presence of nonlinear load and solar irradiation change when system is operating in islanding mode.

It can be seen that the PCC voltage is kept constant and sinusoidal, and the system switches from grid-connected mode to the islanding mode easily without any saturation of the controllers and without large peak current in transition. It can also be observed that power quality is improved in both modes. The source current is balanced and sinusoidal in grid-connected mode and the PCC voltage is sinusoidal and well regulated in islanding mode. One can see clearly that the BESS plays a key role in both modes and during transition.

In Fig. 3.13, the zoomed waveforms of Fig. 3.12 between $t=0.1$ s and $t=0.25$ s are presented. It can be seen that the PCC voltage is well regulated during transition between both operation modes and when the system is operating in islanding mode. The load is supplied without any perturbation during transition and in islanding mode, which confirms the necessity of the storage element in this kind of application.

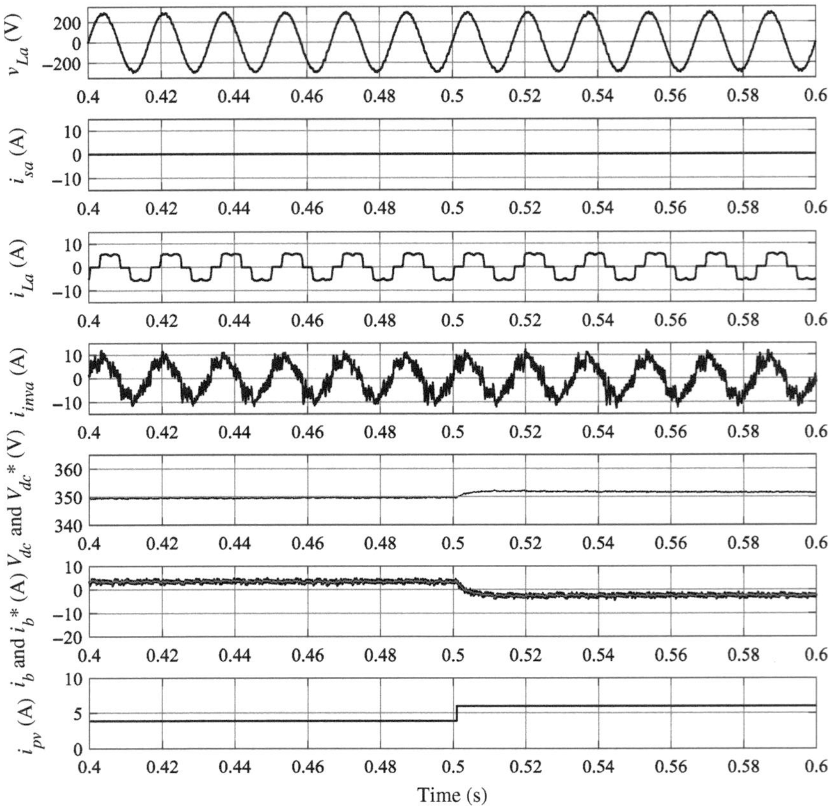

FIG. 3.11 Zoomed waveforms of Fig. 3.10 between $t=0.4$ s and $t=0.6$ s.

3.6.4 BESS Performance Testing Under Load Variation in the Grid-Connected Mode

The waveforms of DC-link voltage (V_{dc}), battery current (i_b), load current (i_{La}), and inverter current (i_{inva}) are shown in Fig. 3.14A and B. One can see clearly that V_{dc} is well regulated and is kept constant during transition at $t=0.4$ s and $t=0.8$ s in Fig. 3.14A, and at $t=0.3$ s, and $t=0.8$ s in Fig. 3.14B. It may be observed that the (i_b) varies with variation of the load to balance the power between generation and load, which confirms the necessity of BESS in this type of application.

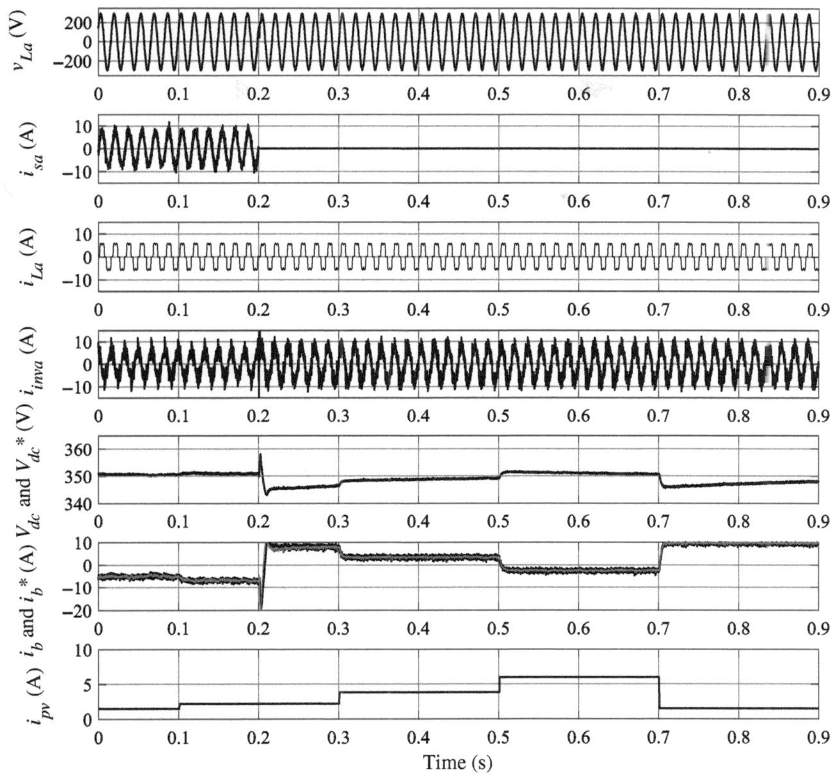

FIG. 3.12 Dynamic performance under presence of nonlinear load and solar irradiation change when system switch from connected mode to islanding mode at $t = 0.2$ s.

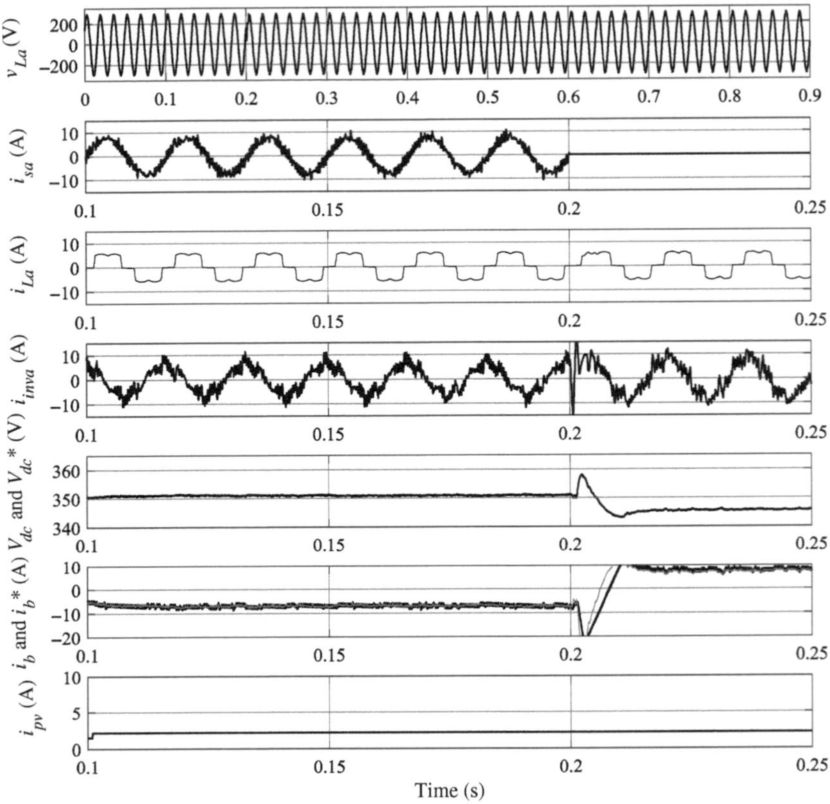

FIG. 3.13 Zoomed waveforms of Fig. 3.12 between $t = 0.1$ s and $t = 0.25$ s.

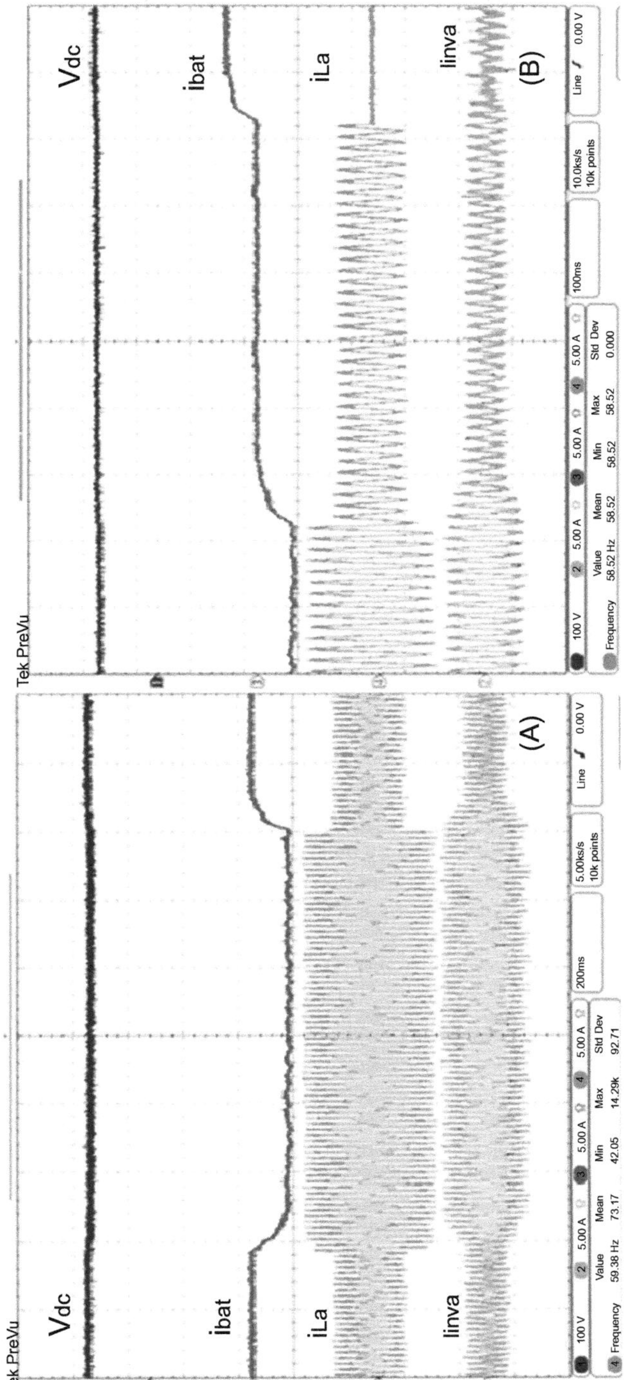

FIG. 3.14 Experimental test under fixed solar irradiation and load change. (A) Sudden increase and decrease of load and (B) sudden decrease and disconnect of load.

3.7 Conclusion

In this chapter, BESS for microgrid configuration based on solar photovoltaic systems in smart grid application was studied. The role and application of BESS by the end-user, as well as policy and recommendations, are given in detail. Control strategies to achieve high performance from the solar photovoltaic system and to take maximum advantage of BESS using such PCC voltage and frequency regulation, balance of power, and power quality are validated. The obtained simulation as well as experimental results show satisfactory performance in the presence of severe conditions. It has been proven that with the help of BESS, it is possible to switch from connected grid to islanding mode without any issue.

References

[1] M. Liserre, T. Sauter, J.Y. Hung, Future energy systems: integrating renewable energy sources into the smart power grid through industrial electronics, IEEE Ind. Electron. Mag. 4 (1) (2010) 18–37.
[2] F.F. Wu, P.P. Varaiya, R.S. Hui, Smart grids with intelligent periphery: an architecture for the energy internet, Engineering 1 (4) (2015) 436–446.
[3] http://swedishsmartgrid.se/globalassets/publikationer/marketanalysis_ger.pdf.
[4] N. Balta-Ozkan, R. Davidson, M. Bicket, L. Whitmarsh, The development of smart homes market in the UK, Energy 60 (2013) 361–372.
[5] G. Xu, W. Yu, D. Griffith, N. Golmie, P. Moulema, Toward integrating distributed energy resources and storage devices in smart grid, IEEE Internet Things J. 4 (1) (2017) 192–204.
[6] R. Bayindir, I. Colak, G. Fulli, K. Demirtas, Smart grid technologies and applications, Renew. Sustain. Energy Rev. 66 (1) (2016) 499–516.
[7] M.L. Tuballa, M.L. Abundo, A review of the development of smart grid technologies, Renew. Sustain. Energy Rev. 59 (1) (2016) 710–725.
[8] A. Bhati, M. Hansen, C.M. Chan, Energy conservation through smart homes in a smart city: a lesson for Singapore households, Energy Policy 104 (2017) 230–239.
[9] R. Zafar, A. Mahmood, S. Razzaq, W. Ali, U. Naeem, K. Shehzad, Prosumer based energy management and sharing in smart grid, Renew. Sustain. Energy Rev. 11 (2017).
[10] A.G. Olabi, Renewable energy and energy storage systems, Energy 136 (2017) 1–6.
[11] M.S. Hossain, N.A. Madlool, N.A. Rahim, J. Selvaraj, A.K. Pandey, A.F. Khan, Role of smart grid in renewable energy: an overview, Renew. Sustain. Energy Rev. 1 (60) (2016) 1168–1184.
[12] Y. Sun, L. Lampe, V.W. Wong, Smart meter privacy: exploiting the potential of household energy storage units, IEEE Internet Things J. 5 (1) (2018) 69–78.
[13] X. Fang, S. Misra, G. Xue, D. Yang, Smart grid—the new and improved power grid: a survey, IEEE Commun. Surv. Tutor 14 (4) (2012) 944–980.
[14] K. Sharma, L.M. Saini, Performance analysis of smart metering for smart grid: an overview, Renew. Sustain. Energy Rev. 49 (2015) 720–735.
[15] https://www.pewtrusts.org.
[16] K. Dehghanpour, Z. Wang, J. Wang, Y. Yuan, F. Bu, A survey on state estimation techniques and challenges in smart distribution systems. IEEE Trans. Smart Grid (2018) https://doi.org/10.1109/TSG.2018.2870600.

[17] Y. Wang, Q. Chen, T. Hong, C. Kang, Review of smart meter data analytics: applications, methodologies, and challenges. IEEE Trans. Smart Grid (2018) https://doi.org/10.1109/TSG.2018.2818167.

[18] J. Ni, K. Zhang, X. Lin, X. Shen, Balancing security and efficiency for smart metering against misbehaving collectors. IEEE Trans. Smart Grid (2017) https://doi.org/10.1109/TSG.2017.2761804.

[19] K.K. Zame, C.A. Brehm, A.T. Nitica, C.L. Richard, G.D. Schweitzer III, Smart grid and energy storage: policy recommendations, Renew. Sustain. Energy Rev. 82 (Part 1) (2017) 1646–1654.

[20] S.O. Geurin, A.K. Barnes, B.J. Carlos, Smart grid applications of selected energy, in: Innovative Smart Grid Technologies (ISGT), IEEE PES, Washington, DC, 2012.

[21] F. Mohamad, J. Teh, C.M. Lai, L.R. Chen, Development of energy storage systems for power network reliability: a review, Energies 11 (9) (2018) 2278.

[22] X. Xu, M. Bishop, D.G. Oikarinen, C. Hao, Application and modeling of battery energy storage in power systems, CSEE J. Power Energy Syst. 2 (3) (2016) 82–90.

[23] M.Y. Suberu, M.W. Mustafa, N. Bashir, Energy storage systems for renewable energy power sector integration and mitigation of intermittency, Renew. Sustain. Energy Rev. 35 (2014) 499–514.

[24] S. Benhalima, R. Miloud, A. Chandra, Real-time implementation of robust control strategies based on sliding mode control for standalone microgrids supplying non-linear loads, Energies 11 (10) (2018) 2590.

[25] J.F. Tsai, Y.P. Chen, Sliding mode control and stability analysis of buck DC-DC converter, Int. J. Electron. 94 (2007) 209–222.

[26] M.A. Reitz, X. Wang, Robust Sliding Mode Control of Buck-Boost DC-DC Converters, in: Proceedings of the ASME 2016 Dynamic Systems and Control Conference, American Society of Mechanical Engineers, 2016. MN, USA, October 12–14.

[27] J.J. Slotine, W. Li, Applied Nonlinear Control, Prentice Hall, Englewood Cliffs, NJ, 1991, pp. 105–116.

[28] R. Gavagsaz-Ghoachani, M. Phattanasak, J.P. Martin, S. Pierfederici, B. Nahid-Mobarakeh, P. Riedinger, A lyapunov function for switching command of a DC–DC power converter with an LC input filter, IEEE Trans. Ind. Appl. 53 (2017) 5041–5050.

Chapter 4

Smart Meters and Advanced Metering Infrastructure

João F. Martins*, Anabela Gonçalves Pronto*, Vasco Delgado-Gomes* and Mihai Sanduleac†

**Faculty of Sciences and Technology, Universidade NOVA de Lisboa, CTS-UNINOVA, Lisbon, Portugal, †Faculty of Power Engineering, Polytechnic University of Bucharest, Bucharest, Romania*

Chapter Outline

4.1 Introduction

The past 50 years have been dominated by the so-called communication revolution. The upcoming years are expected to contemplate what we might call the electric power and energy systems revolution. Concepts such as smart grid, intelligent energy networks, and the "internet of energy" are already in our lexicon. Our modern economy is energy driven, with electricity being the major

Pathways to a Smarter Power System. https://doi.org/10.1016/B978-0-08-102592-5.00004-1

future energy vector among worldwide end-uses of energy. According to the International Energy Agency, electricity will make up 40% of the rise in final consumption by 2040; this corresponds to the same share of growth experienced by oil within the past 25 years [1]. This is mainly due to industrial electric motor systems, which will account for one-third of the increase in power demand, and increasing household cooling systems and electrical appliances (with a rising share of the so-called "smart" connected devices) as a consequence of expected rising incomes.

Nevertheless, regarding the expected future for electric grids, metering has always been a key component in the electricity markets: wherever there is a service, there is the need to measure it. The first known meter dates from 1872. Samuel Gardiner patented a meter that counted the time in which energy was supplied to a set of lamps. This meter became obsolete when lighting circuits became subdivided. Thomas Edison patented his first meter, based on the electrochemical effect of the DC current, in 1881. An initially measured strip of copper, which was decomposed due to the passing current, was measured again later on, with the difference being equivalent to the amount of consumed electricity. These types of meters were still in use by the end of the 19th century. Afterwards, many types of electric meters were developed. Elihu Thomson developed a motor-based DC meter for General Electric in 1889, which had a voltage-excited rotor and a current-excited stator. With the development of AC systems, new meters were needed. In 1889, Ganz employee Otto Titusz Bláthy patented an electric meter for alternating currents, which weighed 23 kg. These induction meters, based on the principles of the Bláthy meter, are still manufactured today in large quantities due to their excellent reliability and low price. In the second half of the 20th century, electronic meters and remote metering were introduced into the measuring market.

A traditional electric grid is a system that provides the power needed by a set of consumers, whether residential, industrial, or commercial. This electric grid is composed by a set of electric power generators, transmission lines, transformers, and substations. Today's electric grid has become aged in most countries and is thus not suited for future challenges [2]. After the climate change agreement at the Paris UN Climate Change Conference, held in Paris, France, from 30 November to 12 December 2015, there was a clear understanding about the actual need of changing the electric grid paradigm [3]. Three major drivers were behind this change:

1) the need to reduce carbon emissions;
2) the decentralization of electrical power and energy production; and
3) the electrification of transport.

The so-called smart grid, as an intelligent electric power infrastructure, has arisen to overcome these challenges. The Energy Independence and Security Act of 2007 was approved by the US Congress in January 2007, and set out one of the first definitions of smart grid [4]. Through proper use of new technologies, the smart grid should become more reliable, secure, flexible, efficient, and sustainable [5]. The smart grid is opposed to the traditional electric grid in several ways. The traditional grid is based mainly on electromechanical devices, while the smart grid encompasses information and communication technologies along with electronic power devices. The traditional grid considers a centralized production system where the production units produce the power that will delivered to the end-users. Smart grids rely on distributed generation, based mainly on renewables. The traditional grid is based on a hierarchical structure while smart grids are networked by nature. The smart grid considers the existence of a huge number of sensors; these are not considered in the traditional grid paradigm.

Among these sensors, it is crucial to consider an advanced metering infrastructure (AMI) based on smart meters [6]. Smart meters are expected to play a key role in the transition from a traditional grid to a smart grid, and they will need to respond to tough challenges [7]. A traditional smart meter can be described as an advanced meter device able electronically to identify and measure energy consumption, and to communicate this information. Making use of bidirectional data communication capabilities, traditional smart meters are also able to collect a huge amount of information regarding end-user electricity consumption, when high reporting rates are selected. Based on this, in addition to metering, they can execute local or remotely received commands for home devices and appliances. While traditional meters are based on manual data collection and billing and electronic/digital meters allow remote communication of indexes, smart meters establish a communication with the utility, allowing automatic billing. Fig. 4.1 presents a basic smart meter architecture. In addition to

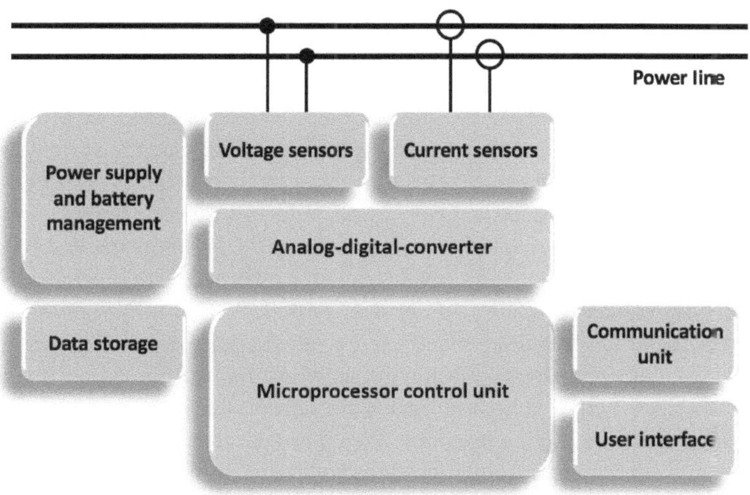

FIG. 4.1 Smart meter architecture.

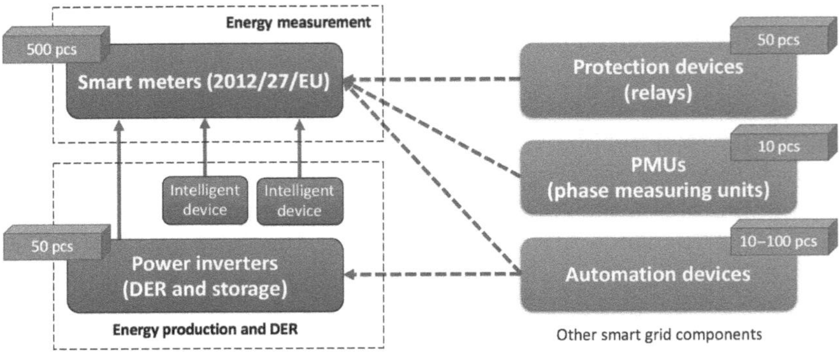

FIG. 4.2 Occurrence of equipment in a smart grid low-voltage network.

the mandatory voltage and current sensors, an analog-digital-converter sends data to the processor control unit, which coordinates data storage, user interface, and communications. (See Fig. 4.2.)

Smart meters present a huge amount of distinct functionalities, providing benefits for both end-users and power service providers. It is important to note that there was no dramatic change from traditional meters to smart meters, because the smart meter development is still a work in progress. It has come a long way from traditional meters, where the first generation of electronic meters with remote reading were only able to provide one-way communication to the power service providers. Nowadays, smart meters can provide a wide variety of functionalities, such as: automatic bidirectional metering and billing; two-way communication and AMI infrastructure; demand response; outage detection and restoration services; distributed control; data storage; fault detection and diagnosis; end-to-end communication; digital interface display; several communication possibilities; and deployment as an IoT device.

Among the huge set of advantages emerging from smart metering deployment, the following ones can be emphasized. For the power service provider, it eliminates the need for manual reading, monitors the electric systems more quickly, allows an optimized resource usage, provides quasi-real-time data (useful for balancing electric loads while reducing power outages), enables dynamic tariffs, avoids capital expense of new grid investments, and helps to optimize income with existing resources. From the end-user's perspective, it provides an enhanced quality of service, provides more informative billing, improves the energy awareness and energy rational use, and provides accurate knowledge about consumption behavior, enabling end-users to adjust their consumption habits. However, some issues are raised with increased smart meter deployment. Power service providers face a higher cost in terms of personnel training and equipment development and production; they also have to deal with some possible negative public reaction when replacing old meters. A long-term financial commitment to the new metering technology and the related software involved is required, and power service providers have to manage and store huge amounts of the metering data and ensure its security and privacy. End-users still face uncertainties in terms of protecting their privacy and personal data collected, as well as some additional fees that may arise (this last issue is usually country related).

The widespread nature of smart meters is now a reality and an unstoppable process. In the USA, it was reported that approximately 37 million smart meters were shipped between 2011 and 2014 0. In Europe, the European Union aims to replace at least 80% of electricity meters with smart meters by 2020 wherever it is cost-effective to do so. Close to 200 million smart meters are expected to be rolled out in the EU by 2020. Fourteen European member countries (Austria, Denmark, Estonia, Finland, France, Greece, Ireland, Italy, Luxemburg, Malta, the Netherlands, Spain, Sweden, and the UK) are currently proceeding with large-scale rollout by 2020 or earlier 0. For all of these, the stated expected penetration by 2020 is 95% or more. Notably, in three of these countries (Finland, Italy, and Sweden), close to 45 million smart meters have already been installed, amounting to almost a quarter of the 2020 total. Several European projects are also responsible for this roll-out. The NOBELGRID project alone is expected to deploy around 500 smart meters over five pilot test sites 0.

4.2 Advanced Metering Infrastructure (AMI)

Nowadays, the AMI is considered a key component of smart grid (SG), integrating software and hardware components, data management and monitoring systems, and smart meters (SM). It enables the bidirectional connection with the SMs, creating home area networks (HANs), neighborhood networks, and even wide area networks (WANs), due to its scalability. Before the AMI, automatic meter reading (AMR) enabled only the reading of the SM. Nowadays, AMI enables the bidirectional communication between the DSO, SMs, and consumers. AMI potentiates other hierarchical levels of complex layers such as ADO (Advanced Distribution Operations), ATO (Advanced Transmission Operations), and AAM (Advanced Asset Management) 0.

Several wired and wireless communication technologies can be used to implement an AMI. Wireless technologies, such as Wi-Fi, GSM, GPRS. ZigBee, and WiMAX, enable the fast and low-cost deployment of the devices that compose an AMI. On the other hand, wired technologies, such as DSL and PLC, provide a more stable communication.

In order to have a proper SG operation, it is necessary to have an almost real-time management of the SG devices and systems. The AMI concept provides real-time readings and allows the DSO to send commands to the power network, thus allowing automatic fault detection and enabling the self-healing features in the power network. Moreover, the end-user has also their consumption information, enhancing consumer empowerment.

4.2.1 Advantages of AMI

There are numerous advantages of using the AMI. From the utilities' point of view, they can prevent illegal electricity usage by analyzing the historical data collected from the SMs. Using the AMI capabilities, the SG is remotely monitored and controlled, improving maintenance, demand management, and planning. Moreover, with a stable SG, electricity outages can be prevented, saving money for electrical companies. The AMI concept eliminates manual monthly meter readings, monitors the electric system much more quickly, makes it possible to use power resources more efficiently, provides real-time data that is useful for balancing electric loads while reducing power outages (i.e., blackouts), enables dynamic pricing (which raises or lowers the cost of electricity based on demand), avoids the capital expense of building new power plants or reinforcing others, and helps to optimize income with existing resources.

AMI consumer benefits inclu7de more precise electrical bills and increased power quality control and awareness. A more informed consumer will be able to make better decisions regarding their consumption profile, reducing the power needed in their homes and, as a result, lowering their house bills. Thus, end-users will have better quality of electric service, adaptive billing mechanisms, more information in energy bills, far greater (and more detailed) feedback regarding energy use, knowledge of consumption behavior, and energy awareness and rational use, enabling consumers to adjust their habits to lower their electricity bills.

4.2.2 Issues and Drawbacks of AMI

Some AMI issues relate to consumer privacy and confidentiality, while the remaining ones relate to unauthorized access to the devices, creating cyber security issues.

On the consumer side, users need to protect their data, creating the need for a proper data management through anonymization of data. End-users may find it hard to verify that the new meter is accurate, there is no way to protect the privacy of the personal data collected, and there may be an additional fee for the installation of the new meter.

The utility companies need to be sure that the data is accurate, and that no physical or cyber-attack to the meter occur. To achieve this, several cyber security measures can be taken, namely: confidentiality, integrity, authorization, and authentication of the exchanged data. These measures can help to prevent some of the following cyber-attacks: eavesdropping, traffic analysis, replay, message modification, impersonation, denial of service, and malware. However, due to the current processing power of some SM and SG systems, most cyber security countermeasures cannot be implemented. Furthermore, the costs in terms of personnel training, equipment development, and production to change to a new technology and new set of processes are higher. Managing negative public reactions and ensuring customer acceptance of the new meters are also difficult matters. In addition, making a long-term financial commitment to the new metering technology and the related software involved raises financial problems, managing and storing vast quantities of the metering data collected raises huge technical problems, and ensuring the security and privacy of metering data is still not a closed issue.

4.3 Unbundled Smart Meter

4.3.1 Why Smart Meters are so Important in a Smart Grid Paradigm and Why They Need a New Architecture

The smart grid is a new/old paradigm that tries to reshape the way power systems should work in a challenging environment, which demands an agenda towards clean energy and communities' resilience. The smart grid can be considered as a new concept because only in the past 10 years has it become possible to ensure an intimate cooperation between power systems and ICT, in order to guarantee an intelligent and efficient balance between production and consumption. It is also an old concept because this cooperation has existed for decades, where at transmission level only a small number of large generators needed to be controlled. The new paradigm intends to implement this procedure at any voltage level, having hundreds of millions of consumers but also the so-called "prosumers," managing both their consumption and production at the same time. In recent times, it was sufficient to deploy remote terminal units, meters, and automation at all power plants, as they were a small number and the associated investment was much lower than the cost of the power plants themselves. However, in the smart grid era, with millions of production units, a new approach is needed. 0 presents a typical low-voltage network diagram, where consumers and producers are considered as distinct entities. For each entity, a meter allowing bidirectional energy measurement is needed.

0 suggests that smart meters will become the most numerous devices in a smart grid, as they will be needed practically at all points of common coupling (between the grid and the several connected entities). For example, merging prosumers and intelligent devices to implement demand response (DR) can only be achieved with meter functionality. Other connected devices, such as protection devices, phase measurement units (PMUs), and automation units, are expected to be less numerous. However, as a way to reduce the number of connected devices, smart meters are appropriate to include most of all other functions, requiring the need for higher flexibility and a new architecture concept at the meter level. The unbundled smart meter offers a flexible solution for those new requirements and allows a multi-actor and multi-protocol operation.

4.3.2 Unbundled Smart Meter Architecture

The **unbundled smart meter** is a specific architecture that simultaneously combines metrology features and high flexibility within smart meters. The unbundled smart meter is a meter systematization where smart meter functionalities are adequately grouped in two separate, unbundled, components as follows:

- one for metrological and hard real-time functions, called the **smart metrology meter** (SMM). This intelligent component has fixed functionality and high security of recorded data (black box-like standard); and
- a **smart meter extension** (SMX) that provides high flexibility needed for new functionalities. This component can be deployed during the meter's lifetime, thus supporting the future evolution of the smart grid and energy services.

Fig. 4.3 presents a simple representation of these two components. The entire communication is completely covered by SMX, addressing both home area network and public IP networks such as ubiquitous internet. See Figs. 4.4–4.7, where Fig. 4.4 shows the smart meters communication architecture and the following show the place of smart meters in the Smart Grid Architecture Model (SGAM).

A key factor in the new smart meter design is the fact that a single piece of equipment, placed at the boundaries between the energy provider and the prosumer, is at the same time, and with the same requested quality standards, serving all possible actors in the energy domain. Integrates in one single equipment features to support all the following functionalities:

- standard billing for legacy energy supply contracts, based on tariffs;
- advanced billing information based on load-profiles of indexes;
- instrumentation values of electrical variables (such as U_{RMS}, I_{RMS}, P, Q, f…) to support real-time observability for smart grid functionality;

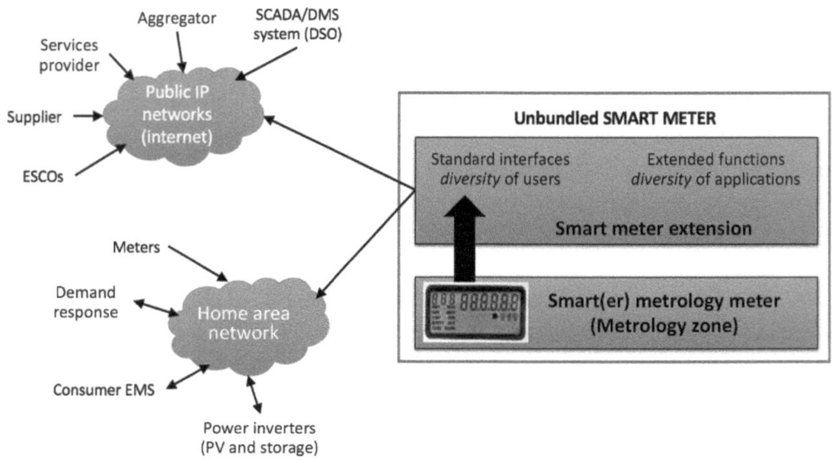

FIG. 4.3 Unbundled smart meter (USM) architecture.

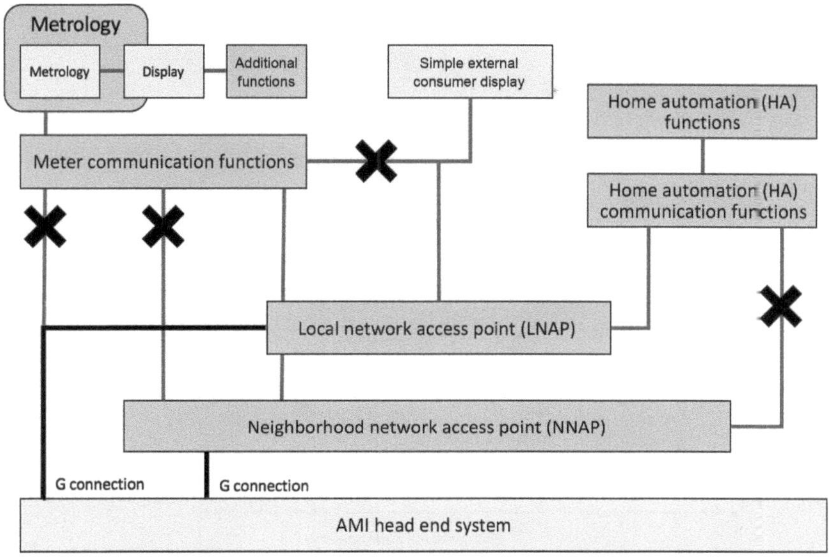

FIG. 4.4 USM—Reference Architecture M/441 (CEN/CLC/ETSI/TR50572, Dec. 2011).

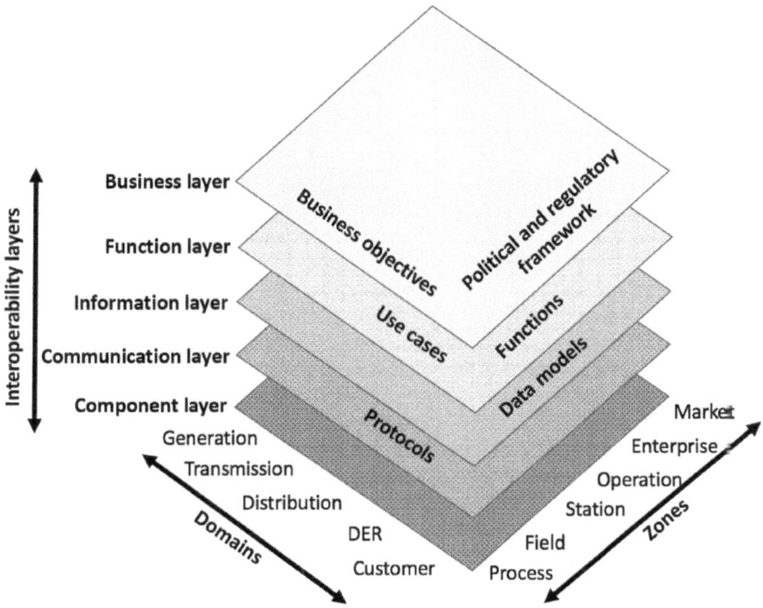

FIG. 4.5 SGAM and its dimensions.

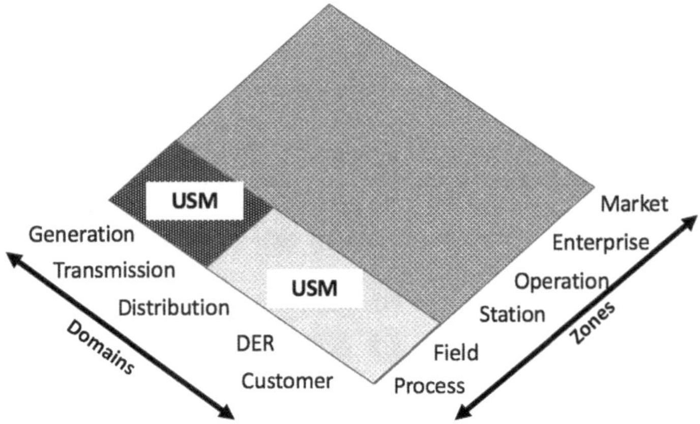

FIG. 4.6 USM position in the SGAM plane.

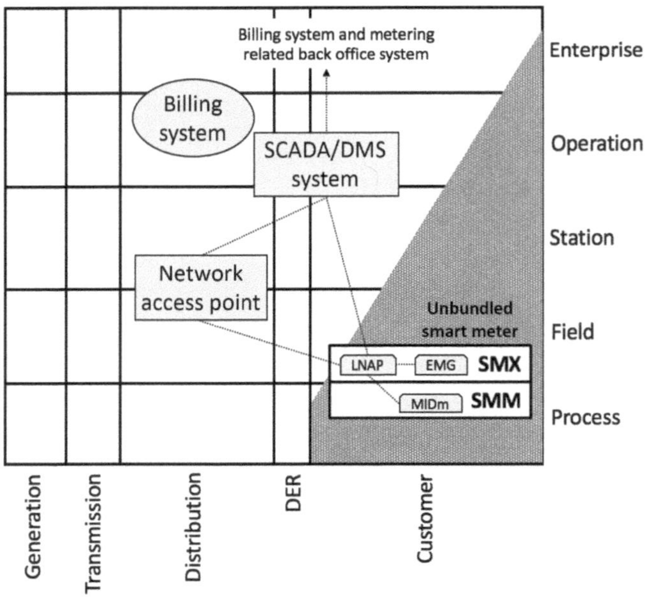

FIG. 4.7 USM mapping in the SGAM architecture.

- support for energy services, through fine-grained load-profiles of consumed and/or produced active and reactive energy at one or five minutes level (LP1 or LP5);
- power quality support;
- support for demand response (DR);
- support for communication with home devices, including DR functionalities;
- support for third party applications;
- interface for local information at the end-user premises; and

- support for controlling intelligent PV and/or storage inverters.
- Besides all the above functionalities, the USM architecture allows support for future SMX functionalities based on third party apps and/or software upgrades of essential SMX functionalities

In the SMM component, hard real-time functions and black-box functionality are essential, while in the SMX component, communication capabilities and flexible local intelligence are essential. The two components do not present any overlapping functionality and have different design, in order to meet their requirements independently.

4.3.3 Communications

When introducing the USM concept, an essential aspect is the compatibility with the recommended architecture of smart metering systems 0. Not meeting such compatibility may bring difficulties when implementing the architecture. 0 presents a simplified representation of the recommended architecture for smart metering systems. This reference architecture is part of the M/441 Mandate 0.

The following points can be observed:

- Each type of connection must be considered and implemented in a specific way. The connections which are discarded are colored in gray and marked with an "X," meaning that they are not used. This is possible because the reference architecture does not require all communication connections – which are giving the all possible ones but they arenot all necessary, thus allowing various particular solutions to be implemented, considering that the functionality is still preserved.
- Remote connections are possible only through the local network access point (LNAP) separated from the metrology part. The component providing the LNAP should be powerful enough to handle more complex functionalities, thus this feature will be assured by the SMX component of the USM.
- The simple external consumer display is changed to a more general "HMI with browser," which can be a local PC, a laptop, a tablet, a smart TV, or a smart phone, as their ubiquitous web browsing facility is needed. This is an improvement over the initial architecture, which opens the possibility for a wide range of functionalities which the end-user can easily access with equipment they already possess.
- Essential functionalities are security and privacy mechanisms, which should be developed in SMX.
- Both direct and neighborhood readout toward AMI head end systems should possible through the G connections.

4.3.4 Compatibility With Smart Grid

As the USM is an essential grid device, it is fundamental to assure its compatibility with smart grid standardization efforts. The European Mandate M/490 defines the SGAM (smart grid architecture model), presented in 0 with its dimensions: domains, one for the zones and interoperability layers 0.

The unbundled smart meter architecture fits well into the SGAM architecture, as presented in 0. The USM position itself in the component layer, covering the process and field zones. In fact, the USM can be simultaneously a metering device (the traditional function) and a supervisory and control device (SCADA related) allowing easy mapping in the SGAM model, as presented in 0 (EMG = energy management gateway; LNAP = local network access point; MID = MID meter).

4.3.5 Data Security in USM

Smart meters are by their nature keepers of data that possess great value. In fact, as already described in the introduction, the basic scope of a meter was initially to measure electrical energy, and that remains the case today. Consumed energy is a commodity that has to be paid to the energy retailer, supplier, or producer. Energy must be measured accurately, which is the role of metrology, but it must also be stored and provided to all interested parties unaltered. Thus, data security is an extremely important aspect, in order to preserve pear to pear confidence and to be able to use it in the relations between the two parts (energy provider and energy user).

Modern smart meters' architectures implement a security method for the communication, especially when the communication path is public, as happens in PLC (power line carrier) communication or in other public IP (Internet Protocol address) networks, such as the internet. However, usually only one trusted communication with one set of credentials is used.

Public key infrastructure (PKI) system is a foundation of hardware, software, procedures, and policies used for creating, distributing, maintaining, storing, and revoking digital signatures. A digital signature associates a specific identity, such as an individual end-user or a company, with a pair of public and private keys. An important challenge of PKIs is to establish a secure and reliable distribution of the keys so no third party can obtain access to private keys. For this reason, a PKI system establishes a trusted third party called a central authority (CA), which is in charge of issuing and verifying digital certificates. The PKI concept provides general concept design and this needs to be adapted for particular use cases and requirements. This situation may ask for a carefully chosen entity that needs to take care of the communication security, restricting the flexibility of a multi-user and multi-actor environment. However, each actor is expected to have its own security policy, so a multicultural security approach is needed. In the USM solution, each partner has the potential to establish its own secure communication environment, meaning that each partner can have its own open VPN (Virtual Private Network), as presented in 0.

An important element of the USM architecture is the security by design, which is part of the USM concept from the beginning. The SMX component has two distinct zones: a "trusted zone" to be handled by a "trusted party," where there is intimate access to powerful functions, and one or more "untrusted zones," where any malicious manipulation does not affect the core of the SMX functionality. In order to make this separation possible, the untrusted zone need to be "sandboxed," meaning that any of its activity has to be inside the walls of a "sandbox," and any interaction with the trusted zone needs to be fully controlled by rules that are defined in the trusted zone and impossible to be changed from outside. Fig. 4.8 suggests Docker 0 as one possible technology for the sandboxing functionality; however, this approach needs a secure access solution to get into the trusted zone.

One good concept has been developed in the German "**smart meter gateway**" (SMG) 0. Germany is the only country officially to support this concept, which acts as a data access point for all customer meters, as well as for the home area network. Even if the German SMG is more related to access to metering data, which mainly relates to data manipulation and to privacy, an important point is that multi-access requires a data access point manager (DAM) 0. This implies a special user which can change data access rights, and which can intervene in the trusted zone. This user needs to be an official trusted part enforced by law and responsible for their behavior also under the law. The DAM access can be solved with another open VPN. As this function is vital, it is advisable to be secured with an even better security mechanism – for instance, by using, in addition to the VPN concept, another mechanism for encrypting the data, based on newer technologies such as the physical unclonable function (PUF) 0. 0 shows how this can be implemented in the USM.

4.3.5.1 Privacy With USM

Privacy is an essential issue on the smart meter, basically because most of the smart meters are targeting citizen end-users whose meter data might show personal behavior. From a sensitivity point of view, the data coming from the meter can be organized into the following categories:

- personal data (data from which citizens' personal behavior can be inferred, directly or by processing);
- critical infrastructure data (data that might be used to attack the electrical grid, seen as critical infrastructure); and
- other data (data commonly shared by all grid actors).

The last data category refers to non-critical and non-personal data that can be easily acquired by anybody, being a sort of public-by-default data. Grid data such as voltage level and frequency are examples of such public-by-default data.

Even if personal data can be considered private information, this does not necessarily need to be secret by default. In fact, some personal data may usefully be shared, for example, with an energy consultant, who can provide important energy saving advice based on the intimate analysis of personal data.

All these requirements need an architectural concept to allow security and privacy by design. In the USM's SMX component, those requirements are fulfilled by using:

- a data-centric architecture (DCS); and
- a powerful role-based access control (RBAC) system, similar to the ones used in critical data applications (such as banks).

The data-centric architecture allows full control of the data with RBAC. It means that any data exchange between parts can be done only through the unique database, which is a combination of a real-time database for high reaction and a mechanism for memorizing this data in a persistent way. 0 shows the SMX design to fulfill privacy by design.

A meter RT (real-time) database, which can exchange data with various entities, should be implemented. This data exchange is done through an RBAC "wall," which uses different privacy profiles for each entity (0 presents, as an example, two different privacy profiles: PP1 and PP2). For instance, application 1 can access the (centric) database only if the corresponding rights for reading and writing are properly described in PP1. No other data and with no other rules can be exchanged with the RT database. Similarly, application 2 – for example, an agent which exchanges data with the central platform of an ESCO (Energy Services Company) – can do this only based on the rules described on PP2. This concept does not allow direct communication between applications 1 and 2, as it is an uncontrollable connection which may jeopardize the privacy of the meter data, and thus that of the end-user which has the rights on it. 0 shows an RBAC implementation that allows different privacy profiles for each type of actor.

The RBAC implementation is based on an MQTT messaging technology, where each external pool of similar actors, having the same privacy profile, has its own MQTT broker. For each MQTT broker, the RBAC implements a specific MQTT client targeting that broker, and a privacy profile PPx describes which data is to be published (sent towards the actor's broker) and to which data the SMX is willing to subscribe.

The published part of the Privacy Profile PPx is a list of rules which describes the following:

- Variables that are allowed to be sent to the actor. This is the most important feature, because with these rules only the data that are agreed with an actor will be sent. As the data is published by the RBAC firewall, there is no possibility for the external actor to get other type of data, as they cannot influence the PP list. With this setup, the meter database is not exposed to being a "server," thus serving the different requests, but instead takes the lead to provide the legally accepted data only, which is decided by both parties.
- Time period when this data can be automatically transmitted (T_ini and T_final). This feature is particularly useful as a service, for example, within a short-term signed contract with an ESCO.

- Metadata related to refresh time periods for the publishing activity (cadence of publishing). This information allows data to be sent according to the contractual agreement.
- Credentials to check the received data against "false data injection" by using, for instance, digital signatures that ensure that the data is coming from the certified sender.
- Credentials to support data decryption of the certified sender data.

4.3.5.2 Giving Consent Automatically, by Using Digital Means

As consent is essential to share personal data provided by the smart meter, the USM also has the ability to support an automatic procedure for this consent. Even if there are many ways to make a contract involving personal data handling, which starts from paper-based signed contracts and finishes at fully digital contracts made with a specific IT model, a solution which connects contract with privacy profiles is presented as an example.

Imagine that an ESCO intends to process power consumption data with low-time granularity, meaning rich data streams showing the power consumption every 10 s. Such a data stream may reveal personal data, starting from habits to come home or leave home, up to habits that may reveal health problems, e.g., going to the toilet too many times during the night (which may be observed through power consumption with light devices). The contract may stipulate that the rich-content data will be sent only for 1 month, at a second-based granularity (meaning power each second), and that the data will be kept by the ESCO for another 15 days, and it will then need to be deleted. Such rich data, for each day, may look graphically like the one presented in 0 (1 day readings with 1 s granularity).

So, for the above example, the privacy profile data is summarized as follows:

- Period: 01.06.2018 to 31.06.2018.
- Type of data: active power P(t) on point of common coupling (PCC).
- Period for the records: 1 s.
- Period of keeping data after finalizing the contract agreement: 15 days.
- Cyber-attack measures: "man in the middle attack" – digital signature; "measure against data theft" – encryption with private keys.
- Data exchange information (DEI): address of the ESCO broker, topics to be received, VPN connection details (including credentials for the ESCO VPN), low-profile agent to be run in the SMX (if needed).

This process, whose concept is presented in 0, can be followed as follows:

- The contract is signed and contains PP data described above (even the contract signature may be made electronically).
- Each part can and will extract digital content which describes the PP data, by producing "electronic PPx rules," as per 0.

- The data owner uploads (can be made also automatically) in the RBAC of the SMX the contract PP. RBAC is profiled to open a new MQTT connection able to make data exchange according to DEI.
- The service provider (ESCO) uploads in their system the PP rules and provides an MQTT broker and an MQTT client for the communication and, if needed, a small application to run in SMX's sandboxed environment.

Each party is running their applications and applies their part of the privacy agreements.

4.3.6 Local data Processing—Obtaining Derived Data From USM Basic Data

Smart meters may need to pre-process data before sending it to different actors. Practically, there is no meter able to provide symmetric components; however, some meters provide all necessary data to compute these values. This type of data is referred as first-order derived data, as they are directly computed from real-time data, after each acquisition cycle based on a full set of new data.

Second-order derived results are also based on a temporal series. One example is the measurement of voltage level, based on the quadratic mean value of voltage over this aggregated interval, as requested on IEC61000–4-30. In 0 an alternative approach is presented, based on the voltage level probability distribution for computing this value. Such a probability distribution can be constructed as shown in 0.

By monitoring the voltage evolution over the longer term, it is possible to obtain the probability distribution for the voltage over an entire commercial interval (e.g., 1 h). 0 presents the superposed probability distributions for different hours, using real measurements on a LV network in Bucharest, Romania.

The probability distribution shows the number of occurrences of a certain voltage level (e.g., between 214 V and 215 V) and gives practical data for coupling consumed energy with the quality of service (in this case, voltage quality). A complete day with voltage evolution processed by the USM is presented in 0.

Some functionalities may need a local agent to provide real-time control, a solution that avoids the danger of making such activity un-operational if it is remotely controlled but the communication is temporarily down.

4.3.7 Running Local Agents to Support Specific Functionalities

In addition to local data processing and sending it customized to different actors, there are functionalities that need local agents to run and work well connected to the USM core. Several types of such agents can be considered, including agents to:

- perform demand response (DR) support for a DR aggregator 0;
- perform secure energy transactions by using blockchain technology;

- act as energy broker for surplus energy of a prosumer. This agent can work for a local (neighborhood) market related to the physical micro grid after an MV/LV transformer or for a more global market;
- provide storage as a service for the grid, or for neighbors that have excess energy and unavailable personal storage means [1, 5];
- supervise and coordinate intelligent PV inverters; and

provide communication with ontology-based intelligent appliances using ontological approach, such as Smart Appliances REFerence (SAREF) 0.

4.3.8 Coordinating Local Resiliency and Immunity

Local resilience and immunity of a prosumer become increasingly important values from the energy supply perspective. Local production and storage along with local consumption coordination are prerequisites for allowing resilience and immunity to the final customer, so that grid perturbations do not affect local energy needs. Such coordination can be done by a special agent that can run in a building management system (BMS). However, as SMX supports local operation of special agents, eventually in a sandboxed solution to protect the main activities, these agents can run on SMX and exchanging data with the smart meter and with local customers. An advanced resilient consumer can control its production, consumption, and storage in such a way that it never needs to send back energy in the system, this being coordinated by a special agent inside the USM's SMX. The S4G project 0 is implementing a solution where an energy router (interconnecting the grid, PV panels, storage devices, and home loads) allows 24/7 prosumer resilience and immunity. This solution (coordinated by an SMX agent) will also supply critical loads during small AC voltage interruptions or during blackouts, as the power inverter connection to the grid will be kept un-operational during this period, to be compatible with existing network codes.

4.3.9 Accommodating and Sharing Data Models as a Multicultural Environment

SMX can organize an internal database model, which can be accomplished by recording data received through MQTT messaging in a different hierarchical model than it has been received in. After the data is received, it is recorded according to the defined internal data model, for example, a DLMS-COSEM style, as the primary data relates to metering. Several possibilities can be considered where the internal model can be converted in different external models, according to the partner/actor needs, as presented in 0. Instead of trying to unify data models between actors, they are recognized and a multicultural communication becomes possible, meaning that each actor can receive or send its specific data model, and a SMX-based converter ensures compatibility between different types of data organizations. This does not mean in any way that the approach is against a unified solution; however, reality shows that domains can differ so much that a unification is not possible, or it distorts some of the specific data exchange structures too much.

This is an important step forward, which overpasses the bottleneck of data models unification, leaving a certain number of market leading solutions to be used, by keeping also the specificity. The data model converter does not impose a solution, but provides different types of solutions for different types of needs, still belonging to the same data roots.

0 shows a conversion from internal data model to a data model that presents data based on the common information model (CIM), which has different types of standard codes compared with OBIS codes. Moreover, the data can be complemented, as an option, with additional data related to the measurement, such as description and unit to be used. Such CIM-oriented organization is useful especially for a SCADA system or for its DMS/EMS components. In particular, 0 presents a conversion example, from an internal SMX data model (DLMS oriented) into a CIM data model (for SCADA applications).

4.4 USM Deployment

Due to its high flexibility, the USM is able to cope with many different requirements, and is thus the repository of new developments in various research projects. Some USM application examples are provided here.

4.4.1 NOBELGRID

NOBELGRID 0 is the European project that pushed the concept and introduced the USM as the main solution for future advanced metering infrastructures. In NOBELGRID, the USM architecture is doubly demonstrated through:

- a new, innovative and affordable smart meter (developed from scratch), based on the unbundled smart meter (USM) concept, which contains two different parts: a smart metrology meter (SMM), with metrology features and fixed functionality, and a smart meter extension (SMX), which has high flexibility to accommodate new functionalities, to support smart grid and various energy services 0; and
- the use of market existing electronic meters, acting as SMM, and a newly designed SMX, which communicates with SMM, and reading all useful data intermediating the multi-user multi-protocol activity with various actors.

In NOBELGRID, the focus is on the following main actors:

- DSO, which receives real-time data from USM, being able to implement meter-based SCADA and future smart grid functionalities;

- aggregator of small demand response services, which receives DR information and energy service measurement (for settlement and quality of service assessment), and sends DR commands at the user level;
- energy service company, which has a direct communication with the USM end-user, thus being able to provide different energy related services such as awareness and advisory services; and
- prosumer having a PV installation, which is integrated and controlled through the SMX.

4.4.2 SUCCESS

SUCCESS 0 is a European project where smart meter data security is the main focus, and where new functionalities are being considered, including:

- advanced cyber security through the physically unclonable function, which significantly improves resilience against cyber-attacks;
- support for data consistency assessment, to implement cyber-security awareness at higher levels, such as local DSO or wider grid level; and
- support for local-level intelligence measures directed at cyber-attacks protection.

4.4.3 NRG5

NRG5 0 is a European project aiming at different targets, including smooth 5G communication accommodation, but also targeting the implementation of blockchain technology for neighborhood markets. The blockchain functionality is designed to be supported by the SMX part of the USM, allowing automatic and secure participation at dynamic short-term local markets (e.g., on time intervals from 15 s to 1 min).

4.4.4 STORAGE4GRID

The STORAGE4GRID European project 0 targets the intelligent use of emerging storage resources in different situations, such as:

- avoiding curtailment in case of renewable energy excess and use of aggregated storage assets over a specific network;
- improving charging services in high EV penetration areas postponing grid reinforcement; and
- using prosumer local storage and hybrid DC/AC network to improve resiliency and immunity against grid outages.

These functionalities are supported by SMX-hosted local agents that handle local storage resources while communicating with higher-level applications coordinating and improving the efficient use of those resources.

4.5 Conclusions and Remarks

According to the Third Energy Package (annex 1.2) and under the EU energy market legislation, member states are required to ensure the implementation of smart metering wherever a long-term cost–benefit analysis (CBA) is positive. In these cases, member states need to replace at least 80% of the traditional electricity meters with smart meters by 2020. The main objectives are reducing gas emissions and annual households' electricity consumption by around 9%. The working document "Cost-benefit analyses & state of play of smart metering deployment in the EU-27" 0 indicated that 20 member states had already conducted their CBA, 13 of them with positive results and 16 with roll-out plans already developed to accomplished the 80% target of smart meters by 2020.

To fulfill the EU Recommendation 2012/148/EU the smart metering systems for electricity should exhibit at least the functional requirements indicated in Table 4.1 0. (See Figs. 4.9–4.18.) Fig. 4.9 presents details on data security issues considering a trusted and some untrusted zones, and also communication through VPN or enhanced communication by adding PUF technology. Fig. 4.10 details the role-based access control (RBAC) and the data-centric architecture which allows an effective RBAC implementation. Fig. 4.11 depicts how RBAC can be implemented with various actors communicating based on MQTT messages. Fig. 4.12 presents an example of smart meter real-time capability used to record fine-grained power load profiles. Fig. 4.13 shows a possible scheme for

TABLE 4.1 List of recommended minimum functional requirements for electricity smart metering systems

Consumer	1. Provide readings directly to the consumer and/or third party. 2. Update readings frequently enough to use energy saving schemes.
Metering operator	3. Allow remote reading by the operator. 4. Provide two-way communication for maintenance and control. 5. Allow frequent enough readings for networking planning.
Commercial aspects of supply	6. Support advanced tariff system. 7. Allow remote ON/OFF control supply and/or flow or power limitation.
Security-data protection	8. Provide secure data communications. 9. Offer fraud prevention and detection.
Distributed generation	10. Provide import/export and reactive metering.

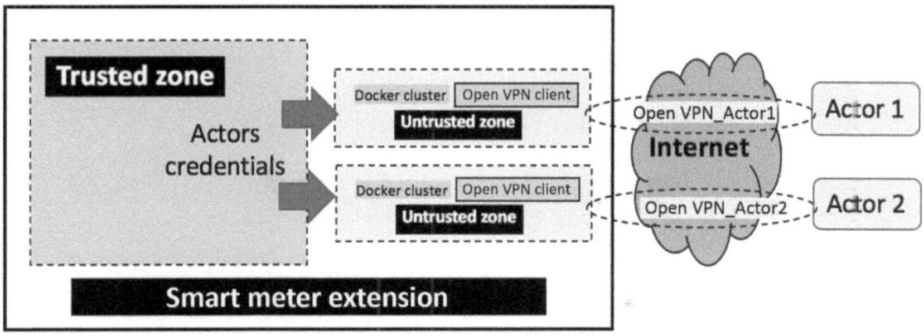

FIG. 4.8 Communicating securely with different actors.

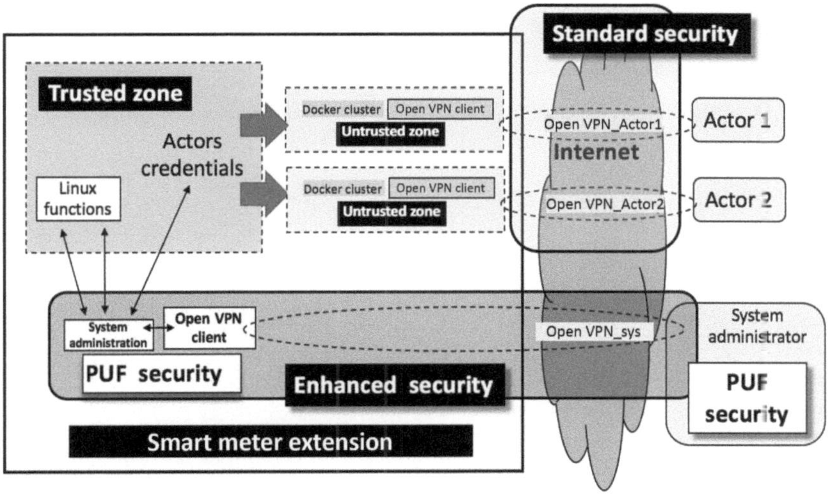

FIG. 4.9 Enhanced access to SMX trusted zone, using new data security mechanism, such as the physical unclonable function (PUF).

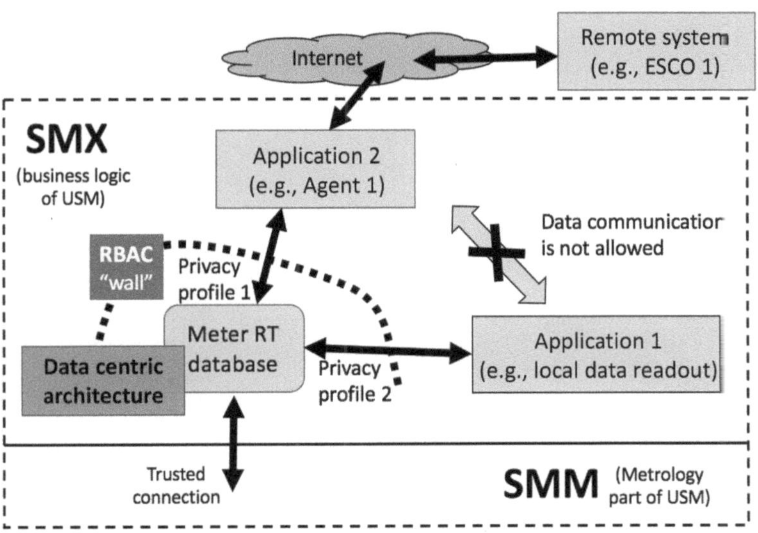

FIG. 4.10 Privacy by design using DCA and RBAC.

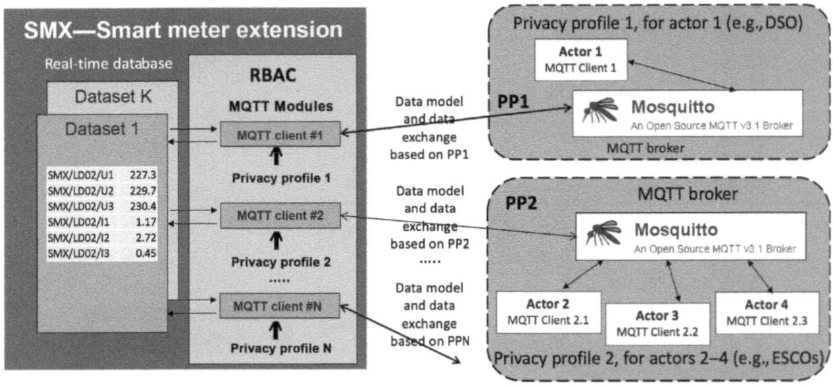

FIG. 4.11 RBAC implementation for ensuring different privacy profiles to different actors.

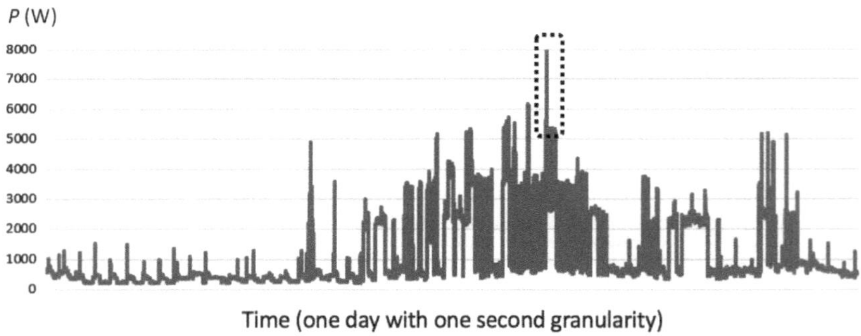

FIG. 4.12 Active power profile for a domestic house, over a day, with second-based series.

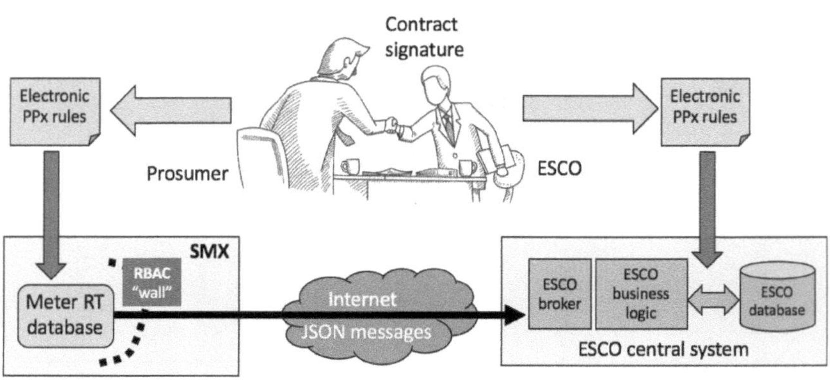

FIG. 4.13 Privacy agreement.

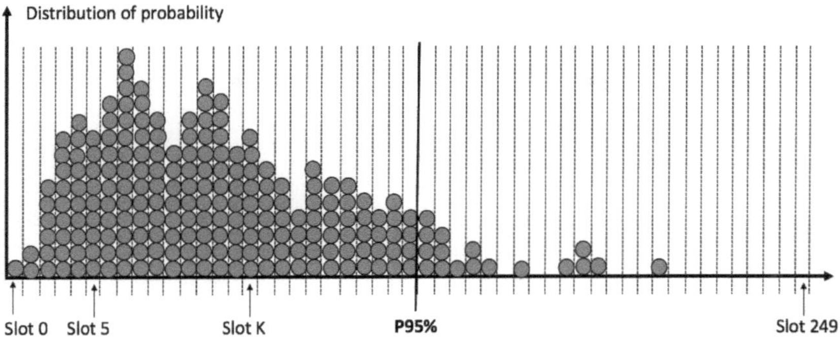

FIG. 4.14 Constructing a distribution of probability for the voltage, as second-order derived results from real-time voltage measurements.

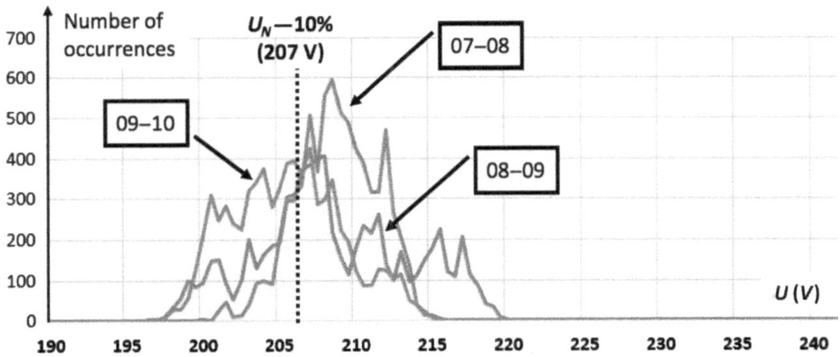

FIG. 4.15 Probability distribution for the voltage in three consecutive hours, on a LV network.

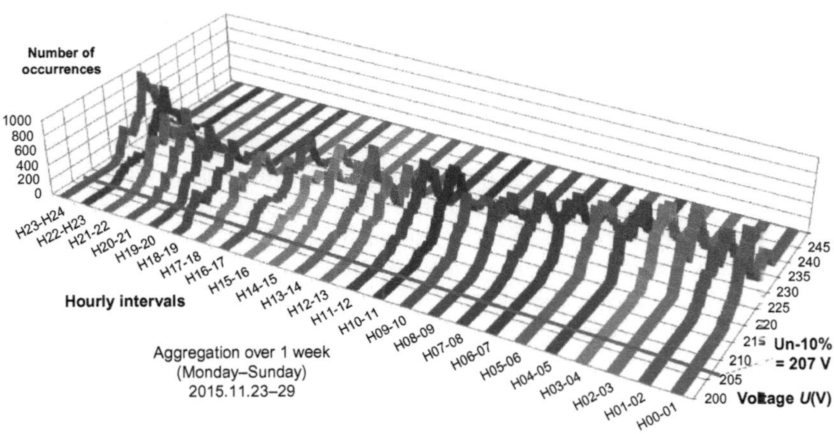

FIG. 4.16 Distribution of probability for each hour over 1 day.

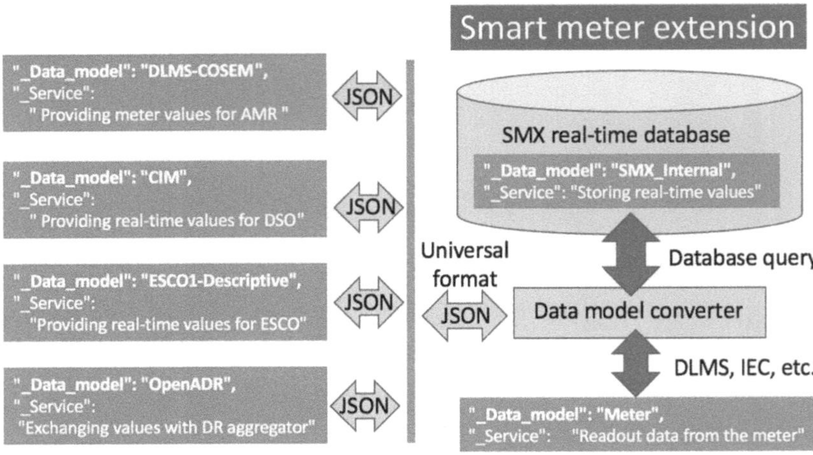

FIG. 4.17 Converting from internal data model to an ESCO specific data model.

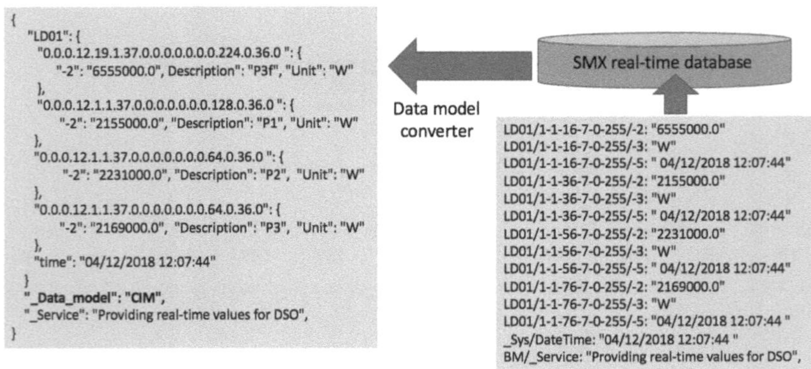

FIG. 4.18 Converting from an internal data model to a common information model (CIM) data model.

energy services transactions by automating the privacy profiles requirements required by the service. Figs. 4.14–4.16 show how voltage level quality indicators can be obtained based on statistical means using the voltage distribution of probability (presented in Fig. 4.14). These results are based on smart meter measurements and respective statistical approach for one day (presented in Fig. 4.15) and for multiple days (one week presented in Fig. 4.16).

As described above, the unbundled smart meter (USM), which simultaneously combines metrology features (through smart metrology meter), high flexibility to accommodate new functionalities (through smart meter extension), and robust data security and protection, assures all these functional requirements and is designed to support future smart grids and energy services developments. The USM should be seen not only as a metering device usually controlled by the DSOs, but also as an intelligent asset used to improve energy efficiency and to benefit both retailers and end-users, contributing to the global accomplishment of the Third Energy Package requirements.

References

[1] World Energy Outlook 2017, International Energy Agency, 2017.

[2] V.C. Gungor, B. Lu, G.P. Hancke, Opportunities and challenges of wireless sensor networks in smart grid, IEEE Trans. Ind. Electron. 57 (10) (2010) 3557–3564.

[3] "Paris Agreement," PARIS2015 UN Climate Change Conference COP21-CMP11, December 2015. http://ec.europa.eu/clima/policies/international/negotiations/paris/index_ en.htm.

[4] Energy Independence and Security Act of, (Pub.L. 110–140), Act of US Congresshttps //www.gpo.gov/fdsys/pkg/PLAW-110publ140/html/PLAW-110publ140.htm, 2007.

[5] X. Fang, S. Misra, G. Xue, D. Yang, Smart grid; the new and improved power grid: a survey, Commun. Surv. Tutorials, IEEE 14 (4) (2012) 944–980.

[6] H. Sui, H. Wang, M.-S. Lu, and W.-J. Lee, "An AMI system for the deregulated electricity markets," IEEE Trans. Ind. Appl., vol. 45, no. 6, pp. 2104–2108, Nov./Dec. 2009.

[7] NIST, Advanced Metering in Smart Distribution Grids, National Institute of Standards and Technology Project, 2012.https://www.nist.gov/programs-projects/advanced-metering-smart-distribution-grids.

Further Reading

[8] AT&T and the Utilities, "An insight into best practice enablers in the U.S. utility sector," M2M Now Magazine, July–August 2015. http://www.m2mnow.biz/2015/07/08/34626-m2m-now-magazine-july-august-2015-edition.

[9] REPORT FROM THE COMMISSION Benchmarking smart metering deployment in the EU-27 with a focus on electricity—COM/2014/0356 final, 2014. http://eur-lex.europa.eu/legal-content/EN/TXT/?uri=COM:2014:356:FIN.

[10] H2020 NOBELGRID project, http://nobelgrid.eu.

[11] R. Rashed Mohassel, A. Fung, F. Mohammadi, K. Raahemifar, A survey on advanced metering infrastructure, Int. J. Electr. Power Energy Syst. 63 (2014) 473–484.

[12] CEN/CLC/ETSI/TR. Functional reference architecture for communications in smart metering systems. s.l.: CEN-CENELEC-ETSI, 2011. 50572:2011 E.

[13] M. Sánchez, Implementation of Smart Electricity Metering in Europe. s.l.: Directorate General for Energy, European Commission, 2014.

[14] Docker Inc, Docker Hub. Docker Hubhttps://hub.docker.com/, 2018https://hub.docker.com.

[15] SMART 2013/0077. Smart appliances reference (SAREF) ontology. https://sites.google.com/site/smartappliancesproject/ontologies/reference-ontology.

[16] H. Kreutzmann, S. Vollmer, Protection Profile for the Gateway of a Smart Metering System (Smart Meter Gateway PP), vol. 1.3, Federal Office for Information Security, 2014 s.l.. BSI-CC-PP-0073.

[17] H2020 Storage4Grid project. n.d. http://www.storage4grid.eu.

[18] M441 Standardization Mandate of 12th March 2009, on development of an open architecture for utility meters. EU mandate M441.

[19] Smart Grid Architecture Model, CEN-CENELEC-ETSI Smart Grid Coordination Group Smart Grid Reference Architecture, CEN-CENELEC-ETSI Smart Grid Coordination Group, (November 2012).

[20] M. Sănduleac, M. Albu, J. Martins, M. Dolores Alacreu, Carmen Stanescu—Power Quality Assessment in LV networks using new Smart Meters design, in: CPEE 2015, June 24–26, 2015, Proceedings 2015 9th International Conference on Campatibility and Power Electronics (CPE), 2015, , pp. 106–112.

[21] M. Sanduleac, I. Ciornei, M. Albu, J. Martins, et al., Resilient prosumer scenario in a changing regulatory environment-the UniRCon solution, ENERGIES 10 (12) (2017) 1941Article Number.

[22] NOBELGRID SLAM, http://nobelgrid.eu/wp-content/uploads/2017/09/NOBEL-GRID-product_SLAM-1.pdf.

[23] H2020 SUCCESS project, https://success-energy.eu.

[24] H2020 NRG5 project, http://www.nrg5.eu.

[25] European Commission, E. (2014). Cost–Benefit Analyses & State of Play of Smart Metering Deployment in the EU-27 Accompanying the Document Report From the Commission Benchmarking Smart Metering Deployment in the EU-27 With a Focus on Electricity. COMMISSION STAFF WORKING DOCUMENT, European Commission, Brussels. http://eur-lex.europa.eu/legal-content/EN/TXT/PDF/?uri=CELEX:52014SC0189&from=EN.

Chapter 5

Energizing Demand Side Participation

Gregor Verbič, Sleiman Mhanna and Archie C. Chapman
The University of Sydney, School of Electrical and Information Engineering, Sydney, NSW, Australia

Chapter Outline

5.1 Introduction

The industry now widely accepts that renewable generation is the cheapest long-term supply option [1]. The increasing penetration of *variable non-synchronous* renewable generation, however, challenges the existing generation-following load operation paradigm based on *dispatchable synchronous* generation. As a result of this, energy security and affordability have taken center stage in

Pathways to a Smarter Power System. https://doi.org/10.1016/B978-0-08-102592-5.00005-3

national political discussions. At the other side of the energy supply chain, residential and small commercial users increasingly opt for solar photovoltaic (PV) and battery storage to reduce their energy expenditure. In addition to this, the technical advancements in sensor, computer, and communication technologies now make it possible to harness the inherent flexibility of these "behind-the-meter" distributed energy resources (DER) in innovative ways [2]. Using prosumers (*pro*ducers-con*sumers*) as a flexible resource creates unique opportunities to address the challenges associated with the increasing uptake of variable renewable energy sources, like wind and solar. In a nutshell, the prosumer-driven decentralized power supply will make the power system more reliable, affordable, and sustainable, thus solving the energy trilemma [3].

5.1.1 Background

The massive uptake of DERs presents the opportunity to manage the power system transformation in a most economical way, and with minimal societal disruption, by exploiting demand-side flexibility [3, 4]. In more detail, at the demand-side, the emergence of cost-effective, behind-the-meter DERs—including on-site generation, energy storage, electric vehicles, and flexible loads—is changing the way that electricity consumers source and consume electric power. To give some context, installed capacity of rooftop PV has increased globally from approximately 4GW in 2003 to nearly 227GW in 2015, driven by government incentives, electricity price increases, and decreasing PV capital cost [5]. A similar trend is observed in Australia, where the installed capacity of rooftop PV has grown from 0.8GW in 2011 to over 8GW in 2018 (c.f. peak demand of 35GW). Also, improvements in small-scale battery storage technologies are continuing, and it is anticipated that economic storage options will be soon available. Indeed, recent studies suggest that rooftop PV-battery systems will reach retail price parity from 2020 in the US grids and the Australian National Electricity Market [6, 7].

A recent forecast by Morgan Stanley suggests that the uptake might be even faster—boldly predicting that up to 2 million Australian households could install battery storage by 2020 [8]. This has been confirmed by the Australian Energy Networks Association (ENA) and the Commonwealth Scientific and Industrial Research Organisation (CSIRO), who estimate the projected uptake of solar PV and battery storage in 2050 to be 80GW and 100GWh [4], representing between 30% and 50% of total demand at that time. In their prediction of the future grid, the CSIRO considered a scenario called "Rise of the prosumer" [9]. They refer to small-scale (residential, commercial, and small industrial) electricity consumers with onsite generation as prosumers. The opportunities offered by demand-side resources have also been recognized by the Australian Energy Market Commission (AEMC) [10], who argue that consumers with PV-battery systems can provide network support and ancillary services that were traditionally confined to the domain of large generators. A successful transition to the future electricity grid will thus require a more decentralized power supply, underpinned by a paradigm shift in power system operation from generation-following-load to load-following-generation. The key to the transition is to tap the inherent flexibility of the demand-side. But before we discuss that in more detail, we first give an overview of the "traditional" demand response.

5.1.2 Overview of Traditional Demand Response Programs

The flexibility of the demand-side has been recognized as a valuable system resource for a long time. Historically, various demand response programs have been used as a reliability and capacity resource, to manage emergency and peak load events, respectively. In 2006, the Department of Energy in the United States formally defined demand response as [11]: Changes in electric usage by end-use customers from their normal consumption patterns in response to changes in the price of electricity over time, or to incentive payments designed to induce lower electricity use at times of high wholesale market prices or when system reliability is jeopardized.

Demand response programs take many forms. The North American Electric Reliability Corporation (NERC) categorized different forms of demand response in relation to overall demand-side management strategies, as illustrated in Fig. 5.1 [12]. As you can see, demand response is part of a broader suite of measures for demand-side management that also includes energy efficiency. Energy efficiency focuses on users and their behavioral changes to achieve more efficient energy usage. Demand response, on the other hand is a tool used by utilities to improve reliability and economic performance. It encompasses a range of activities undertaken by electricity users, or other agents acting on their behalf, to temporarily change their grid consumption in response to some form of financial incentive. Demand response is categorized based on the *level of control* (dispatchable or non-dispatchable), *purpose* (economic, reliability), and the *type of resource* provided (energy, capacity, reserves, regulation).

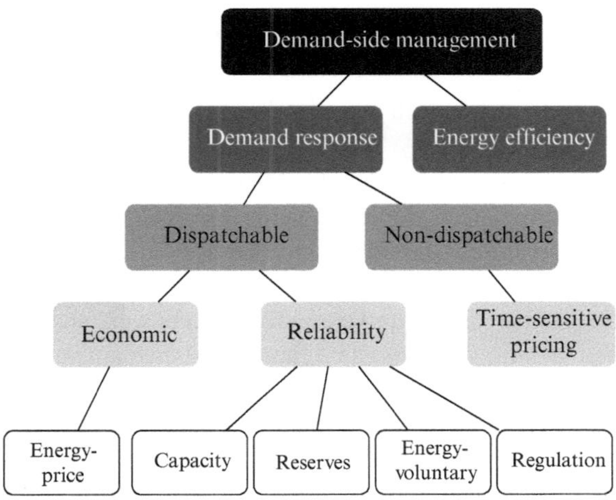

FIG. 5.1 Classification of demand response programs [12].

Existing demand response programs predominantly rely on large industrial customers with electricity-intensive processes that can be used as curtailable loads. Under the old *generation-following-load* operation paradigm, demand response was an effective way of managing predictable discrete events. The increasing variability of the supply, however, will require a more generation-like resource to provide balancing services, which can be offered by behind-the-meter DERs.

5.1.3 Demand Response Using Behind-the-Meter DERs

Large industrial and commercial loads are relatively easy to aggregate, but their large individual size is not well suited for continuous load balancing. On the other hand, fast, flexible and continuously responsive DERs, connected behind-the-meter in residential, small commercial, and industrial buildings, are highly granular and, therefore, well suited for balancing services. To be able to harness their flexibility we need to aggregate them, so that they behave like a single dispatchable resource.

Conventional aggregation techniques are, however, unsuitable for large-scale DER aggregation. Technical advancements in sensor, computer, wireless communication, and energy management technologies now make it possible to harness the inherent flexibility of these behind-the-meter DERs in innovative ways. However, a number of fundamental technical challenges need to be solved first:

- (i) DERs are highly diffuse;
- (ii) the nature of their energy usage is stochastic and inherently task-oriented;
- (iii) their number is enormous, which renders their treatment as conventional power plants unfeasible; and
- (iv) the existing wireless network standards do not cater for two-way communication flows with heterogeneous latency, packet size, and reliability requirements.

As a result of this, two decoupled research avenues have emerged:

- (i) residential demand response aggregation on one side, and
- (ii) home energy management on the other.

However, there is a gap between these two branches of work, which affects the usefulness of research in both.

Research on home energy management has consistently addressed more and more difficult optimization problems, using increasingly sophisticated techniques and incorporating a broader scope of activities and phenomena [13, 14]. On the other hand, the techniques used for aggregating electricity consumer entities have focused on tariff, market, and auction design, and are now making use of advanced techniques from game theory and mechanism design [15, 16]. Unfortunately, this represents a divergence between the two branches of research. From one perspective, the techniques often used for energy management are not amenable to aggregation using standard techniques from mechanism design. Likewise, the models employed by researchers using mechanism design are qualitatively different to those proposed in state-of-the-art energy management systems, as we have explored in [17]. The home energy management system typically considers a single residential building, which allows capturing the nature of residential energy usage with sufficient complexity. Residential demand response aggregation schemes, on the other hand, require a significantly simplified representation of the DERs, which typically results in unrealistic solutions, for example, by representing a household's valuation for electricity by a concave increasing function [18], which is simple enough to be conducive to convex analysis, but fails to capture the discrete states of appliances. Also, the DER models are typically deterministic, which fails to capture the stochastic nature of energy consumption. The existing techniques for prediction of energy usage in residential buildings provide a suitable avenue to address this problem, but they typically focus on individual appliances [19] and are thus unsuitable for online applications.

Finally, a significant prosumer penetration at low voltage (LV) levels will push electricity grids and communications networks to the brink of their technical limits. Therefore, the DER aggregation problem should explicitly consider network power flow and communication constraints, which has not yet been sufficiently addressed in the existing literature. There does, however, exist a large body of work on optimal power flow formulations of LV grids with renewables [20, 21]. The DERs considered typically include rooftop PV, which admits a computationally conducive convex (or convexified) formulation. However, explicitly considering battery storage produces a—computationally much harder—mixed-integer formulation, which is yet to receive any attention in the available literature. Additionally, an explicit network representation introduces non-convex constraints stemming from the AC load flow equations. Moreover, unlike network power flow constraints, communication constraints have not yet received significant attention in the residential demand response aggregation context—sufficient communication capacity is either assumed [22], or is only mentioned implicitly [23]. The emergence of the internet of things paradigm is bound to increase the demand for communication resources, which require an explicit representation.

In summary, given these observations, this chapter argues that it is essential that demand-side participation is actively coordinated, in order for distributed generation and behind-the-meter resources to contribute the most value to distribution networks and the broader power system.

The rest of the chapter progresses as follows. Section 5.2 introduces the home energy management problem, which serves as the fundamental building block for demand response aggregation. Uncoordinated control of DER can, however, have an adverse impact on the operation of LV networks, which is the subject of Section 5.3. Section 5.4 then discusses active management of DERs, including peer-to-peer trading with a DSO permission structure and distributed optimal power flow. Section 5.5 discusses the distributed optimal power flow in more detail. Section 5.6 discusses a generic aggregate prosumer model for future grid studies. Finally, Section 5.7 concludes.

5.2 Home Energy Management

Buildings consume 40% of the world's energy, so they present the largest untapped pool of DR, which makes them a key enabler of future grids. Load flexibility stems from different sources. Water use, for example, requires water pumps and intermediate storage, which allows an operator to schedule their operation strategically. Similarly, the thermal inertia of the building's envelope can serve as a virtual battery, which offers a level of flexibility in scheduling the heating, ventilation, and air conditioning (HVAC) systems; and the same can be said for thermal water storage, both chilled and heated. Next, rooftop PV coupled with battery storage allows an operator to reduce peak demand and thereby minimize distribution network expenditures. Moreover, an efficient aggregation of DERs can turn a precinct, or a building, into a virtual power plant, which allows the owner to achieve multiple objectives, from cost reduction, to more sustainable use of resources and improved user comfort.

Currently, there are few mechanisms in place for load aggregation and integration into existing energy markets [24]. The advancements in smart grid technologies, however, are bound to change that. In addition, battery storage is becoming financially attractive, which opens new possibilities for using residential and small commercial loads for demand response. This is because battery storage effectively decouples energy and power requirements, so user comfort does not need to be compromised. The key here is proper management of energy flows.

As an example, consider a residential building equipped with the PV-battery system shown in Fig. 5.2, and the power flows for a typical sunny day illustrated in Fig. 5.3. Observe that, during the day when PV generation exceeds the demand, the energy management system charges the battery. In the evening, when electricity prices increase,[1] the power demand is supplied by the battery, which in turn reduces the grid demand. Note that the device-level management (e.g., HVAC or a hot-water system) is not visible in the grid demand, which

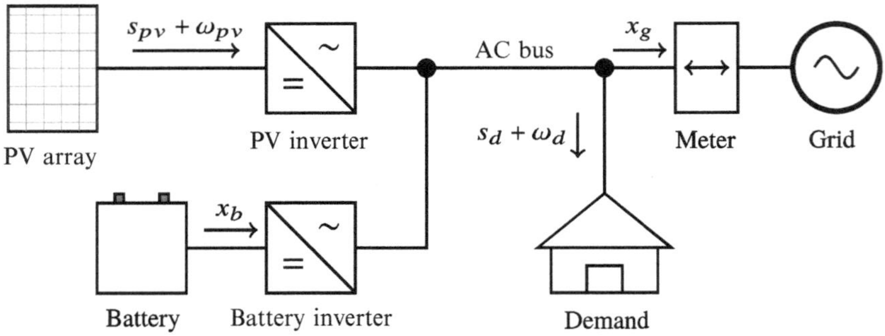

FIG. 5.2 A typical configuration of residential building equipped with a PV-battery system.

1. This is typical in a so-called time of use (ToU) pricing.

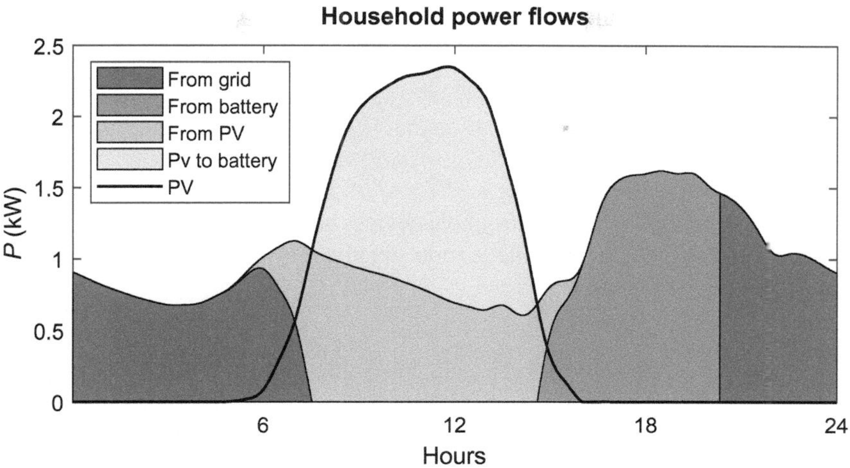

FIG. 5.3 Power flows in a residential building equipped with a PV-battery system.

preserves users' privacy. The battery thus effectively serves as an interface between the grid and the house.

The home energy management problem can be thought of as a sequential decision making process under uncertainty. We first, in Section 5.1.1, formulate the problem as a *Markov Decision Process* (MDP), and then follow this with a discussion of alternative solution techniques in Section 5.2.2.

5.2.1 General Formulation as a Markov Decision Process

A sequential stochastic optimization problem consists of:

- A sequence of *time-steps*, $\mathcal{T}=\{1\ldots t\ldots T\}$, where T and t represent the total number of time-steps, and a particular time-step in the decision horizon, respectively;
- A set of *non-controllable inputs*, $\mathcal{B}=\{1\ldots b\ldots B\}$, where each b is represented using:
  - A state variable, $s_b^t\in\mathcal{S}$.
  - A random variable, $\omega_b^t\in\Omega$, that captures perturbations or exogenous information.
- A set of *controllable devices*, $\mathcal{A}=\{1\ldots a\ldots A\}$, where each a is represented using:
  - A state variable, $s_a^t\in\mathcal{S}$;
  - A decision variable, $x_a^t\in\mathcal{X}$ for every control action;
  - Constraints for the control and state variables;
  - A transition function $s_a^{t+1}=s^M(s_a^t,x_a^t,\omega_b^t)$, describing the evolution of a state at time t to $t+1$. Here $s^{\mathrm{M}}(\cdot)$ is the system model that consists of the

operational constraints of controllable device a, such as power flow limits, efficiencies and losses.

Given this, let: $\mathbf{s}^t = [s_a^t \ldots s_A^t, s_b^t \ldots s_b^t]^{\mathrm{T}}$, $\mathbf{x}^t = [x_a^t \ldots x_A^t]^{\mathrm{T}}$, and $\boldsymbol{\omega}^t = [\omega_b^t \ldots \omega_B^t]^{\mathrm{T}}$. The state variables contain the information that is necessary and sufficient to make the decisions and compute rewards, costs and transit ions. The compact form of the transition functions is given as $\mathbf{s}^{t+1} = \mathbf{s}^M(\mathbf{s}^t, \mathbf{x}^t, \boldsymbol{\omega}^t)$. Note that we only need transition functions for the controllable devices and the combined random variables vector of the noncontrollable inputs is given by ω^t (without a subscript).

An *objective function*:

$$F = \mathbb{E}\left\{\sum_{t=1}^{T} C^t(\mathbf{s}^t, \mathbf{x}^t, \boldsymbol{\omega}^t)\right\}, \tag{5.1}$$

where $C^t(\mathbf{s}^t, \mathbf{x}^t, \boldsymbol{\omega}^t)$ is the contribution (i.e., reward or cost of energy, or a discomfort penalty) incurred at time-step t, which accumulates over time.

5.2.1.1 Instantiation

The aim of home energy management is to minimize the household's energy costs over a decision horizon. In general, stochastic optimization techniques solve this sequential stochastic optimization problem before the beginning of each day, using either a daily or a 2 day decision horizon. For illustration purposes, we only consider a PV unit and a battery, as depicted in Fig. 5.2, however the formulation is general and can seamlessly include other DERs as well, for example, fuel cells [25] or hot water systems [26]. An effective home energy management system needs to incorporate variations in electrical demand and PV output of the building. Assuming that they can be predicted with reasonable accuracy, we can model the stochastic variables using their respective means as state variables and the variations as random variables. In general, electricity prices have to be considered stochastic when the exact prices are unavailable. However, for most residential applications, electricity prices are available, for example, as time-of-use prices. Given this, there is no random variable associated with the electricity tariff.

Now we cast the home energy management problem as a sequential stochastic optimization problem as follows. The daily decision horizon is 1 day, divided into suitable time intervals, typically 30 min, giving $T = 48$ time-steps. The noncontrollable inputs are the electrical demand, PV output, and electricity tariff (note that there is no random variable associated with the electricity price), which are represented using:

- State variables for the mean electrical demand, $s_{d,\mathrm{e}}^t$, mean PV output, s_{pv}^t, and electricity tariff, s_p^t.
- Random variables for the variations in electrical demand, $\omega_{d,e}^t$, and variations in PV output, ω_{pv}^t.

The controllable device is the battery, which is represented using:

- State variables for the battery SOC, s_b^t.
- Control variables for charge and discharge rates of the battery, x_b^t. Given this, $\mathbf{x}^t = x_b^t$, $\mathbf{s}^t = [s_b^t, s_{d,e}^t, s_{pv}^t, s_p^t]$, and, $\boldsymbol{\omega}_k = [\omega_{pv}^t, \omega_{d,e}^t]$, are defined for each time-step, t, in the decision horizon, as depicted in Fig. 5.2.
- The energy balance constraint is given by:

$$s_{d,e}^t + \omega_{d,e}^t = \eta_i x_i^t + x_g^t, \tag{5.2}$$

where $x_i^t = s_{pv}^t + \omega_{pv}^t - \eta_b x_b^t$ is the inverter power at the DC side (positive value means power into the inverter); η_i is the efficiency of the inverter (note that the efficiency is $1/\eta^\text{i}$ when the inverter power is negative); η_b is the efficiency of the battery action corresponding to either charging or discharging; and x_g^t is the electrical grid power. The charge rate of the battery is constrained by the maximum charge rate $x_b^{t,+} \leq \gamma_c$ and discharge rate of the battery is constrained by the maximum discharge rate $x_b^{t,-} \geq \gamma_d$. The energy stored in the battery should be within the limits $\underline{s}_b \leq s_b^t \leq \bar{s}_b$.

- Transition functions govern how the state variables evolve over time. The battery SOC, denoted $s_b^t \in [\underline{s}_b, \bar{s}_b]$, progresses by:

$$s_b^{t+1} = \left(1 - l_b\left(s_b^t\right)\right)\left(s_b^t - x_b^{t,-} + \eta_b^+\left(x_b^{t,+}\right)x_b^{t,+}\right), \tag{5.3}$$

where $l^b\ (s_b^t)$ models the self-discharging process of the battery. The battery SOC transition function is a non-linear function of state and charge rate. Note that the discharge efficiency of the battery, the efficiency of the inverter and the maximum possible charge rate with respect to the battery SOC are nonlinear, so they need to be suitable linearized in certain formulations (i.e., in a mixed-integer linear formulation, as discussed in Section 5.2.2).

The optimal policy, π^*, is a choice of action for each state $\pi: \mathcal{S} \rightarrow \mathcal{X}$, that minimizes the expected sum of future costs over the decision horizon; that is:

$$F^{\pi^*} = \min_{\pi} \mathbb{E}\left\{\sum_{t=0}^{T} C^t(\mathbf{s}^t, \pi(\mathbf{s}^t), \boldsymbol{\omega}^t)\right\}, \tag{5.4}$$

where $C^t(\boldsymbol{s}^t, \boldsymbol{x}^\text{t}, \boldsymbol{\omega}^t)$ is the cost incurred at a given time-step, which is given by:

$$C^t(s^t, x^t, \omega^t) = s_p^t\left(s_{d,e}^t + \omega_{d,e}^t - \eta_i x_i^t\right). \tag{5.5}$$

Here the problem is formulated as an optimization of the expected contribution because the contribution is generally a random variable, due to the effect of $\boldsymbol{\omega}^t$. In a home energy management problem, we obtain the battery decisions x_b^t, depending on the state variables $\mathbf{s}^t = [s_b^t, s_{d,e}^t, s_{pv}^t, s_p^t]$, and realizations of random variables $\boldsymbol{\omega}^t = [\omega_{pv}^t, \omega_{d,e}^t]$ at each time-step.

5.2.2 Home Energy Management Solution Techniques

The home energy management problem can be solved using different optimization methods. We first discuss mixed-integer linear programming (MILP), which is the most popular due to the ubiquitous solvers (e.g., Gurobi, CPLEX, Mosek) and modeling languages (e.g., AMPL, GAMS). The MILP formulation requires linearized models of components and transition functions, which can result in lower-quality solutions (see [26] for details). Furthermore, MILP cannot natively handle stochasticity, for which a stochastic MILP formulation is required. Dynamic programming (DP), on the other hand, can seamlessly handle both nonlinearity and stochasticity, but it is computationally burdensome due to the *curse of dimensionality*—that is, the size of the state, action, and outcome spaces, which can be overcome using *approximate dynamic programming* (ADP) [26].

The quality of the schedules from ADP, DP, and stochastic MILP depend on the quality of the PV and electrical demand predictions. Moreover, even though ADP is computationally efficient compared to other methods, computing the value function approximations in the off-line planning stage is computationally expensive and has high memory requirements. In order to overcome these difficulties, we have proposed an algorithm based on policy function approximations using machine learning that generates fast offline models using training data, which is then used to make fast real-time decisions [27].

5.2.2.1 Mixed-Integer Linear Programming

The home energy management problem is cast as a mixed-integer linear programming problem of the general form:

$$\begin{aligned} \underset{x_r, x_b}{\text{minimize}} \quad & c' x \\ \text{subject to} \quad & \underline{x}_r \leq x_r \leq \overline{x}_r \\ & Ax \leq b \\ & x_b \in \{0, 1\}^{n_b} \\ & x_r \in \mathbb{R}_+^{n_r} \end{aligned}$$

where x_b and x_r are binary, and continuous decision variables of the problem, respectively. Note that both the objective function and the constraints (many corresponding to the transition functions in the MDP formulation) are linear.

The home energy management objective is to minimize the energy expenditure over the decision horizon (1 day), which for a PV-battery system takes the form:

$$\text{minimize} \sum_{t=1}^{T} \left(x_g^{+,t} c_g^t - x_g^{-,t} c_{\text{fit}} \right), \tag{5.6}$$

where T is the number of time steps (48 for 1 day with half-hourly resolution), $x_g^{+,t}$ is power taken from the grid, $x_g^{-,t}$ is power returned to the grid, c_g^t is the retail tariff, and c_{fit} is the feed-in-tariff, all at time step t. The decision vector consists of the following continuous variables: power flow from/to grid $x_g^{+/-,t}$, battery charge/discharge power $x_b^{+/-,t}$, power flow from battery to grid $x_b^{g,t}$, power flow from battery to demand $x_b^{d,t}$, proportion of PV power flow $d_{pv}^t \in [0,1]$ (0: all PV power flows to grid, 1: all PV power flows to demand), and battery state of charge e_b^t (note here we use e_b rather than s_b as in the MDP formulation above to indicate this is the battery linearized model). The binary variables include: battery charging status $d_b^t \in \{0,1\}$ (0: discharge, 1: charge), and direction of grid power flow $d_g^t \in \{0,1\}$ (0: demand→grid, 1: grid→demand).

The optimization problem is subject to the following equality constraints (5.7)–(5.11), inequality constraints (5.12)–(5.15), upper and lower limits on the continuous variables (5.16)–(5.20), and binary constraints (5.21):

$$x_g^{+,t} = s_d^t - \eta_i\left(\eta_b x_b^{d,t} + d_{pv}^t s_{pv}^t - x_b^{+,t}\right) \tag{5.7}$$

$$x_g^{-,t} = \eta_i\left(\eta_b x_b^{g,t} + \left(1 - d_{pv}^t\right) s_{pv}^t\right) \tag{5.8}$$

$$x_b^{-,t} = x_b^{d,t} + x_b^{g,t} \tag{5.9}$$

$$e_b^{t+1} = e_b^t + \Delta t\left(x_b^{+,t} - x_b^{-,t}\right) \tag{5.10}$$

$$e_{b,1/n} = e_b^{\text{start/end}} \tag{5.11}$$

$$x_g^{+,t} \le \overline{x}_g d_g^t \tag{5.12}$$

$$x_g^{-,t} \le \overline{x}_g\left(1 - d_g^t\right) \tag{5.13}$$

$$x_b^{+,t} \le \overline{x}_b d_b^t \tag{5.14}$$

$$x_b^{-,t} \le \overline{x}_b\left(1 - s_b^t\right) \tag{5.15}$$

$$0 \le x_g^{+/-,t} \le \overline{x}_g \tag{5.16}$$

$$0 \le x_b^{+/-,t} \le \overline{x}_b^{+/-} \tag{5.17}$$

$$0 \le x_b^{g,d,t} \le \overline{x_b} \tag{5.18}$$

$$0 \le d_{pv}^t \le 1 \tag{5.19}$$

$$\underline{e}_b \le e_b \le \overline{e}_b \tag{5.20}$$

$$d_g, d_b \in \{0, 1\}. \tag{5.21}$$

The equality constraints in (5.7)–(5.9) are balance constraints, ensuring the energy flows within the home are conserved. The positive and negative flows from the grid and battery are coupled using binary variables to ensure only one of the positive or negative variables is not equal to zero in (5.12)–(5.15).

The battery operation is modeled in (5.10) and (5.11). The former is the (linearized) state-of-charge transition function, linking present and future battery states via the energy drawn from them. The latter ensures the state of charge at the start and end of the day are maintained at a useful level. The remaining expressions are variable value constraints.

5.2.2.2 Stochastic MILP

In order to incorporate stochasticity into the deterministic version of the home energy management problem presented above, a large set of scenarios need to be generated to cover the space of possible realizations of the uncertain variables. A larger number of scenarios should improve the solution, but this imposes a greater computational burden. Therefore, heuristic scenario reduction techniques are employed to obtain a scenario set of size N, which can be solved within a given time with reasonable accuracy. Given this, a scenario-based stochastic MILP formulation of the problem is described by:

$$\text{minimize} \sum_{j=1}^{J} P_j(\mathbf{s}_j) \sum_{t=1}^{T} \left(x_g^{+,t} c_g^t - x_g^{-,t} c_{\text{fit}} \right), \tag{5.22}$$

where $P_j(\mathbf{s}_j)$ is the probability of a particular scenario j corresponding to realizations of stochastic variables $\mathbf{s}_j$, subject to $\sum_{j=1}^{J} P_j(\mathbf{s}_j) = 1$.

For each realized scenario, the optimization problem is solved for the whole horizon at once, using a standard MILP solver, so the solution time grows exponentially with the length of the horizon. In the existing literature, a 1 day optimization horizon is typically assumed. The solutions are of lower quality because of the linear approximations made and the inability to incorporate all the probability distributions. In response to these limitations, *dynamic programming* was proposed in [25] to improve the solution quality.

Dynamic Programming

The problem in (5.4) is easily cast as an MDP, due to the separable objective function and Markov property of the transition functions. Given this, DP solves the MDP form of (5.4) by computing a value function $V^{\pi}(\mathbf{s}^t)$. This is the expected future cost of following a policy, π, starting in state, $\mathbf{s}^t$, and is given by:

$$V^{\pi}(s^t) = \sum_{s' \in \mathcal{S}} \mathbb{P}(\mathbf{s}' | \mathbf{s}^t, \pi(\mathbf{s}^t), \boldsymbol{\omega}^t)[C(\mathbf{s}^t, \pi(\mathbf{s}^t), \mathbf{s}') + V^{\pi}(\mathbf{s}')], \tag{5.23}$$

where $\mathbb{P}(\mathbf{s}' | \mathbf{s}^t, \boldsymbol{x}^t\, \omega^{t)}$ is the transition probability of landing on state $\mathbf{s}'$ from $\mathbf{s}^t$ if we take action $\boldsymbol{x}^t$. An optimal policy, π^*, is one that minimizes (5.4), and which also satisfies Bellman's optimality condition:

$$V^{\pi^*}(s^t) = \min_{\pi^*} \left(C^t(\mathbf{s}^t, \pi(\mathbf{s}^t)) + \mathbb{E}\{ V^{\pi^*}(\mathbf{s}') | \mathbf{s}^t \} \right). \tag{5.24}$$

The expression in (5.24) is typically computed using backward induction, a procedure called *value iteration*, and then an optimal policy is extracted from the value function by selecting a minimum value action for each state. This is the key functional point of difference between DP and stochastic MILP. DP enables us to plan offline by generating value functions for every time-step. Once we have the value functions, we can make faster online solutions using (5.24). Note that a value function at a given time-step consists of the expected future cost from all the states. This process of mapping states and actions is not possible with stochastic MILP.

5.2.2.3 Approximate Dynamic Programming

ADP, also known as *forward DP*, is an algorithmic strategy for approximating a value function, which steps forward in time, compared to backward induction, used in *value iteration*. Policies in ADP are extracted from these value function approximations (VFA) [28]. Similar to DP, ADP operates on an MDP formulation of the problem, so all the non-liner constraints and transition functions can be incorporated with the same computational burden as modeling linear transition functions and constraints. ADP is an anytime optimization solution technique,[2] so we always obtain a solution regardless of the constraints and the inputs. In this instance, the problem is formulated as a maximization problem for convenience.

VFAs are obtained iteratively, and here the focus is on approximating the value function around a post decision state vector, $\mathbf{s}_x^t$, which is the state of the system at discrete time, t, soon after making the decisions but before the realization of any random variables [28]. This is because approximating the expectation within the max or min operator in (5.24) is difficult in large practical applications, as transition probabilities from all the possible states are required. Pseudo-code of the method used to approximate the value function is given in Algorithm 1, which is a double pass algorithm referred to as *temporal difference learning* with a discount factor $\lambda = 1$ or TD(1).

Given this, the original transition function $s^{t+1} = s^M(s^t, x^t, \omega^t)$ is divided into the post-decision state:

$$\mathbf{s}_x^t = \mathbf{s}^{M,x}(\mathbf{s}^t, \mathbf{x}^t), \tag{5.25}$$

and the next pre-decision state:

$$\mathbf{s}^{t+1} = \mathbf{s}^{M,\omega}\left(\mathbf{s}_x^t, \boldsymbol{\omega}^t\right), \tag{5.26}$$

which are used in Line 12 of Algorithm 1.

An example of the new modified MDP is illustrated in Fig. 5.4, which uses the mean and variation of the stochastic variables to obtain the post-decision, and next pre-decision states, respectively. In more detail, at s^1, there are three

2. An anytime algorithm is an algorithm that returns a feasible solution even if it is interrupted prematurely. The quality of the solution, however, improves if the algorithm is allowed to run until the desired convergence.

Algorithm 1 ADP using temporal difference learning TD(1)

Initialize $\overline{V}_0^t$, $t \in T$,
Set $r = 1$ and $t = 1$,
Set $\mathbf{s}^1$.
while $r \leq R$ **do**
  Choose a sample path ω_r.
  for $t = 0, \dots, T$ **do**
    Solve the deterministic problem.
    for $a = 1, \dots, A$ **do**
      Find right and left marginal contributions.
    end for
    if $t < T$ **then**
      Find the post-decision states (5.25) and the next pre-decision states (5.26).
    end if
  end for
  for $t = T, \dots, 0$ **do**
    Calculate the marginal values.
    Update the estimates of the marginal values.
    Update the VFAs using CAVE algorithm [28].
    Combine value functions of each controllable device (5.30).
  end for
  $r = r + 1$.
end while
Return the value function approximations $\overline{V}_R^t$ $\forall t$.

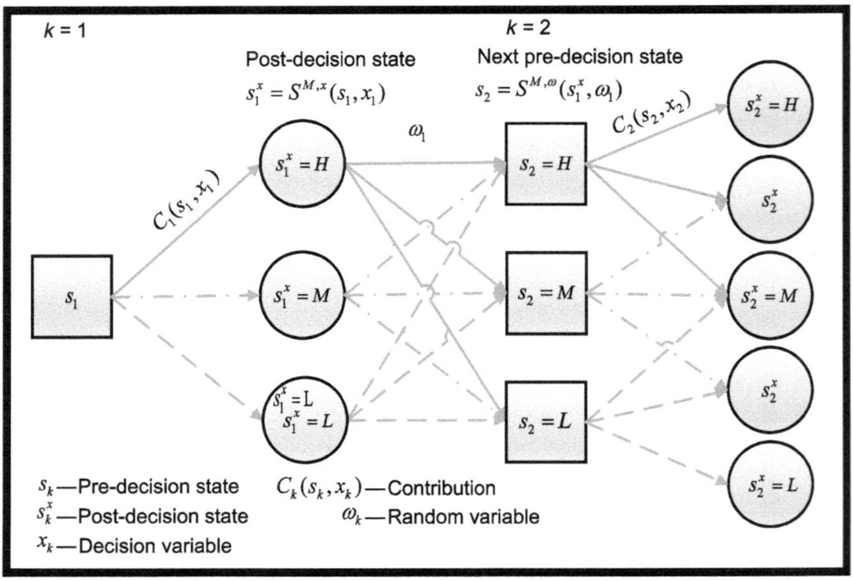

FIG. 5.4 Illustration of the modified Markov decision process, which separates the state variables into post decision states and pre-decision states.

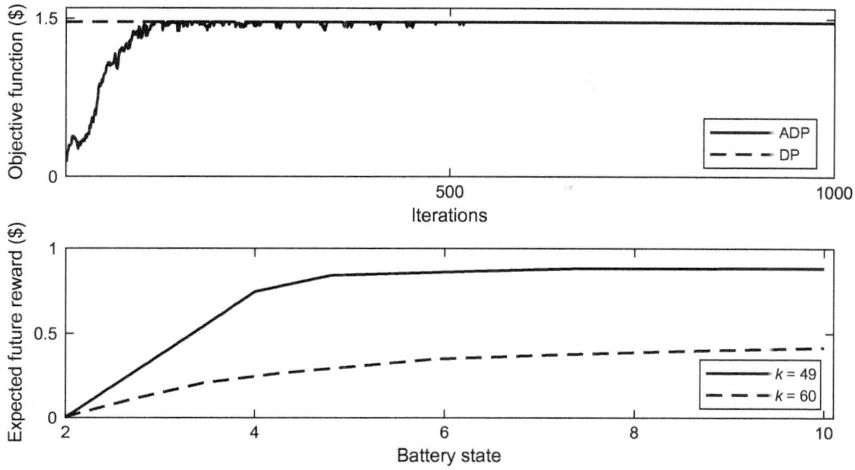

FIG. 5.5 Expected future reward or VFA (5.28) for following the optimal policy vs. state of the battery for time-steps $t=49$ and $t=60$ (top), and value of the objective function (i.e., reward) vs. iterations for ADP approach and the expected value from DP (bottom).

possible decisions that take us to three post-decision states, which correspond to high, middle, and lowest states. However, the next pre-decision state s^2 depends on the random variables ω^1.

Given this, the new form of the value function is written as:

$$\overline{V}^{\pi}(\mathbf{s}^t) = \max_{x^t}\left(\mathbf{C}^t(\mathbf{s}^t, \mathbf{x}^t) + \overline{V}^{\pi,x}(\mathbf{s}_x^t)\right), \tag{5.27}$$

where $\overline{V}^{\pi,x}(s_x^t)$ is the VFA around the post-decision state s_x^t, given by:

$$\overline{V}^{\pi,x}(\mathbf{s}_x^t) = \mathbb{E}\left\{V^{\pi}(\mathbf{s}^{t+1}) \mid \mathbf{s}_x^t\right\}. \tag{5.28}$$

This method is computationally feasible because $\mathbb{E}\left\{V^{\pi}(\mathbf{s}^{t+1}) \mid \mathbf{s}_x^t\right\}$ is a function of the post-decision state $\mathbf{s}_x^t$, that is a deterministic function of $\boldsymbol{x}^t$. However, in order to solve (5.27), we still need to calculate the value functions in (5.28) for every possible state $\mathbf{s}_x^t$ for all t. This can be computationally difficult since $\mathbf{s}_x^t$ is continuous and multidimensional, so we approximate (5.28).

Two strategies are employed in Algorithm 1. First we construct lookup tables for VFAs in (5.28) that are concave and piecewise linear in the resource dimension of all the state variables of controllable devices [29]. For example, in the VFA for $t=49$, which is depicted in Fig. 5.5(A), the expected future rewards stay the same after approximately 7 kWh; so if we are at 7 kWh in the time-step $t=48$, charging the battery further will have no future rewards and will only incur an instantaneous cost if the electricity has to come from the grid. However, if the electricity price or demand is high then we can discharge the battery as the expected future rewards will only decrease slightly. Given this, we never charge the storage when there is no marginal value so the slopes of the VFA are always greater than or equal to zero. Accordingly, the VFA is given by:

$$\overline{V}^a\left(s^t_{a,x}\right) = \sum_{a=1}^{A^t} \overline{v}^t_a z^t_a, \tag{5.29}$$

where $\sum_a z^t_a = s^t_{i,x}$ and $0 \leq z^t_a \leq \overline{z}^t_a$ for all a; z^t_a is the resource coordinate variable for segment $a \in (1 \ldots A^t), A^t \in \mathcal{A}$; $\overline{z}^t_a$ is the capacity of the segment, and; $\overline{v}^t_a$ is the slope. Other strategies that could be used for this step are parametric and non-parametric approximations of the value functions [28].

Second, we handle the multidimensional state space by generating independent VFAs for each controllable device, which are then combined to obtain the optimum policy. The separable VFA is given by:

$$\overline{V}_k\left(\mathbf{s}^x_k\right) = \sum_{i=1}^{I} \overline{V}^i_k\left(s^{i,x}_k\right). \tag{5.30}$$

For example, for a home energy management system including both a battery and thermal energy storage (TES), it is possible to separate the VFAs for battery and TES because their state transitions are independent. Instead, the interdevice coupling between the battery and the TES only arise through their effect on energy costs, so it is done in the contribution function in (5.5). If the state transition functions of the controllable devices depend on each other, then the VFAs are not separable and we have to use multidimensional value functions. In such situations the number of iterations needed for VFA convergence will increase and concavity needs to be generalized as well.

Note that when solving a deterministic home energy management problem using ADP, the post decision and next pre decision states are the same because the random variables will be zero. However, the remaining steps in Algorithm 1 stay the same. This means, with ADP, there is no noticeable computational burden for considering variation in the stochastic variables compared to solving the deterministic problem. On the other hand, with DP we have to loop over all the possible combinations of realization of random variables, which significantly increases the computational burden.

Home Energy Management Based on Policy Function Approximation

An alternative solution that does not require optimization in the execution stage is based on policy function approximation (PFA), which uses machine learning to effectively control PV-storage systems [27]. The algorithm uses an offline policy planning stage and an online policy execution stage. In the planning stage, a suitable machine learning technique is used to generate models that map states (inputs) and decisions (outputs) using training data. The training data set is generated by solving a deterministic home energy management problem using a suitable optimization technique (e.g., mathematical programming or

Algorithm 2 PFA algorithm

1: Generate empirical data set by solving the perfect foresight deterministic home energy management problem over a day.
2: Separate empirical data into seasons and/or weekdays/weekends.
3: Cluster empirical data into different day types using k-means.
4: Train non-parametric models for each time-step in every cluster.
5: **for** $k = 1, \ldots, K$ **do**
6: Compute the electrical grid power using the trained non-parametric model.
7: Calculate the battery decision that satisfies (5.5).
8: Find the next pre-decision state using (5.3).
9: **end for**

dynamic programming). In the execution stage, the model generated by the machine learning algorithm is then used to generate fast real-time decisions. Since the decisions can be made in real-time, the policy can rely on up-to-date information on PV output, electrical demand, and battery state of charge. Moreover, we can use PFA models over a long period of time (i.e., months) without having to update them but still obtain similar quality solutions. The implementation details of the proposed PFA algorithm are summarized in Algorithm 2, which proceeds as follows:

1. Generate the training data set by solving the perfect foresight deterministic DP-based home energy management over a day to obtain battery SOC (input) and electrical grid power (output) for corresponding electrical demand and PV output values (inputs) from the SGSC data set (Line 1).
2. Separate empirical data into seasons and weekends/weekdays. This step is only required when we have a large empirical data set or the number of possible clusters (i.e., different day types) that customers can choose are limited (i.e., depends on customers comfort) (Line 2). The empirical data set used to train the nonparametric model consists of input states, such as battery SOC, electrical demand, PV output and electricity tariff; and output decision is the electrical grid power, as shown in Fig. 5.6.
3. Cluster empirical data into different day types using k-means algorithm. The empirical data is clustered according to the total daily PV output and electrical demand so the resulting clusters correspond to different day types, such as sunny day with low demand, sunny day with high demand, cloudy day with high demand or cloudy day with low demand etc. (Line 3).
4. Train non-parametric models for every time-step in every cluster. We can train these non-parametric models offline or even real time since it is computationally efficient (Line 4).

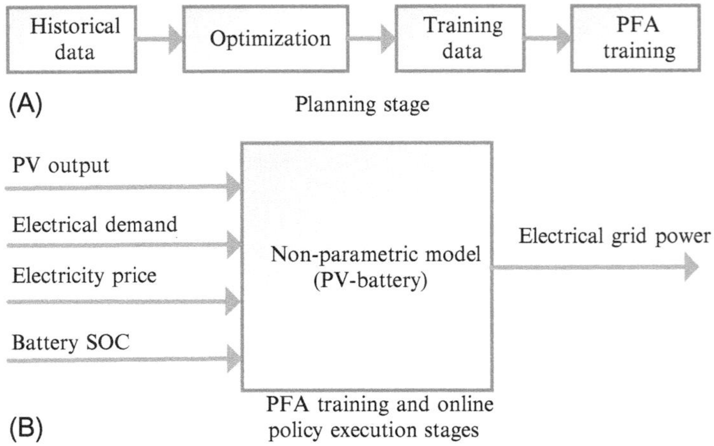

FIG. 5.6 (A) Procedure used to obtain PFA training models and (B) PFA model that maps inputs and outputs for a PV-battery system. When training PFAs, battery SOC and electrical grid power are obtained from the optimization, and PV output and electrical demand are perfect foresight inputs. During the online execution stage, PV output and electrical demand are predictions for the next time-step while the battery SOC is the current value.

5. In real-time, provide inputs to the trained non-parametric model to obtain the outputs (Line 6). Output is the electrical grid power.
6. Calculate the battery decisions that satisfy (5.5).
7. Obtain the next pre-decision states of the controllable device using the transition function in (5.3) (Line 8).

In summary, reasonably accurate day-type forecasts allow us to choose the correct PFA model in real-time operation. PFA models can be deployed on low cost computation devices, and don't rely on large amounts of computation to plan a policy each period. Once we generate the PFA models, they can be used over months without updating and still obtain quality solutions.

When historical data are not available, which is typically the case for new installations, we can use typical day types, as illustrated in Fig. 5.7. Upon installation, the users only need to choose the correct day type. When new data become available, the day type estimate can be updated using machine learning.

5.3 Impact of Passive DER on LV Distribution Networks

When penetration levels of solar PV are small, the impact on the networks in negligible. For these settings, solely focusing on home energy management is warranted, and local optimization-based approaches like those described above can be beneficial to customers and the power system.

However, many countries have already reached levels where solar PV has a material impact on the performance of distribution networks, in particular at LV

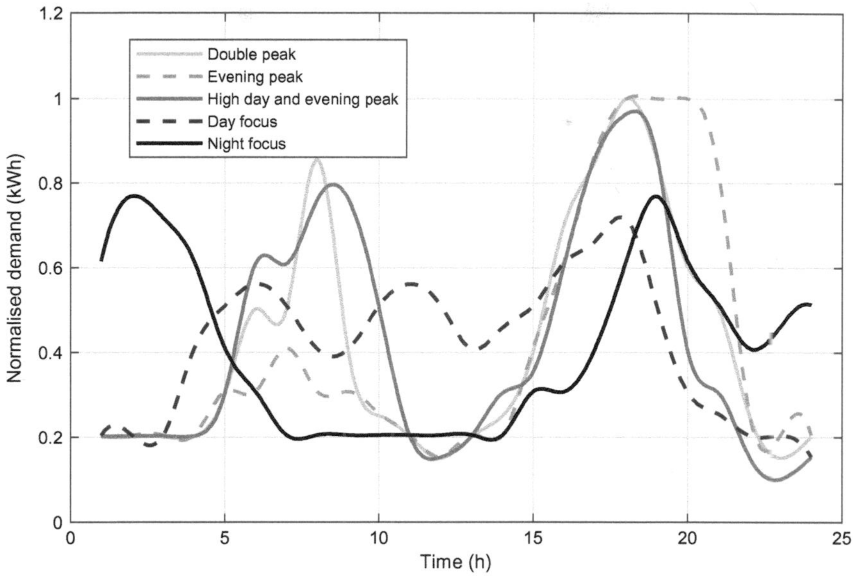

FIG. 5.7 Generic load profiles.

levels. In Australia, for example, state-wide PV penetration in Queensland and South Australia is already above 30%, with some feeders well above 50%. As an illustration, consider a low-voltage distribution feeder with different penetration levels of rooftop PV. Fig. 5.8 illustrates the number of households with overvoltage problems as a function of PV penetration and the size of the PV system [30]. As you can see, problems start to emerge as early as at 20% PV penetration and can affect all users at 60% penetration for PV system sizes between 5 and 7 kW.[3]

Different strategies have been proposed to mitigate the emerging LV network issues. While some methods leave the responsibility to the distribution system operator (DSO; e.g., grid reinforcement, and active transformers with on-load tap changers [31, 32]), other strategies consider a direct participation of end users. For example, PV generation can be curtailed proportionally to avoid voltage problems using a *dynamic curtailment method* [33]. This method brings benefits to weak nodes that could be highly restricted due to their location in the network, but requires designing a cost-sharing model among prosumers to guarantee fair conditions in the curtailment.

At the same time, the increasing uptake of residential batteries has led to suggestions that the prevalence of batteries on LV networks will serendipitously mitigate the technical problems induced by PV installations. However, in general, the effects of PV-battery systems on LV networks have not been well

3. In Australia, the average size of new installations in 2018 is around 5.5 kW.

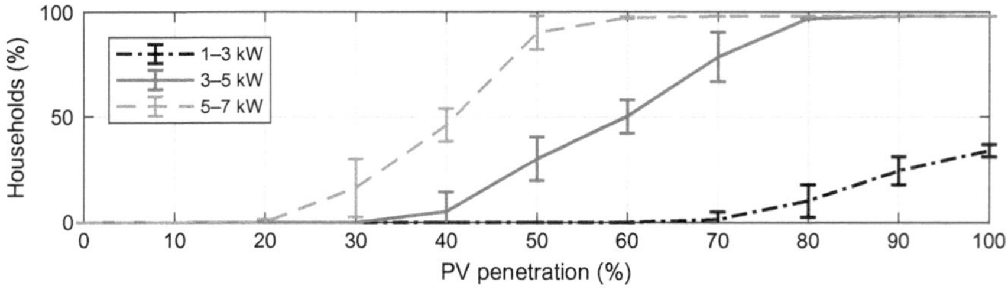

FIG. 5.8 The effect of PV penetration on LV network voltage.

studied. Therefore, we first look at the effect of uncoordinated penetration of DER on the performance of LV networks.

To quantify the effects of uncoordinated DER use on LV networks we use the methodology proposed in [34]. In short, the methodology incorporates *home energy management* operational decisions within a *Monte Carlo* (MC) power flow analysis, comprising three parts. First, due to the unavailability of large numbers of load and PV traces required for MC analysis, we used a *maximum a-posteriori Dirichlet process* to generate statistically representative synthetic profiles [35]. Second, a *policy function approximation* (PFA) that emulates the outputs of the home energy management solver is implemented to provide battery scheduling policies for a pool of customers, making simulation of optimization-based home energy management feasible within MC studies [27]. Third, the resulting net loads are used in an MC power flow time series study.

We split the analysis into two parts: (i) the impact of uncoordinated penetration of PV-battery system for different levels of battery penetration, and (ii) the impact of network tariffs in LV distribution networks.

5.3.1 Uncoordinated Penetration of PV-Battery Systems

We show results for two representative suburban (~5 km long, ~250 customers) 0.4 kV LV feeders, an Australian one and a UK one. Typically, Australian LV networks are designed to have higher capacity than the UK ones, mainly due to much larger air-conditioning loads. In the analysis, the *R*/*X* of the UK feeder is three times higher than the Australian one, so we expect the UK feeder to be more prone to overvoltage problems at higher PV penetration levels. The results are summarized in Fig. 5.9, which shows the percentage of cases with overvoltage and overloading problems, respectively, for three different battery penetration levels: 0%, 50%, and 100%.

The frequency of voltage problems with respect to the increasing PV and battery penetration on the LV feeders is shown in Fig. 5.9, row one. The percentage of customers with a voltage problem follows an increasing trend across both test feeders with respect to rising PV penetration, especially from 30% to

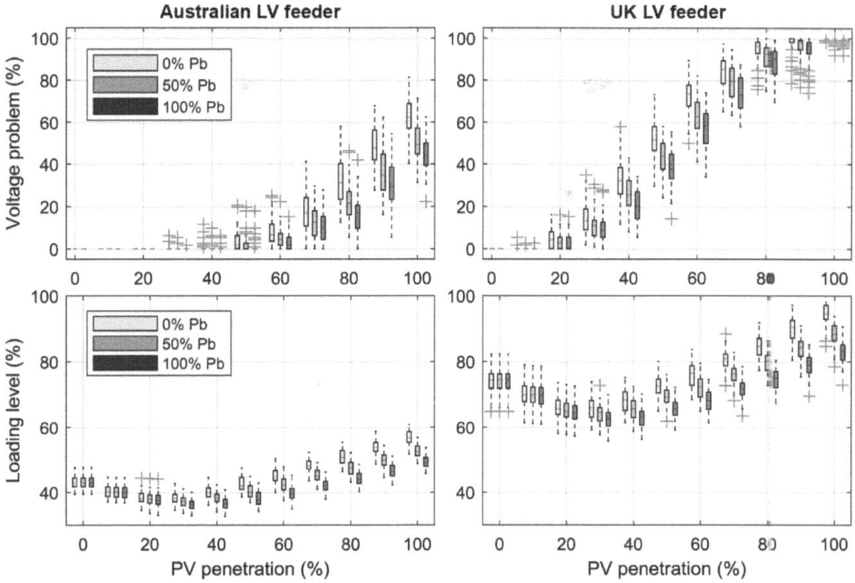

FIG. 5.9 Percentage of customers with voltage problems (first row) and transformer loading level (second row) for three different battery penetration levels: 0%, 50%, and 100%. Each bar from top to bottom shows the maximum value, 75 percentile, median value, 25 percentile, and minimum values.

100%, while the UK feeder presents more voltage problems due to higher line impedances. Voltage problems can be reduced by 10%–20% across all test feeders using home energy management under time-of-use (ToU) tariff. This scheduling strategy encourages batteries to charge when the electricity price is low, and discharge when the price is high (during peak hours). However, the timespan for high PV outputs can extend and even overlap with peak demand, especially in summer. This is illustrated for some specific case in Fig. 5.10, in which the peak demand occurs between 4 and 6pm, causing the battery to discharge during high PV output. This reduces the grid power supply (p_g), when compared to the case without the battery ($\hat{p}_g$). As a result, p_g and $\hat{p}_g$ cross at around 4:30pm, where the voltages become the same (as highlighted in the black boxes). Furthermore, at 4:30pm, the rising demand causes the battery to decrease its charging power at high PV output, which keeps the voltage high. In these scenarios, home energy management under ToU is less effective at reducing over-voltage problems.

The ToU tariff is rather ineffective in reducing the peak demand. Surprisingly, the performance of home energy management with self-consumption maximization (SCM)[4] is only marginally worse, as demonstrated in

4. Under SCM, the battery is charged as soon as PV generation exceeds the demand and is discharged when the generation drops below the demand.

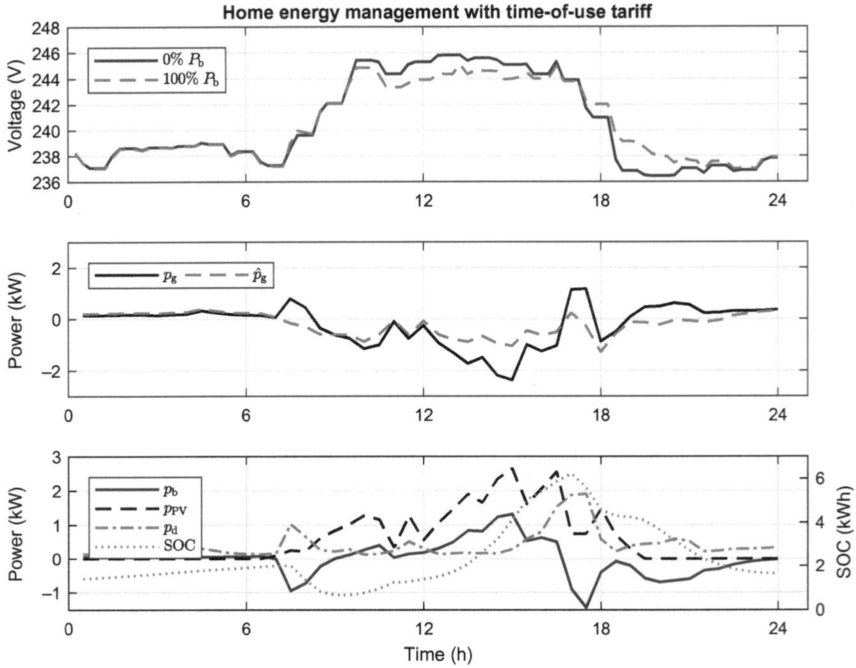

FIG. 5.10 Voltage profiles (top), grid power (p_g and $\hat{p}_g$ denote grid power with and without battery, respectively) (middle), battery scheduling (p_b), PV (p_{pv}), demand (p_d), and the SOC (bottom) of a customer with 3 kW PV and 6.5 kWh battery installed on the Australian feeder on a particular summer day for home energy management with time-of-use tariff.

Fig. 5.11. An illustration for a specific case shown in Fig. 5.12 shows that SCM forces the battery to charge with excess PV output to reach the battery's full capacity, which usually occurs before the sun goes down; as a result, the battery fails to consistently reduce voltage problems across the entire PV generation period. Observe in Fig. 5.12 that the battery reaches its full capacity at 3 pm. As a result, all excess solar generation after 3 pm is exported to the grid, as illustrated by the section where $\hat{p}_g$ overlaps $\hat{p}_g$, keeping the voltage high between 3 and 6 pm. On the other hand, the HEM under ToU considers the price of electricity, which is an indication of the timely demand and PV generation. In this method (Fig. 5.10), the charging profile is more evenly distributed throughout the PV generation period, hence a more consistent reduction of the problems is expected. Although the HEM with ToU helps reduce voltage problems, it is far from the panacea for over-voltage problems. This goes to show that a different approach is needed to mitigate the adverse impact of DER penetration. Electricity tariffs with a demand charge might be an option, which we discuss next.

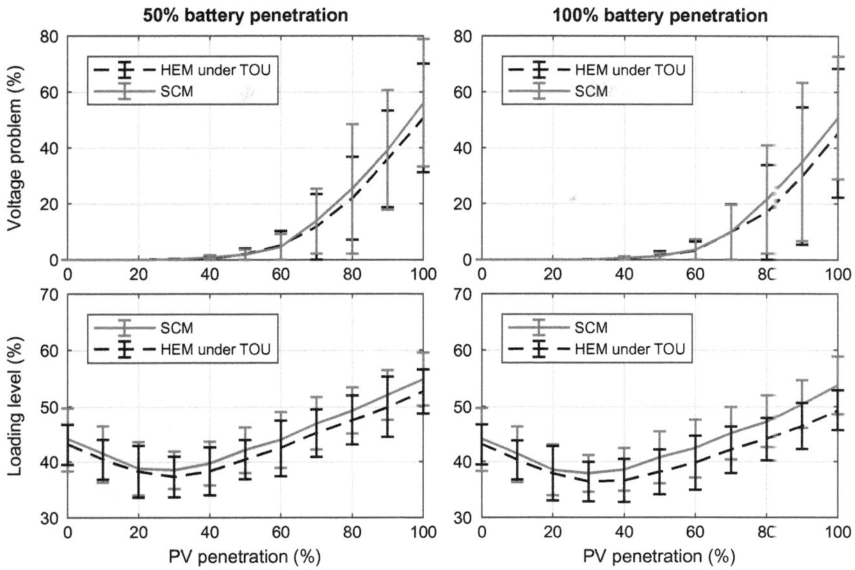

FIG. 5.11 Comparisons between HEM under ToU and SCM for the Australian feeder.

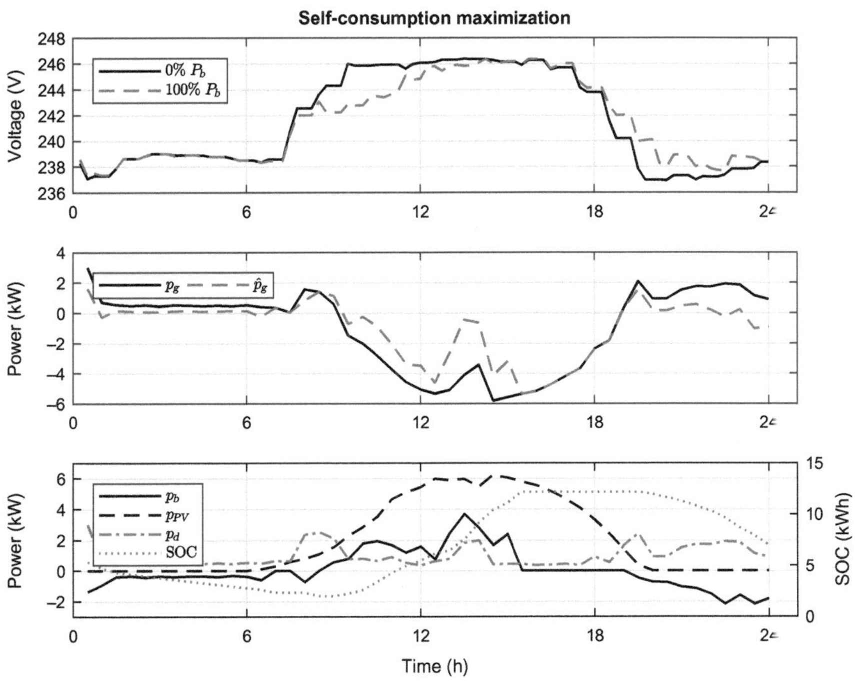

FIG. 5.12 Voltage profiles (top), grid power profiles (middle), battery scheduling, PV, demand, and the SOC (bottom) of a customer with 7kW PV and 14kWh battery installed on the Australian feeder on a particular summer day for self-consumption maximization.

5.3.2 Tariff Design as a Means to Mitigate Adverse Network Impact

In theory, cost-reflective tariffs charge consumers according to the extent to which they contribute to network peak demand and other long-run network cost drivers. In practice, however, things are a bit more complicated because "one size fits all" approaches seldom work, and tariff design is no exception. Nevertheless, *demand-based tariffs* have been proposed by network service providers as an interim solution to inequitable pricing and cross-subsidies (between consumers and prosumers) existing as a result of purely volumetric tariffs.

Given this background, we now examine residential customers' response to energy and demand-based network tariffs, in particular the impact of demand-based charges on peak demand. We consider consumers with solar PV, thermostatically controlled loads and battery storage, all managed by a home energy management system, formulated as a mixed integer linear program (see Section 5.2.2) aiming to minimize electricity cost. The formulation of the home energy management problem differs from the existing approaches [25, 26] in that it also considers the demand charge. Incorporating the demand charge, however, is challenging. Whilst modeling *demand-based tariffs*, if the demand charge is included explicitly in the optimization model as a constraint, the problem can become intractable, because a min-max problem results even with a MILP formulation. To overcome this, we implicitly model the peak demand charge as a linear term in the objective function, with an additional inequality constraint which limits the monthly demand according to the set demand charge [36, 37]. In this way, we achieve the same results as with the explicit modeling of the peak demand constraint but with a lesser computational burden.

For customers facing an *energy-based tariff* (*Flat* or *ToU*) the objective of the daily optimization model is given as, for all $h \in \mathcal{H}$:

$$\underset{p_h^{g+}, p_h^{g-}, p_h^{b+}, p_h^{b-}, p_h^{d}}{\text{minimize}} \left[\sum_{h \in \mathcal{H}} T^{\text{flt/tou}} p_h^{g+} - T^{\text{flt}} p_h^{g-} \right], \tag{5.31}$$

where $T^{\text{flt/tou}}$ is flat/time-of-use energy charge, and $T^{\text{fix/fit}}$ is fixed daily charge/feed-in-tariff (FiT).

For customers facing a *demand-based tariff* (*FlatD* or *ToUD*), the daily optimization model is given as, for all $h \in \mathcal{H}$:

$$\underset{p_h^{g+}, p_h^{g-}, p_h^{b+}, p_h^{b-}, p_h^{d}, \hat{p}}{\text{minimize}} \quad T^{\text{pk}} \hat{p} + \left[\sum_{h \in \mathcal{H}} T^{\text{flt/tou}} p_h^{g+} - T^{\text{flt}} p_h^{g-} \right], \tag{5.32}$$

where $T^{\text{pk}}/p^{\text{pk}}$ is peak demand charge/monthly peak demand. In addition to the existing MILP constraints, we effectively limit peak demand by imposing:

$$p_h^{g+} \leq \hat{p}, \tag{5.33}$$

which is used to limit the grid import according to the demand charge component, $T^{\text{pk}} \hat{p}$ in (5.32).

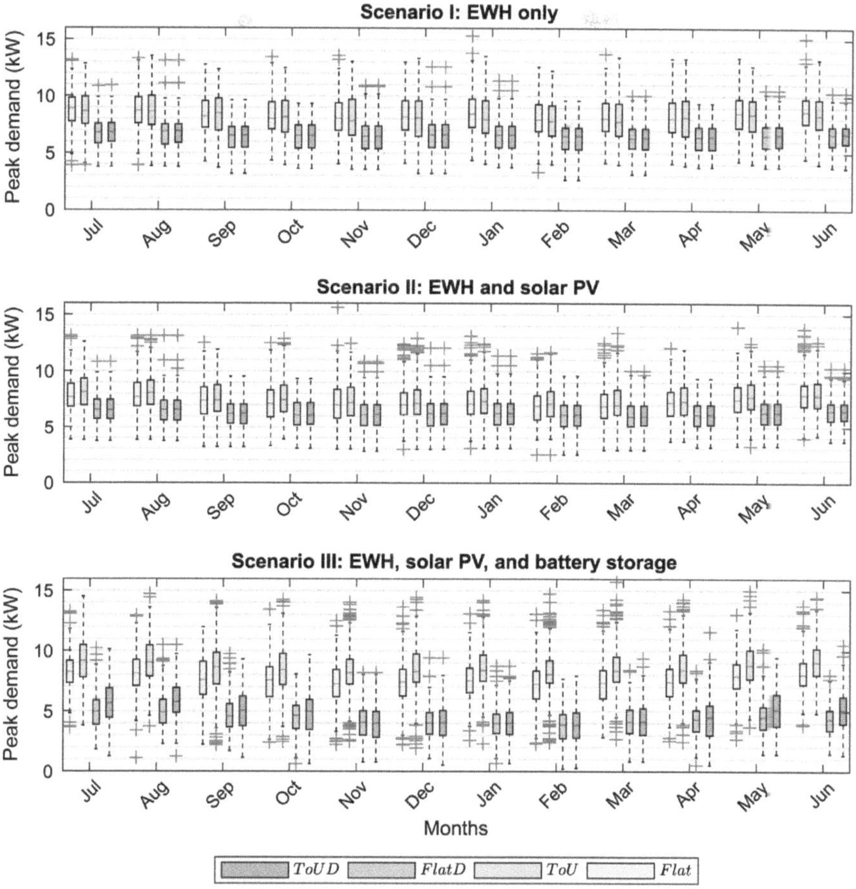

FIG. 5.13 The impact of different tariff structures on peak demand.

We now assess the impacts of different network tariffs on the consumption pattern of electricity consumers with distributed energy resources, such as thermostatically controlled loads and battery storage, and the resultant effects on the LV distribution network. Results presented in Fig. 5.13 show that tariffs with a peak-demand component perform better in terms of electricity cost reduction for the customer, as well as in reducing average daily peak demand and line loading for networks with high PV-battery penetration levels, but do little to reduce seasonal peak loads.

5.4 Active Operation of DER in Distribution Networks

In contrast to the uncoordinated operation of DER on networks, orchestrated responses by DERs to low voltage network conditions have demonstrated excellent prospects for improving embedded PV-battery and network value. Essential to this framework is the *distribution system operator* (DSO). As a concept, a DSO is expected to act analogously to an independent system operator for transmission and bulk generation assets. In practice, however, the methods used by a DSO to coordinate and orchestrate DER are likely to be quite different to those used at high voltage levels, due to the difference in device and network characteristics, and most importantly, constraints applied to them by their owners, including, for example, comfort levels for air-conditioning and timing for washers and dryers.

In this section, two DSO-administered approaches are covered. First, the effects of operating a dispatch-based DSO are assessed in [38], where a fleet of distributed battery energy storage systems successfully moderated losses, voltage variations, and phase unbalance on LV networks. In practice, this can be scaled to thousands of customers using advanced techniques, such those in [39] or [40]. Indeed, the CONSORT project demonstrates the potential of this type of orchestrating technology [41], which could sit behind a future DSO, as flagged by [42]. Moreover, by embedding this type of dispatch engine into an appropriate auction format, whole-of-feeder dispatchable energy and service markets, potentially arranged hierarchically into wholesale energy and services markets, become a technical possibility. An example of this style of approach to DER coordination is given in Section 5.5.

Second, an alternative DSO approach involves overlaying a permission structure on peer-to-peer energy markets that allows only trades that do not violate network technical constraints. This represents a less active role for the DSO, and relies on a greater degree of active participation by prosumers (or their agents), but nonetheless, our results demonstrate that it is sufficient to keep the LV network operating within its technical limits as PV penetrations increase, while also improving outcomes for PV and battery owners and regular customers alike. Building on [43], we model a local peer-to-peer energy market in the presence of different curtailment schemes, and assess the performance of the trading mechanism on the distribution network. Similarly, we have assessed the effects of different curtailment regimes on network loadings and also their impacts on customers with PV units. Again, this shows that trade may be beneficial to energy prosumers, but there is no guarantee of a positive effect on the network.

5.4.1 DSO-Operated Storage

A first alternative for active DSO management of LV networks is for the DSO to invest in and operate their own utility battery systems. These battery systems can be used as nonnetwork solutions, as alternatives to conventional network augmentation, or reconfigurations, thereby adding value by avoiding or deferring large capital investment in place of a smaller capital and operating cost, and improving the efficiency with which electricity is delivered to customers [44]. Specifically, battery systems can be used to ensure voltage and thermal constraints are satisfied, while also moderating power losses, improving rooftop PV hosting capacity, and rectifying network unbalance. In particular, they can mitigate over voltages by absorbing excessive renewable generation in LV networks, and release the stored energy whenever it is needed [45]. Considering these benefits, the value of battery systems has risen significantly, especially in Australia, where PV penetration is significantly higher than the rest of the world [46].

This section describes a method for managing such DSO-operated utility battery systems in unbalanced LV networks. A *mixed integer quadratic programmming* (MIQP) model is used to minimize annual energy losses and determine the sizing and placement of battery systems, while satisfying voltage constraints. A real unbalanced LV UK grid is adopted to examine the effects of battery systems under two scenarios: the installation of one community battery system (CBS) and multiple distributed battery systems (DBSs), and the impacts of battery systems in power losses, the hosting capacity and network unbalance in LV networks are investigated. Specifically, two scenarios are examined: (i) the installation of a single CBS, and (ii) the installation of multiple DBS with the same aggregated size. In this section, in order to determine the optimal location and the associated capacity of the battery systems in both scenarios, a linearized approximation of AC *optimal power flow* (OPF) is applied (cf. the full AC OPF model that is considered in later sections). The optimization model aims to minimize annual energy losses while satisfying voltage constraints. Given the results of simulations, we assess, quantify, and compare the effectiveness of these two ESS configurations.

However, there is still considerable debate as to whether a CBS or DBSs is more valuable to our networks, and currently few studies have compared their benefits and drawbacks. In [47], a DC OPF model is used to determine the optimal battery systems sizing and placement. The DC OPF is a linearization of AC OPF, which is use because it has limited computational complexity, but this type of approximation is better suited to transmission networks.

In contrast, a full AC OPF can produce more accurate results for distribution networks, but this representation's nonlinearities often make the computational problem intractable. This issue is overcome with convex relaxations, as suggested in [48, 49]. Relaxations are widely used for linearizing higher-order terms in objective functions and constraints, which often result in a *second order conic programming* (SOCP) model. For example, the study in [43] uses SOCP to show how battery storage can provide voltage regulation, congestion management, and power loss reduction. However, this method is not applied to LV networks. The study in [50] describes a linearized AC OPF for optimal battery system sizing and location on a LV feeder based on cost minimization. It demonstrated the great potential of battery systems in boosting PV hosting capacity with the aid from active power curtailment (APC). However, it does not consider power loss minimization, and the network is assumed to be balanced.

Recently, [51] studied the differences between a CBS and multiple DBSs in a LV network. A forward-backward sweep linearized AC OPF model was proposed for determining the optimal location and sizing for the battery systems. However, this study was conducted on a balanced network, and it failed to investigate the benefits of an optimally placed and sized centralized battery system. None of the papers above have explored the differences between a CBS and multiple DBSs when considering their impacts on power losses, hosting capacity and network unbalance as a whole.

The study that follows uses an unbalanced AC OPF model formulation to determine the optimal location and the associated battery system capacity. Our results show that the DBS configuration has an edge over the CBS configuration in accomplishing the tasks described above, although both of them are highly effective. The description begins with the battery system model, before considering the optimal location and sizing problem formulation, and then discussing simulation results.

5.4.1.1 Battery System Model

Key to this problem is modeling the performance and intertemporal couplings of the battery systems, which are slightly different to those used in the home energy management problem. To begin, a battery system's *state of charge*, at time t, is a function of the state of charge at time $t-1$ and its charging and discharge rates during this time interval. The charging and discharging rates stated are for one phase only, so they must be multiplied by 3, giving the following state-of-charge transitions for a battery placed at bus i:

$$e_{t,i} = e_{t-1,i} + 3\left(\eta^{+} p_{t,i}^{+} - \frac{1}{\eta^{-}} p_{t,i}^{-}\right)\Delta t. \tag{5.34}$$

For convenience, the initial state of charge after each cycle (typically 1 day) is fixed, and given by:

$$e_{\Omega} = e_{0}. \tag{5.35}$$

To ensure that the above constraint can be satisfied, the total charged and discharged power for one cycle must be the same:

$$\sum_{t \in \mathcal{T}} \left(\eta^{+} p_{t,i}^{+} - \frac{1}{\eta^{-}} p_{t,i}^{-}\right) = 0. \tag{5.36}$$

In addition, the state of charge cannot exceed the minimum and maximum states of charge at all times, and similarly, charging and discharging rates at each time interval Δt are constrained by the battery systems characteristics:

$$0 \leq 3p_{t,i}^{+} \leq \overline{p}^{+} \tag{5.37}$$

$$0 \leq 3p_{t,i}^{-} \leq \overline{p} \tag{5.38}$$

$$\underline{e} \leq e_{t,i} \leq \overline{e}. \tag{5.39}$$

Last, we note that in this study, the charging and discharging efficiencies in the model are neglected.

5.4.1.2 Optimization Problem Formulation

The optimization model aims to place and size a CBS (or multiple DBS) in an unbalanced LV network in order to minimize annual system losses. Let $\mathcal{B}$ be the set of buses, $\mathcal{L}$ be the set of total network lines, indexed lm where l and m are elements of $\mathcal{B}$. Denote the single-phase resistance and impedance $R_{f,lm}$ and $X_{f,lm}$, respectively, for phase $f \in \mathcal{F}$.

The CBS objective is to minimize system losses, is formally stated as:

$$\underset{p_l^{+,t}, p_l^{-,t}, B_l^{\text{loc}}, e^{\text{com}}}{\text{minimize}} \sum_{d \in \mathcal{D}} \sum_{t \in \mathcal{T}} \sum_{lm \in \mathcal{L}} p_{lm}^{\text{loss},t}, \tag{5.40}$$

where $\mathcal{D}$ are the days in a year, $\mathcal{T}$ are the hours in a day, and $p_{lm}^{\text{loss},t}$ is given by:

$$p_{lm}^{\text{loss},t} = \sum_{f \in \mathcal{F}} R_{f,lm} \frac{\left(p_{f,lm}^t\right)^2 + \left(q_{f,lm}^t\right)^2}{v_{\text{sub}}^2} \forall f \in \mathcal{F}. \tag{5.41}$$

This objective function indicates that the overall model is a *quadratic* problem.

For the DBS case, the single variable e^{com} in (5.40) is replace with N variables e^{dist_n}. Similarly, the battery storage location constraint, corresponding to the binary variables B_l^{loc} and given by:

$$\sum_{l \in \mathscr{E}} B_l^{\text{loc}} = N. \tag{5.42}$$

is adjusted by varying N from 1 (for the CBS case) to the largest number of distributed battery storage installations under consideration, where $\mathscr{E} \subseteq \mathcal{B}$ are the set of buses suited to a battery installation.

The study reported below compares the impacts between a CBS and multiple DBS. Therefore, the battery location model above is used to determine the optimal locations and capacities for multiple DBS, assuming they have the same aggregated capacity as the CBS. In this scenario, N in (5.42) must be considered as a variable instead of a problem parameter. At the same time, a constraint equating the aggregated capacity of the DBS to the size of the CBS is added to the optimization model:

$$\sum_{j \in \mathscr{E}} e_j^{\text{dist}} = e^{\text{com}}. \tag{5.43}$$

In addition to the battery characteristics and battery system location variables and constraints described above, the model is also subject to branch flow constraints and voltage constraints. Load flows in LV networks with numerous branches can be difficult to solve, especially when the network is unbalanced and each phase must be taken care of individually. Here, we use the *branch flow* model introduced in [52, 53], which has been shown to be accurate in modeling power flows in radial distributed networks [54]. The equations in this model show that power flows on each phase from bus l to bus m, can be formulated as a function of power flows and power losses entering bus m, and the generation and consumption at bus m. The modified power flow equations with battery storage placed on bus m, phase $f \in \mathcal{F}$ are presented below (dropping the time index t as there are no intertemporal couplings in the network constraints):

$$\begin{aligned} p_{f,lm} = &\sum_{lm \in \mathcal{L}} \left(p_{f,mn} + R_{f,lm} \frac{p_{f,lm}^2 + q_{f,lm}^2}{v_{\text{sub}}^2} \right) \\ &+ p_m^{\text{load}} - p_m^{\text{pv}} + \sum_{j \in \varepsilon} \left(p_m^+ - p_m^- \right), \end{aligned} \tag{5.44}$$

$$\begin{aligned} q_{f,lm} = &\sum_{m,n \in \mathcal{L}} \left({}_{f,mn} + X_{f,lm} \frac{p_{f,lm}^2 + q_{f,lm}^2}{v_{\text{sub}}^2} \right) \\ &+ q_j^{\text{load}} - \mathrm{q}_j^{\text{pv}}, \end{aligned} \tag{5.45}$$

$$\begin{aligned} v_{f,m}^2 = &v_{f,1}^2 - 2\left(p_{f,lm} R_{f,lm} + q_{f,lm} X_{f,lm} \right) \\ &+ \left(R_{f,lm}^2 + X_{f,lm}^2 \right) \frac{p_{f,lm}^2 + q_{f,lm}^2}{v_{f,m}^2}. \end{aligned} \tag{5.46}$$

However, here only the active power flow of battery storage is considered. It has been clearly illustrated by the literature that the nonlinear terms in (5.44) and (5.45), representing power losses, can be neglected, as the impacts of these terms on the objective function are limited [54]. Under this assumption, the branch flow equations are given by:

$$\begin{aligned} p_{f,lm} \approx &\sum_{mn \in \mathcal{L}} p_{f,mn} + p_m^{\text{load}} - p_m^{\text{pv}} \\ &+ \sum_{j \in \varepsilon} \left(p_m^+ - p_m^- \right) \forall f \in \mathcal{F}. \end{aligned} \tag{5.47}$$

$$q_{f,lm} \approx \sum_{m,n \in \mathcal{L}} q_{f,mn} + q_m^{\text{load}} - q_m^{\text{pv}} \forall f \in \mathcal{F}. \tag{5.48}$$

Based on [54, 55], voltage expressions can be linearized by assuming voltage variations at an arbitrary bus m are close to 0 at all times, for example, $v_{f,m} - v_{\text{sub}} \approx 0$. This assumption leads to the following expression:

$$v_{f,m}^2 \approx v_{\text{sub}}^2 + \mathbf{2} v_{\text{sub}} \left(v_{f,m} - v_{\text{sub}} \right) \; \forall f \in \mathcal{F}. \tag{5.49}$$

The linearized version of the voltage equation can be developed by substituting (5.49) to (5.46):

$$v_{f,m} \approx v_{f,l} - \frac{p_{f,lm}R_{f,lm} + q_{f,lm}X_{f,lm}}{v_{\text{sub}}} \; \forall f \in \mathcal{F}. \tag{5.50}$$

The accuracy of this equation has been justified by the authors in [53]. In addition, the model should satisfy the voltage limits for all three phases in presence of loads and PV generation:

$$\underline{v} \leq v_{f,l} \leq \overline{v} \; \forall f \in \mathcal{F}. \tag{5.51}$$

The battery system formulations from (5.34) to (5.39) and the branch flow (5.47), (5.48), and (5.50), along with voltage and battery storage constraints in (5.42) and (5.43) formulate a mixed integer quadratic programming model for determining the optimal location and the associated capacity for the CBS and the DBS. We now report on a comparison of the effects using the two different installation patterns.

Test Case

The test LV feeder is from the United Kingdom, which consists of 175 single phase consumers with a total length of 4.3 km. This is an unbalanced network with 61, 60, and 54 consumers sitting on phase A, B, and C, respectively. Each load, which follows a specific load profile, adopts a single phase PV system. This feeder covers a large area and it is reasonably loaded, presenting great potential for over-voltages. The network diagram is shown in Fig. 5.14.

Load and PV profiles for 1 year are generated using the CREST model, which allows one to generate high resolution PV and load profiles [57]. For this study, each individual load and PV system is assigned with unique daily load, and PV profiles with 1 h resolution, respectively. When performing simulations, the base voltage is 0.4 kV (1 pu) and the phase voltage of transformer secondary winding is fixed at 0.23 kV.

The two performance indices are power losses and the hosting capacity, which are explained next. First, the primary task of the optimization model

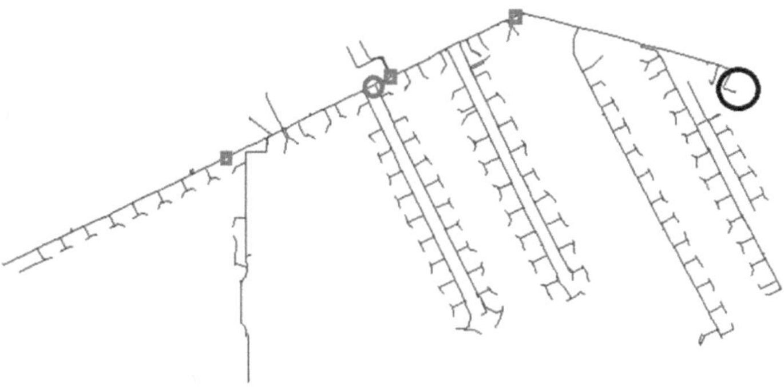

FIG. 5.14 The unbalanced LV network. The black and red circles denote the substation transformer and CESS, while the three squares represent DESSs [56].

is to reduce annual network energy losses. For the base scenario, without any battery storage, simulations are performed for a year. The total network losses in summer, autumn, winter, and spring are approximately 2530, 2820, 4660, and 2190kWh, respectively. It can be seen that the total energy loss in winter is significantly higher than other seasons due to the employment of heating facilities. The annual network loss is 12,200kWh, which is expected to be decreased by implementing the two battery storage configurations.

Second, in order to determine the hosting capacity, the rated power generation of PV systems on the entire feeder is gradually increased until the voltage threshold is met. This is most likely to occur at loads towards the end of the feeder. Specifically, the installed PV capacity is 175kVA in the case that all PV system is rated at 1kW. Iterative power flow simulation is performed, which automatically increases PV generation until the voltage limit is exceeded. A typical summer day is tested using the algorithm, since summer usually presents less network loading and reasonably high PV generation. The outcomes indicate that the upper limit is exceeded at bus 2266 at 1pm when each rated PV generation is raised to 2.1kW. This means the 100% penetration of this network without battery storage is 368kW.

Community Battery Storage

The results of installing a CBS in the LV network are shown and analyzed in this section.

Optimization-based simulations are performed for 1year, and the outcomes suggest an installation at bus 702 with the capacity being 942kWh. The initial state of charge for each cycle is 282kWh. The annual network energy loss in this case drops to 9410kWh, which shows a 22.9% reduction compared with the base scenario. The detailed charging and discharging patterns of a sample day for each season are observed and presented in Fig. 5.15. It can be seen that summer presents the highest charging rates around mid-day due to high PV penetration.

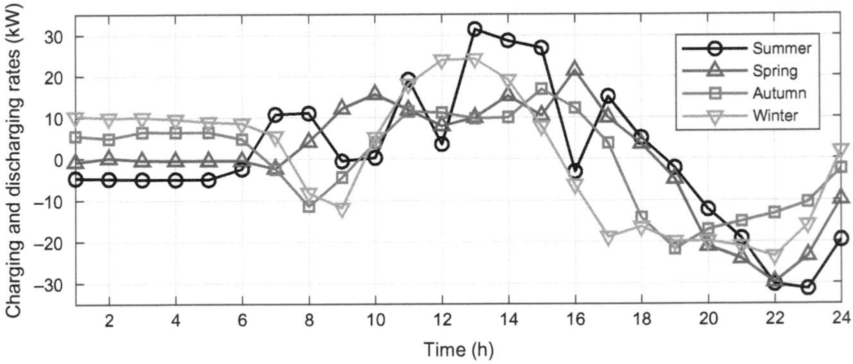

FIG. 5.15 Charging and discharging rates on a typical day in each season.

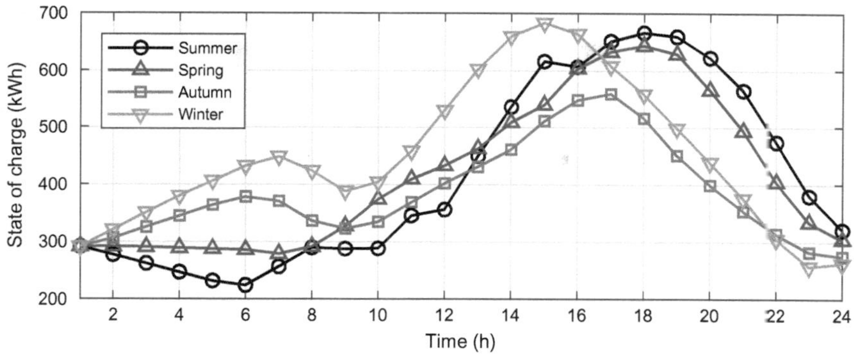

FIG. 5.16 State of charge on a typical day in each season.

Meanwhile, the discharging period comes at around 4pm in winter as night falls faster, while this time is delayed in summer, as expected. Due to the long discharging period in winter, the battery storage must charge in early mornings to maintain the state of charge on the pre-defined level at the beginning of each cycle. The corresponding state of charge is illustrated by Fig. 5.16, from which we observe that the winter curve reaches its peak at 850kWh at around 5pm, earliest and highest among all, then drops rapidly due to the high energy consumption.

The battery system with the same charging and discharging patterns is implemented in Open DSS to quantify the hosting capacity in this scenario. Applying the same method as in the base scenario, an over voltage is discovered at bus 2266 on a typical summer day at 1pm, with PV systems at all loads being rated at 2.71kW. This shows that the CBS is capable of increasing the hosting capacity of the network by 27.8%.

Distributed Battery Storage

In this section, the results of installing multiple DBSs—which have the same aggregated size as the CBS—in the network are detailed and compared with the performance indices and the first scenario.

Fig. 5.17 illustrates the improvements in annual energy losses and the hosting capacity with respect to different numbers of DBS. An obvious decrease in energy losses is shown as the CBS is implemented. However, as we continue to raise the number of storage batteries, the reduction immediately plateaus. When three batteries are installed, energy loss is decreased only by 190kWh, to 9220kWh, from the previous scenario. After this, the reduction becomes subtle. This is because the aggregated size is fixed, which limits the performance of the DBS. Thus, larger numbers of batteries require further increase in the aggregated capacity to become more effective.

Greater number and size of DBS potentially provide greater power loss reductions; however, these also lead to higher installation cost. Usually, if the installation cost is considerable and the utilities expect the battery storage

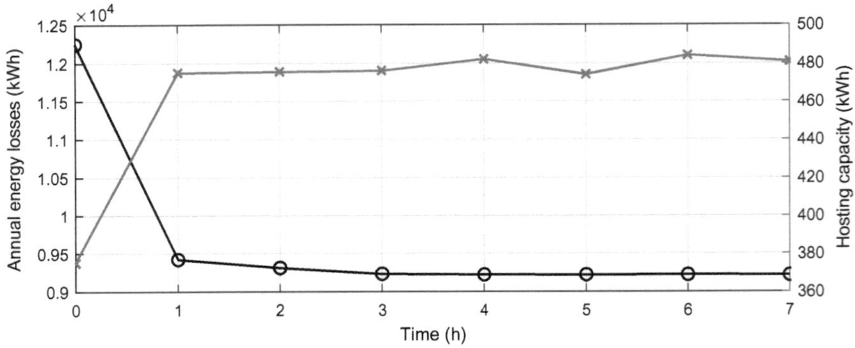

FIG. 5.17 Annual energy losses (left) and hosting capacity (right) with regard to number of batteries.

to be up-gradable for greater performance in the future, multiple DBS can be applied. Further, the locations for DBS are fairly uniformly distributed on the main branch of the feeder, which allows the DBS to cover larger areas and become more effective. An example is shown in Fig. 5.14 where the optimal locations of three DBS are indicated by green squares.

The results presented in Table 5.1 and Fig. 5.17 show that there is an obvious increase, 27.8%, in the hosting capacity after the CBS is installed. However, this rate reduces when the large battery storage is divided into DBSs. Specifically, after two installations, the curve experiences minor fluctuations and the increase becomes subtle. The utilities must consider this, with the trade-off between the battery storage costs and performance introduced previously, for deciding the most beneficial number of installations.

Comparisons of Community and Distributed Battery Storage

Speaking overall, both CBS and DBS can accomplish the tasks described in this study. Minor improvement can be made when attempting to decrease the VUF by replacing the CBS with multiple DBS, while this improvement is larger regarding power losses and the hosting capacity. It is important to realize that the performance can be improved significantly when increasing the aggregated capacity of distributed battery storage. However, this is not the case for CBS, for which a further increase in size will result in more energy losses. Therefore, in general, DBS can be more effective than CBS if the correct number of DBS is placed at optimal locations with reasonable sizes, and consequently, they can make distribution upgrade deferral and increased PV generation a reality.

5.4.2 Peer-to-Peer Energy Trading

Local energy trading between consumers and prosumers is one of the new scenarios of growing importance in the domain of distribution networks. Local

TABLE 5.1 Improvement in energy losses and the hosting capacity with respect to different numbers of batteries

No of ESSs	Locations	Sizes (kWh)	Power loss reduction (%)	Increase in hosting capacity (%)
1	702	942	22.9	27.8
2	234, 1704	527, 414	23.6	29.1
3	153, 676, 1786	322, 451, 169	24.5	29.2
4	170, 652, 996, 1786	322, 427, 114, 78	24.6	31.1
5	153, 429, 728, 996, 1786	179, 269, 312, 63, 117	24.7	29.2
6	234, 652, 820, 1484, 1851, 2191	320, 346, 114, 57, 49, 54	24.8	31.4
7	153, 429, 728, 996, 1484, 1786, 1924	179, 269, 312, 39, 34, 17, 89	24.8	30.6

distribution markets have been proposed as means of efficiently managing the uptake of DERs [10, 58]. This involves the creation of new roles and market platforms that allow the active participation of end-users and the direct interaction between them. This scenario brings potential benefits for the grid and users, by facilitating: (i) the efficient use of demand-side resources, (ii) the local balance of supply and demand, as well as, (iii) opportunities for users to receive economic benefits through sharing and using clean and local energy.

Given this context, a decentralized peer-to-peer (P2P) architecture has been proposed to implement local energy trading. Unlike the traditional scheme, under a P2P scheme, prosumers can trade their energy surplus with neighboring users. Currently, the implementation of decentralized market platforms is possible due to new advances in information and communication technology, such as *blockchain* and other *distributed ledger technologies* (DLTs), which support transparent and decentralized transactions. Many studies have already considered DLTs as the base of their P2P energy trading platforms [59, 60]. For example, [61] proposed a P2P energy trading model for electrical vehicles, showing the potential of blockchain to enhance cybersecurity on P2P transactions. Similarly, the work in [62] demonstrates the benefits of a

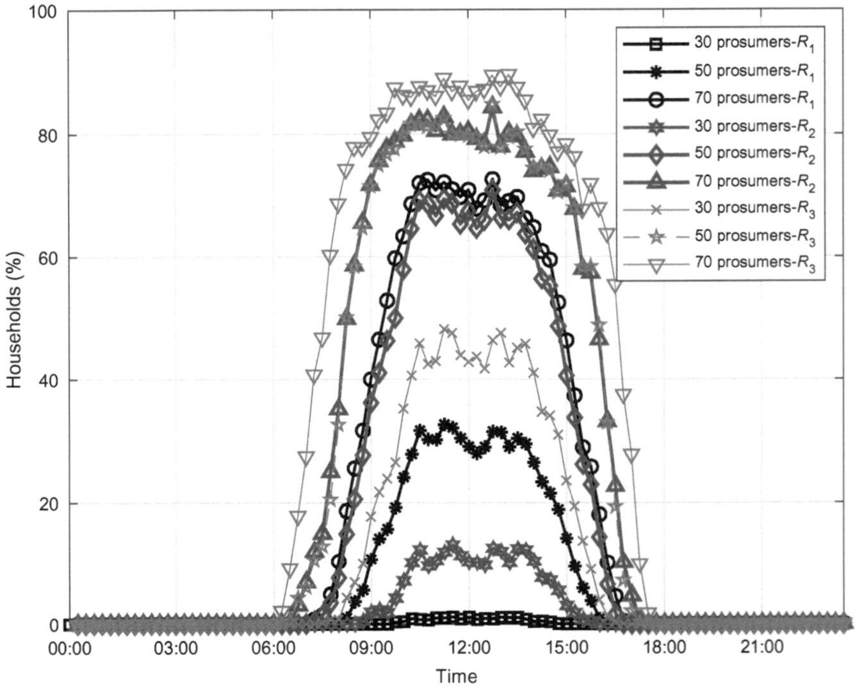

FIG. 5.18 Percentage of households with voltage problems [30].

blockchain-based microgrid energy market using smart contracts. Additionally, commercial P2P trading pilot projects have also been implemented recently. Most of these create a cryptocurrency that is used to trade energy between users.[5]

However, electricity exchange is different from any other exchange of goods. Residential users are part of an electricity network, which imposes hard technical constraints on the energy exchange. Completely decentralized energy trading, without any coordination, compromises the operation of the network within its technical limits. Therefore, physical network constraints must be included in energy trading models. For example, let us assume that a group of end-users in a typical UK LV network [56] participates in P2P energy trading without considering the network constraints in their trading mechanism, and the energy traded is supplied by nondispatchable generation, such as PV systems. Based on the probabilistic impact assessment methodology proposed in [63], we evaluated the voltage issues at different levels of PV penetrations. Fig. 5.18 illustrates the impact of PV penetration using different types of conductors in the network, showing that voltage issues may be worst for networks with

5. Examples of DLTs in P2P energy trading include PowerLedger (https://powerledger.io), Enosi (https://enosi.io) and LO3 Energy (https://lo3energy.com).

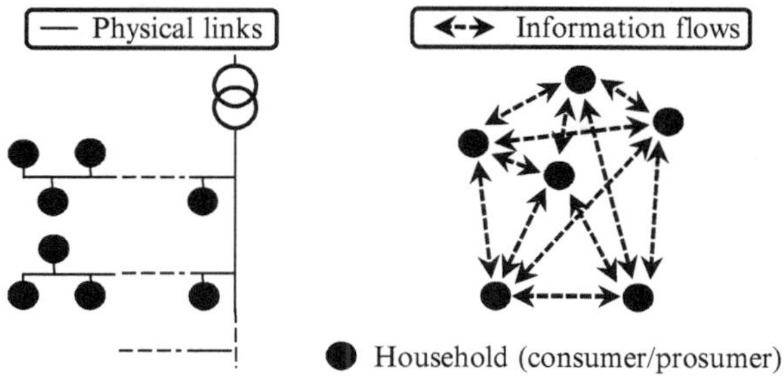

FIG. 5.19 Model of information flows and physical links between households under a P2P scheme [30].

greater resistance values. Throughout the day (Fig. 5.18), the most critical situation happens around midday (peak of PV generation). Similarly, the situation is worst when there are more prosumers in the network.

P2P Model

We consider a smart grid system for a P2P energy trading in a low-voltage (LV) network, under a decentralized scheme proposed in [30], where residential users interact through an online platform. The P2P scheme is illustrated in Fig. 5.19. The information flows between peers in a decentralized manner. As such, every peer can interact, through financial flows, with the others; users can sell and buy energy to/from their neighbors or a retailer. We consider this a realistic assumption, since currently there are pilot projects based on this concept, and it does not interfere with existing institutional arrangements (retail).[6] A general P2P scheme is a method by which households interact directly with other households. Users are self-interested and have complete control of their energy used (different to centralized direct load control structures, in which some entity may have control of some appliances).

Let $\mathcal{H}=\{1,2,\ldots,H\}$ be the set of all *households* in the local grid. The time is divided into time slots $t\in\mathcal{T}$, where $\mathcal{T}=\{1,2,\ldots,T\}$ and T is the total number of time slots. The set of all households $\mathcal{H}$ is composed of the union of two sets: consumers $\mathcal{P}$ and prosumers $\mathcal{C}$ (i.e.,$\mathcal{H}=\mathcal{P}\cup\mathcal{C}$). We assume that all households are capable of predicting their levels of demand and generation for electrical energy for a particular time slot t. Specifically, we assume consumers bid in the market based on their demand profiles. As such, a demand profile is not divided into tasks or device utilization patterns, so that the demand levels represent the total energy consumption over time. Prosumers are classified into two

6. Examples of pilot projects include Decentralized Energy Exchange (deX) Project, available at https://arena.gov.au/projects/decentralised-energy-exchange-dex/; and White Gum Valley energy sharing trial, available at https://westernpower.com.au/energy-solutions/projects-and-trials/white-gum-valley-energy-sharing-trial/.

types. Type 1 prosumers include those that have only PV systems; Type 2 includes prosumers that have PV systems, battery storage, and home energy management systems. Prosumers have two options to sell their energy surplus: (i) they can sell to the retailer and receive a payment for the amount of energy (e.g., feed-in tariff), or (ii) they can sell on the local market to consumers who participate in the P2P energy trading process.

Household Agent Model

A household $h \in \mathcal{H}$ uses d_t^h units of electrical energy in slot t. Likewise, a household $h \in \mathcal{H}$ has w_t^h units of energy surplus in slot t. The total quantity of electrical energy purchased in a slot t is given by x_t^+, and its price is denoted by s_t^+. The total energy consumption x_t^+ includes the amount of electrical energy purchased from the grid and from the local market. Similarly, the quantity of electrical energy sold in a slot t is given by x_t^-, and its price is denoted by s_t^-. While the energy surplus of Type 1 prosumers in $\mathcal{P}$ comes entirely from the PV system, each prosumer Type 2 in $\mathcal{P}$ uses its home energy management system to optimize its self-consumption, considering their demand and energy surplus by solving the following mixed-integer linear programming (MILP) problem [26]:

$$\begin{aligned} \underset{x \in \mathcal{X}}{\text{minimize}} & \sum_{t \in \mathcal{T}} \left(s_t^+ x_t^+ - s_t^- x_t^-\right) \\ \text{s.t.} \ & \text{device operation constraints,} \\ & \text{energy balance constraints, } \forall t \in \mathcal{T}, \end{aligned} \tag{5.52}$$

where $\mathcal{X}$ is the set of decision variables $\{x_t^+, x_t^-\}$. State variables in the model are s_k^+ and s_t^-. The former is associated with the price of energy in time slot t, and the latter with the incentive received for the contribution to the grid. In other words, s_t^+ and s_t^- are related to import tariffs (e.g., flat, time-of-use) or export tariffs (e.g., feed-in-tariff). The outcome of this process provides *net load profiles* for users with a home energy management system. After their self-optimization, prosumers can export their energy surplus to the grid.

The market mechanism for a P2P energy trading consists of three components [30, 43]: (i) a *continuous double auction* (CDA), (ii) the agents' bidding strategies, and (iii) the network permission structure, as described below.

Continuous Double Auction

A CDA matches buyers and sellers in order to allocate a commodity. It is widely used, including in major stock markets like the NYSE. A CDA is a simple market format that matches parties interested in trading, rather than holding any of the traded commodity itself. This makes it very well suited for P2P exchanges. Bids into a CDA indicate the prices at which participants are willing to accept a trade, and reflect their desire to improve their welfare. As such, the CDA tends towards a highly efficient allocation of commodities [64]. In more detail, a CDA comprises:

- A set of *buyers* $\mathcal{B}$, where each $b \in \mathcal{B}$ defines its trading price π_b and the amount of energy to purchase σ_b.
- A set of *sellers* $\mathcal{S}$, where each $s \in \mathcal{S}$ defines its trading price π_s and the amount of energy to sell σ_s.
- An *order book*, with bids o_b (b, π_b, σ_b, t), made by buyers $\mathcal{B}$, and asks o_s (s, π_s, σ_s, t), made by sellers $\mathcal{S}$.

Pseudo-code of the matching process in a CDA is given in Algorithm 3. A CDA is run for each time slot separately. Any intertemporal couplings that arise on a customer's side from using batteries, or loads with long minimum operating times, are not passed up to the market clearing entity. Once the market is open, arriving orders are queued in the *order book* for trades during a fixed interval t_d (Lines 2–8), which is limited by the start time t_d^{st} and the trading end time t_d^{end} (i.e., $t_d^{end} = t_d^{st} + t_d$). During the trading period, orders are submitted for buying or selling units of electrical energy in time-slot t. At the end of the trading period, the market closes—thereby no more offers are received. We assume the orders arrive according to a Poisson process with mean arrival rate λ. The current best bid (ask) is the earliest bid (ask) with the highest (lowest) price. A bid and an ask are matched when the price of a new bid (ask) is higher than or equal to the price of the best ask $o_s{}^*(s^*, \pi_s{}^*, \sigma_s{}^*, t^*)$ (the best bid $o_b{}^*(b^*, \pi_b{}^*, \sigma_b{}^*, t^*)$) in the order book (Line 9). However, if a new bid (ask) is not matched, then it is added to the order book, recording its arrival time and price. Note that after matching, an order may be only partially covered. If this is the case, it will remain at the top of the order book waiting for a new order. This process is executed continually during the trading period as new asks and bids arrive.

Algorithm 3 Matching process in a CDA with ZIP traders

1: **while** market is open **do**
2: randomly select a new ZIP trader
3: **if** buyer **then**
4: new o_b (b, π_b, σ_b, t)
5: **else**
6: new o_s (s, π_s, σ_s, t)
7: **end if**
8: allocate a new order in the *order book* ▹ Evaluate matching process with best bid and ask
9: **if** $\pi_b^* \geq \pi_s^*$ **then**
10: clear orders o_b^* and o_s^* at a price π_t and amount σ_t
11: **end if**
12: Update values of profit margins for buyers and sellers (Algorithms 4 and 5, respectively.)
13: **end while**

Bidding Strategies

Conventionally, market participants (buyers and sellers) define their asks and bids based on their preferences and the associated costs. The home energy management systems act as agents for the customers, and are continually responding to new stochastic information. As such, they appear very unpredictable from the outside. Moreover, because the market is thin, this can produce large swings in available energy and prices. In this context, constructing an optimal bidding strategy is futile, but simple bidding heuristics are still valuable. In particular, in our study, the agents are *zero intelligence plus* (ZIP) traders [43, 65]. ZIP traders use an adaptive mechanism which can give performance very similar to that of human traders in stock markets. Agents have a profit margin which determines the difference between their limit prices and their asks or bids. Under this strategy, traders adapt and update their margins based on the matching of previous orders (Algorithms 4 and 5, respectively). Indeed, the participation of ZIP traders in a CDA allows us to assess the economic benefits of the market separate from that of a particular bidding strategy. Specifically, ZIP traders are subject to a budget constraint (L_{max} and L_{min} are the maximum and minimum price respectively) which forbids the trader to buy or sell at a loss. Then, buyers and sellers select their bids or asks uniformly at random between these limits.

Network Permission Structure

The outline of the mechanism is presented in Fig. 5.20. A third party entity (e.g., DSO) validates the transactions using a network permission structure based on the network's features and sensitivity coefficients. Specifically, we incorporate the following sensitivities in the market mechanism [30]:

Algorithm 4 Update ZIP buyers' profit margins

if the last order was matched at price π_t **then**
  all buyers for which $\pi_b \geq \pi_t$, raise their margins;
  if the last trader was a seller **then**
    any active buyer for which $\pi_b \leq \pi_t$,
    lower its margin;
  end if
else
  if the last trader was a buyer **then**
    any active buyer for which $\pi_b \leq \pi_t$,
    lower its margin;
  end if
end if

Algorithm 5 Update ZIP sellers' profit margins

if the last order was matched at price π_t **then**
  all sellers for which $\pi_s \leq \pi_t$, raise their margins;
  if the last trader was a buyer **then**
    any active seller for which $\pi_s \geq \pi_t$,
    lower its margin;
  end if
else
  if the last trader was a seller **then**
    any active seller for which $\pi_s \geq \pi_t$,
    lower its margin;
  end if
end if

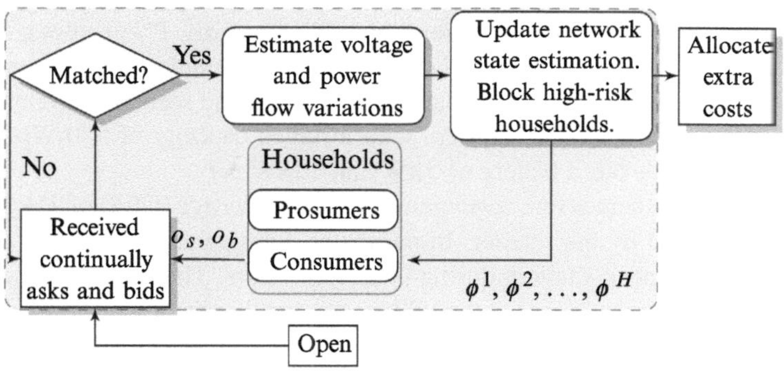

FIG. 5.20 Schematic of the P2P trading under network constraints.

- *Voltage sensitivity coefficients* (VSCs): Through VSCs, we can estimate the variation in the voltages as a function of the power injections in the network;
- *Power transfer distribution factors* (PTDFs): These reflect the changes in active power line flows due to an exchange of active power between two nodes;
- *Loss sensitivity factors* (LSFs): These reflect the portion of system losses due to power injections in the network.

Every time one ask and one bid are matched, voltage variation and line congestion are evaluated. All households receive a signal (ϕ^h) which informs them if they can still participate in the market without causing problems in the network. For instance, one prosumer could be blocked from injecting power into the grid at a certain time due to the high risk of causing voltage problems in the network. This is achieved using the VSCs and PTDFs. If the transaction is approved, the extra cost associated with the network constraints are allocated to the users involved in the matched transaction.

Importantly, power curtailment is implicitly incorporated in the trading. Thus, this method may bring extra benefits in comparison to other curtailment methods. For example, users at the worst node location still have the opportunity to participate if their order can be matched and if the mechanism allows the trade. This improves the efficiency by allowing greater participation of consumers and a better reflect of network conditions.

Case Study

As an illustration, we consider a smart grid system for energy trading at a local level. The methodology is applied to the UK LV network shown [56], comprising one feeder and 100 single phase households. The simulations are carried out with $T=24\,\text{h}$, $\Delta\tau=15\,\text{min}$, and up to 100 agents. There are 50 consumers and 50 prosumers, 40 for Type 1 (PV) and 10 for Type 2 (PV, battery, and a home energy management system). Each household has a stochastic load consumption profile, with load profiles using the tool presented in [66]. Similarly, PV profiles are generated considering sun irradiance data, capturing the sunniest days in order to evaluate the method on the most challenging yet realistic scenarios. We assume that all prosumers have a PV system with installed capacity of 5.0kWp. Each Type 2 households has a battery of 3kW and 10kWh.

Additionally, there is one *community electricity storage* (CES) of 25kW and 50 kWh operated by the retailer. In particular, the operation objective of the CES is to apply peak shaving during peak load hours. The CES strategy is to buy only the energy to charge in the P2P market to other prosumers around midday (when there are low rates and a high number of prosumers with energy surplus) and resell the energy during peak demand hours to the consumers. Like the prosumers behavior, the CES is modeled as a ZIP trader. We define the price constraints L_{max} and L_{min} based on the values of import and export electricity tariffs through the day. L_{max} depends on the time-of-use tariff (ToU) and L_{min} on the feed-in-tariff (FiT). These definitions are consistent in the sense that no buyer would pay more than the tariff of a retailer (ToU), and no seller would sell their units cheaper than the export tariff (FiT).

Fig. 5.21 shows the average transaction price (ATP) and the amount of energy purchased from the grid or in the P2P market during 1 day. The transaction prices remain in the range of ToU and FiT rates because of the ZIP limits L_{max} and L_{min}. Hence, both prosumers and consumers obtain monetary benefits by participating in P2P trading. Most of the energy is traded during 8:00 and 14:00. During that time, there is an excess of energy due to PV generation. Notably, there is a peak of energy sold in the market around 11 am because of the charging strategy of the CES. There is some energy traded after 18:00 due to the CES and the prosumers who kept some energy in the battery. Once the peak time ends (20:00), the ZIP maximum limit (L_{max}) is low. As a consequence, no

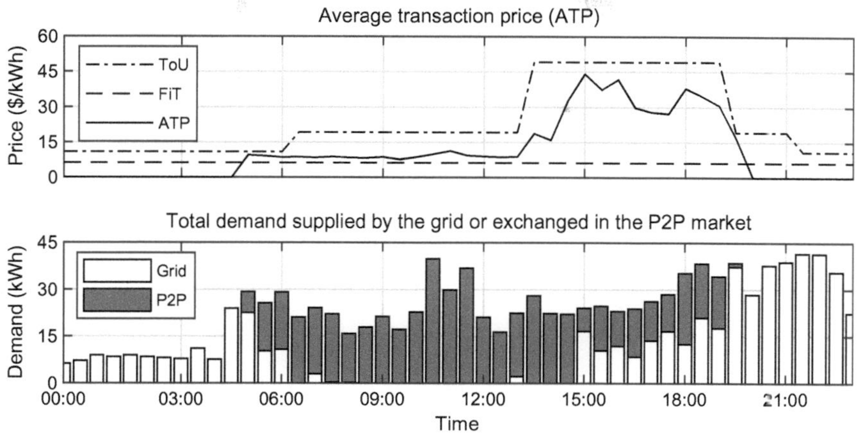

FIG. 5.21 Average transaction prices (top), demand and generation levels (bottom).

prosumers submit any new asks to trade in the market. Moreover, in this case, when the total energy surplus from prosumers is greater than the total demand of consumers (e.g., around midday), some prosumers (those who do not match their asks with consumers' bids) have to curtail their power generation.

Fig. 5.22 presents a histogram of voltages at all users' nodes during 1 day of simulation. There are no cases of overvoltage. The voltages varied between 0.945 and 1.022 pu. Around 55% of the voltages are between 0.99 and 1 pu. As such, all exchanges respect the network constraints, and the external costs were attributed among the households involved in each transaction.

As demonstrated in Fig. 5.23, P2P with network permission consistently outperforms alternative power curtailment methods used to manage adverse impacts of DER penetration, including capacity reduction (Red. Cap), tripping, and

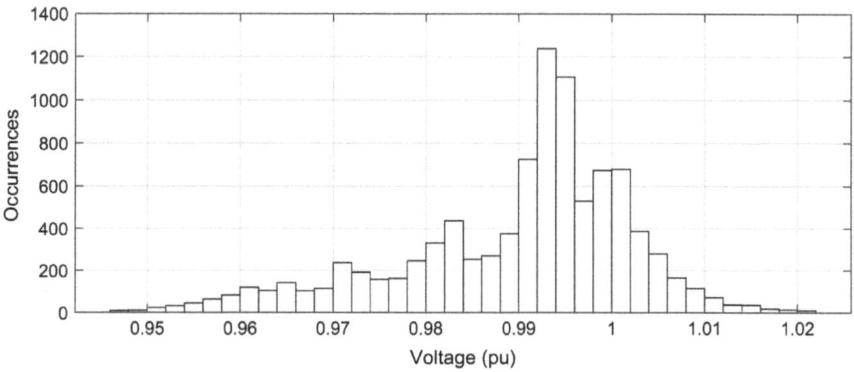

FIG. 5.22 Histogram of voltages at users' nodes—number of occurrences in 1 day period at a certain voltage [pu].

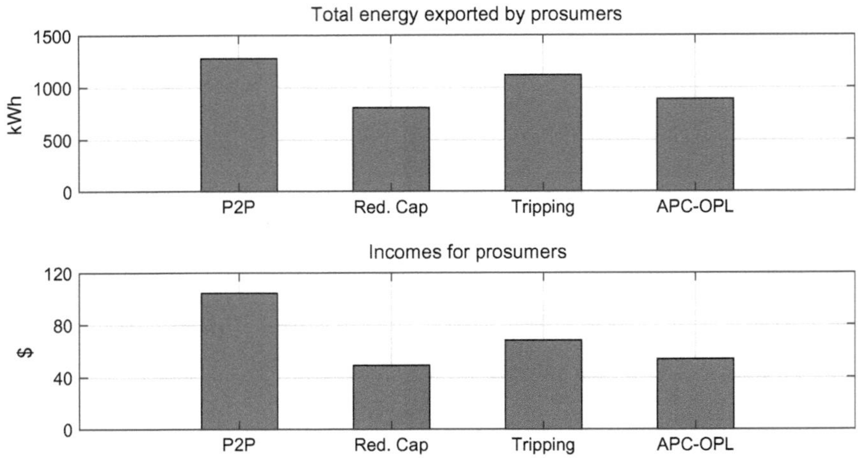

FIG. 5.23 P2P vs alternative solutions: prosumers' total energy supplied and income received.

droop-based active curtailment (APC-OLP). The results show that in the P2P case there is more energy traded, and the prosumers' revenue is bigger in comparison with the other methods. In particular, the drawback of the power curtailment methods is that they do not consider the impact on the end-users' revenue. In contrast, the P2P scheme offers greater economic benefits to all users.

5.5 Large-Scale Demand Response as Distributed Optimal Power Flow

A principled way of addressing the problem of *DER dispatch subject to network constraints* is to use distribution optimal power flow (DOPF) [67], which is formulated as welfare maximization as follows:

$$\begin{aligned} \underset{p^g, p^c}{\text{maximize}} \quad & \sum_{i \in \mathcal{I}} c_i^g p_i^g - \sum_{i \in \mathcal{I}} c_i^c p_i^c \\ \text{s.t.} \quad & \text{power flow constraints,} \\ & \text{power balance constraints, and} \\ & \textit{DER} \text{ operational constraints,} \end{aligned} \tag{5.53}$$

where c_i^c is marginal benefit of consumption at node i, c_i^g is marginal cost of generation at node i, p_i^c is the real power consumption, and p_i^g is the real power generation.[7]

DOPF produces distribution locational marginal prices (DLMPs), which can be used to attribute the network cost to the market. In doing so, a central entity (e.g., DSO) solves the optimization problem across the scheduling horizon with the goal of minimizing the total cost of supplying power to the consumers,

7. The formulation in (5.53) is general and can vary depending on the application.

subject to network constraints. As a result, real power losses, and (binding) capacity and voltage constraints result in DLMPs being different across the network. Conceptually, DOPF is the same as the optimal power flow (OPF) problem used in the wholesale market.

The DOPF implementation is, however, riddled with technical and market design barriers. First, the number of market agents (consumers) is significantly larger than in the conventional OPF problem, so that a centralized DOFF computation can be challenging, if not intractable. To this end, distributed optimization approaches have been considered [39, 40, 68] to ensure scalability and to preserve consumers' prerogative and privacy [17, 68]. Thirdly, household consumption patterns are stochastic, so proper mechanisms are required to ensure that the customers follow the allocated power profiles [17, 23, 69]. Finally, a DOPF implementation would require a complete redesign of the tariff structures, so it cannot be easily incorporated into the existing market framework. A possible solution would be to use price network services externally, as proposed in [70].

Section 5.5.1 first discusses fast distributed optimization techniques for largescale demand response aggregation without network constraints. Section 5.5.2 then discusses distributed OPF using ADMM. Section 5.5.3 discusses a mechanism design for a faithful implementation of demand response aggregation (DRA).

5.5.1 Network-Agnostic Demand Response Aggregation

In the absence of network constraints, the DSO can find the minimum cost of purchasing electricity for a set of households $\mathcal{I}: = \{1, 2, \ldots, I\}$, over a scheduling horizon $\mathcal{T}: = \{t, t+\Delta\tau, \ldots, t+T-\Delta\tau\}$ (typically 1 day) comprising T time slots, by solving the following problem:

$$\underset{x_i\in\mathcal{X}_i}{\text{minimize}} \sum_{t\in\mathcal{T}} C^t\left(p_0^t\right), \tag{5.54a}$$

$$\text{subject to} \sum_{i\in\mathcal{T}} p_i^t = p_0^t, \quad t\in\mathcal{T} \tag{5.54b}$$

$$\sum_{a\in\mathcal{A}_i} p_{i,a}^t - P_i^{PV,t}\Delta\tau = p_i^t, \quad t\in\mathcal{T} \tag{5.54c}$$

$$0 \le p_i^t \le P_i^{\max}\Delta\tau, \quad t\in\mathcal{T} \tag{5.54d}$$

where $C^t(p_0^t)$ is the cost of withdrawing from the grid p_0^t units of energy at timeslot t and

$$\boldsymbol{x}_i := \left(\left\{\{p_{i,a}^t\}_{t\in\mathcal{T}}\right\}_{a\in\mathcal{A}_i}, \{p_i^t\}_{t\in\mathcal{T}}\right).$$

In this setting, it is assumed that each household agent $i\in\mathcal{I}$ controls a set of flexible devices $\mathcal{A}_i$ with $p_{i,a}^t \in \mathcal{X}_{i,a}$ being the energy consumption variable of device

$a \in \mathcal{A}_i := \{1, \ldots, A_i\}$ at time-slot t [68]. The control of these devices requires the use of binary variables which makes the feasible sets $\mathcal{X}_i$ disjoint, and consequently, the DR model in (5.54) is a nonconvex problem that belongs to the class of mixed-integer nonlinear programming (MINLP) problems, which are NP-hard and notorious for tending to be intractable when they grow in size, if solved centrally.

However, by relaxing the coupling constraints (5.54b), the Lagrangian relaxation bestows a separable structure on problem (5.54). Formally, the (partial) Lagrangian of (5.54) is given by

$$L(x, \lambda) := \sum_{t \in \mathcal{T}} \left(C^t(p_0^t) + \lambda^t \left(\sum_{i \in \mathcal{I}} p_i^t - p_0^t \right) \right),$$

where $\lambda := [\lambda^\tau, \ldots, \lambda^{\tau+T-\Delta\tau}]^T$ is the vector of Lagrange multipliers. Now, the Lagrange dual function

$$D(\boldsymbol{\lambda}) := \underset{x_i \in \mathcal{X}_i}{\text{minimize}}\ L(x, \boldsymbol{\lambda}). \tag{5.55}$$

can be decomposed into $I+1$ independent subproblems that can be solved in parallel. In more detail, (5.55) can be rewritten as follows

$$D(\boldsymbol{\lambda}) := D_0(\boldsymbol{\lambda}) + \sum_{i \in \mathcal{I}} D_i(\boldsymbol{\lambda}), \tag{5.56}$$

where the DSO solves

$$D_0(\boldsymbol{\lambda}) := \inf_{x_0 \in \mathcal{X}_0} \left\{ \sum_{t \in \mathcal{T}} \left(C^t(p_0^t) - \lambda^t p_0^t \right) \right\}, \tag{5.57}$$

and the household agents solve

$$D_i(\boldsymbol{\lambda}) := \inf_{x_i \in \mathcal{X}_i} \left\{ \sum_{t \in \mathcal{T}} \lambda^t p_i^t \right\}. \tag{5.58}$$

Physical system limits restrict the power that can be drawn from the grid to $G^{\max}$, and therefore $\boldsymbol{x}_0 := \{p_0^t\}_{t \in \mathcal{T}} \in \mathcal{X}_0 := \{p_0^t \in \mathbf{R}_+ | \, 0 \le p_0^t \le G^{\max} \Delta\tau, t \in \mathcal{T}\}$.

Finally, the dual problem is given by

$$\max_{\lambda} D(\boldsymbol{\lambda}). \tag{5.59}$$

However, in this DR scenario, the concave dual function $D(\boldsymbol{\lambda})$ is typically nondifferentiable. Specifically, as the subproblems in (5.57) and (5.58) can have multiple optimal solutions for a given vector $\boldsymbol{\lambda}$, the subdifferentials $\partial D(\boldsymbol{\lambda})$ may be not be unique and the dual function $D(\boldsymbol{\lambda})$ can be nonsmooth.[8] Consequently, applying the conventional gradient ascent method to this problem would most likely exhibit very slow convergence.

8. If a function $f(\boldsymbol{x})$ is smooth, its subdifferential contains only one point and therefore $\partial f(\boldsymbol{x}) = \nabla f(\boldsymbol{x})$.

One way to obtain a smooth approximation of $D(\boldsymbol{\lambda})$ is to modify the subproblems in (5.58) to ensure a unique optimal solution for every $\boldsymbol{\lambda}$ as follows

$$D_\mu(\lambda) = D_0(\lambda) + \sum_{i\in\mathcal{I}} D_{i,\mu}(\lambda), \tag{5.60}$$

where

$$D_{i,\mu}(\lambda) = \inf_{x_i\in X_i} \left(\sum_{t\in\mathcal{T}} \lambda^t \mathrm{p}_i^t + \frac{\mu}{2}\|x_i\|^2 \right), \; i\in\mathcal{I} \tag{5.61}$$

and $\mu > 0$ is a smoothness parameter [71]. Since the aim is not only to efficiently solve the dual problem but also to recover a feasible solution to the primal, a second smoothing is applied to the dual function to make it strongly concave. The resulting modified dual function,

$$D_{\mu,k}(\lambda) = D_0(\lambda) + \sum_{i\in\mathcal{I}} D_{i,\mu}(\lambda) - \frac{\kappa}{2}\|\lambda\|^2, \tag{5.62}$$

is now strongly concave with parameter $\kappa > 0$. Applying a fast gradient method now ensures the same rate of convergence for $\|\nabla D_{\mu,\,\kappa}(\lambda)\|$ as for $D_{\mu,\,\kappa}(\lambda^*) - D_{\mu,\,\kappa}(\lambda^{(k)})$, where k is the iteration number. This property is essential for recovering a near-optimal solution for the primal in fewer iterations, compared to just applying a single smoothing [72]. A fast gradient algorithm for solving (5.62) is described in [68].

Other methods can be used to solve the nonsmooth dual in (5.55) but with extra care in handling the binary variables. Potential candidates are the cutting plane method [73], the bundle method [74], the alternating direction method of multipliers (ADMM) [75], and disaggregated bundle methods [76].

5.5.2 Network-Aware Demand Response Aggregation

In a network-aware demand response scenario, a multi-period OPF problem with demand response consists of finding the lowest cost of dispatching power from generators to satisfy the flexible loads at all buses over a scheduling horizon $\mathcal{T}$, in a way that is governed by physical laws, such as Ohm's Law and Kirchhoff's Law, and other technical restrictions, such as transmission line thermal limit constraints.

By letting

$$w_l = |V_l|^2, w_{l,m}^r = \Re\{V_l V_m^*\}, w_{l,m}^i = \Im\{V_l V_m^*\}, \tag{5.63}$$

the multi-period OPF with demand response can be written as[9]

9. See [77] for this *alternative formulation* of the OPF problem.

$$\underset{\substack{x_{i,l}\in\mathcal{X}_{i,l},p^t_{g,l},q^t_{g,l},w^t_l,w^{r,t}_{l,m},w^{i,t}_{l,m},\\ \theta^t_l,p^t_{l,m},q^t_{l,m},p^t_{m,l}q^t_{m,l}}}{\text{minimize}} \sum_{t\in\mathcal{T}}\sum_{(gl)\in\mathcal{G}} f_{g,l}\left(p^t_{g,l}\right) \tag{5.64a}$$

subject to

$$\underline{p}_{g,l} \le p^t_{g,l} \le \overline{p}_{g,l},\quad (g,l)\in\mathcal{G}, t\in\mathcal{T} \tag{5.64b}$$

$$\underline{q}_{g,l} \le q^t_{g,l} \le \overline{q}_{g,l},\quad (g,l)\in\mathcal{G}, t\in\mathcal{T} \tag{5.64c}$$

$$-p^r_{g,l} \le p^t_{g,l} - p^{t-1}_{g,l} \le p^r_{g,l},\ (g,l)\in\mathcal{G}, t\in\mathcal{T} \tag{5.64d}$$

$$\sum_{a\in\mathcal{A}_i} p^t_{i,l,a} - P^{pv,}_{i,l} t\Delta\tau = p^t_{i,l}\quad l\in\mathcal{B}, t\in\mathcal{T} \tag{5.64e}$$

$$0 \le p^t_{i,l} \le P^{\max}_{i,l}\Delta\tau,\qquad l\in\mathcal{B}, t\in\mathcal{T} \tag{5.64f}$$

$$|\underline{V}_l|^2 \le w^t_l \le |\overline{V}_l|^2,\qquad l\in\mathcal{B}, t\in\mathcal{T} \tag{5.64g}$$

$$\underline{\theta}^{\Delta}_{l,m} \le \theta^t_l - \theta^t_m \le \overline{\theta}^{\Delta}_{l,m},\qquad (l,m)\in\mathcal{L}, t\in\mathcal{T} \tag{5.64h}$$

$$\sum_{(g,l)\in\mathcal{G}} p^t_{g,l} - p^{d,t}_l = \sum_{m\in\mathcal{B}_l} p^t_{l,m} + \sum_{(i,l)\in\mathcal{I}} p^t_{i,l} + g^{sh}_l w^t_l,\quad l\in\mathcal{B}, t\in\mathcal{T} \tag{5.64i}$$

$$\sum_{(g,l)\in\mathcal{G}} q^t_{g,l} - q^{d,t}_l = \sum_{m\in\mathcal{B}_l} q^t_{l,m} - b^{sh}_l w^t_l,\qquad l\in\mathcal{B}, t\in\mathcal{T} \tag{5.64j}$$

$$p^t_{l,m} = g^c_{l,m} w^t_l - g_{l,m} w^{r,t}_{l,m} + b_{l,m} w^{i,t}_{l,m},\qquad (l,m)\in\mathcal{L}, t\in\mathcal{T} \tag{5.64k}$$

$$q^t_{l,m} = b^c_{l,m} w^t_l - b_{l,m} w^{r,t}_{l,m} - g_{l,m} w^{i,t}_{l,m},\qquad (l,m)\in\mathcal{L}, t\in\mathcal{T} \tag{5.64l}$$

$$p^t_{m,l} = g^c_{m,l} w^t_m - g_{m,l} w^{r,t}_{l,m} - b_{m,l} w^{i,t}_{l,m},\qquad (lm)\in\mathcal{L}, t\in\mathcal{T} \tag{5.64m}$$

$$q^t_{m,l} = b^c_{m,l} w^t_m - b_{m,l} w^{r,t}_{l,m} + g_{m,l} w^{i,t}_{l,m},\qquad (l,m)\in\mathcal{L}, t\in\mathcal{T} \tag{5.64n}$$

$$\left(w^{r,t}_{l,m}\right)^2 + \left(w^{i,t}_{l,m}\right)^2 = w^t_l w^t_m,\qquad (l,m)\in\mathcal{L}, t\in\mathcal{T} \tag{5.64o}$$

$$\theta^t_m - \theta^t_l = \operatorname{atan2}\left(w^{i,t}_{l,m}, w^{r,t}_{l,m}\right),\qquad (l,m)\in\mathcal{L}, t\in\mathcal{T} \tag{5.64p}$$

$$\sqrt{\left(p^t_{l,m}\right)^2 + \left(q^t_{l,m}\right)^2} \le \overline{S}_{l,m},\qquad (lm)\in\mathcal{L}\cup\mathcal{L}_r, t\in\mathcal{T} \tag{5.64q}$$

where $f_{g,l}(p^t_{g,l}) := c_{2,g,l}(p^t_{g,l})^2 + c_{1,g,l}(p^t_{g,l}) + c_{0,g,l}$.[10] The multi-period OPF with DR problem in (5.64) has a nonconvex relaxation (of the binary variables) and belongs to the class of mixed-integer nonlinear programming (MINLP) problems which are proven to be NP-hard.

10. See [39] for more details on the parameters of the network.

There is a plethora of existing works on distributed OPF. These can be broadly classified into four categories: dual decomposition methods, optimality conditions decomposition (OCD) methods, sparse semidefinite programming (SDP) decomposition methods, and primal decomposition methods (see [39, 78] for a review). The dual decomposition techniques underlying the dual-decomposition-based distributed OPF methods in the literature can, in turn, be classified into two categories: region-based decompositions and component-based decompositions. The focus of this study revolves around the latter decomposition techniques, because they can distribute the computation across *every* component in the network (generators, transformers, loads, buses, transmission lines, etc.) and are flexible enough to incorporate discrete decision variables to suit a wide variety of optimization applications in power systems and future grids. The result of the component-based dual decomposition is a consensus problem that can be solved in a distributed fashion using the alternating direction of multipliers method (ADMM).

In its native form in (5.64), this problem is not separable in terms of components, as relaxing the coupling constraints in (5.64i) and (5.64j) is not enough to bestow separability. This is because branch power flow constraints in (5.64k–5.64n) are dependent on bus variables (voltage and angle variables). A separable structure can therefore be bestowed by creating copies of the following variables,

$$p_{i,l}^t = p_{i,l_{(l)}}^t, \qquad (i,l) \in \mathcal{I}, t \in \mathcal{T} \tag{5.65}$$

$$p_{g,l}^t = p_{g,l_{(l)}}^t, \qquad (g,l) \in \mathcal{G}, t \in \mathcal{T} \tag{5.66}$$

$$q_{g,l}^t = q_{g,l_{(l)}}^t, \qquad (g,l) \in \mathcal{G}, t \in \mathcal{T} \tag{5.67}$$

$$p_{l,m}^t = p_{l,m_{(l)}}^t, \qquad (l,m) \in \mathcal{L} \cup \mathcal{L}_r,\ t \in \mathcal{T} \tag{5.68}$$

$$q_{l,m}^t = q_{l,m_{(l)}}^t, \qquad (l,m) \in \mathcal{L} \cup \mathcal{L}_r,\ t \in \mathcal{T} \tag{5.69}$$

$$w_{l_{(l,m)}}^t = w_l^t, \qquad (l,m) \in \mathcal{L} \cup \mathcal{L}_r,\ t \in \mathcal{T} \tag{5.70}$$

$$\theta_{l_{(l,m)}}^t = \theta_l^t, \qquad (l,m) \in \mathcal{L} \cup \mathcal{L}_r,\ t \in \mathcal{T} \tag{5.71}$$

and making sure that these variables are in *consensus*. The multi-period OPF with DR problem now becomes

$$\underset{x,z,w_{ij}^r,w_{ij}^i}{\text{minimize}} \sum_{t \in \mathcal{T}} \sum_{(g,l) \in \mathcal{G}} f_{g,l}\left(p_{g,l}^t\right) \tag{5.72a}$$

$$\text{subject to (5.64b)–(5.64f), (5.64q), and (5.65)–(5.71)} \tag{5.72b}$$

$$|\underline{V}_l|^2 \leq w_{l_{(l,m)}}^t \leq |\overline{V}_l|^2, \quad l \in \mathcal{B}, t \in \mathcal{T} \tag{5.72c}$$

$$\underline{\theta}_{l,m}^{\Delta} \leq \theta_{l_{(l,m)}}^t - \theta_{m_{(m,l)}}^t \leq \overline{\theta}_{l,m}^{\Delta}, \quad (l,m) \in \mathcal{L}, t \in \mathcal{T} \tag{5.72d}$$

$$\sum_{(g,l)\in\mathcal{G}} p^t_{g,l_{(l)}} - p^{d,t}_l = \sum_{m\in\mathcal{B}_l} p^t_{l,m_{(l)}} + \sum_{(i,l)\in\mathcal{I}} p^t_{i,l_{(l)}} + g^{sh}_l w^t_l, l\in\mathcal{B}, t\in\mathcal{T} \tag{5.72e}$$

$$\sum_{(g,l)\in\mathcal{G}} q^t_{g,l_{(l)}} - q^{d,t}_l = \sum_{m\in\mathcal{B}_l} q^t_{l,m_{(l)}} - b^{sh}_l w^t_l,\ \ l\in\mathcal{B}, t\in\mathcal{T} \tag{5.72f}$$

$$p^t_{l,m} = g^c_{l,m} w^t_{l_{(l,m)}} - g_{l,m} w^{r,t}_{l,m} + b_{l,m} w^{i,t}_{l,m},\ \ (l,m)\in\mathcal{L}, t\in\mathcal{T} \tag{5.72g}$$

$$q^t_{l,m} = b^c_{l,m} w^t_{l_{(l,m)}} - b_{l,m} w^{r,t}_{l,m} - g_{l,m} w^{i,t}_{l,m},\ \ (l,m)\in\mathcal{L}, t\in\mathcal{T} \tag{5.72h}$$

$$p^t_{m,l} = g^c_{m,l} w^t_{m_{(m,l)}} - g_{m,l} w^{r,t}_{l,m} - b_{m,l} w^{i,t}_{l,m},\ \ (l,m)\in\mathcal{L}, t\in\mathcal{T} \tag{5.72i}$$

$$q^t_{m,l} = b^c_{m,l} w^t_{m_{(m,l)}} - b_{m,l} w^{r,t}_{l,m} + g_{m,l} w^{i,t}_{l,m},\ \ (l,m)\in\mathcal{L}, t\in\mathcal{T} \tag{5.72j}$$

$$\left(w^{r,t}_{l,m}\right)^2 + \left(w^{i,t}_{l,m}\right)^2 = w^t_{l_{(l,m)}} w^t_{m_{(m,l)}},\ \ (l,m)\in\mathcal{L}, t\in\mathcal{T} \tag{5.72k}$$

$$\theta^t_{m_{(m,l)}} - \theta^t_{l_{(l,m)}} = \text{atan2}\left(w^{i,t}_{l,m}, w^{r,t}_{l,m}\right),\ \ (l,m)\in\mathcal{L}, t\in\mathcal{T} \tag{5.72l}$$

where

$$\boldsymbol{x} := \begin{bmatrix} \left\{\left\{p^t_{g,l}, q^t_{g,l}\right\}_{t\in\mathcal{T}}\right\}_{(g,l)\in\mathcal{G}} \\ \left\{\left\{p^t_{i,l}\right\}_{t\in\mathcal{T}}\right\}_{(i,l)\in\mathcal{I}} \\ \left\{\left\{p_{l,m}, q_{l,m}, w_{l_{(l,m)}} \theta_{l_{(l,m)}}\right\}_{t\in\mathcal{T}}\right\}_{(l,m)\in\mathcal{L}\cup\mathcal{L}r} \end{bmatrix},$$

and

$$\boldsymbol{z} := \begin{bmatrix} \left\{\left\{p^t_{g,l_{(l)}}, q^t_{g,l_{(l)}}\right\}_{t\in\mathcal{T}}\right\}_{(g,l)\in\mathcal{G}} \\ \left\{\left\{p^t_{i,l_{(l)}}\right\}_{t\in\mathcal{T}}\right\}_{(i,l)\in\mathcal{I}} \\ \left\{\left\{p^t_{l,m_{(l)}}, q^t_{l,m_{(l)}}\right\}_{t\in\mathcal{T}}\right\}_{(l,m)\in\mathcal{L}\cup\mathcal{L}_r} \\ \left\{\left\{w^t_l, \theta^t_l\right\}_{t\in\mathcal{T}}\right\}_{l\in\mathcal{B}} \end{bmatrix}.$$

Let $N_{\mathbf{x}} = N_\lambda = (2|\mathcal{G}| + |\mathcal{I}| + 4|\mathcal{L}\cup\mathcal{L}_r|) \times |\mathcal{T}|$ and $N_{\mathbf{z}} = (2|\mathcal{G}| + |\mathcal{I}| + 2|\mathcal{L}\cup\mathcal{L}_r| + 2|\mathcal{B}|) \times |\mathcal{T}|$. Problem (5.72) is now of the general form

$$\underset{x\in\mathcal{X}, z\in\mathcal{Z}}{\text{minimize}}\ \ f(\mathbf{x}) + g(\mathbf{z}) \tag{5.73a}$$

$$\text{subject to}\ \ A\mathbf{x} + B\mathbf{z} = \mathbf{c}, \tag{5.73b}$$

where $f: \mathbf{R}^{N_x} \rightarrow \mathbf{R}$ and $g: \mathbf{R}^{N_z} \rightarrow \mathbf{R}$ are closed convex functions, $A = I \in N_\lambda \times N_x$, $B \in \mathbf{R}^{N_\lambda \times N_z}$, and $c \in \mathbf{R}^{N_\lambda}$.[11] Constraints (5.73b) are defined by (5.65)–(5.71), $\mathcal{X}$ is

11. Note that $\boldsymbol{c} = \mathbf{0}$ in this OPF case.

the feasible set defined by constraints (5.64b)–(5.64f), (5.64q), (5.72c), (5.72d), (5.72g)–(5.72l) and $\mathcal{Z}$ is the feasible set defined by constraints (5.72e) and (5.72f). The augmented (partial) Lagrange function of (5.73) is written as

$$L_\rho(\boldsymbol{x}, z, \boldsymbol{\lambda}) := f(\boldsymbol{x}) + g(z) + \boldsymbol{\lambda}^T(A\boldsymbol{x} + Bz - \boldsymbol{c}) + \frac{\rho}{2}\|A\boldsymbol{x} + Bz - \boldsymbol{c}\|^2, \tag{5.74}$$

where $\rho > 0$ is a penalty parameter and $\lambda \in \mathbf{R}^{N_\lambda}$, is the vector of dual variables associated with coupling constraints (5.73b). The augmented Lagrangian in (5.74) is not separable in terms of sets of variables ($\mathcal{X}$ and $\mathcal{Z}$). Nonetheless, ADMM can be used to decouple these sets of variables, by using alternate minimizations over these sets. In particular, given the current iterates ($\boldsymbol{x}^k$, z^k, $\boldsymbol{\lambda}^k$), ADMM generates a new iterate ($\boldsymbol{x}^{k+1}$, z^{k+1}, $\boldsymbol{\lambda}^{k+1}$) as follows

$$\boldsymbol{x}^{k+1} := \underset{x \in \mathcal{X}}{\operatorname{argmin}} L_\rho(\boldsymbol{x}, z^k, \boldsymbol{\lambda}^k), \tag{5.75a}$$

$$z^{k+1} := \underset{z \in \mathcal{Z}}{\operatorname{argmin}} L_\rho(\boldsymbol{x}^{k+1}, z, \boldsymbol{\lambda}^k), \tag{5.75b}$$

$$\boldsymbol{\lambda}^{k+1} := \boldsymbol{\lambda}^k + \rho(A\boldsymbol{x}^{k+1} + Bz^{k+1} - \boldsymbol{c}). \tag{5.75c}$$

In general, if the sets $\mathcal{X}$ and $\mathcal{Z}$ are convex and the primal problem is feasible, ADMM is guaranteed to converge to an optimal point [79]. However, the multi-period OPF with DR problem at hand is nonconvex and, therefore, ADMM is no longer theoretically guaranteed to converge.

The advantage of the component-based decomposition is smaller subproblems that can be solved efficiently. In fact, by adopting the above formulation, which uses the *alternative formulation* of the OPF problem, bus subproblems now have closed-form solutions (see [39]).

The primal residuals are defined as

$$\boldsymbol{r}^{k+1} = \left[r_1^{(k+1)}, \ldots, r_{N_\lambda}^{(k+1)}\right] = A\boldsymbol{x}^{k+1} + Bz^{k+1} - \boldsymbol{c}, \tag{5.76}$$

and the dual residuals as

$$\boldsymbol{s}^{k+1} = \left[s_1^{(k+1)}, \ldots, s_{N_\lambda}^{(k+1)}\right] = \rho A^T B(z^{k+1} - z^k). \tag{5.77}$$

The algorithm in (5.75) is terminated when

$$\|\boldsymbol{r}^k\| \le \epsilon^{\text{pri}} \text{ and } \|\boldsymbol{s}^k\| \le \epsilon^{\text{dual}}, \tag{5.78}$$

where ϵ^{pri} and ϵ^{dual} are feasibility tolerances which are chosen using an absolute and relative criterion as follows

$$\epsilon^{\text{pri}} = \sqrt{N_\lambda}\epsilon^{\text{abs}} + \epsilon^{\text{rel}} \max\left\{\|A\boldsymbol{x}^k\|, \|Bz^k\|, \|\boldsymbol{c}\|\right\}, \tag{5.79}$$

$$\epsilon^{\text{dual}} = \sqrt{N_x}\epsilon^{\text{abs}} + \epsilon^{\text{rel}}\|A^T\boldsymbol{\lambda}^k\|, \tag{5.80}$$

where $\epsilon^{\text{abs}} > 0$ and $\epsilon^{\text{rel}} > 0$ are absolute and relative tolerances, respectively (see [79]). The values of $\boldsymbol{r}^k$ and $\boldsymbol{s}^k$ indicate how distant the iterates are from a solution. More specifically, $\boldsymbol{r}^k$ approaches zero when the iterates accurately satisfy the coupling constraints, and $\boldsymbol{s}^k$ approaches zero as the iterates approach a minimizer of the objective.

At each iteration of the distributed algorithm in (5.75), the DSO sends price signals, to which the agents respond by computing the local demand schedule that minimizes their payments. These demand schedules are then communicated back to the DSO and the prices are updated. This process is repeated until the algorithm converges to a near-optimal price vector that minimizes the total cost of purchasing electricity.

Despite the absence of theoretical guarantees of convergence and the presence of binary variables, ADMM applied to a problem of the form (5.72) has been empirically shown in [80] to converge to near optimal solutions on a suburb-sized microgrid within computation times that are practical for receding-horizon control.

The convergence of ADMM is in practice sensitive to the choice of penalty parameter and is therefore problem dependent [79]. A natural extension is to use variable penalty parameter methods such as residual balancing [81] and adaptive penalty schemes [82], which allow this penalty parameter to vary at each iteration k.

5.5.3 Faithful Mechanism Design for Demand Response Aggregation

Any economic mechanism consists of an *allocation rule* and a *payment rule* [83], which are used to implement a particular objective function. For example, an English auction for a single indivisible item has an allocation rule of "highest bidder wins" and a payment rule of "winner pays their bid" that together implement the social objective of allocating the item to the bidder that values it the most (the *Pareto-efficient* outcome). The same procedure can be followed for more general settings, in which any particular objective function is being implemented. However, in the presence of self-interested actors, a corresponding payment rule needs to be defined to ensure that the actors truthfully implement their assigned protocols, which are based on the preferences of the users themselves, and their allocated schedules. The problem of deriving such a payment rule is the problem addressed in this section.

If the distributed algorithm in (5.75) is to be considered as a distributed mechanism for residential electricity pricing, its outcome would be an allocation vector $\hat{\boldsymbol{x}}_i$ and a payment vector $\wp_i\left(\hat{\boldsymbol{\lambda}}_i\right)$ for each agent $i \in \mathcal{I}$. In more detail, the near-optimal allocation vector $\hat{\boldsymbol{x}}_i$ and Lagrange multipliers $\hat{\boldsymbol{\lambda}}_i$ (associated with consensus constraints (5.65)) are iteratively computed as in (5.75), in which the agents are assumed to faithfully implement their prescribed protocols (or algorithms)

$$\boldsymbol{x}_i^{(k)} := \underset{\boldsymbol{x}_i \in \mathcal{X}_i}{\operatorname{argmin}} \left\{ \sum_{t \in \mathcal{T}} \left(\hat{\lambda}_i^{t,(k)} p_i^t \right) \right\}, \quad i \in \mathcal{I}, \tag{5.81}$$

at each iteration k. Additionally, the utility of agent i at time-slot t would be

$$U_i^t\left(w_i^t, \lambda_i^t\right) := -\wp_i^t\left(w_i^t, \hat{\lambda}_i^t\right) = -w_i^t \hat{\lambda}_i^t,$$

where w_i^t is the *actual* consumption of agent i at time-slot t, revealed by the smart meters after time-slot t has elapsed.

There are two ways the agents can manipulate the outcome of a distributed algorithm if it leads to a decrease in their payment. The agents can either tamper with their assigned computation in (5.81) and/or misreport the result of this computation. Specifically, one obvious way for agent i to maximize $U_i^t\left(w_i^t, \hat{\lambda}\right)$ is to minimize $\hat{\lambda}_i^t$. This can be easily done by reporting any value of $\hat{p}_i^t$ less than the true value p_i^t, which leads the mechanism to choose $\hat{\lambda}_i^t$ less than the true λ_i^t. In the extreme case, the agents can collude and report $\hat{\boldsymbol{x}}_i = \boldsymbol{0}$, at each iteration k, which results in $\hat{\boldsymbol{\lambda}}_i = \boldsymbol{0}$ and therefore a total utility of 0 for each agent $i \in \mathcal{I}$. Therefore, without a carefully designed payment scheme, the distributed algorithm in (5.75) would be prone to manipulation by self-interested and strategic agents, who aim to maximize their utilities. Consequently, a mechanism directly based on a distributed dual-decomposition algorithm, like the one in (5.75), is not efficient. In other words, because it can be easily manipulated, it would not be able to find an outcome that maximizes the social welfare (the sum of the utilities of the agents based on their true information).

To prevent this gaming behavior, a distributed mechanism should be designed to incentivize the agents to *faithfully* implement their prescribed computation and *truthfully* report their negotiated total hourly demands at each iteration of the distributed algorithm. More specifically, the payment function should be designed to bring *faithful* computation to a *dominant-strategy* equilibrium. In a dominant-strategy equilibrium, it is in the best interest of every agent to follow all aspects of the algorithm—computation and message-passing—whatever the private types of the other agents, *even if* the other agents do not follow the prescribed protocol [84]. This latter requirement makes a dominant-strategy equilibrium stronger than an *ex post* Nash equilibrium, in which the prescribed protocol is the best protocol for every agent *as long as* the other agents also follow the prescribed protocol. Another desired property is *collusion-proofness* in a dominant-strategy equilibrium, in which no deviation by a coalition of agents can strictly improve the outcome *of even a single agent* in that coalition.

One way to ensure the desired properties of faithfulness and collusion proofness is to modify the payment from each agent i at time-slot t as follows

$$\wp_i^t\left(w_i^t; \hat{p}_i^t, \vartheta^t\right) := \kappa_0 + \frac{\kappa_1 w_i^t}{\kappa_2 e^{\left(-\frac{\vartheta}{2}\left(w_i^t - \hat{p}_i^t\right)^2\right)} + \kappa_3} - \kappa_4 w_i^t, \tag{5.82}$$

where $\kappa_0{:=}w_i^t\hat{\lambda}_i^t, \kappa_1{:=}\lambda_{w,i}^t S, \kappa_2{:=}\left(S^{\max}-S^{\min}\right), \kappa_3{:=}S^{\min}, \kappa_4{:=}\kappa_1/S^{\max}$, $S^{\max}$, and $S^{\min}$ are scaling parameters and $S{:=}\ (S^{\max}-S^{\min})/(S^{\max}-S^{\min})$ is a scaling parameter for $\lambda_{w,i}^t$.[12] The payment function in (5.82) *couples* agent i's actual consumption w_i^t with its report $\hat{p}_i^t$ directly, and indirectly through $\lambda_{w,i}^t$, which is the electricity price (Lagrange multiplier) obtained from the actual consumption $\boldsymbol{w}_i=\left\{w_i^t\right\}_{t\in\mathcal{T}}$. More specifically, $\boldsymbol{\lambda}_{w,i}{:=}\left\{\lambda_{w,i}^t\right\}_{t\in\mathcal{T}}$ is the Lagrange multiplier vector associated with constraint (5.65) of problem (5.72), but with (fixed) $\boldsymbol{w}_i$ instead of variables $\boldsymbol{x}_i$.

Since this electricity pricing scheme is the result of an efficient coordinated negotiation between the DSO and the prosumers, it does not suffer from rebound peaks in the aggregate residential demand, which is a major advantage over existing time-varying electricity rate plans like time-of-use (TOU) pricing and critical-peak pricing (CPP). These existing time-varying prices (TOU and CPP) are not a result of coordination among the agents, so it should not be surprising that they have been shown to create rebound peaks that stem from the synchronized responses of the agents, when deferring their flexible loads [85]. These rebound peaks can be higher and steeper than the original peaks that TOU and CPP electricity pricing structures were intended to eliminate.

The payment function in (5.82) can also be thought of as a hedging against strategic uncertainty that arises in this setting. In fact, there are two types of uncertainties in this setting: strategic uncertainty and demand/generation uncertainty. Strategic uncertainty arises when self-interested agents perceive an opportunity to manipulate a DR scheme for their own benefit, either through manipulating their assigned protocols or by deviating from their allocated DER policy. The chain of reasoning is as follows: the negotiated consumption policies are based on the preferences of the users and the mechanism's payment rule incentivizes the users to be truthful in implementing their assigned protocols, which are based on these preferences. This allows the DSO to iteratively compute the cost minimizing policies for each user. The payment rule then incentivizes the users to accurately follow the resulting policy profiles, instead of trying to manipulate these outcomes if it leads to higher monetary savings; and this thereby reduces strategic uncertainty. Demand/generation uncertainty is addressed by embedding the mechanism in a receding-horizon control process, which allows time-coupled devices to be accurately controlled in an uncertain environment. At first glance, this receding-horizon control mechanism might seem like a highly dynamic pricing, which might be hard for users to follow [86], but since the households are (assumed) equipped with an energy management system, all the users have to do is follow the schedule suggested by the energy management system and this guarantees the minimum cost of electricity.

12. See [69] for the proofs.

5.6 Generic Aggregate Prosumer Model for Future Grid Studies

The increasing uptake of residential PV-battery systems is bound to significantly change demand patterns of future power systems and, consequently, their dynamic performance. In this section, we describe a generic demand model that captures the aggregated effect of a large population of price-responsive users equipped with small-scale PV-battery systems, called *prosumers*, for market simulation in future grid scenario analysis [87].

5.6.1 Modeling Assumptions

The main purpose of developing generic demand models is to provide accurate dispatch decisions for balancing and stability analysis of future grid scenarios. Given the uncertainty associated with future grid studies, the modeling framework should be market structure agnostic, and capable of easy integration of various types and penetrations of emerging demand-side technologies. To this effect, we make the following assumptions:

1. The loads are modeled as price anticipators. It is well understood that pricetaking load behavior, which simply responds to a given price profile, can result in load synchronization, i.e., all users move their consumption to a low-price period, resulting in an inefficient market outcome. In contrast, price anticipatory loads influence the electricity price by playing "a game" with the wholesale market. The game is captured by a *bi-level model*. Specifically, the price-anticipating assumption implies that the aggregate effect of the prosumers is to change the market clearing prices and quantities, and, moreover, that the prosumers have a model of this effect. Given this, the prosumer aggregation bids follow an equilibrium strategy, with an accurate expectation about the response of the market; this is the standard reasoning behind a Cournot or Stackelberg game formulation corresponding to a bilevel optimization problem

   This equilibrium strategy can be thought of as being generated by the following iterative process: First, the market operator creates a price profile by clearing the market based on the predicted demand. Second, the prosumers (and other price-anticipatory participants) respond by shifting their consumption to cheaper time slots. This gives a new demand profile, and the market operator clears the market again, and the process repeats until convergence. Note that the proposed model does not specify an iterative mechanism, but rather, it encodes the optimality conditions for any price profile for the prosumers in Karush-Kuhn-Tucker (KKT) conditions, and in this way it captures the outcome resulting from the price-anticipatory prosumer behavior

   In contrast, we can model the loads as price-takers, inspired by the smart home concept [25] where the loads respond to the electricity price to minimize energy expenditure. With a large penetration of price-taking prosumers the marginal benefit might become negative when secondary peaks are created due to the load synchronization (as an illustration, see

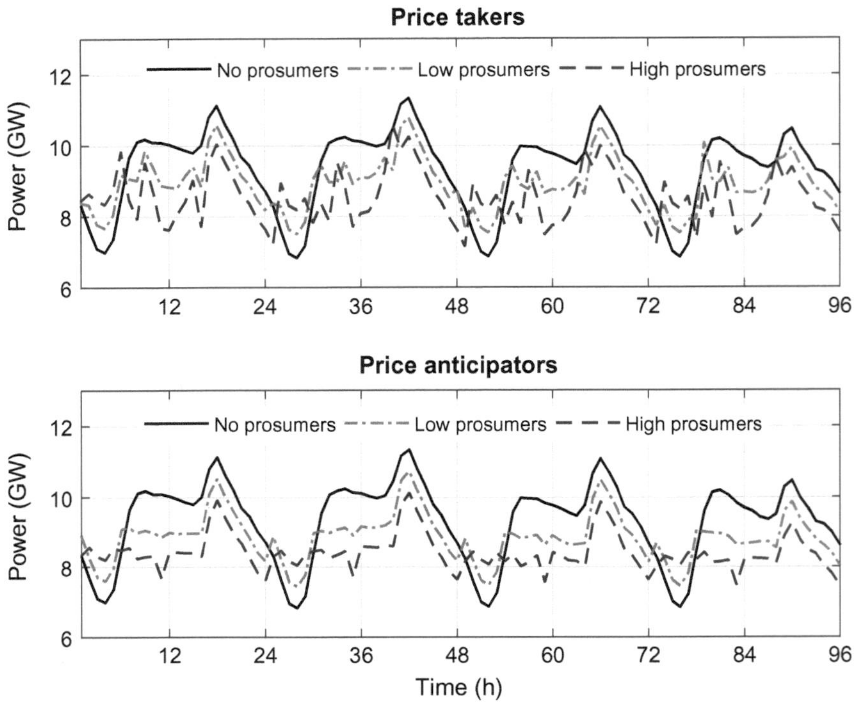

FIG. 5.24 Operational demand with different penetrations of price-taking (top) and price-anticipating prosumers (bottom).

Fig. 5.24 showing the operational demand with different penetrations of prosumers). Therefore, demand response aggregators (henceforth simply called aggregators) have started to emerge to fully exploit the demand-side flexibility. To that effect, the model implicitly assumes an efficient mechanism for demand response aggregation (see [17] for a discussion on practical implementation issues). However, specific implementation details, like price structure or the division of the profit earning by the aggregated collection of prosumers, are not of explicit interest in the proposed model.

2. The demand model representing an aggregator consists of a large population of prosumers connected to an unconstrained distribution network, who collectively maximize self-consumption made possible by an efficient internal trading and balancing mechanism, like peer-to-peer energy trading (Section 5.4.2) or large-scale demand response aggregation (Section 5.5).
3. Aggregators do not alter the underlying power consumption of the prosumers. That is, except for battery losses, the total power consumption before and after aggregation remains the same; however, the grid power intake profile does change, as a result of the prosumers using batteries to maximize self-consumption.

4. Prosumers have smart meters equipped with home energy management systems for scheduling of the PV-battery systems. Also, a communication infrastructure is assumed that allows a two-way communication between the grid, the aggregator and the prosumers, facilitating energy trading between prosumers in the aggregation.

These assumptions appear to be appropriate for scenarios arising in time frame of several decades into the future.

5.6.2 Bi-level Optimization Framework

In the model, we are specifically interested in the aggregated effect of a large prosumer population on the demand profile, assuming that the prosumers collectively maximize their self-consumption. Given that the objective of the wholesale market is to minimize the generation cost, the problem exhibits a bi-level structure. In game theory, such hierarchical optimization problems are known as Stackelberg games. They can be formulated as bi-level mathematical programs of the form [88]:

$$\begin{aligned} \underset{\mathbf{x},\mathbf{y}}{\text{minimize}} \quad & \Phi(\mathbf{x},\mathbf{y}) \\ \text{subject to} \quad & (\mathbf{x},\mathbf{y}) \in \mathcal{Z} \\ & \mathbf{y} \in \mathcal{S} = \arg\min_{\mathbf{y}} \{\Omega(\mathbf{x},\mathbf{y}) : \mathbf{y} \in \mathcal{C}(\mathbf{x})\} \end{aligned}$$

where $\mathbf{x} \in \mathbb{R}^n$, $\mathbf{y} \in \mathbb{R}^m$, are decisions vectors, and Φ $(\mathbf{x}, \mathbf{y})$: $\mathbb{R}^{n+m} \rightarrow \mathbb{R}$ and Ω $(\mathbf{x}, \mathbf{y})$: $\mathbb{R}^{n+m} \rightarrow \mathbb{R}$ are the objective functions of the upper and the lower-level problems, respectively. $\mathcal{Z}$ is the joint feasible region of the upper-level problem and C($\mathbf{x}$) the feasible region of the lower-level problem induced by $\mathbf{x}$. In the existing market models that adopt a hierarchical approach, the coupling variable $\mathbf{y}$ is the electricity price (e.g., [88, 89]). That is, the upper-level (the wholesale market in our case) determines the price schedule, while the lower-level (the aggregator acting on behalf of the prosumers), optimizes its consumption based on this price schedule.

Fig. 5.25 shows the structure of the proposed modeling framework. The demand model consists of two parts: (i) *inflexible* demand, $p_{\mathrm{d}}^{\mathrm{inf},m}$, with a fixed demand profile, representing large industrial loads and loads without flexible resources; and (ii) *flexible* demand, $p_{\mathrm{d}}^{\mathrm{flx},m}$, comprising a large population of prosumers who collectively maximize self-consumption. Note that not every bus in the system has a load connected to it, hence the distinction between an aggregator $m \in \mathcal{M}$ and bus $i \in \mathcal{B}$. Unlike in most existing studies, the interaction between the wholesale market and the aggregators in our model is through the demand profile of the aggregator, $p_{\mathrm{d}}^{\mathrm{flx},m}$. Note, also, that in contradistinction to the price-taking assumption when the electricity price is known in advance, now the collective action of the prosumers affects the wholesale market dispatch, which is the salient feature of the proposed model.

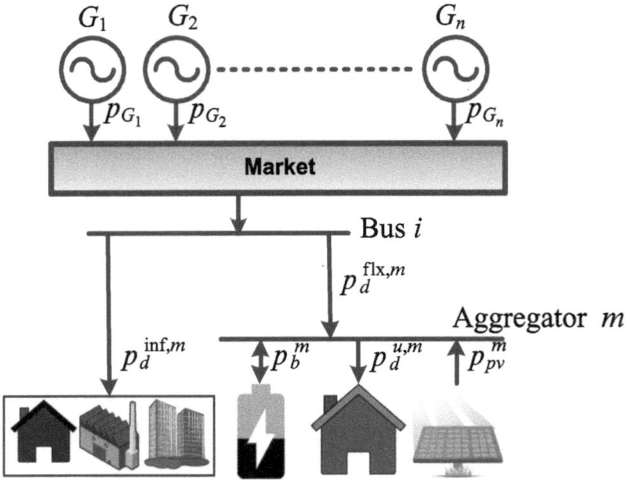

FIG. 5.25 Structure of the bi-level market model.

5.6.2.1 Upper-Level Problem (Wholesale Market)

To emulate the market outcome, the upper-level problem is cast as a unit commitment (UC) problem aiming to minimize the generation cost:

$$\underset{s,u,d,p,\theta}{\text{minimize}} \sum_{h\in\mathcal{H}} \sum_{g\in\mathcal{G}} \left(c_g^{\text{fix}} s_{g,h} + c_g^{\text{su}} u_{g,h} + c_g^{\text{sd}} d_{g,h} + c_g^{\text{var}} p_{g,h}\right), \tag{5.83}$$

where $s_{g,h}$, $u_{g,h}$, $d_{g,h} \in \{0, 1\}$, $p_{g,h} \in \mathbb{R}_+$, $\theta_{i,h} \in \mathbb{R}$ are the decision variables of the problem. The problem is subject to the following constraints:

$$\sum_{g\in\mathcal{G}_i} p_{g,h} = \sum_{m\in\mathcal{M}_i} \left(p_{d,h}^{\text{inf},m} + p_{d,h}^{\text{flx},m}\right) + \sum_{l\in\mathcal{L}_i} (p_{l,h} + \Delta p_{l,h}), \tag{5.84}$$

$$\left|B_{i,j}\left(\theta_{i,h} - \theta_{j,h}\right)\right| \leq \overline{p}_l, \tag{5.85}$$

$$\underline{p}_g s_{g,h} \leq p_{g,h} \leq \overline{p}_g s_{g,h}, \tag{5.86}$$

$$u_{g,h} - d_{g,h} = s_{g,h} - s_{g,h-1}, \tag{5.87}$$

$$\sum\nolimits_{g_{\text{synch}}\in\mathcal{R}} \left(\overline{p}_g s_{g,h} - p_{g,h}\right) \geq p_{r,h}^{\text{res}}, \tag{5.88}$$

$$u_{g,h} + \sum\nolimits_{\tilde{h}=0}^{\tau_g^{\text{u}}-1} d_{g,h+\tilde{h}} \leq 1, \tag{5.89}$$

$$d_{g,h} + \sum\nolimits_{\tilde{h}=0}^{\tau_g^{\text{d}}-1} u_{g,h+\tilde{h}} \leq 1, \tag{5.90}$$

$$-r_g^- \leq p_{g,h} - p_{g,h-1} \leq r_g^+, \tag{5.91}$$

$$\arg\min_{p_d^{\text{flx}}, p_b, e_b} \left\{ \sum_{h \in \mathcal{H}} p_{d,h}^{\text{flx},m} \text{subject to} (1.94) - (1.97) \right\}, \tag{5.92}$$

where (5.84) is the power balance equation at each bus i in the system,[13] with $\mathcal{G}_i$, $\mathcal{M}_i$, $\mathcal{L}_i$ representing respectively the sets of generators, aggregators and lines connected to bus i, and $p_{d,h}^{\text{inf},m}$, $p_{d,h}^{\text{flx},m}$, $p_{l,h}^{i,j}$, and $\Delta p_{l,h}^{i,j}$ representing, respectively, the inflexible and flexible demand of aggregator m, line power and line power loss (assumed to be 10% of the line flow) on each line connected to bus i; (5.85) represents line power limits; (5.86) limits the dispatch level of a generating unit between its respective minimum and maximum limits; (5.87) links the status of a generator unit to the up and down binary decision variables; (5.88) ensures sufficient spinning reserves are available in reach region of the grid; (5.89) and (5.90) ensure minimum up and minimum down times of the generators; (5.91) are the generator ramping constraints; and (5.92) is the constraint resulting from the prosumer aggregation optimization problem, as explained in the next section.

5.6.2.2 Lower-Level Problem (Aggregators)

The prosumer aggregation is formulated in the lower-level problem. The loads within an aggregator's domain are assumed homogeneous, which allows us to represent the total aggregator's demand with a single load model. The flexibility provided by the battery is only used to maximize self-consumption, with grid supply readily available. This implies that the end-users' power consumption pattern is left unaltered, so that their comfort is not jeopardized.

The coupling between the upper-level (wholesale) and the lower-level (retail) problem in the proposed model is through the power demand. This removes the need to define the market in terms of a pricing mechanism or rule, and it follows that the electricity price is not explicitly shown in the optimization problem. This is why it is called a *generic model.*

This approach stands in contrast to other existing bi-level formulations [88, 89], in which the loads and the wholesale market are coupled through the electricity price. The prices generated by any market depend inherently on the specific pricing mechanism adopted. However, in practice, several different pricing rules can implement a desired outcome, including the widely used uniform price reverse auction or nodal pricing mechanisms. For example, the electricity price could comprise the dual variables associated with the power balance constraint (5.84) and power flow constraints (5.85) of the upper-level problem, plus retail/aggregator and network charges. Moreover, a range of different pricing rules result in different retail/aggregator and network charges, with all supporting an efficient outcome. However, by instead coupling the upper- and lower-level

13. Note that the flexible demand of each aggregator m, $p_{\text{d},h}^{\text{flx},m}$ couples the upper-level (wholesale market) problem with each of the m lower-level (aggregator) problems.

problems via power demand, our proposed model avoids the need to specify a particular pricing rule, which makes it market-structure agnostic. Nonetheless, it is fair to assume that in any practical system, users will have access to a price forecast[14] and other bidders' historical behavior—but these are all market design-specific. In the absence of pricing rule details, the proposed optimization model does not require such information.

Specifically, the lower-level problem is formulated as follows for each demand aggregator $m \in \mathcal{M}$:

$$\underset{p_d^{\text{flx}}, p_b, e_b}{\text{minimize}} \sum_{h \in \mathcal{H}} p_{d,h}^{\text{flx},m}, \tag{5.93}$$

where the decision variables of the problem are flexible demand $p_d^{\text{flx},m}$, battery power p_b, and battery capacity e_b. The problem is subject to the following constraints:

$$p_{d,h}^{\text{flx},m} = p_{d,h}^{\text{u},m} - p_{pv,h}^{m} + p_{b,h}^{m}, \tag{5.94}$$

$$e_{b,h}^{m} = \eta_b^m e_{b,h-1}^{m} + p_{b,h}^{m}, \tag{5.95}$$

$$\underline{p}_b^m \leq p_{b,h}^m \leq \overline{p}_b^m, \tag{5.96}$$

$$\underline{e}_{\text{b}}^m \leq e_{\text{b},h}^m \leq \overline{e}_{\text{b}}^m, \tag{5.97}$$

where (5.94) is the power balance equation; and (5.95)–(5.97) are the battery storage constraints. Power $p_{d,h}^{u,m}$ is the underlying power demand of the prosumers. Note that except for battery losses, the underlying power demand does not change, however the power intake from the grid can. Finally, the KKT optimality conditions of the lower-level problem are added as the constraints to the upper-level problem, which reduces the problem to a single mixed integer linear program that can be solved using of-the-shelf solvers. Note that because the two levels interacts through a power, not through a price, unlike in [88], no linearization is required.

5.6.3 Case Study

As an illustration, we consider a future grid scenario for the Australian National Electricity Market (NEM) with a 40% penetration of variable renewable generation (wind and solar PV). Inspired by two recent Australian 100% renewables studies [90, 91], part of coal generation is replaced with wind and utility PV, which results in 28.94 GW coal, 5.22 GW gas, 2.33 GW hydro, 21 GW wind, and 12 GW utility PV. Given the deterministic nature of the model, we assume 10% reserves for each region in the system to cater for

14. Note that in a HEM problem [25], a HEM system is an agent acting on behalf of a prosumer, and the electricity price is known ahead of time, resulting in a price-taking behavior.

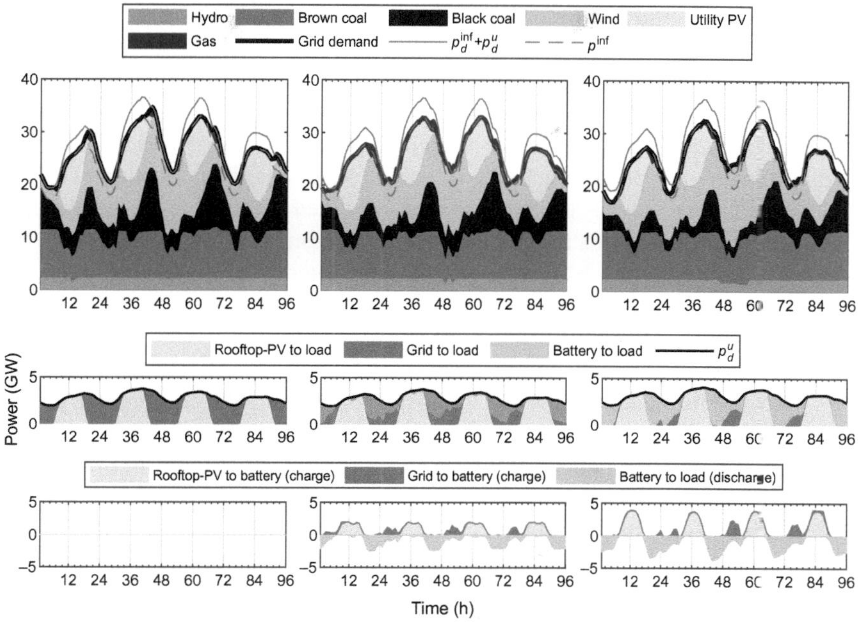

FIG. 5.26 Dispatch results for a typical summer week with high demand for a medium prosumer penetration with different amounts of storage: zero (left), 2h (middle), and 4h (right).

demand and errors in forecasts of renewable energy sources. In market simulations, generators are assumed to bid according to their short-run marginal costs, while renewable energy sources bid at zero cost. Simulations are performed using a rolling horizon approach with hourly resolution assuming a perfect foresight. The optimization horizon is 3 days with a 2-day overlap.

We assume four different prosumer penetrations: zero, low, medium, and high. With no prosumer penetration, the demand is assumed inflexible. For the other three scenarios, we assume that part of the demand is equipped with small-scale (residential and small commercial) PV-battery systems. The PV capacities are respectively 5GW, 10GW, and 20GW for the low, medium, and high uptake of prosumers (the existing penetration in the NEM is 3GW). We consider three different amounts of storage: zero, 2 kWh, and 4 kWh of storage for 1kW of rooftop PV.[15]

Dispatch results for a typical summer week with high demand for a medium prosumer penetration with, respectively, zero, 2h, and 4h hours of storage are shown in Fig. 5.26. The figure shows, respectively, generation dispatch results (top row), combined flexible demand of all aggregators

15. A typical ratio in the NEM today is 2 h of storage [7], however, in the future, this will likely increase due to the anticipated cost reduction.

(middle row), and a combined battery charging profile of all aggregators (bottom row). Observe that in all six cases peak demand occurs at mid-day due to a high air-conditioning load. After sunset, however, the demand is still high, so gas generation is needed to cover the gap. In the available generation mix, gas has the highest shortrun marginal cost, which increases the electricity price in late afternoon/early evening. The balancing results over the simulated year have revealed that the increased renewable penetration in the renewable scenarios requires more energy from gas generation compared to business as usual. This is due to the intermittency of renewables, and the ramp limits of conventional coal-fired generation. An increased penetration of prosumers with higher amounts of storage, however, reduces the usage of gas due to a flatter demand profile.

Observe in Fig. 5.26 how an increasing amount of storage increases prosumers' self-sufficiency. Without storage, the load is supplied by PV during the day, and the rest is supplied from the grid. When storage is added to the system, batteries are charged when electricity is cheap (mostly from rooftop PV during the day and from wind during the night) and discharged in late afternoon to offset the demand when the electricity is most expensive. Note that the plots in the bottom two rows show a combined load profile of all aggregators in the system, which explains why storage is seemingly charged and discharged simultaneously. Observe also how high amounts of storage (rightmost column in Figs. 5.26) flatten the demand profile. During the day, the flexible demand is supplied by rooftop PV, which *reduces* the grid demand, while during the night, with sufficient wind generation, batteries are charged, which *increases* the grid demand. A flatter demand profile also has a significant beneficial effect on loadability and static voltage stability.

5.7 Conclusion

A combination of government support and falling technology cost has resulted in a rapid uptake of behind-the meter distributed energy resources. For a while, the networks were free-ride beneficiaries of that wave, mainly because of reduced peak demands. That era, however, is all but over. In many countries (e.g., Germany and Australia) the uptake has reached a point where high and uncoordinated penetration of behind-the meter distributed energy resources has started to cause network issues, so alternative solutions for the operation of LV networks need to be found.

In this chapter, we have shown that not all is doom and gloom. On the contrary, properly coordinated prosumer-owned DERs can offer system services while minimizing the cost for the owners. However, this needs to be done in a way that preserves prosumer's prerogative and privacy. To that end, we argue that the problem needs to be tackled bottom up, starting with the home energy management problem as the fundamental building block. Next, the aggregation should be done at the *household level* not at the device level, as done in the existing approach. A key enabler in this household level approach is battery storage, which is very close to becoming economically viable.

We have shown that a possible way of solving the large-scale demand response aggregation problem that explicitly considers network technical constraints is to formulate it as a distributed optimal power flow where each prosumers solves its own energy management problem. Conceptually, the solution procedure can be thought of as bargaining between prosumers and a third-party entity responsible for aggregation. In this chapter, we have shown what is technically possible, but have left open the question who this third-party entity should be and how it can be integrated into the existing market framework, which is subject for future work.

References

[1] International Renewable Energy Agency (IRENA), The power to change: solar and wind cost reduction potential to 2025, Technical Report, 2016.

[2] International Energy Agency (IEA), Digitalization & energy, Technical Report, 2017.

[3] World Energy Council (WEC), Changing dynamics—using distributed energy resources to meet the trilemma challenge, The World Energy Trilemma 2017 Report, 2017.

[4] CSIRO and Energy Networks Australia, Electricity network transformation roadmap, Final Report, 2017.

[5] International Energy Agency (IEA) Trends in photovoltaic applications 2017, Technical Report, 2017.

[6] Electric Power Research Institute (EPRI), The integrated grid: Realizing the full value of central and distributed energy resources, Technical Report, 2014.

[7] Australian Energy Market Operator (AEMO) Emerging technologies information paper, Technical Report, 2015.

[8] Morgan Stanley Research, Australia utilities asia insight: solar & batteries, Alphawise report, 2016.

[9] Commonwealth Scientific and Industrial Research Organisation (CSIRO), Change and choice—the future grid forum's analysis of Australia's potential electricity pathways to 2050, Technical Report, 2013.

[10] Australian Energy Market Commission (AEMC), Distribution market model, Technical Report, (2016).

[11] U.S. Department of Energy, Benefits of demand response in electricity markets and recommendations for achieving them, A Report to the United States Congress Pursuant to Section 1252 of the Energy Policy Act of 2005, 2006.

[12] Federal Energy Regulatory Commission (FERC), Assessment of demand response and advanced metering, Staff Report, 2006.

[13] M.A.A. Pedrasa, T.D. Spooner, I.F. MacGill, Coordinated scheduling of residential distributed energy resources to optimize smart home energy services, IEEE Trans. Smart Grid 1 (2) (2010) 134–143.

[14] M.C. Bozchalui, S.A. Hashmi, H. Hassen, C.A. Canizares, K. Bhattacharya, Optimal operation of residential energy hubs in smart grids, IEEE Trans. Smart Grid 3 (4) (2012) 1755–1766.

[15] Z. Ma, D.S. Callaway, I.A. Hiskens, Decentralized charging control of large populations of plug-in electric vehicles, IEEE Trans. Control Syst. Technol. (1) (2013) 67–78.

[16] A. C. Chapman, G. Verbič, and D. J. Hill, A healthy dose of reality for game-theoretic approaches to residential demand response, in 2013 IREP Symposium Bulk Power System Dynamics and Control—IX Optimization, Security and Control of the Emerging Power Grid. IEEE, 2013.

[17] A.C. Chapman, G. Verbič, D.J. Hill, Algorithmic and strategic aspects to integrating demand-side aggregation and energy management methods, IEEE Trans. Smart Grid 7 (6) (2016) 2748–2760.

[18] M. Fahrioglu, F. Alvarado, Using utility information to calibrate customer demand management behavior models, IEEE Trans. Power Syst. 16 (2) (2001) 317–322.

[19] N. C. Truong, J. Mcinerney, L. Tran-Thanh, E. Costanza, and S. D. Ramchurn, Forecasting multi-appliance usage for smart home energy management, in Proceedings IJCAI, 2013.

[20] E. Dall'Anese, S.V. Dhople, G.B. Giannakis, Optimal dispatch of photovoltaic inverters in residential distribution systems, IEEE Trans. Sustain. Energy 5 (2) (2014) 487–497.

[21] X. Su, M.A.S. Masoum, P.J. Wolfs, Optimal PV inverter reactive Power control and real Power curtailment to improve performance of unbalanced four-wire LV distribution networks, IEEE Trans. Sustain. Energy 5 (3) (2014) 967–977.

[22] A.H. Mohsenian-Rad, V.W.S. Wong, J. Jatskevich, R. Schober, A. Leon-Garcia, Autonomous demand-side management based on game-theoretic energy consumption scheduling for the future smart grid, IEEE Trans. Smart Grid 1 (3) (2010) 320–331.

[23] S. Mhanna, G. Verbič, A.C. Chapman, A faithful distributed mechanism for demand response aggregation, IEEE Trans. Smart Grid 7 (3) (2016) 1743–1753.

[24] D.S. Callaway, I.A. Hiskens, Achieving controllability of electric loads, Proc. IEEE 99 (1) (2011) 184–199.

[25] H. Tischer and G. Verbič, Towards a smart home energy management system—A dynamic programming approach, in 2011 IEEE Innovative Smart Grid Technologies—Asia (ISGT-Asia), Perth, Australia, 2011.

[26] C. Keerthisinghe, G. Verbič, A.C. Chapman, A fast technique for smart home management: ADP with temporal difference learning, IEEE Trans. Smart Grid 9 (4) (2018) 3291–3303.

[27] C. Keerthisinghe, G. Verbič, A.C. Chapman, Energy management of PV-storage systems: Policy approximations using machine learning, IEEE Trans. Ind. Informat. 15 (2018) 257–265.

[28] W.B. Powell, Approximate Dynamic Programming—Solving the Curses of Dimensionality, John Wiley, 2007.

[29] D. F. Salas and W. B. Powell, Benchmarking a scalable approximate dynamic programming algorithm for stochastic control of multidimensional energy storage problems, Technical report, Working Paper, Department of Operations Research and Financial Engineering, Princeton, NJ, 2013.

[30] J. Guerrero Orbe, A. C. Chapman, and G. Verbič Decentralized P2P energy trading under network constraints in a low-voltage network, 2018, arXiv:1809.06976.

[31] S. Hashemi, J. Østergaard, Methods and strategies for overvoltage prevention in low voltage distribution systems with PV, IET Renew. Power Generation 11 (2) (2017) 205–214.

[32] K.E. Antoniadou-Plytaria, I.N. Kouveliotis-Lysikatos, P.S. Georgilakis, N.D. Hatziargyriou, Distributed and decentralized voltage control of smart distribution networks: models, methods, and future research, IEEE Trans. Smart Grid 8 (6) (2017) 2999–3008.

[33] R. Tonkoski, L.A.C. Lopes, T.H.M. El-Fouly, Coordinated active power curtailment of grid connected PV inverters for overvoltage prevention, IEEE Trans. Sustainable Energy 2 (2) (2011) 139–147.

[34] Y. Ma, G. Verbič, and A.C. Chapman Probabilistic assessment of PV-battery system impacts on LV distribution networks, 2018, arXiv:1809.07488.

[35] T. Power, G. Verbič, and A. C. Chapman, A nonparametric bayesian model for synthesising residential solar generation and demand, 2018, arXiv:1808.00615.

[36] D. Azuatalam, G. Verbič, and A. C. Chapman, Impacts of network tariffs on distribution network power flows, in Australasian Universities Power Engineering Conference (AUPEC). Melbourne, Australia: IEEE, 2017.

[37] D. Azuatalam, A. C. Chapman, and G. Verbič. Probabilistic impact assessment of network tariffs in low voltage distribution networks, 2018, arXiv:1810.02013.

[38] Y. Ma, M. S. S. Abad, D. Azuatalam, G. Verbič, and A. C. Chapman, "Impacts of community and distributed energy storage systems on unbalanced low voltage networks," in Australasian Universities Power Engineering Conference (AUPEC). Melbourne, Australia: IEEE, 2017.

[39] S. Mhanna, A.C. Chapman, G. Verbič, Component-based dual decomposition methods for the OPF problem, Sustainable Energy, Grids Networks (2017)arXiv:1704.03647.

[40] P. Scott and S. Thiébaux, "Distributed multi-period optimal Power flow for demand response in microgrids," ACM e-Energy'15—Proceedings of the 2015 ACM Sixth International Conference on Future Energy Systems Pages 17–26, 2015.

[41] E. Franklin, D. Gordon, D. Jones, P. Scott, L. Blackhall, and S. Thiébaux, "Peak demand management on distribution networks using coordinated behind-the-meter pv/battery systems: The Bruny Island battery trial," in Asia-Pacific Solar Research Conference 2016, 2016.

[42] AEMO and ENA, "Open energy networks, consultation paper," Technical Report, 2018.

[43] J. Guerrero, A. C. Chapman, and G. Verbič, "A study of energy trading in a low-voltage network: centralised and distributed approaches," in Australasian Universities Power Engineering Conference (AUPEC). Melbourne, Australia: IEEE, 2017.

[44] S. Chu, A. Majumdar, Opportunities and challenges for a sustainable energy future, Nature 488 (7411) (2012) 294–303.

[45] J. Eyer, G. Corey, Energy storage for the electricity grid: benefits and market potential assessment guide, Sandia Natl. Lab. 20 (10) (2010) 5.

[46] D. Condon, D. McPhail, and D. Ingram, "Customer and grid impacts of solar, storage and cost reflective tariffs," in Australasian Universities Power Engineering Conference (AUPEC). Brisbane, Australia: IEEE, 2016.

[47] H. Pandžić, Y. Wang, T. Qiu, Y. Dvorkin, D.S. Kirschen, Near-optimal method for siting and sizing of distributed storage in a transmission network, IEEE Trans. Power Syst. 30 (5) (2015) 2288–2300.

[48] M. Nick, R. Cherkaoui, M. Paolone, Optimal allocation of dispersed energy storage systems in active distribution networks for energy balance and grid support, IEEE Trans. Power Syst. 29 (5) (2014) 2300–2310.

[49] S.H. Low, Convex relaxation of optimal power flow—Part I: Formulations and equivalence, IEEE Trans. Control Network Syst. 1 (1) (2014) 15–27.

[50] V. Poulios, E. Vrettos, F. Kienzle, E. Kaffe, H. Luternauer, G. Andersson, Optimal placement and sizing of battery storage to increase the PV hosting capacity of low voltage grids, in: International ETG Congress 2015; Die Energiewende-Blueprints for the new energy age; Proceedings of VDE, 2015, pp. 1–8.

[51] P. Fortenbacher, M. Zellner, and G. Andersson, Optimal sizing and placement of distributed storage in low voltage networks. in Power Systems Computation Conference (PSCC), 2016 IEEE, 2016, pp. 1–7.

[52] M. Farivar, R. Neal, C. Clarke, and S. Low, "Optimal inverter var control in distribution systems with high PV penetration," in Power and Energy Society General Meeting. IEEE, 2012.

[53] M. Baran, F.F. Wu, Optimal sizing of capacitors placed on a radial distribution system, IEEE Trans. Power Delivery 4 (1) (1989) 735–743.

[54] H.-G. Yeh, D.F. Gayme, S.H. Low, Adaptive VAR control for distribution circuits with photovoltaic generators, IEEE Trans. Power Syst. 27 (3) (2012) 1656–1663.
[55] Y. Xu, Z.Y. Dong, R. Zhang, D.J. Hill, Multi-timescale coordinated voltage/var control of high renewable-penetrated distribution systems, IEEE Trans. Power Syst. (2017)
[56] Electricity North West Limited, "Low voltage network solutions closedown report," 2014.
[57] I. Richardson, M. Thomson, D. Infield, C. Clifford, Domestic electricity use: a highresolution energy demand model, Energ. Buildings 42 (10) (2010) 1878–1887.
[58] F. Moret, P. Pinson, Energy collectives: a community and fairness based approach to future electricity markets. IEEE Trans. Power Syst. (2018)https://doi.org/10.1109/TPWRS.2018.2808961.
[59] G. Zizzo, E.R. Sanseverino, M.G. Ippolito, M.L.D. Silvestre, P. Gallo, A technical approach to P2P energy transactions in microgrids, IEEE Trans. Ind. Informat. (2018) 1.
[60] A. Goranović, M. Meisel, L. Fotiadis, S. Wilker, A. Treytl, and T. Sauter, "Blockchain applications in microgrids: An overview of current projects and concepts," in IECON 2017—43rd Conference of IEEE Industrial Electronics Society, 2017, pp. 6153–6158.
[61] J. Kang, R. Yu, X. Huang, S. Maharjan, Y. Zhang, E. Hossain, Enabling localized peer-to- peer electricity trading among plug-in hybrid electric vehicles using consortium blockchains, IEEE Trans. Ind. Informat. 13 (6) (2017) 3154–3164.
[62] E. Münsing, J. Mather, and S. Moura, "Blockchains for decentralized optimization of energy resources in microgrid networks," in 2017 IEEE Conference on Control Technology and Applications (CCTA), 2017, pp. 2164–2171.
[63] A. Navarro-Espinosa, L.F. Ochoa, Probabilistic impact assessment of low carbon technologies in LV distribution systems, IEEE Trans. Power Syst. 31 (3) (2016) 2192–2203.
[64] D.K. Gode, S. Sunder, Allocative efficiency of markets with zero-intelligence traders: Market as a partial substitute for individual rationality, J. Polit. Econ. 101 (1) (1993) 119–137.
[65] D. Cliff, J. Bruten, Minimal-intelligence agents for bargaining behaviors in market-based environments, Technical Report HPL-97-91, 1997.
[66] E. McKenna, M. Thomson, High-resolution stochastic integrated thermal electrical domestic demand model, Appl. Energy 165 (2016) 445–461.
[67] A. Papavasiliou, Analysis of distribution locational marginal prices, IEEE Trans. Smart Grid 9 (5) (2018) 4872–4882.
[68] S. Mhanna, A.C. Chapman, G. Verbič, A fast distributed algorithm for large-scale demand response aggregation, IEEE Trans. Smart Grid 7 (4) (2016) 2094–2107.
[69] S. Mhanna, A.C. Chapman, G. Verbič, A faithful and tractable distributed mechanism for residential electricity pricing, IEEE Trans. Power Syst. 33 (4) (2017) 4238–4252.
[70] A. C. Chapman, S. Mhanna, and G. Verbič, "Cooperative game theory for non-linear pricing of load-side distribution network support," in 2017 IREP Symposium Bulk Power System Dynamics and Control –X (IREP), 2017.
[71] Y. Nesterov, Smooth minimization of non-smooth functions, Math. Program. 103 (1) (2005) 127–152.
[72] O. Devolder, F. Glineur, Y. Nesterov, Double smoothing technique for large-scale linearly constrained convex optimization, SIAM J. Optim. 22 (2) (2012) 702–727.
[73] N. Gatsis, G.B. Giannakis, Decomposition algorithms for market clearing with large-scale demand response, IEEE Trans. Smart Grid 4 (4) (2013) 1976–1987.
[74] S.-J. Kim, G. Giannakis, Scalable and robust demand response with mixed-integer constraints, IEEE Trans. Smart Grid 4 (4) (2013) 2089–2099.
[75] D. Bertsekas, Convex Optimization Algorithms, Athena Scientific, 2015.

[76] Y. Zhang, N. Gatsis, and G. B. Giannakis, "Disaggregated bundle methods for distributed market clearing in power networks," in Proceedings of 2013 IEEE Global Conference on Signal and Information Processing—GlobalSIP 2013. IEEE, 2013, pp. 835–838.
[77] A.G. Expósito, E.R. Ramos, Reliable load flow technique for radial distribution networks, IEEE Trans. Power Syst. 14 (3) (1999) 1063–1069.
[78] D.K. Molzahn, F. Dorfler, H. Sandberg, S.H. Low, S. Chakrabarti, R. Baldick, J. Lavaei, A survey of distributed optimization and control algorithms for electric Power systems, IEEE Trans. Smart Grid 3053 (C) (2017) 1.
[79] S. Boyd, N. Parikh, E. Chu, B. Peleato, J. Eckstein, Distributed optimization and statistical learning via the alternating direction method of multipliers, Foundations Trends Mach. Learn. 3 (1) (2011) 1–122.
[80] P. Scott and S. Thiébaux, "Distributed multi-period optimal Power flow for demand response in microgrids," in ACM e-Energy, 2015.
[81] S. Mhanna, G. Verbič, and A. C. Chapman, "Accelerated methods for the SOCP-relaxed component-based distributed optimal Power flow," in 20th Power Systems Computation Conference, PSCC 2018. IEEE, 2018, pp. 1–7.
[82] Z. Xu, G. J. Taylor, H. Li, M. A. T. Figueiredo, X. Yuan, and T. Goldstein, "Adaptive consensus ADMM for distributed optimization," in International Conference on Machine Learning (ICML), 2017.
[83] Y. Shoham, K. Leyton-Brown, Multiagent Systems: Algorithmic, Game-Theoretic, and Logical Foundations, Cambridge University Press, New York, 2008.
[84] A. Petcu, B. Faltings, D.C. Parkes, M-DPOP: faithful distributed implementation of efficient social choice problems, Auton. Agent. Multiagent Syst. (2006) 1397–1404.
[85] M. Muratori, G. Rizzoni, Residential demand response: Dynamic energy management and time-varying electricity pricing, IEEE Trans. Power Syst. 31 (2) (2016) 1108–1117.
[86] E. Dütschke, A.-G. Paetz, Dynamic electricity pricing—Which programs do consumers prefer, Energy Policy 59 (2013) 226–234.
[87] S. Riaz, H. Marzooghi, G. Verbič, A. Chapman, D. Hill, Generic demand model considering the impact of prosumers for future grid scenario analysis, IEEE Trans. Smart Grid 10 (1) (2019) 819–829.
[88] M. Zugno, J. Miguel Morales, P. Pinson, H. Madsen, A bilevel model for electricity retailers' participation in a demand response market environment, Energy Econ. 36 (2013) 182–197.
[89] M. Gonzalez Vaya, G. Andersson, Optimal bidding strategy of a plug-in electric vehicle aggregator in day-ahead electricity markets under uncertainty, IEEE Trans. Power Syst. 30 (5) (Sep 2015) 2375–2385.
[90] M. Wright and P. Hearps, "Zero carbon Australia stationary energy plan," The University of Melbourne Energy Research Institute, Technical Report, 2010.
[91] Australian Energy Market Operator (AEMO), "100 per cent renewable study—modelling outcomes," Technical Report, 2013.

Chapter 6

Evolving New Market Structures

Amin Shokri Gazafroudi*, Miadreza Shafie-khah[†], Francisco Prieto-Castrillo*[,‡,§], Saber Talari[†], Juan Manuel Corchado*[,¶] and João P.S. Catalão[†,‖,#]

[*]BISITE Research Group, University of Salamanca, Salamanca, Spain, [†]C-MAST, University of Beira Interior, Portugal, [‡]Media Laboratory, Massachusetts Institute of Technology, Cambridge, MA, United States, [§]Harvard T.H. Chan School of Public Health, Harvard University, Boston, MA, United States, [¶]Osaka Institute of Technology, Osaka, Japan, [‖]INESC TEC and the Faculty of Engineering of the University of Porto, Porto, Portugal, [#]INESC-ID, Instituto Superior Técnico, University of Lisbon, Lisbon, Portugal

Chapter Outline

6.1 Introduction

6.1.1 Motivation

Conventional electricity markets usually consist of day-ahead and real-time markets. The day-ahead market is required in order to allocate generating units that have slow ramp-rate and need advance planning. The clearing of the real-time market allows the energy to maintain a balance between the supply and the demand during the decision-making period. In addition, the real-time market is needed because of quick output and stochastic producers, for example, wind farms.

Pathways to a Smarter Power System. https://doi.org/10.1016/B978-0-08-102592-5.00006-5

In conventionally structured electricity markets, day-ahead and balancing markets are cleared sequentially and independently. However, the participation of stochastic generation of non-dispatchable renewable energy resource injects power generation uncertainty, causing an electricity market problem. In this way, new services called ancillary services, for example, operating reserves, are required to achieve balancing in the real-time markets. Hence, the simultaneous clearing of joint energy and reserve day-ahead and real-time markets makes this process more efficient.

In addition, central energy management systems are not good strategies for resolving issues related to distributed energy resources' real and fair price in medium- and low-voltage distribution network locations. Finally, centralized energy management strategies in electricity markets are transferred to decentralized approaches.

6.1.2 Literature Review and Background

Electricity markets have experienced many changes over the past 30 years; their evolution has been aimed at increasing the efficiency of the power system [1]. These changes have formed the foundation of the restructured electricity market in terms of design and architecture. However, rapid technological development in the area of renewable energy generation caused these resources to become cost-competitive in comparison to conventional energy. This was due to lower variable cost in the electricity markets, in addition to providing clean and eco-friendly power [2, 3]. On this basis, in many power systems, an essential evolution has been formed. It should be noted that the high penetration of renewable energy can have a negative effect on the operation and planning of power systems [1]. In addition, the implementation of several environmental policies combined with renewable energies have contributed to considerable changes [4]. The share of renewable resources, for example, wind and solar energy, has been advancing, while the thermal units have been losing their contributions in power systems [5, 6]. The thermal units can be replaced with these resources, which leads to a decrease in short-term market prices.

Despite the benefits of renewable energy, their high penetration can jeopardize the secure operation of the power system because of their intermittent output and uncertain nature [7, 8]. The effect of different renewable support mechanisms on the performance of the power market was investigated in Ref. [9]. Similarly, Ref. [10] proposed a green power system and designed an electricity market that would support renewable energies. In Ref. [11], the integration of large-scale renewable resources in the electricity market was analyzed.

Moreover, as a result of the changes in the power system, it is necessary to make changes in the electricity markets as well. Some proposals in the state of the art have already made an effort to redesign the market. J.L. Sawin et al. [12] studied the changes in the electricity markets due to the increase in renewable energies. In Ref. [13], the capacity market was modified to make the generation of renewable energy dependent on weather conditions. In Refs. [14] and [15], the Flexiramp market was introduced to decrease the negative impact of solar generation on California ISO.

To overcome the insecurity of renewable-based power systems, a larger number of backup units are needed, as well as some flexible units to supply ancillary services, for example, regulation and reserve markets. These flexible units can cover the uncertainty of renewable resources and ensure the balance between supply–demand in real-time [16, 17]. It should be noted that by increasing the share of renewable energy resources, the demand for the described regulation and reserve services increases [18].

The main flexibility services are currently provided by the thermal units, and there will be no major changes to this situation in the future. However, a large part of the profit obtained by thermal units comes from participation in the energy markets, not from the ancillary services. Consequently, the thermal units prefer to supply energy, not to deliver regulation or reserve services. Since the requirement for energy is much greater than the ancillary service, the profit of thermal units resulting from a regulation/reserve service is approximately 1% of the total profit [19].

However, a higher penetration of renewable resources can cause a drastic drop in energy prices, which will result in less motivation to invest in backup plants. This would be similar to the current situation in the Danish electricity market [20], where the energy prices may have zero or negative values. Therefore, conventional thermal units should participate more in the regulation and reserve markets in order to gain more incomes. This will allow them to survive in the renewable-based electricity markets and compete with renewable resources, which are supported by a variety of policies [21].

In Ref. [22], the critical steps to create a flexible power system preserving the required stability and reliability in the presence of renewable energy resources are presented. To this end, the policies, market schemes, and technical necessities are reviewed, leading to a power system with only renewable resources. In Ref. [23], the tasks of conventional units are presented to enable the units to manage the inconsistency of renewable resources, such as wind ramping, to identify the importance of each unit characteristics to operate the system reliably. On this basis, the market policies to employ the current generation units are presented to decrease the negative impacts of renewable resources.

The impacts of uncertainty on the investment in renewable resources that are supported by governments are studied [24] by assessing the effect of various support schemes and flexibilities. On this basis, a framework of the investment decisions is modeled and the investors' decisions are investigated by employing the net present value. In Ref. [25], a power system model is developed to determine the optimal generation mix, taking the operational constraints into account. It is shown that operational constraints of generation units, particularly the constraints associated with the unpredictable nature of renewable energies, significantly affect the energy mix. Relating to the Greek power system, in [26], a multi-period approach is modeled for short-term operational scheduling considering high penetration of renewables. In Ref. [27], the unit commitment program is extended to longer horizons such as a few days, by considering a rolling horizon concept to keep the solution for only the next day. In Ref. [28], a medium-term market-based framework is proposed that finds the energy mix and the market prices for the Greek electricity market.

Some articles in the literature consider the evolution of market design. In Ref. [29], which was published in 2009, the electricity market was reviewed in terms of architecture and design. In that year, the main market design issues related to electricity price forecast, bilateral contracts, and auction designs. Therefore, Ref. [29] did not study the effects of the upcoming power system on electricity markets. In Refs. [30] and [31], a market-splitting framework was proposed for future integration in day-ahead markets in Europe. In Ref. [32], a model was proposed for the electricity markets' clearing process, and had high computational efficiency. This model enabled the system operator to consider the supply orders and ramping limits. In Ref. [33], the efficiency of the balancing market in Germany was studied in terms of electricity market design. The authors of Ref. [34] investigated the problem of market design from a conventional thermal power plant's point of view. The work examined various market designs in order to achieve the optimal participation and success of such power plants.

6.1.3 Contributions

In this chapter, sequential and simultaneous approaches are used to solve and analyze the stochastic market-clearing problem of joint energy and reserve. Then, the influences of electrical consumers' flexibility behaviors on our proposed market-clearing model is evaluated. The rest of the chapter is organized as follows. In Section 6.2, the proposed restructured electricity market model is presented. Simulation results are described in details in Section 6.3. Section 6.4 concludes this chapter.

6.2 Restructured Electricity Market Model

6.2.1 Nomenclature

A. Indices and Numbers

n	Index of system buses, from 1 to N_B.
i	Index of conventional generating units, from 1 to N_G.
j	Index of loads, from 1 to N_L.
t	Index of time periods, from 1 to N_T.
m	Index of energy blocks offered by conventional generating units, from 1 to N_{Oit}.
ω	Index of wind power, electrical load and power grid scenarios, from 1 to Ω.

A. Indices and Numbers

n	**Index of system buses, from 1 to N_B.**

B. Continuous Variables

C_{it}^{SU}	Scheduled start-up cost ($).
$P^S{}_{it}$	Power output of units in the day-ahead market (MW).
$p_{itm}{}^G$	Power output from the *m*-th block of energy offered by unit in day-aheac market (MW).
$L^S{}_{jt}$	Power consumed of load in day-ahead market (MW).
$P_t^{S,\ WP}$	Wind power in day-ahead market (MW).
$C^A{}_{it\omega}$	Start-up cost due to change in commitment status of units in day-ahead market and balancing market ($).
$P^G{}_{it\omega}$	Power output of unit in balancing market (MW).
$L^C{}_{jt\omega}$	Electrical consumed in balancing market (MW).
$r^U{}_{it\omega}$	Up-spinning reserve in balancing market (MW).
$r^D{}_{it\omega}$	Down-spinning reserve in balancing market (MW).
$r^{NS}{}_{it\omega}$	Non-spinning reserve in balancing market (MW).
$r^U{}_{jt\omega}$	Up-spinning reserve from demand-side in balancing market (MW).
$r^U{}_{jt\omega}$	Down-spinning reserve from demand-side in balancing market (MW).
$r_{itm\omega}{}^G$	Reserve deployed from the *m*-th block of energy offered in balancing market (MW).
$L_{jt\omega}{}^{shed}$	Load shedding (MW).
$S_{t\omega}$	Wind power generation spillage (MW).
$f_{t\omega(n,r)}$	Power flow through line (n, r) (MW).
$\delta_{t\omega n}$	Voltage angle at node.

C. Binary Variables

u_{it}	Commitment status of units in day-ahead market.
$v_{it\omega}$	Commitment status of units in balancing market.

D. Random Variables

$P^{WP}{}_{t\omega}$	Wind power generation in balancing market (MW).

E. Constants

$\lambda_{it}{}^{SU}$	Start-up offer cost of unit ($).
$\lambda_{itm}{}^G$	Marginal cost of the *m*-th block of energy offered ($/MWh).
$\lambda_{jt}{}^L$	Utility of electrical load ($/MWh).
$\lambda^{WP}{}_t$	Marginal cost of the energy offer submitted by the wind producer ($/MWh).
$VOLL_{jt}$	Value of loss load for load ($/MWh).
$V^S{}_t$	Wind spillage cost ($/MWh).
π_ω	Probability of scenarios.
$\overline{P}_i$	Maximum capacity of units (MW).
P_i	Minimum power output of generation units (MW).
$B_{(n,r)}$	Absolute value of the imaginary part of the admittance of line (n, r) (p.u.).
$\overline{f}_{(n,\ r)}$	Maximum capacity of line (n, r) (MW).

6.2.2 Modeling Description

This section is an introduction to the restructured electricity market. As stressed earlier, electricity markets include new services called ancillary services. In this chapter, only spinning and non-spinning reserves are modeled in the proposed market-clearing problem. In addition, two different approaches are used to solve the two-stage stochastic market-clearing problem. The first stage represents a day-ahead market-clearing problem, and the balancing market-clearing problem is described in the second stage. As mentioned before, two approaches were utilized to model the electricity market-clearing problem. In the first one, the market-clearing problem is solved sequentially. In this way, the day-ahead electricity market is cleared first, then the balancing market is cleared according to the outputs of the day-ahead market. In the second one, the day-ahead and balancing markets are cleared simultaneously. It should be noted that the uncertainty of decision-making variables is seen only in the second (balancing) stage. In the following subsections, the day-ahead and balancing stages of the market-clearing problem are described.

6.2.3 Day-Ahead Stage

In the proposed day-ahead electricity market model, only energy is cleared between market players as an electricity commodity. Besides, uncertainty is not seen in the day-ahead stage. Thus, an objective function is defined as the social welfare's expected cost, which should be minimized in the day-ahead market-clearing problem.

$$EC_{da} = \sum_{t=1}^{N_T}\sum_{i=1}^{N_G} C_{it}^{SU} + \sum_{t=1}^{N_T}\left[\sum_{i=1}^{N_G}\sum_{m=1}^{N_{Oit}} \lambda_{itm}^{G}.p_{itm}^{G} - \sum_{j=1}^{N_L}\lambda_{jt}^{L}.L_{jt}^{S} + \sum_{k=1}^{N_W}\lambda^{WP}{}_{kt}.P_{kt}^{S,WP}\right] \tag{6.1}$$

The expected cost of the day-ahead market is expressed in Eq. (6.1) in four terms. The first term represents the start-up cost of units and the second the energy cost of units. The utility of electricity customers and the cost of wind farm energy generation are expressed in the third and fourth terms, respectively. In addition, there are constraints related to different players of the electricity market that are represented in the following.

Eqs. (6.2)–(6.4) represent constraints linked to the power generation of the conventional generation units in the day-ahead electricity market. Specifically, Eq. (6.2) states maximum and minimum limitations of conventional units' power scheduling. The constraints related to the generation units' energy blocks are expressed in Eq. (6.3). Moreover, the total scheduled power of a conventional unit in each time period is represented in Eq. (6.4); its power is equal to the sum of its energy blocks.

$$\underline{P}_i.u_{it} \le P^S{}_{it} \le \overline{P}_i.u_{it},\ \forall i, \forall t \tag{6.2}$$

$$0 \le p^G_{itm} \le \overline{p}^G_{itm}, \ \forall m, \forall i, \forall t \tag{6.3}$$

$$P^S{}_{it} = \sum_{m=1}^{N_{Oit}} p^G_{itm}, \forall i, \forall t. \tag{6.4}$$

Eqs. (6.5)–(6.7) present constraints linked to the start-up cost of the conventional units.

$$C^{SU}_{it} \ge \lambda^{SU}_{it}.\left(u_{it} - u_{i(t-1)}\right), \ \forall i, \forall t > 1 \tag{6.5}$$

$$C^{SU}_{i(t=1)} \ge \lambda^{SU}_{i(t=1)}.\left(u_{i(t=1)} - u_{i(0)}\right), \ \forall i, t = 1 \tag{6.6}$$

$$C^{SU}_{it} \ge 0, \ \forall i, \ \forall t. \tag{6.7}$$

As seen in Eqs. (6.2)–(6.7), conventional units only provide energy as a commodity that can be cleared in the day-ahead market. Moreover, these constraints show that conventional units can be committed by the market operator. Hence, these units are called dispatchable generation units. Eq. (6.8) expresses the balancing equation between conventional generation units, wind farms, and electrical loads.

$$\sum_{i=1}^{N_G} P^S{}_{it} + \sum_{k=1}^{N_W} P^{S,WP}_t = \sum_{i=1}^{N_L} L^S{}_{jt}, \forall t. \tag{6.8}$$

As highlighted before, the uncertainty of stochastic variables such as wind power generation is not considered in the day-ahead stage. Therefore, the scheduled power of the wind farm is modeled in a way that limits its maximum and minimum power generation, as represented in Eq. (6.9).

$$\underline{P}^{WP}_{kt} \le P^{S,WP}_{kt} \le \overline{P}^{WP}_{kt}, \ \forall k, \forall t. \tag{6.9}$$

6.2.4 Balancing Stage

In this stage, both energy and operating reserve services are provided. Energy is supplied by wind farms. However, operating reserves are provided by generation units and electrical customers. In this chapter, only spinning and non-spinning reserves are modeled. Spinning reserves are classified as upward and downward spinning reserves that can be provided by generation-side and demand-side. However, the non-spinning reserve can be provided only by generation units. Moreover, the uncertainty of stochastic parameters is considered in the balancing stage. Eq. (6.10) represents the balancing stage's objective function. This objective function expresses the expected social welfare cost of the system in the balancing electricity market.

$$EC_b = \sum_{\omega=1}^{N_\Omega} \pi_\omega.\left\{\sum_{t=1}^{N_T}\sum_{i=1}^{N_G} C^A{}_{it\omega} + \sum_{t=1}^{N_T}\sum_{i=1}^{N_G}\sum_{m=1}^{N_{Oit}}\left(C^{R^U}{}_i.r^U{}_{it\omega} + C^{R^D}{}_i.r^D{}_{it\omega} + C^{R^{NS}}{}_i.r^{NS}{}_{it\omega}\right)\right.$$
$$\left. + \sum_{j=1}^{N_L}\left(C^{R^U}{}_j.r^U{}_{jt\omega} + C^{R^D}{}_j.r^D{}_{jt\omega} + VOLL_{jt}.L^{shed}_{jt\omega}\right) + \sum_{k=1}^{N_W} V^S{}_t.S_{kt\omega}\right\} \tag{6.10}$$

As expressed in Eq. (6.10), the social welfare's expected cost in the balancing stage includes eight terms. The first line represents the cost caused by changes in the start-up states of generation units in day-ahead and balancing stages. The second line states costs linked to the generation sides upward, downward spinning, and non-spinning reserves, respectively. Finally, the third line lists the costs of upward and downward spinning reserves from the demand-side, the load shedding cost, and the spillage cost of wind power generation.

The power generation constraint of generation units in the balancing market is expressed by Eq. (6.11).

$$\underline{P}_i.v_{it\omega} \le P^G{}_{it\omega} \le \overline{P}_i.v_{it\omega}, \quad \forall i, \forall t, \forall \omega. \tag{6.11}$$

Eq. (6.12) expresses the balancing of allocated energy in the generation units in the day-ahead and balancing electricity markets, and the operating reserves from the generation-side in the balancing market. As seen in Eq. (6.12), if the power provided by the generation units in the balancing market is greater than their committed power in the day-ahead market, upward spinning, or non-spinning reserves should be committed in the balancing stage. Otherwise, downward spinning reserve should be provided by generation units in the balancing stage. It should be noted that a non-spinning reserve can be committed only from units that are "OFF" in the day-ahead market, as represented in Eq. (6.15). In other words, spinning reserves can be dispatched when the commitment status of generation units is "ON." Eqs. (6.15)–(6.17) state the constraints related to the operating reserve of generation units in the balancing stage.

$$P^G{}_{it\omega} - P^S{}_{it} = r^U{}_{it\omega} + r^{NS}{}_{it\omega} - r^D{}_{it\omega}, \quad \forall i, \forall t, \forall \omega. \tag{6.12}$$

$$0 \le r^U{}_{it\omega} \le \overline{R^U}_i.u_{it}, \quad \forall i, \forall t, \forall \omega. \tag{6.13}$$

$$0 \le r^D{}_{it\omega} \le \overline{R^D}_i.u_{it}, \quad \forall i, \forall t, \forall \omega. \tag{6.14}$$

$$0 \le r^{NS}{}_{it\omega} \le \overline{R^{NS}}_i.(1 - u_{it}), \quad \forall i, \forall t, \forall \omega. r^U{}_{it\omega} + r^{NS}{}_{it\omega} - r^D{}_{it\omega} = \sum_{m=1}^{N_{Oit}} r^G_{itm\omega}, \quad \forall i, \forall t, \forall \omega. \tag{6.15}$$

$$r^G_{itm\omega} \le \overline{p}^G_{itm} - p^G_{itm}, \quad \forall i, \forall t, \forall \omega. \tag{6.16}$$

$$r^G_{itm\omega} \ge -p^G_{itm}, \quad \forall m, \forall i, \forall t, \forall \omega. \tag{6.17}$$

In the balancing stage, the uncertainty of the power system causes generation units to make new commitments, which increase start-up costs in the system. The start-up equation and limitations in the balancing stage are represented by Eqs. (6.18)–(6.21).

$$C^A{}_{it\omega} = C^{SU}{}_{it\omega} - C^{SU}{}_{it}, \quad \forall i, \forall t, \forall \omega. \tag{6.18}$$

$$C^{SU}{}_{it\omega} \geq \lambda_{it}^{SU} \cdot \left(v_{it\omega} - v_{i(t-1)\omega}\right), \quad \forall i, \forall t > 1, \forall \omega. \tag{6.19}$$

$$C_{i(t=1)\omega}^{SU} \geq \lambda_{i(t=1)}^{SU} \cdot \left(v_{i(t=1)\omega} - u_{i(0)}\right), \quad \forall i, t = 1 \tag{6.20}$$

$$C^{SU}{}_{it\omega} \geq 0, \quad \forall i, \forall t, \forall \omega. \tag{6.21}$$

In Eq. (6.22), the power balance equation in the balancing stage is represented considering line power flow. In this chapter, the market-clearing is modeled according to the DC optimal power flow problem. In this way, Eqs (6.23) and (6.24) express constraints related to obtaining power flow and its transmission line's capacity, respectively.

$$\sum_{i:(i,n)} P^{G}{}_{it\omega} - \sum_{j:(j,n)} \left(L^{C}{}_{jt\omega} - L_{jt\omega}^{shed}\right) + \sum_{k:(k,n)} \left(P^{WP}{}_{kt\omega} - S_{kt\omega}\right) - \sum_{r:(n,r)} f_{t\omega(n,r)} = 0, \quad \forall n, \forall t, \forall \omega. \tag{6.22}$$

$$f_{t\omega(n,r)} = B_{(n,r)} \cdot \left(\delta_{t\omega n} - \delta_{t\omega r}\right), \quad \forall t, \forall \omega. \tag{6.23}$$

$$-\overline{f}_{(n,r)} \leq f_{t\omega(n,r)} \leq \overline{f}_{(n,r)}, \quad \forall t, \forall \omega. \tag{6.24}$$

As pointed out, wind farms are renewable energy resources, which are modeled as stochastic power generation units. Although wind power generation in the day-ahead market is modeled based on its maximum and minimum limitation, wind power generation is modeled as a stochastic parameter in the balancing market. Besides, wind power can be spilled in the balancing stage due to economic and technical concerns, as expressed in Eq. (6.25).

$$0 \leq S_{kt\omega} \leq P^{WP}{}_{kt\omega}, \quad \forall k, \forall t, \forall \omega. \tag{6.25}$$

In the balancing electricity market, electrical loads can act as interruptible agents. In this case, they present their flexible behavior to decrease or increase their consumption in the balancing stage. Hence, if customers increase their consumption in the balancing market, they act as virtual generation units, which decrease their power generation. Hence, this flexible behavior of electrical loads is called downward spinning reserve from demand-side. On the other hand, if they decrease their electrical demand in the balancing market, they provide upward spinning reserve from the demand-side. The above definitions are represented in Eqs. (6.26)–(6.28).

$$0 \leq r^{U}{}_{jt\omega} \leq \overline{R^{U}}_{j}, \quad \forall j, \forall t, \forall \omega. \tag{6.26}$$

$$0 \leq r^{D}{}_{jt\omega} \leq \overline{R^{D}}_{j}, \quad \forall j, \forall t, \forall \omega. \tag{6.27}$$

$$L^{C}{}_{jt\omega} - L^{S}{}_{jt} = r^{D}{}_{jt\omega} - r^{U}{}_{jt\omega}, \quad \forall j, \forall t, \forall \omega. \tag{6.28}$$

Moreover, the portion of loads that is decreased non-voluntarily in the balancing market is called the shed load. The load shedding limitation is represented in Eq. (6.29).

$$0 \leq L_{jt\omega}^{shed} \leq L^{C}{}_{jt\omega}, \quad \forall j, \forall t, \forall \omega. \tag{6.29}$$

6.3 Simulation Results

This section presents the two approaches that were used to solve the proposed two-stage stochastic market-clearing problem: a sequential approach and a simultaneous approach. A modified three-bus test system from Refs. [35, 36] is used to evaluate our study as shown in Fig. 6.1. Table 6.1 presents system data. Table 6.2 shows transmission lines' power capacity. The day-ahead scheduled load is presented in Table 6.3. Wind power generation's scenarios and their corresponding probabilities are outlined in Tables 6.4 and 6.5, respectively. It should be noticed that uncertainty of the power grid is considered in this case study that its scenarios come from ORR, which equals 0.02 for conventional generation units and 0.01 for transmission lines. Besides, the Value Of Lost Load (VOLL) of consumers is supposed to equal 1000. Our stochastic market-clearing problem is model by Mixed Integer Linear Programming (MILP) to solve in GAMS 24.7.4 [38] that has been linked with MATLAB software [39].

6.3.1 Case 1: Sequential Market-Clearing Model

In case 1, the day-ahead and balancing electricity markets are cleared sequentially. In this way, the day-ahead market-clearing problem is solved independently. Then, the balancing market is cleared according to the outputs of the day-ahead market-clearing problem as shown in Fig. 6.2. Table 6.6 presents

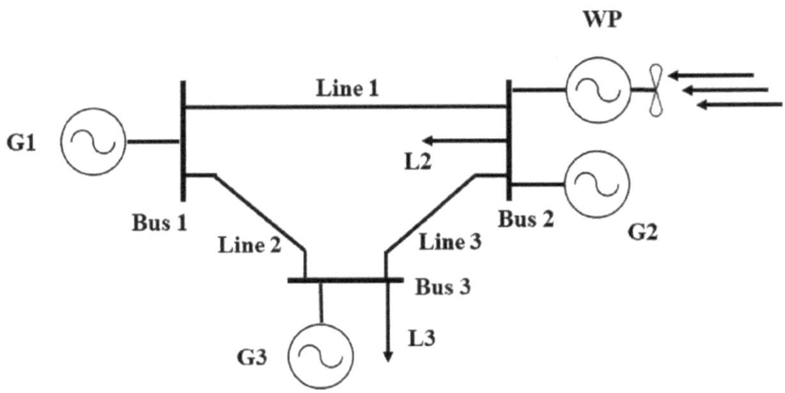

FIG. 6.1 The three-bus test system [35, 36].

TABLE 6.1 System Data in a 3-Bus Test System [35–37]

$C^{R^U}{}_{jt}$ (\$/MWh)	70
$C^{R^D}{}_{jt}$ (\$/MWh)	70
$VOLL_{base}$ (\$/MWh)	1000
Lines reactance (p.u.)	0.13
Lines capacity (MW)	55
P_{base} (MW)	41
V_{base} (kV)	120

TABLE 6.2 Line Capacities [35–37]

Transmission lines	Capacity (MW)
Line (1,2)	10
Line (1,3)	28
Line (2,3)	24

TABLE 6.3 Day-Ahead Electrical Demand of Consumers [35–37]

Consumer (MW)		Time (h)			
		t1	*t2*	*t3*	*t4*
L2	$L^S{}_{jt}$	20	60	90	30
L3	$L^S{}_{jt}$	30	80	110	40

TABLE 6.4 Scenarios of Wind Power Generation [35–37]

Period *t*	$P^{WP}{}_{t\omega}$ (MW)		
	As forecasted	*High*	*Low*
1	6	9	2
2	20	30	13
3	35	50	25
4	8	12	6

TABLE 6.5 Scenarios Probabilities of Wind Power Generation [35–37]

	$P^{WP}_{t\omega}$ (MW)		
	As forecast	*High*	*Low*
Probability	0.6	0.2	0.2

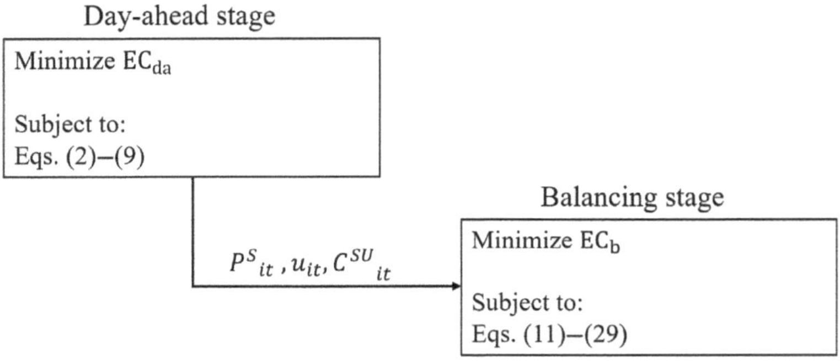

FIG. 6.2 Stochastic market-clearing procurement in the sequential approach.

TABLE 6.6 Expected Demand, Load Shedding, Spinning Reserves of Consumers in the Balancing Market of the Sequential Market-Clearing Model

Consumer (MW)		Time (h)			
		t1	*t2*	*t3*	*t4*
L2	L^C_{jt}	18.399	54	81	32.349
	L_{jt}^{shed}	0	0	0	0
	$r^D_{jt\omega}$	0	0	0	2.375
	$r^U_{jt\omega}$	1.601	6	9	0.025
L3	L^C_{jt}	33	87.909	101.973	44
	L_{jt}^{shed}	0	0.080	0.355	0
	$r^D_{jt\omega}$	3	7.952	0	4
	$r^U_{jt\omega}$	0	0.044	8.027	0

the expected electrical consumption, load shedding, and downward and upward spinning reserves of consumers in the balancing market. As seen in Tables 6.3 and 6.6, the expected consumption of L2 in t1, t2, and t4 in the balancing market is lower than its scheduled load in the day-ahead market. However, for L3 the day-ahead scheduled demand is greater than its expected consumption in the balancing stage only in t3. If the consumer decreases its demand in the balancing market voluntarily, this quantity of decrement acts as an upward spinning reserve from the demand-side. As stated in Table 6.6, this decrement plays only as the upward spinning reserve form L2. However, load shedding occurs in t2 and t3 for L3. On the other hand, if the balancing expected consumption of the electrical consumers is greater than their day-ahead scheduled demand, they act as virtual downward spinning reserve providers in the market. As seen in Table 6.6, while L2 provides downward spinning reserve only in t4, the downward spinning reserve is provided by L3 in t1, t2 and t4.

Table 6.7 shows the scheduled power and the expected balancing power generation of the generation units. The difference between the dispatched power of generation units in the day-ahead and balancing markets is allocated to the spinning and non-spinning reserves of the conventional generation units in the balancing stage, as expressed in Eq. (6.12). As seen in Table 6.8, G1 provides both upward and downward spinning reserves in the fourth time period. At first, it seems that these results are not true because the generation units can only provide upward or downward spinning reserves. However, the results are the expected reserves that are supplied in different scenarios. This means that G1 produces only the upward spinning reserve in one scenario, and it provides the downward spinning reserve in another one. This occurs in t1,

TABLE 6.7 Dispatched Power of Generation Units in the Day-Ahead Market and Their Expected Balancing Power Generation in the Sequential Market-Clearing Model

		Time (h)			
Gen. unit (MW)		*t1*	*t2*	*t3*	*t4*
G1	P^S_{it}	0	70	100	12
	P^G_{it}	0.063	33.944	31.935	12.191
G2	P^S_{it}	0	0	15	0
	P^G_{it}	0.027	37.484	64.982	8.156
G3	P^S_{it}	44	50	50	50
	P^G_{it}	45.529	49.841	49.721	49.602

TABLE 6.8 Expected Allocated Operating Reserves of Generation Units in the Sequential Market-Clearing Model

		Time (h)			
Gen. unit (MW)		*t1*	*t2*	*t3*	*t4*
G1	$r^{U}{}_{it}$	0	0	0	0.199
	$r^{D}{}_{it}$	36.056	0	68.065	0.008
	$r^{NS}{}_{it}$	0.063	0	0	0
G2	$r^{U}{}_{it}$	0	0	49.982	0
	$r^{D}{}_{it}$	0	0	0	0
	$r^{NS}{}_{it}$	0.027	37.484	0	8.156
G3	$r^{U}{}_{it}$	1.597	0	0	0
	$r^{D}{}_{it}$	0	0	0	0
	$r^{NS}{}_{it}$	0	0	0	0

where G1 provides both spinning and non-spinning reserves. In this case, G1 provides only a non-spinning reserve in one scenario, and a downward spinning reserve in another one.

6.3.2 Case 2: Simultaneous Market-Clearing Model

In case 2, the joint energy and reserve stochastic market-clearing problem is solved in the day-ahead and balancing electricity markets, simultaneously. The objective function of the simultaneous market-clearing will be total expected cost—sum of day-ahead and balancing stages' objective functions—that should be minimized as represented in Eq. (6.30).

$$EC = EC_{da} + EC_{b} \tag{6.30}$$

Hence, the simultaneous market-clearing problem is presented in the following:

Min. EC
S.t.
Eqs. (6.2)–(6.9) and (6.11)–(6.29).

Table 6.9 shows the expected load, load shedding and downward and upward spinning reserves of consumers in the balancing stage.

As seen in Tables 6.6 and 6.9, there is no difference between consumers' flexibility behavior in sequential and simultaneous market-clearing models.

TABLE 6.9 Expected Electrical Demand, Load Shedding and Downward and Upward Spinning Reserves of Consumers in the Balancing Market of the Simultaneous Market-Clearing Model

Consumer (MW)		Time (h) t1	t2	t3	t4
L2	$L^C{}_{jt}$	18.399	55.59	81.014	32.349
	L_{jt}^{shed}	0	0	0	0
	$r^D{}_{jt\omega}$	0	0.398	0	2.375
	$r^U{}_{jt\omega}$	1.601	4.807	8.986	0.025
L3	$L^C{}_{jt}$	33	87.909	101.973	44
	L_{jt}^{shed}	0	0.080	0.355	0
	$r^D{}_{jt\omega}$	3	7.952	0	4
	$r^U{}_{jt\omega}$	0	0.044	8.027	0

The dispatched power of the generation units in the day-ahead and balancing electricity markets is represented in Table 6.10. In addition, spinning and non-spinning reserves that are provided by generation units are outlined in Table 6.11. As seen in Tables 6.7 and 6.10, the difference between the day-ahead and balancing power generation of the conventional units in the simultaneous model is lower than the sequential one. Hence, generation units show smoother behavior to provide operating reserves in the simultaneous market-clearing model, as shown in Table 6.11. Hence, this smoother behavior decreases the balancing stage's expected cost in the simultaneous market-clearing model.

Furthermore, as shown in Table 6.12, the total expected cost in the simultaneous market-clearing model is lower than the sequential model, proving that the simultaneous model is more efficient than the sequential model.

6.3.3 Case 3: Uncertainty analysis

In this case, the impact of wind power generation's uncertainty is evaluated where joint energy and electricity markets are cleared simultaneously. As seen in Table 6.13, the EC is higher without wind power generation. Hence, considering wind power generation has a positive effect of the EC of the system. On the other hand, uncertainty of the wind power generation has a negative effect on the EC and increases the total EC of the system. Fig. 6.3 shows the impact of

TABLE 6.10 Dispatched Power of Generation Units in the Day-Ahead Market and Their Expected Balancing Power Generation in the Simultaneous Market-Clearing Model

Gen. unit (MW)		Time (h) t1	t2	t3	t4
G1	P^S_{it}	0	34	49	12
	P^G_{it}	0.063	33.944	31.935	12.191
G2	P^S_{it}	0	36	66	0
	P^G_{it}	0.027	39.074	67.996	8.156
G3	P^S_{it}	44	50	50	50
	P^G_{it}	45.509	49.801	49.701	49.601

TABLE 6.11 Expected Allocated Operating Reserves of Generation Units in the Simultaneous Market-Clearing Model

Gen. unit (MW)		Time (h) t1	t2	t3	t4
G1	r^U_{it}	0	0	0	0.199
	r^D_{it}	0	0.056	17.065	0.008
	r^{NS}_{it}	0.063	0	0	0
G2	r^U_{it}	0	3.074	1.996	0
	r^D_{it}	0	0	0	0
	r^{NS}_{it}	0.027	0	0	8.156
G3	r^U_{it}	1.597	0	0	0
	r^D_{it}	0	0	0	0
	r^{NS}_{it}	0	0	0	0

wind power generation's prediction accuracy on the total EC. As seen in Fig. 6.3, there is a linear relationship between the prediction accuracy of the wind power generation and the total EC. In this way, increasing its prediction accuracy has a positive influence on the total EC and decreases the EC of the market-clearing problem.

TABLE 6.12 Expected Costs (ECs) in the Sequential and Simultaneous Market-Clearing Models

	Sequential market-clearing model	Simultaneous market-clearing model
EC ($)	11,428.803	11,243.460
EC_{da}($)	10,240.000	11,110.000
EC_b ($)	1188.803	133.460

TABLE 6.13 Impact of Wind Power Generation's Uncertainty on the Total Expected Cost (EC) in the Simultaneous Market-Clearing Model

	Without Wind Power Generation	With Uncertain Wind Power Generation	With Certain Wind Power Generation
EC ($)	13,920.459	11,243.460	11,228.733

6.3.4 Case 4: Flexibility Analysis

In this case, the impact of electrical consumers' flexible behavior is assessed in the simultaneous stochastic joint energy and reserve market-clearing problem. Three different scenarios are studied. In Scenario 1, consumers do not join the balancing market in order to provide spinning reserves. In Scenario 2, consumers act as interruptible loads, in the same way as they did in Case 2. In Scenario 3, we consider that consumers act as shiftable loads in the balancing market, as expressed in Eq. (6.31).

$$\sum_{t=1}^{N_T}\sum_{\omega=1}^{N_\Omega}\pi_\omega\left({r^D}_{jt\omega}-{r^U}_{jt\omega}\right)=0,\quad \forall j,\forall t,\forall\omega. \tag{6.31}$$

Thus, in Scenario 3 the market-clearing model is represented by the following:

Min. EC
S.t.
Eqs. (6.2)–(6.9), (6.11)–(6.29), and (6.31).

On the other hand, interruptible loads can be constrained over all interruptible loads in each time step, as represented in Eq. (6.32). In this way, Eq. (6.32) increases the self-sustainability of the electrical loads in the balancing market.

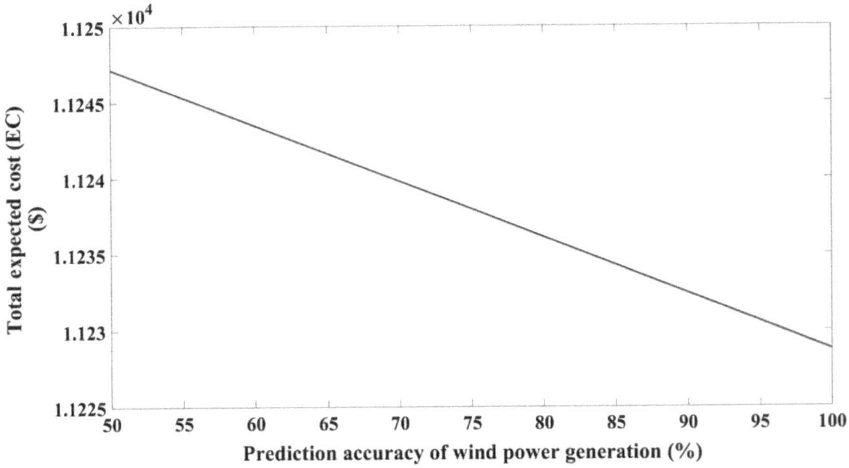

FIG. 6.3 Impact of prediction accuracy of wind power generation on the total expected cost (EC) in the simultaneous market-clearing model.

$$\sum_{j=1}^{N_L}\sum_{\omega=1}^{N_\Omega}\pi_\omega\left({r^D}_{jt\omega}-{r^U}_{jt\omega}\right)=0,\quad \forall j, \forall t, \forall \omega. \tag{6.32}$$

In this way, the market-clearing model of Scenario 4 is represented by the following:

Min. EC
S.t.
Eqs. (6.2)–(6.9), (6.11)–(6.29), and (6.32).

As shown in Table 6.14, the worst scenario is Scenario 1, where the total EC of the system is the greatest one. Also, the total expected costs in Scenarios 3 and 4 are greater than the EC in Scenario 2. Besides, the EC in Scenario 3 is greater than the EC in Scenario 4 because the shiftable load constraint gives less freedom to electrical consumers to show their desired flexibility behavior.

However, the proposed market-clearing models are solved in a centralized manner by the system operator who is in charge of making policies regarding Scenarios 1–4. In this way, it can be concluded that efficient decision-making in

TABLE 6.14 Impact of Customers' Flexibility Behavior on the Total Expected Cost (EC) in the Simultaneous Market-Clearing Model

	Scenario 1	Scenario 2	Scenario 3	Scenario 4
EC ($)	19,837.361	11,243.460	11,707.427	11,574.521

electricity markets can be improved if consumers as active agents can decide autonomously based on a decentralized way.

6.4 Conclusion

In this chapter, the two-stage stochastic joint energy and reserve market-clearing problem has been solved. The two stages of the problem are called day-ahead and balancing. Two different approaches—sequential and simultaneous—were used in the problem. In the sequential model, the day-ahead stage was solved independently, and the balancing stage was solved based on the first stage's outputs. However, the day-ahead and balancing stages were solved together according to the coupling constraints in the simultaneous market-clearing problem. Furthermore, the impacts of electrical consumers' flexibility programs have been assessed in this chapter. According to our study, the simultaneous market-clearing model is more efficient than the sequential one. This is because the total expected cost in the simultaneous model is lower than the sequential one. Moreover, from the analysis of the flexibility programs, the shiftable load constraint increases the total expected cost of the system because it decreases the electrical consumers' freedom to have their desired flexibility behavior. However, both market models, with or without shiftable load constraints, improve the efficiency of the electricity market and decrease the total expected cost of the system.

Acknowledgments

Amin Shokri Gazafroudi, Francisco Prieto-Castrillo, and Juan Manuel Corchado acknowledge the support of the European Commission H2020 MSCA-RISE-2014: Marie Sklodowska-Curie project DREAM-GO Enabling Demand Response for short and real-time Efficient And Market Based Smart Grid Operation—An intelligent and real-time simulation approach, ref. 641794. Moreover, Amin Shokri Gazafroudi acknowledges the support of the Ministry of Education of the Junta de Castilla y Leon and the European Social Fund through a grant from predoctoral recruitment of research personnel associated with the research project "Arquitectura multiagente para la gestión eficaz de redes de energía a través del uso de tecnicas de inteligencia artificial" of the University of Salamanca. The work of João P.S. Catalão was supported by FEDER funds through COMPETE 2020 and by Portuguese funds through FCT, under Projects SAICT-PAC/0004/2015—POCI-01-0145-FEDER-016434, POCI-01-0145-FEDER-006961, UID/EEA/50014/2013, UID/CEC/50021/2013, and UID/EMS/00151/2013. In addition, the research leading to these results has received funding from the EU Seventh Framework Programme FP7/2007–2013 under grant agreement no. 309048.

References

[1] P. Zou, Q. Chen, Y. Yu, Q. Xia, C. Kang, Electricity markets evolution with the changing generation mix: an empirical analysis based on China 2050 high renewable energy penetration roadmap, Appl. Energy 185 (2017) 56–67. Part 1.

[2] Project Partner Paul Scherrer Institute of World Energy Council. World Energy Scenarios: composing energy futures to 2050. Switzerland; Technical Report; 2013.

[3] Energy Research Institute of National Development and Reform Commission. China 2050 High Renewable Energy Penetration Scenario and Roadmap Study. Beijing; Technical Report; 2015.

[4] A.J. Conejo, R. Sioshansi, Rethinking restructured electricity market design: lessons learned and future needs, Int. J. Electr. Power Energy Syst. 98 (2018) 520–530.

[5] T.R. Karl, J.M. Melillo, T.C. Peterson (Eds.), Global Climate Change Impacts in the United States, Cambridge University Press, 2009.

[6] Ardani K, Margolis R. 2010 Solar Technologies Market Report, Technical Report DOE/GO-102011-3318, United States Department of Energy; November 2011.

[7] E.A. DeMeo, W. Grant, M.R. Milligan, M.J. Schuerger, Wind plant integration, IEEE Power Energy Mag. 3 (2005) 38–46.

[8] J.C. Smith, M.R. Milligan, E.A. DeMeo, B. Parsons, Utility wind integration and operating impact state of the art, IEEE Trans. Power Syst. 22 (2007) 900–908.

[9] E.A. DeMeo, G.A. Jordan, C. Kalich, J. King, M.R. Milligan, C. Murley, et al., Accommodating wind's natural behavior, IEEE Power Energy Mag. 5 (2007) 59–67.

[10] M.R. Milligan, K. Porter, E.A. DeMeo, P. Denholm, H. Holttinen, B. Kirby, et al., Wind power myths debunked, IEEE Power Energy Mag. 7 (2009) 89–99.

[11] Denholm P, O'Connell M, Brinkman G, Jorgenson J. Overgeneration from Solar Energy in California: A Field Guide to the Duck Chart, Technical Report NREL/TP-6A20-65023, National Renewable Energy Laboratory; 2015.

[12] Sawin JL, Sverrisson F, Chawla K, Lins C, McCrone A, Musolino E, et al. Renewables 2014 Global Status Report, Technical Report, Renewable Energy Policy Network for the 21st Century; 2014.

[13] H. Holttinen, M. Milligan, E. Ela, N. Menemenlis, J. Dobschinski, B. Rawn, et al. Methodologies to determine operating reserves due to increased wind power IEEE Trans. Sustain. Energy, 3 (October) (2012), pp. 713–723.

[14] Monitoring Analytics, LLC, State of the Market Report for PJM, Independent Market Monitor, 2015.

[15] Nord Pool Spot. n.d. Europe's leading power market [Online]. Available: <http://www.nordpoolspot.com>.

[16] C. Weiller, Plug-in hybrid electric vehicle impacts on hourly electricity demand in the United States, Energy Policy 39 (2011) 3766–3778.

[17] R. Sioshansi, Evaluating the impacts of real-time pricing on the cost and value of wind generation, IEEE Trans. Power Syst. 25 (2010) 741–748.

[18] Ela E, Milligan M, Bloom A, Botterud A, Townsend A, Levin T. Evolution of Wholesale Electricity Market Design with Increasing Levels of Renewable Generation, Technical Report. NREL/TP-5D00-61765, National Renewable Energy Laboratory; 2014.

[19] Bushnell J, Harvey SM, Stoft S, Hobbs BF. Final Opinion on Payment for Provision of Flexible Ramping, Technical Report; 16 August 2011.

[20] B. Wang, B.F. Hobbs, A flexible ramping product: can it help real-time dispatch markets approach the stochastic dispatch ideal? Electr. Pow. Syst. Res. 109 (2014) 128–140.

[21] P.N. Biskas, D.I. Chatzigiannis, A.G. Bakirtzis, European electricity market integration with mixed market designs—Part II: Solution algorithm and case studies, IEEE Trans. Power Syst. 29 (2014) 466–475.

[22] G. Papaefthymiou, K. Dragoon, Towards 100% renewable energy systems: uncapping power system flexibility, Energy Policy 92 (2016) 69–82.

[23] M.L. Kubik, P.J. Coker, J.F. Barlow, Increasing thermal plant flexibility in a high renewables power system, Appl. Energy 154 (2015) 102–111.

[24] A. Welling, The paradox effects of uncertainty and flexibility on investment in renewables under governmental support, Eur. J. Oper. Res. 251 (2016) 1016–1028.

[25] A. Belderbos, E. Delarue, Accounting for flexibility in power system planning with renewables, Int. J. Electr. Power Syst. 71 (2015) 33–41.

[26] E.A. Bakirtzis, P.N. Biskas, D.P. Labridis, A.G. Bakirtzis, Multiple time resolution unit commitment for short-term operations scheduling under high renewable penetration, IEEE Trans. Power Syst. 29 (2014) 149–159.

[27] P. Andrianesis, G. Liberopoulos, P. Biskas, A. Bakirtzis, 2011. Medium-Term Unit Commitment in a Pool Market, 2011 in: Proceedings of the 8th International Conference on the European Energy Market (EEM), 461–466.

[28] N.E. Koltsaklis, A.S. Dagoumas, M.C. Georgiadis, G. Papaioannou, C. Dikaiakos A mid-term, market-based power systems planning model, Appl. Energy 179 (2016) 17–35.

[29] L.C. Chien, W. Lin, Y.C. Yin, Enhancing frequency response control by DFIGs in the high wind penetrated power systems, IEEE Trans. Power Syst. 26 (May) (2011) 710–718.

[30] R.G. Almeida, J.A. Lopes, Participation of doubly fed induction wind generators in system frequency regulation, IEEE Trans. Power Syst., 22 (August) (2007), pp. 944–950.

[31] N. Aparicio, I. MacGill, J.R. Abbad, H. Beltran, Comparison of wind energy support policy and electricity markets design in Europe, the United States, and Australia, IEEE Trans. Sustain. Energy, 3 (October) (2012), pp. 809–818.

[32] Y. Lin, J.X. Johnson, J.L. Mathieu, Emissions impacts of using energy storage for power reserves, Appl. Energy, 168 (January) (2016), pp. 444–456.

[33] B. Kyle, P. Lincoln, P. Dalia, Economic viability of energy storage systems based on price arbitrage potential in real-time electricity markets, Appl. Energy, 114 (October) (2013), pp. 512–519.

[34] G.M. Casolino, G. Liuzzi, A. Losi, Combined cycle unit commitment in a changing electricity market scenario, Int. J. Electr. Power Energy Syst. 73 (2015) 114–123.

[35] A. Shokri Gazafroudi, K. Afshar, N. Bigdeli, "Assessing the operating reserves and costs with considering customer choice and wind power uncertainty in pool-based power market," Int. J. Electr. Power Energy Syst., Vol. 67, May 2015, pp. 202–215.

[36] A. Shokri Gazafroudi, M. Shafie-khah, M. Abedi, S.H. Hosseinian, G.H.R. Dehkordi, L. Goel, P. Karimyan, J.M. Corchado, J.P.S. Catalão, A novel stochastic reserve cost allocation approach of electricity market agents in the restructured power systems, Electr. Pow. Syst. Res. 152 (C) (2017) 223–236.

[37] A.J. Conejo, M. Carrion, J.M. Morales, Decision making under uncertainty in electricity markets, in: International Series in Operations Research & Management Science, Springer, 2010.

[38] GAMS Release 2.50. A User's Guide; GAMS Development Corporation: 1999. Available online: http://www.gams.com Accessed 30 January 2018, 2018.

[39] The MathWorks, MATLAB. Available online: http://www.mathworks.com. Accessed 30 January 2018, 2018).

Chapter 7

Better Transmission Networks for a Smarter Global System

Sara Lumbreras and Andres Ramos
Universidad Pontificia Comillas, Madrid, Spain

Chapter Outline

7.1 Introduction: The Current Need for Transmission

We can define the power network as the set of assets that are used to transfer the energy produced at the generating units to reach the consumption nodes. These assets are categorized attending their functionality, and can be separated into transmission and distribution. Transmission lines span long distances and support large amounts of electricity. Distribution lines are in charge of the final delivery to the consumers. In Europe, the transmission network is made mainly of high voltage AC and DC lines of 132, 220, and 400 kV, while the distribution network includes voltages below 36 kV.

The shape of the future energy is expected to change dramatically in the near future. These changes will be reflected in the power grids. Although it seems that renewable penetration is set to increase greatly, there is an ongoing debate about whether this increase will take form in a centralized or a decentralized manner. However, it is clear that renewable integration will be one of the main

Pathways to a Smarter Power System. https://doi.org/10.1016/B978-0-08-102592-5.00007-7

drivers for the development of the network. In particular, the two main needs for transmission are as follows:

- The fulfillment of energy policy targets and the large-scale integration of renewables: policy targets have been set by countries in order to satisfy the global climate-change goals committed in the Paris Agreement. In Europe, the so-called Winter Package reflects the compromise of achieving a 40% emission reduction, 30% improvement in energy efficiency, and 27% increase in renewables by 2030 [1]. In the United States, Renewable Portfolio Standards play a similar role by introducing some renewable generation targets in different years depending on the state [2]. Renewable generation at utility-scale is becoming a cost-competitive technology and, therefore, is becoming widespread in many areas and countries. However, this generation is usually far away from loading centers, and its integration into the system requires the development of new transmission lines.
- The creation of regional electricity markets: larger electricity markets allow for increasing competition among the different agents located in different countries and for mitigating market power improving competition. The creation of a single European market is a goal for achieving higher economic efficiencies.

Other additional drivers that guide the development of the transmission network are as follows:

- *Reliability*: the transmission network can increase the security of supply by allowing alternative paths for the electricity to supply the demand. Deploying a messed network increases the reliability of the grid.
- *Reduction of network losses*: although ohmic losses are small, sometimes lines can be economically efficient just for the impact in loss reduction that they have.
- *Connection of new generating power plants*: the evacuation of the energy produced by large-scale power plants can condition network development, especially if the new lines are close to local congestions.
- *Solve local capacity constraints*: congestions located in certain areas can be solved by new network additions.

Because of the previous drivers, network expansion planning is going to play a major role in the transition to a future carbon-free energy sector. This will happen not only because of the high impact in the system operation, but also because of the extended permitting processes they require, which often demands more than a decade to build a new overhead line. Therefore, the network planning exercise becomes more relevant as a way to determine this influence.

This chapter is organized as follows. Section 7.2 discusses the new technologies that are altering the network landscape. Section 7.3 then presents a general formulation for the TEP problem that can be adapted to most problem instances.

Section 7.4 describes the most important modeling choices that can be made when performing transmission planning. This is complemented by Section 7.5, which details the most popular solution methods for this problem. Section 7.6 illustrates the chapter with a case of large-scale planning in an optimization setting. Finally, Section 7.7 extracts conclusions.

7.2 A New Network: Game-Changing Technologies

The challenges above are leading to a paradigm shift in the transmission system: we are working towards a smarter transmission network. The proliferation of distributed energy resources (DER) has invalidated the *trickling-down* model, where power flew from higher voltage levels in transmission to lower ones in distribution, from generation to demand. This makes it necessary to review our assumptions about how the network should be built, as well as the principles for the operation of the system. In addition, it is necessary to rethink how to define the boundary between transmission and distribution now that distribution will likely incorporate elements like locational marginal prices or the management of congestions [3]. DERs can be understood as a source of local flexibility that can be activated to solve problems in the system and, among other things, delay investments in transmission [4], but there are others. Several recent projects and task forces have focused on the design of a *smart transmission network* [5, 6]. The majority of works places its emphasis on distribution and demand-side management, but the big picture with respect to transmission is already emerging. Most views agree on the key role of new technologies. Long-distance, efficient transmission will be accomplished by a combination of AC and DC technologies. Advanced conductors present advantages with respect to current-carrying capability, and in the last few years there has been a number of high-temperature options made available, as well as new transmission configurations (such as 6 or 12 phases) that allow for greater power transmission in the same right-of-way due to greater phase cancellation [6].

Underground cables will be especially important. This is the case of urban areas and also of underwater transmission (which is particularly relevant in the case of offshore wind farms (OWFs), which will play a very important role in the future energy system). A very high proportion of these underground facilities will be HVDC, as shown by the long list of recent projects that rely on this technology [7]. HVDC is a more efficient alternative for long-distance transmission (usually, 500km is considered as the breakeven distance [8]). In addition, it can be the only feasible alternative for asynchronous connections and long submarine cable crossings. Because DC lines can be controlled independently of voltage angles at the converter stations, it is also possible to use them to bypass network congestions. The number of HVDC projects has increased dramatically in the last few decades. The core HVDC technologies are LCC (line-commutated current source converters) and VSC (self-commutated voltage source converters). LCC require a synchronous voltage source and can only operate with the AC current lagging the voltage. This means that the process demands some reactive power, which must be provided by the system. On the contrary, VSC uses pulse-width modulation (PWM) and insulated-gate bipolar transistor (IGBT) valves. This technology can efficiently control both reactive and active power independently. Multiterminal HVDC allows the creating of DC networks that can be especially attractive in an offshore context, where they can link several OWFs among them and with the neighboring countries [9].

Another technology that provides enhanced flexibility is the phase-shifting transformer (PST). PSTs have become instrumental in a system with increasing cross-border transactions. Previously, countries used to be relatively isolated, maybe coupled to their neighbors only in times of need. The deregulation of markets led to energy exchanges between countries, but the physical reality of flows rarely follows those contractual paths. The uneven loading of lines due to Kirchhoff's laws can be alleviated by the use of PSTs, which control strictly the flow through a given line. Several configurations of PST exist with different tradeoffs of flexibility and cost [10]. The joint use of PSTs and HVDC can greatly improve the dynamic performance of the overall system [11].

The distance operation of substations offers the possibility of opening and closing some lines to better adapt to the current conditions of the system. This is especially interesting when some lines become congested due to Kirchhoff's second law, but there are parallel paths available for the energy transfer between two zones. Opening the congested path can lead to a feasible solution for the transfer. The consideration of transmission switching considerably increases the complexity of the TEP problem [12–14], but can provide a lower-cost alternative to a massive deployment of PSTs.

Another key factor that seems to appear in most visions is the upgrading of control centers and substations. Current control algorithms should be greatly improved by the use of state variables obtained from state measurement and GIS data, and more accurate stability analyses. Smart substations will provide autonomous responses tailored to the state of the grid. In addition, a cost-effective distributed line condition monitoring will optimize the utilization of lines with dynamic rating [15] based on overhead conductor sags, temperature and wind profiles, and the detection of potential issues like vegetation or ice.

All these developments enable the development of a smart transmission grid. However, they also greatly increase the complexity of the planning problem, as can be seen in the formulation below.

7.3 Formulation

7.3.1 Overview

This section presents a general mathematical formulation of the transmission expansion planning (TEP) problem. It can be easily adapted to most problem instances and can be solved using MIP.

7.3.2 Indices

y	Year
p	Period
s	Sub-period
n	Load level
g	Thermal unit, hydro plant or intermittent generator
t	Thermal generator
h	Storage hydro or pumped-storage hydro plant
i, j	Node
ij	Line
E, C	Sets of existing and candidates lines

7.3.3 Parameters

Costs		
α, β, γ	Weights of the different components of the objective function	
Demand		
D_{ypsni}	Demand in each node	MW
DUR_{psn}	Duration	h
R_{ps}	Operating reserve	MW
$CENS$	Cost of not served energy, value of lost load (VoLL)	EUR/MWh
$CPNS$	Cost of not served power	EUR/MW
Generation system		
$\underline{GP}_g, \overline{GP}_g$	Minimum load and maximum output of generator	MW
$\overline{GC}_h$	Maximum consumption of a pumped-storage hydro	MW
FC_t, VC_g	Fixed and variable cost of generator, variable cost includes fuel, O&M, and emission cost	EUR/h, EUR/MWh
SU_t	Startup cost of thermal unit	EUR
η_h	Efficiency of pumped-storage hydro plant	p.u.
I_{ph}	Inflows of hydro reservoir	hm^3
$\underline{R}_h, \overline{R}_h$	Minimum and maximum hydro reservoir levels	hm^3
Transmission system		
FCT_{ij}	Annualized fixed cost of a transmission line	EUR
$\overline{F}_{ij}$	Transfer capacity of a transmission line	MW
$\overline{F'}_{ij}$	Upper bound of the disjunctive constraint of a transmission line	MW
R_{ij}, X_{ij}	Resistance and reactance of a transmission line	p.u.

7.3.4 Variables

Demand		
ens_{ypsni}	Energy not served	MW
pns_{yps}	Power not served	MW
Generation system		
gp_{ypsng}, gc_{ypsng}	Generator output and pump consumption	MW
$u_{ypst}, su_{ypst}, sd_{ypst}$	Commitment, startup and shutdown of thermal unit [0,1]	p.u.
r_{yph}	Hydro reservoir level	hm^3

Continued

—cont'd

s_{yph}	Water spillage	hm^3
Transmission system		
ic_{yij}	Indicator of cumulative installed capacity of candidate line in each year {0,1}	p.u.
f_{ypsnij}	Flow through a line	MW
θ_{ypsni}	Voltage angle of a node	rad

7.3.5 Equations

The objective function reflects the minimization of total costs for the scope of the model, usually a long-term horizon with many years. Generally, three kinds of costs are included: investment costs in candidate transmission lines, system operation costs, and reliability costs. These costs are weighted with α, β, and γ, respectively, for assigning different importance to each one.

The transmission investment cost is the sum for all the years of the fixed annual cost FCT_{ij} of each candidate line times the investment decision ic_{yij}:

$$\alpha\sum_{yij} FCT_{ij} ic_{yij} \tag{7.1}$$

The system variable operation cost includes fixed, variable, and startup cost of the generating units plus the penalties associated to energy not served and deficit of operating reserve:

$$\beta\left[\begin{array}{c}\sum_{ypsnt} DUR_{psn}VC_t gp_{ypsnt} + \sum_{ypsnt} DUR_{psn}FC_t u_{ypst} + \sum_{ypst} SU_t su_{ypst} + \\ \sum_{ypsni} DUR_{psn}CENS\, ens_{ypsni} + \sum_{yps} CPNS\, pns_{yps}\end{array}\right] \tag{7.2}$$

The reliability costs are evaluated for N-1 generation and transmission contingencies:

$$\gamma\sum_{ypsni} DUR_{psn}CENS\, ens_{ypsni} \tag{7.3}$$

Balance of generation and demand for each node including the generators and the pumped hydro units located in that node and the flow injected or extracted through the lines:

$$\sum_{g\in i} gp_{ypsng} - \sum_{h\in i}\frac{gc_{ypsnh}}{\eta_h} + ens_{ypsni} = D_{ypsni} - \sum_j f_{ypsnji} + \sum_j f_{ypsnij}\ \forall ypsni \tag{7.4}$$

Operation reserve for the first load level follows. Intermittent generation does not contribute to the reserve margin

$$\sum_t \overline{GP}_t u_{ypst} + \sum_h \overline{GP}_h + pns_{ys} \geq \sum_i \left[D_{yps1i} + R_{ps}\right]\ \forall yps \tag{7.5}$$

The logical relation between commitment, startup status, and shutdown status of the thermal generation units is as follows:

$$u_{ypst} - u_{yps-1t} - su_{ypst} + sd_{ypst} = 0 \quad \forall ypst \tag{7.6}$$

For hydro power plants, a reservoir inventory constraint is formulated:

$$r_{yp-1h} + I_{ph} - \sum_{n} DUR_{psn}\left[gp_{ypsnh} - gc_{ypsnh}\right] - s_{yph} = r_{yph} \quad \forall yph \tag{7.7}$$

The second Kirchhoff law is formulated for existing and candidate lines, if they are installed:

$$\begin{aligned} f_{ypsnij} = \left(\theta_{ypsni} - \theta_{ypsnj}\right)\frac{S_B}{X_{ij}} \qquad & \forall ypsnij, ij \in E \\ \left|f_{ypsnij} - \left(\theta_{ypsni} - \theta_{ypsnj}\right)\frac{S_B}{X_{ij}}\right| \leq \overline{F'}_{ij}\left(1 - ic_{yij}\right) \quad & \forall ypsnij, ij \in C \end{aligned} \tag{7.8}$$

The maximum transfer capacity in existing and candidate transmission lines is as follows:

$$\begin{aligned} \left|f_{ypsnij}\right| \leq \overline{F}_{ij} \quad & \forall ypsnij, ij \in E \\ \left|f_{ypsnij}\right| \leq \overline{F}_{ij} ic_{yij} \quad & \forall ypsnij, ij \in C \end{aligned} \tag{7.9}$$

The relation between the indicators of cumulative installed capacity in consecutive years is as follows:

$$ic_{yij} \leq ic_{y'ij} \quad \forall yy'ij, ij \in C, y' > y \tag{7.10}$$

Bounds on generation and consumption variables, reservoir volume, and energy not served are as follows:

$$\begin{aligned} \underline{GP}_g u_{ypsg} \leq gp_{ypsng} \leq \overline{GP}_g u_{ypsg} \quad & \forall ypsng \\ 0 \leq gc_{ypsnh} \leq \overline{GC}_h \quad & \forall ypsnh \\ \underline{R}_h \leq r_{yph} \leq \overline{R}_h \quad & \forall yph \\ 0 \leq ens_{ypsni} \leq D_{ypsni} \quad & \forall ypsni \end{aligned} \tag{7.11}$$

Although in the previous formulation, we have assumed only AC transmission lines, also HVDC lines are considered, given the current deployment limitations that AC lines are observing in many countries. In addition, new network devices as PSTs play a role in making a flexible network operation and also need to be included in the expansion problem.

For clarification purposes, we have decided to present in this chapter a general formulation for a TEP model. However, recent developments need to be considered, especially those oriented to improve the assessment of the main drivers; see [2]. For example, the following are very relevant:

- Renewable generation and their variability along the year is represented as a conventional generation with a maximum capacity that varies hourly and with zero variable cost; see [1].

- Uncertainty in renewable generation (mainly hydro, solar, and wind) is included via several operation scenarios.
- Capacity constraints between countries (or regions) are introduced as lower and/or upper bounds on the flows connecting those countries that could be released by future transmission lines.

7.4 A Complex Problem: Modeling Choices

The formulation above can be adapted to the wide variety of situations that can be presented in a real-life instance of transmission planning. TEP is notoriously difficult to solve, and so the planner must make the exercise of selecting the problem features that he or she will deem relevant for a particular instance of planning. This section reviews the main modeling choices that can be made when performing this exercise.

First, it is necessary to establish whether transmission will be planned in an isolated manner, or jointly with the expansion of generation. In deregulated contexts, it is not possible to establish a centralized plan for generation; each company makes their own investment decisions independently. However, solving the two problems simultaneously can give interesting information, particularly in the cases where large amounts of renewable generation must be placed far from current demand centers [16–18].

This renewable generation is also the cause of one of the key factors behind TEP's complexity: uncertainty. Generation expansion is the single most important source of uncertainty, only followed by generation costs. These types of uncertainty, which cannot be described by a probability distribution, are known as nonrandom. Random uncertainties, on the contrary, can be described by means of a short-term probability distribution. Demand, renewable generation, hydro inputs, or element contingencies are all examples of this sort of uncertainty. The latter is especially important for transmission expansion, as reinforcements are often selected with the objective of improving reliability. An N-1 criterion is the usual modeling choice [19, 20], although it is also possible to predefine a list of contingencies. Most works have disregarded uncertainty, due to the added difficulty of solving the problem in its uncertain version. When it has been included, three main techniques have been used. Stochastic optimization minimizes the expected value of the total cost related to the expansion plan [19, 21–23]. On the contrary, robust optimization focuses on the worst-case scenario [24–26]. Last, fuzzy decision analysis considers the outcome associated to different scenarios simultaneously, as a multi-criteria decision problem would do with different objectives [27, 28].

TEP is naturally a multi-stage problem: investments are made in discrete moments where decisions are evaluated using the most recent information. However, given its complexity, only a few works solve it as a multi-stage problem [23, 29]. Some works consider several time horizons, but without considering their relationship to each other—that is, in a sequential-static manner

[19, 30–32]. Finally, most studies consider only one point in the future as their single time horizon. This is known as the static TEP problem [33–35].

Another important factor is whether market considerations are important for a given instance of planning. In most cases, transmission is operated in a centralized manner while generation operates in a market. Given the complexities of TEP, most works ignore the market for generation and instead assume cost-based operation [33, 36]. Some limited cases study the impact of generation markets [33, 36]. A few works study decentralized expansion with different agents acting independently [37–39] and the possibility of forming coalitions to build transmission lines in the interest of generation companies or demand [40].

The criteria that can be considered in TEP include investment and operation cost (which considers not only generation costs but also penalties for ENS or carbon emissions), and, in a market context, aggregate social welfare and the support of competition. Most works focus on a single added objective, but some authors have applied tools such as fuzzy decision theory [23], goal programming (GP) [41], and analytic hierarchy process (AHP) [42].

The level of detail on system operation is also a key modeling decision. Namely, transportation models (which only consider energy balances) [43], DCPF models [44, 45], and ACPF models can be used [46]. DCPF seem to offer a good compromise between accuracy and simplicity and are the most vastly used. ACPF are, in general, too difficult to solve due to nonlinearity, although they would have the capability of incorporating stability considerations.

Last, most works do not consider the new technologies described above; only a few do incorporate HVDC [47–50] or PSTs explicitly [50]. How to identify candidate investments with a potential to improve the transmission networks is particularly difficult, as it is not possible to consider all the possible transmission lines. We refer the reader to reference [51] for a detailed presentation of a method to deal with this issue.

7.5 Solution Methods

The previous transmission expansion planning problem has been stated as an optimization problem. Under this framework, classical and nonclassical optimization methods have been widely used.

7.5.1 Classical Methods

Linear programming (either simplex or interior point) (LP) methods are powerful and robust algorithms able to solve large-scale optimization problems. In this case, the discrete nature of the investments is ignored. Linear formulations can accommodate transportation power flow models [43].

Mixed integer programming (MIP) acknowledges the discrete nature of investment (a transmission line can either be installed or not). In addition, a

DC load flow can be introduced by using a disjunctive constraint (as Eq. 7.10) to improve the representation of the network operation. A linear approximation of ohmic losses can also be considered. This option is the preferred one in an optimization context, and can be found in references such as [52, 53].

Nonlinear programming (NLP) and mixed-integer nonlinear programming (MINLP) are used to describe the nonlinearities of an AC power flow. The computational burden associated to this can only be solved in relatively small networks, and is only justified when some effects that cannot be accurately captured with the DC power flow need to be taken into account (e.g., reactive power). Only a few works follow this route [54].

As the size of the electric system and number of operation hours/scenarios to consider grows, stochastic programming (SP) solved via decomposition algorithms becomes the preferred technique. A good review of different techniques that can be applied to increase the computation speed of the problem can be found in reference [55].

7.5.2 Nonclassical Methods

Nonclassical methods do not guarantee the global optimum nor give an estimation of the closeness to it. However, they may have affordable computation times to reach a certain optimal solution.

Metaheuristic algorithms improve a solution iteratively, usually including some form of random evolution. Genetic algorithms replicate the principles of Darwinian evolution to solve optimization problems. They are currently a widely popular approach when solving combinatorial problems, and have been applied to TEP in a long list of studies [56].

Other types of algorithms are based on heuristic guided searches, such as [45]. No a-priori guarantee of the goodness of the final solution or the soundness of the rules applied can be tracked. In particular, greedy local searches directed by sensitivity analyses are rather popular. Expert systems can extract expansion rules by generalizing information from a set of sample systems and were applied in works such as [57].

7.5.3 Iterative Methods With Human Interaction

The previous optimization methods cannot represent all the details from a practical point of view. One way to reconcile it is to iterate manually between two different functional modules:

- an expansion planning module that provide one or several transmission expansion plans and
- an evaluation module that improves the evaluation of the system operation by introducing additional consideration not previously taken into account in the expansion module.

By coordinating both modules, it is possible to find a satisfactory and practical expansion plan fulfilling reasonable optimality conditions and detailed evaluation of the system operation.

7.6 Designing Transmission for a High Renewable Penetration: A Planning Exercise

This section presents a transmission planning exercise, which can be found in more detail in the paper [58]. It uses the tool TEPES (transmission expansion planning for an electrical system), which was developed at the Institute for Research in Technology at Universidad Pontificia Comillas. A complete description of this tool, which features stochastic decomposition and modeling of HVDC and PSTs, can be found on its website (https://www.iit.comillas.edu/aramos/TEPES.htm).

The planning exercise presented here should be understood as an illustration of the application of optimization-based approaches in realistic context, even in large systems with a reasonably detailed model for uncertainty. The case is based on the south-west region in Europe, composed of the countries France, Spain, and Portugal, in the medium-term future (2030) and considering current RES targets. The rest of the continent is considered only in a simplified manner, with a single node representing each country. The data were collected from publicly available information from the relevant TSOs. NTCs between countries were taken from Entso-e. This initial network is represented in Fig. 7.1 and Table 7.1.

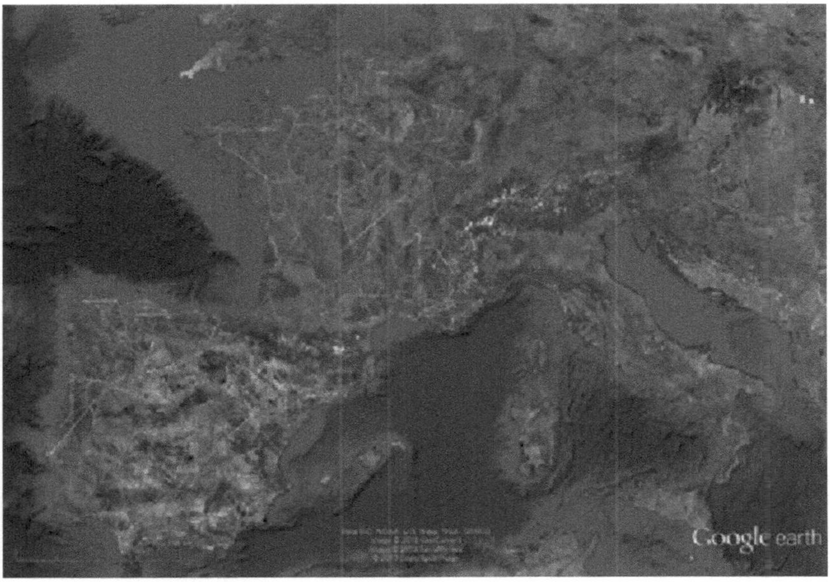

FIG. 7.1 Initial network. *Green (light grey)* is used for 220 kV (only for Spain, France, and Portugal) and *red (dark grey)* for 400 kV *lines* and HVDC.

TABLE 7.1 System Summary

	Number
Nodes	1413
Existing lines	2661
Thermal units	349
Hydro plants	116
Intermittent generators	566

TABLE 7.2 Demand and Peak Demand for the Year 2030

	PT	ES	FR	Rest	Total
Demand (TWh)	64	364	558	2677	3663
Peak demand (GW)	12	63	103	424	595

Uncertainty has been modeled by means of four operation situations corresponding to peak demand, maximum cross-border flows and two other intermediate states selected via clustering techniques. RES have been distributed based on resource potentials Tables 7.2 and 7.3.

The model considers the installation of AC overhead and underground lines, transformers, converter stations, and HVDC overhead, underground, or submarine cables. In particular, it selects 1406 candidate investments, including reinforcements and new transmission lines.

TEPES found the optimal solution for this cases study in less than 1 h in an ordinary desktop computer. A summary of the results is presented in Figs. 7.2 and 7.3 and Table 7.4.

As can be seen, most of the expansion occurs in Spain, where a large share of RES was located. The transmission plan has reinforced the south and northeast of Spain considerably. In some scenarios, there are still imports of energy from France into Spain and Portugal, given that there is much more conventional generation located in France. The wide variation of flows for the different operation situations considered highlights the importance of considering uncertainty, as well as keeping the study region large enough to capture the large cross-border flows that are bound to happen in the energy system of the future. Even in these conditions (uncertainty and large systems), the case study presented demonstrates how it is possible to tackle this large optimization problem successfully.

FIG. 7.2 Optimal network, with 220kV *lines* represented in *green (light grey)* and 400kV or HVDC represented in *red (dark grey)*.

TABLE 7.3 Installed Capacity of Conventional Generation and RES (Peak and Energy) Expected for the Year 2030

	PT	ES	FR	Rest	Total	%
Gas (MW)	0	814	1,318	86,065	88,197	19
CCGT (MW)	2,791	29,283	8,808	103,771	144,653	31
Coal (MW)	5,68	9,180	957	94,465	105,170	23
OCGT (MW)	2,467	0	2,456	0	4,923	1
Oil (MW)	774	3,714	5,829	17,821	28,138	6
Nuclear (MW)	0	2,875	44,696	41,938	89,509	19
Total (MW)	6,600	45,866	64,064	344,060	460,590	
RES peak expected (GW)	17	98	75	314	435	
RES energy expected (TWh)	60	339	315	1,679	2,393	

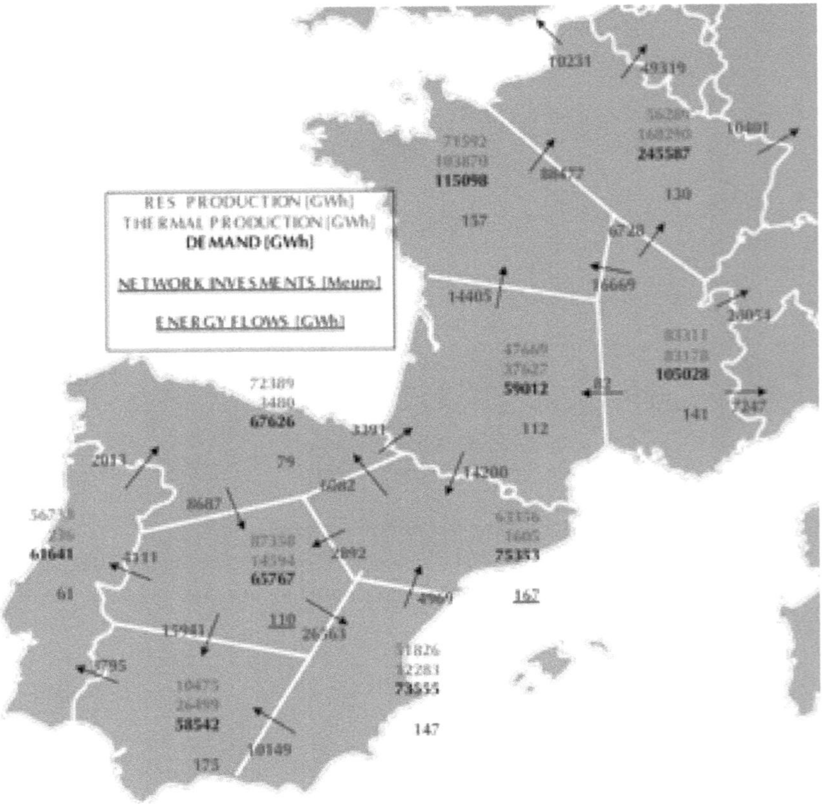

FIG. 7.3 Resulting cross-border flows in the optimal solution.

TABLE 7.4 Characterization of the Optimal Solution

Investment cost (M€/y)	1,279
Operation cost (M€/y)	11,513
Total cost (M€/year)	12,792
Length built (km)	27,200
Capacity built (GW)	588
GW*km built (GW*km)	24,729

7.7 Conclusions

The transmission network will need to be improved in order to keep up with the changes that are affecting the global energy system. The current need for transmission is the result of renewable production targets—which stress the system with nondispatchable generation that calls for improved network support—and the creation of regional electricity markets that demand higher cross-border flows. In addition, transmission must respond to the need for reliability, the reduction of losses, the connection of new generation, or the resolution of local capacity constraints.

Several developments have recently altered the network landscape. The proliferation of DERs has invalidated the *trickling-down* model, where power flew from higher voltage levels in transmission to lower ones in distribution, from generation to demand. This questions the existing assumptions on valid network architectures and system operation. In addition, it blurs the boundary between transmission and distribution.

DERs should be understood as a local source of flexibility that can postpone the need for transmission investment. Although most smart-grid projects focus on distribution and demand-side management, the view with respect of transmission seems to emphasize the use of technology to improve the management of existing and new transmission assets. Long-distance flows will benefit from advanced conductors such as high-temperature cables or new phase configurations. Underground cables will be necessary to cater to urban areas and offshore wind farms. HVDC, and in particular VSC, makes it possible to transfer large amounts of power to greater distances with added benefits for controllability. PSTs will be instrumental to overcome congestion and control flows in a regional system with increasing cross-border transactions. Transmission switching might provide a lower-cost alternative with the same aim. Smart control centers and substations will make use of state measurements and GIS to perform more accurate stability analyses, optimize the utilization of lines with dynamic rating, and detect potential issues such as vegetation or ice.

TEP, as featured above, can be expressed as a MIP optimization problem amenable to classical decomposition algorithms. This chapter has presented a structured formulation that can be adapted to the wide variety of existing problem instances; given the complexities inherent to the problem, the planner must specify the scope of the planning exercise very carefully. The consideration of market structures, mainly for generation, must be decided. Next, the inclusion of uncertainties is one of the most complicating factors that have a deep impact on the resulting expansion plan. The higher shares of renewable penetration are the single most important factor in this respect, followed by the inclusion of contingency scenarios with the aim of capturing reliability. While most works do not consider the new available technologies, it seems that the incorporation of HVDC and PSTs will be key to obtaining efficient expansion plans in the current energy context.

The relevancy and difficulty of TEP has motivated the application of a wide array of techniques to its resolution, classical, and nonclassical. MIP dominates the classical programming spectrum, often in conjunction with stochastic programming tools. The formulation provided in this chapter falls in this category. In the nonclassical domain, a long list of different methods has been tested, sometimes striking a good compromise between affordable times and solution accuracy. Iterative methods with human interaction complement optimization with external analyses that might include additional considerations (such as stability tests) that are too complex to be modeled in an optimization setting. This hybrid method iterates between optimization and evaluation analyses, and can find satisfactory planning solutions that do not violate any important constraints in the subsequent system operation.

This chapter illustrates TEP with a case study based on the European Continental South-West Region, composed of France, Spain, and Portugal. The case study considers the high levels of renewable targets imposed for 2030. Although the system is large (over 1000 nodes and 2000 lines, and considering four different operation scenarios), the model TEPES finds an optimal solution is less than 1 h using an ordinary desktop computer. The solution obtained shows the resulting investments, which reinforce strategically the zones where large amounts of RES are located. Cross-border flows are also presented and analyzed. This exercise demonstrates the possibility of performing large-scale transmission planning using optimization.

The challenges of the future global system, sustainable and connected, can only be addressed by using all the available tools. We need the transmission network to be a backbone that provides support to increasing amounts of renewables, and to endure the long-distance cross-border flows that will be the basis for the future regional markets. For this, it is imperative to use all the means at our disposal, new technologies, as well as smart planning methods. We need better transmission networks for a smarter global system.

References

[1] EC–European Commission, 2016 "Clean energy for all Europeans," COM (2016), vol. 860.

[2] G. Barbose, "US Renewables Portfolio Standards: 2016 Annual Status Report," 2016.

[3] I.J. Perez-Arriaga, The transmission of the future: the impact of distributed energy resources on the network, IEEE Power Energy Mag. 14 (4) (2016) 41–53.

[4] P. Vasquez, F. Olsina, Valuing flexibility of DG investments in transmission expansion planning, in: 2007 IEEE Lausanne Power Tech, 2007, pp. 695–700.

[5] MIT, Utility of the Future, project report, 2016. http://energy.mit.edu/wp-content/uploads/2016/12/Utility-of-the-Future-Full-Report.pdf.

[6] F. Li, et al., Smart transmission grid: vision and framework, IEEE Trans. Smart Grid 1 (2) (2010) 168–177.

[7] R. Rudervall, J. Charpentier, R. Sharma, High voltage direct current (HVDC) transmission systems technology review paper, Energy Week 2000 (2000) 1–19.

[8] M.P. Bahrman, B.K. Johnson, The ABCs of HVDC transmission technologies, IEEE Power Energy Mag. 5 (2) (2007) 32–44.

[9] J.G. Dedecca, S. Lumbreras, A. Ramos, R.A. Hakvoort, P.M. Herder, Expansion planning of the North Sea offshore grid: simulation of integrated governance constraints. Energy Econ. 72 (2018) 376–392, https://doi.org/10.1016/j.eneco.2018.04.037.

[10] J. Verboomen, et al., Phase shifting transformers: principles and applications, in: 2005 International Conference on Future Power Systems, 2005, p. 6.

[11] J. Paserba, Recent power electronics/FACTS installations to improve power system dynamic performance, in: 2007 IEEE Power Engineering Society General Meeting, 2007, pp. 1–4.

[12] K.W. Hedman, et al., Optimal transmission switching with contingency analysis, IEEE Trans. Power Syst. 24 (3) (2009) 1577–1586.

[13] E.B. Fisher, R.P. O'Neill, M.C. Ferris, Optimal transmission switching, IEEE Trans. Power Syst. 23 (3) (2008) 1346–1355.

[14] H. Ying, G. Andersson, R.N. Allan, Determining optimum location and number of automatic switching devices in distribution systems, in: 1999 International Conference on Electric Power Engineering, PowerTech Budapest 99, 1999, p. 259.

[15] Y. Yang, et al., Power line sensornet-a new concept for power grid monitoring, in: 2006 IEEE Power Engineering Society General Meeting, 2006, p. 8.

[16] B. Alizadeh, S. Jadid, Reliability constrained coordination of generation and transmission expansion planning in power systems using mixed integer programming, IET Gener. Transm. Dis. 5 (9) (2011) 948–960.

[17] D. Pozo, E.E. Sauma, J. Contreras, A three-level static MILP model for generation and transmission expansion planning, IEEE Trans. Power Syst. 28 (1) (2013) 202–210.

[18] A. Liu, et al., Co-optimization of transmission and other supply resources, in: Prepared for the Eastern Interconnection States' Planning Council, National Association of Regulatory Utility Commissioners, Washington, DC, 2013.

[19] C. Serna, J. Duran, A. Camargo, A model for expansion planning of transmission systems a practical application example, IEEE Trans. Power App. Syst. PAS-97 (2) (1978) 610–615.

[20] G.A. Orfanos, et al., Transmission expansion planning by enhanced differential evolution, in: 2011 16th International Conference on Intelligent System Application to Power Systems (ISAP), 2011, pp. 1–6.

[21] G. Latorre, et al., PERLA: a static model for long-term transmission planning: modeling options and suitability analysis, in: Proceeding of the 2nd Conferences on Electrical Engineering Spanish–Portuguese, 1991. https://www.iit.comillas.edu/aramos/papers/COIMBRA.pdf.

[22] T. Akbari, S. Zolfaghari, A. Kazemi, Multi-stage stochastic transmission expansion planning under load uncertainty using benders decomposition, Int. Rev. Electr. Eng. 4 (5) (2009) 976–984.

[23] M. Moeini-Aghtaie, A. Abbaspour, M. Fotuhi-Firuzabad, Incorporating large-scale distant wind farms in probabilistic transmission expansion planning—Part I: theory and algorithm, IEEE Trans. Power Syst. 99 (2012) 1.

[24] V. Miranda, L.M. Proenca, Probabilistic choice vs. risk analysis-conflicts and synthesis in power system planning, IEEE Trans. Power Syst. 13 (3) (1998) 1038–1043.

[25] M.O. Buygi, et al., Market based transmission planning under uncertainties, in: 2004 International Conference on Probabilistic Methods Applied to Power Systems, 2004, pp. 563–568.

[26] P. Maghouli, et al., A scenario-based multi-objective model for multi-stage transmission expansion planning, IEEE Trans. Power Syst. 26 (1) (2011) 470–478.

[27] H. Sun, D.C. Yu, A multiple-objective optimization model of transmission enhancement planning for independent transmission company (ITC), in: 2000 IEEE Power Engineering Society Summer Meeting, vol. 4, 2000, , pp. 2033–2038.

[28] A.A. El-Keib, J. Choi, T. Tran, Transmission expansion planning considering ambiguities using fuzzy modeling, in: 2006 IEEE PES Power Systems Conference and Exposition, PSCE '06, 2006, pp. 207–215.

[29] K. Yoshimoto, K. Yasuda, R. Yokoyama, Transmission expansion planning using neuro-computing hybridized with genetic algorithm, in: 1995 IEEE International Conference on Evolutionary Computation, 1995, p. 126.

[30] C. Dechamps, E. Jamoulle, Interactive computer program for planning the expansion of meshed transmission networks, Int. J. Electr. Power Energy Syst. 2 (2) (1980) 103–108.

[31] C. Barbulescu, et al., Congestion management driven transmission expansion planning, in: Proceedings of 2011 46th International Universities' Power Engineering Conference (UPEC), 2011, pp. 1–6.

[32] M. Costeira, J.T. Saraiva, Discrete evolutionary particle swarm optimization for multiyear transmission expansion planning, in: 17th Power Systems Computation Conference, 2011. Stockholm, Sweden.

[33] L.L. Garver, Transmission network estimation using linear programming, IEEE Trans. Power App. Syst. PAS-89 (7) (1970) 1688–1697.

[34] R.J. Bennon, J.A. Juves, A.P. Meliopoulos, Use of sensitivity analysis in automated transmission planning, IEEE Trans. Power App. Syst. PAS-101 (1) (1982) 53–59.

[35] P. Buijs, R. Belmans, Transmission investments in a multilateral context, IEEE Trans. Power Syst. 27 (1) (2012) 475–483.

[36] N. Gupta, R. Shekhar, P.R. Kalra, A Probabilistic Transmission Expansion Planning Methodology Based on Roulette Wheel Selection and Social Welfare, ArXiv Computer Science, 2012.

[37] J. Contreras, A Cooperative Game Theory Approach to Transmission Planning in Power Systems, Department of Electrical Engineering and Computer Sciences, University of California at Berkeley, http://www.uclm.es/area/gsee/JavierC/tesis/thesis_JC.pdf, 1997.

[38] J. Contreras, F.F. Wu, A kernel-oriented algorithm for transmission expansion planning, IEEE Trans. Power Syst. 15 (4) (2000) 1434–1440.

[39] C. Cagigas, M. Madrigal, Centralized vs. competitive transmission expansion planning: the need for new tools, in: 2003 IEEE Power Engineering Society General Meeting, vol. 2, 2003, p. 1017.

[40] J. Contreras, F.F. Wu, Coalition formation in transmission expansion planning, in: 1999 IEEE Power Engineering Society Winter Meeting, vol. 2, 1999, p. 887.

[41] J. Alseddiqui, R.J. Thomas, Transmission expansion planning using multi-objective optimization, in: 2006 IEEE Power Engineering Society General Meeting, 2006, p. 8.

[42] E.A.C.A. Neto, et al., An AHP multiple criteria model applied to transmission expansion of a Brazilian southeastern utility, in: 2010 IEEE/PES Transmission and Distribution Conference and Exposition: Latin America (T&D-LA), 2010, pp. 575–580.

[43] A. Marin, J. Salmeron, Electric capacity expansion under uncertain demand: decomposition approaches, IEEE Trans. Power Syst. 13 (2) (1998) 333–339.

[44] M.V.F. Pereira, et al., A decomposition approach to automated generation/transmission expansion planning, IEEE Trans. Power App. Syst. PAS-104 (11) (1985) 3074–3083.

[45] S. Binato, G.C. de Oliveira, J.L. de Araujo, A greedy randomized adaptive search procedure for transmission expansion planning, IEEE Trans. Power Syst. 16 (2) (2001) 247–253.

[46] A.P. Meliopoulos, et al., Optimal long range transmission planning with AC load flow, IEEE Trans. Power App. Syst. PAS-101 (10) (1982) 4156–4163.

[47] A.O. Ekwue, B.J. Cory, Transmission system expansion planning by interactive methods, IEEE Trans. Power App. Syst. 103 (7) (1984) 1583–1591.

[48] J. O. G. Tande, M. Korpås, L. Warland, K. Uhlen, F. Van Hulle, "Impact of TradeWind Offshore Wind Power Capacity Scenarios on Power Flows in the European HV Network," 2008.

[49] A. Garzillo, et al., Offshore grids in Europe: the strategy of Ireland for 2020 and beyond, in: 2010 9th IET International Conference on AC and DC Power Transmission, ACDC, 2010, pp. 1–7.

[50] G. Blanco, et al., Real option valuation of FACTS investments based on the least square monte carlo method, IEEE Trans. Power Syst. 26 (3) (2011) 1389–1398.

[51] S. Lumbreras, A. Ramos, P. Sánchez, Automatic selection of candidate investments for transmission expansion planning. Int. J. Electr. Power Energy Syst. 59 (2014) 130–140, https://doi.org/10.1016/j.ijepes.2014.02.016.

[52] S.T.Y. Lee, K.L. Hicks, E. Hnyilicza, Transmission expansion by branch-and-bound integer programming with optimal cost—capacity curves, IEEE Trans. Power App. Syst. 93 (5) (1974) 1390–1400.

[53] S. Lumbreras, A. Ramos, Transmission expansion Planning Using an efficient version of benders' decomposition. A case study. in: IEEE Power Tech, 2013https://doi.org/10.1109/PTC.2013.6652091. Grenoble, France.

[54] M. Rahmani, et al., Efficient method for AC transmission network expansion planning, Electr. Pow. Syst. Res. 80 (9) (2010) 1056–1064.

[55] S. Lumbreras, A. Ramos, How to solve the transmission expansion planning problem faster: acceleration techniques applied to benders' decomposition. IET Gener. Transm. Dis. 10 (2016) 2351–2359, https://doi.org/10.1049/iet-gtd.2015.1075.

[56] H.A. Gil, E.L. da Silva, A reliable approach for solving the transmission network expansion planning problem using genetic algorithms, Electr. Pow. Syst. Res. 58 (1) (2001) 45–51.

[57] R.K. Gajbhiye, et al., An expert system approach for multi-year short-term transmission system expansion planning: an Indian experience, IEEE Trans. Power Syst. 23 (1) (2008) 226–237.

[58] S. Lumbreras, A. Ramos, F. Banez-Chicharro, Optimal transmission network expansion planning in real-sized power systems with high renewable penetration. Electr. Pow. Syst. Res. 149 (2017) 76–88, https://doi.org/10.1016/j.epsr.2017.04.020.

Chapter 8

Power Quality in Smart Grids

Miloud Rezkallah*, Ambrish Chandra*, Abdelhamid Hamadi*, Hussein Ibrahim† and Mazen Ghandour‡

**École de Technologie Supérieure (ETS), Montréal, QC, Canada, †Institut Technologique de Maintenance Industrielle (ITMI), Cégep de Sept-Îles, Sept-Îles, QC, Canada, ‡Faculty of Engineering, Lebanese University, Beirut, Lebanon*

Chapter Outline

8.1 Introduction

The power quality and harmonics issues in future modernized electrical networks' so-called smart grid, which is based on advanced information and communication technology (ICT), can affect the span life and efficiency of electrical network, and can also reduce system performance. Generally, the smart grid will consist of many microgrids connected to the main grid, as they can operate as standalone systems (off grid) in the case of faults in the grid, or as grid-connected systems. The microgrid is defined as a small-scale low voltage AC grid. It employs a combination of distributed energy sources, such as solar photovoltaic panels, a wind turbine, microhydro power, and/or a diesel generator [1]. It also contains a storage system, controlled/uncontrolled power electronics devices, and loads. Generally, the indexes of power in grid or in microgrid are

Pathways to a Smarter Power System. https://doi.org/10.1016/B978-0-08-102592-5.00008-9

the frequency and voltage variations, harmonics, power factor, flicker, etc. [2]. Mostly, the power quality at a point of common coupling (PCC) is affected due to the power converters device used to control the system elements, such as energy sources, battery energy storage system, etc., and due to the presence of polluting loads. Therefore, it is necessary to solve this issue locally at the level of each microgrid. According to the IEEE 519 and IEEE 1159 standards, total harmonic distortion (THD) should be maintained below 5% and the amplitude voltage variation should not exceed $\pm 5\%$ [3, 4]. Regarding harmonics mitigation, power factor correction, and unbalance, passive harmonics filters and active filters (so-called shunt filters) are installed in parallel to the power lines [5], and for voltage sag and swell compensation, series power converters are suggested [6]. To achieve these tasks of series and shunt power converters and provide active power to the connected load, simultaneously, a single unit series of active power filters, called a unified power quality conditioner (UPQC), is advised in [7, 8]. According to [9], UPQC is classified in two groups: (1) based on physical structure (technology of storage used, number of phases, and physical location of series and parallel power converters); and (2) voltage sag approach used. Generally, for active and reactive power compensation, UPQS-S is selected.

Regarding the control of UPQC, two strategies of control are required to operate series/parallel power converters. In [10], many control strategies for UPQCs, such as d-q theory, P-Q theory, and I cos φ algorithm, are detailed.

Recently, UPQC integrated with DC microgrid (UPQC-DC) was proposed in [11] as solution for the smart grid to achieve objectives such as smart load management and plug-in electric vehicles. According to [12], DC microgrid also suffers from power quality issues due to many factors such as: (1) sudden variation of voltage at utility grid; (2) generating of harmonics by power converters; (3) voltage disturb in bipolar DC bus; (4) sudden variation of the DC loads; (5) DC bus faults; and (6) circulating current.

In this chapter, low-voltage UPQC-S is proposed for the smart grid, which is based on n-microgrids as shown in Fig. 8.1. These microgrids, based on solar photovoltaic systems, can operate independently to the grid (islanding mode) and can also connect easily to the grid (grid-connected mode).

The stable and clean energy can be transferred in the case of the grid-connected mode from the utility grid and solar photovoltaic panels to the connected loads, and from the solar photovoltaic panels to the grid during peak hours. In the case of the off-grid mode (islanding mode), solar photovoltaic panels supported by battery energy storage system provide a clean, stable, and uninterruptible power to the connected loads.

8.2 Configuration of UPQC-S Based Microgrid and Operation Modes

Fig. 8.2 shows the configuration of the proposed UPQC-S based microgrid for smart grid application. It consists of two parts: series branch and shunt branch. The parallel power converter (VSC1) in the shunt branch is connected in

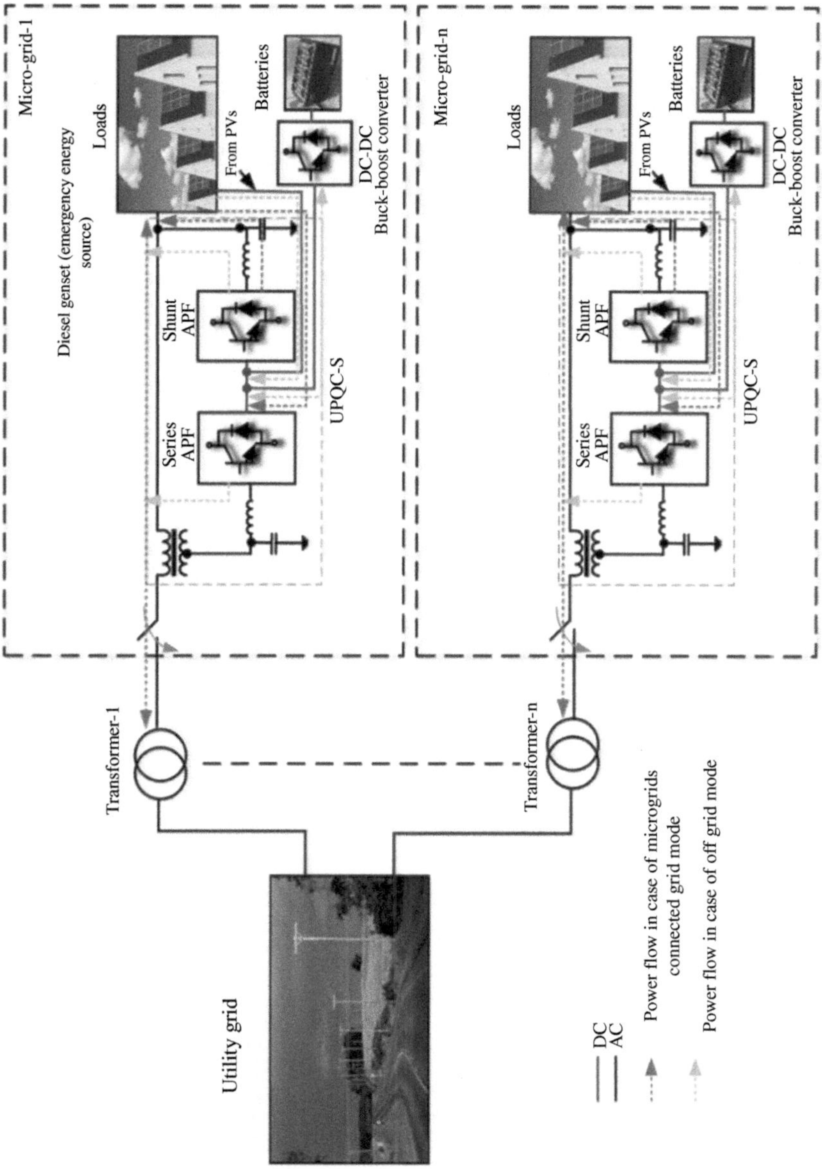

FIG. 8.1 UPQC-S for smart grid based on microgrids.

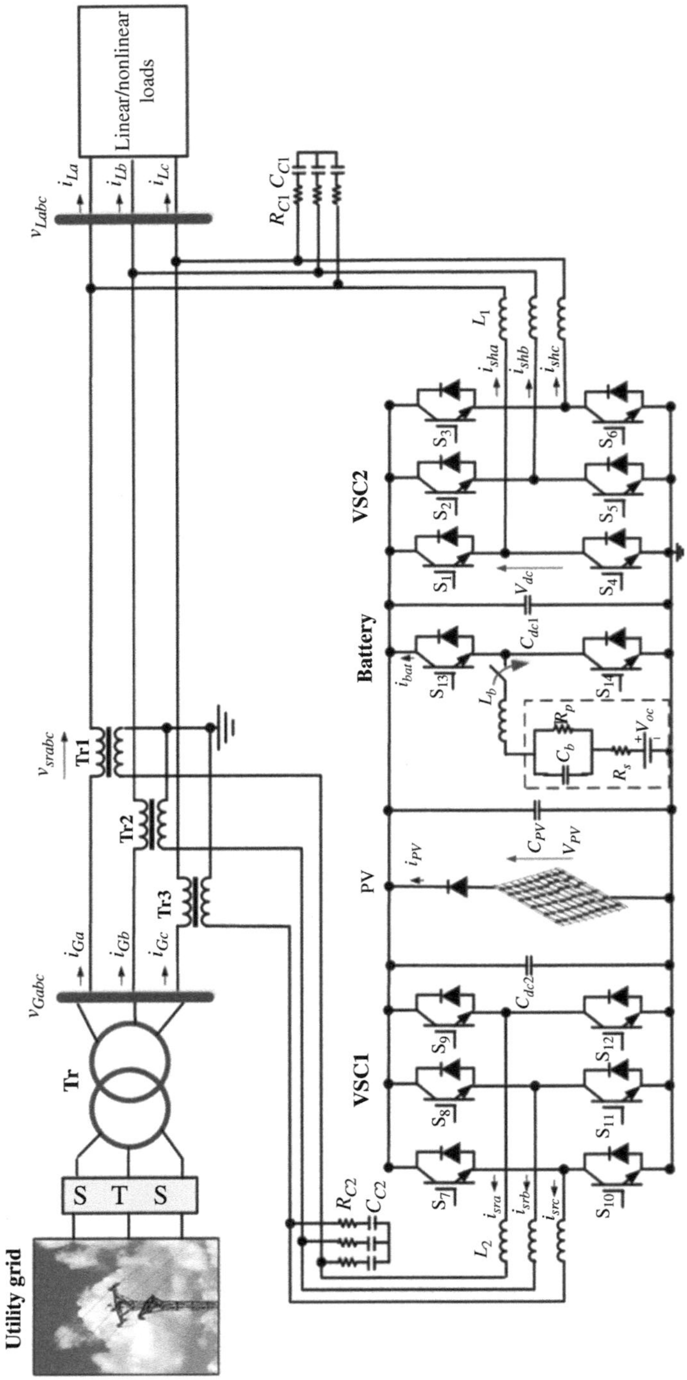

FIG. 8.2 Detailed scheme of UPQC-S based microgrid configuration.

parallel; it is controlled to mitigate harmonics and compensate the reactive power, load unbalance, and supply active power to the load and utility grid in peak hours.

Regarding the series power converter (VSC2), which is controlled to compensate voltage sag/swell in grid-connected mode and to regulate the load voltage at a constant rate, as well as the system frequency in off-grid mode, it is connected in series with the power lines through a transformer. The solar photovoltaic panels are installed on the roofs of houses and are connected directly to the common DC bus. Furthermore, the UPQC-S configuration based on microgrid configuration is reinforced by a battery storage system connected to the common DC-bus through a controlled DC-DC buck/boost converter to balance the power in the system when the system operates in the off-grid mode. RC passive filters are installed at the output of the both power converters to eliminate the high frequency noise coming from the switching frequencies. In grid-mode operation, the DC-DC buck/boost is completely isolated from the system and only the VSC2 ensures the regulation of the DC-bus and power quality at PCC, as well as extracting the maximum of power from solar photovoltaic panels. In off-grid operation, DC-DC buck/boost converter ensures regulation of DC-link voltage, MPPT tracking, and control of the battery storage system.

8.3 Control of UPQC-S Based Microgrid

In this section, the developed control strategies for the series and shunt power converters, as well as for the DC-DC buck/boost converter, are detailed.

8.3.1 Control Strategy for the Series Power Converter (VSC1)

In Fig. 8.3, the developed control strategy to compensate voltage imbalance (sag and swell) when the system operates in grid-connected mode, as well as to regulate the voltage at a constant rate and frequency at the PCC when the system is operated in off-grid mode, is presented.

The two-layer Adaline neural network (ANN) algorithm, which consists of an input and output layer, is employed to estimate the load harmonics currents ($i_h(k)$). As presented in Fig. 8.4, the inputs (x_i), which represent in our study the measured load currents (i_L) and in-phase unit vectors templates, are multiplied by the modifiable weight (Wi) and then summed by summation units (output neuron). The output, which represents the estimated current (i_{est} (k)), is compared to load currents and the resulting error is used by the LMS learning algorithm to train the network weights.

The load harmonics current ($i_h(k)$), are calculated as follows:

$$i_L(\mathrm{k}) = i_{Ldc} + i_h(k) \quad (8.1)$$

where i_{Ldc} (k) denote the fundamental component and i_h (k) presents harmonic component, which is obtained as follows [13]:

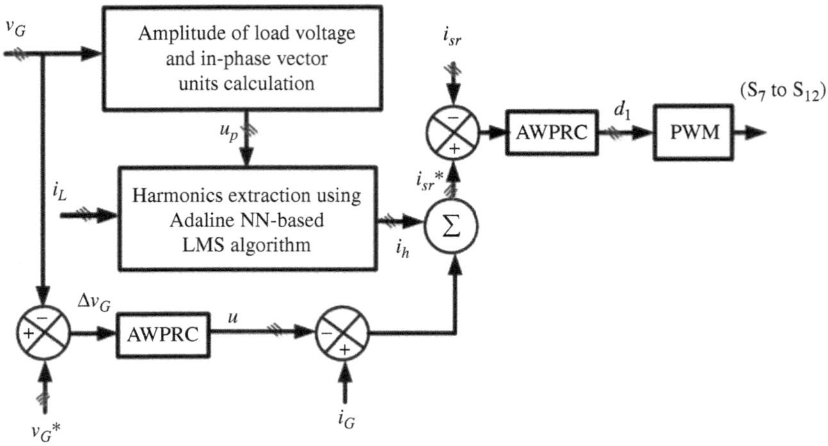

FIG. 8.3 Control strategy for the series power converter.

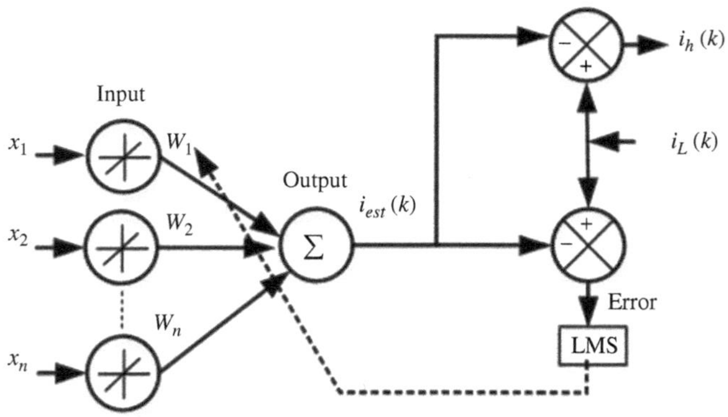

FIG. 8.4 Adaline NN-based LMS algorithm for load harmonics estimation.

$$i_h(k) = i_L(k) - W\sin(k\omega t) \tag{8.2}$$

where $W\sin(k\omega t)$ represents the load current fundamental sine part, which is obtained as is presented in Fig. 8.3 by calculating the in-phase unit vectors templates as follows:

$$u_{pa} = \frac{v_{Ga}}{V_{GP}}, u_{pb} = \frac{v_{Gb}}{V_{GP}}, u_{pc} = \frac{v_{Gc}}{V_{GP}} \tag{8.3}$$

where v_{Ga}, v_{Gb}, and v_{Gc}, u_{pa}, u_{pb}, and u_{pc}, and V_{GP} represent the grid phase voltages, in-phase unit vector templates, and the amplitude of the grid voltage, respectively, which is calculated as follows:

$$V_{GP} = \sqrt{\frac{2}{3}\left(v_{Ga}^2 + v_{Gb}^2 + v_{Gc}^2\right)} \tag{8.4}$$

The reference grid voltages are calculated:

$$\begin{cases} v_{Ga}^{*} = V_{GP}\sin(\omega t) \\ v_{Gb}^{*} = V_{GP}\sin(\omega t - 2\pi/3) \\ v_{Gc}^{*} = V_{GP}\sin(\omega t + 2\pi/3) \end{cases} \tag{8.5}$$

where ω is the pulsation and is equal to $2\pi f_s t$.

The measured grid voltage (v_G) is compared with its references ($v_G{}^*$), and its errors (Δv_G) are fed to proportional resonant controllers with antiwindup (AWPRCs):

$$\begin{cases} i_{Ga}^{*} = G_{PRC}(S)\left(v_{Ga}^{*} - v_{Ga}\right) \\ i_{Gb}^{*} = G_{PRC}(S)\left(v_{Gb}^{*} - v_{Gb}\right) \\ i_{Gc}^{*} = G_{PRC}(S)\left(v_{Gc}^{*} - v_{Gc}\right) \end{cases} \tag{8.6}$$

where $G_{PRC}(S)$ represents the transfer function of the AWPRC controller where its mathematical model is defined in Eq. (8.6) and presented in Fig. 8.5 [14].

$$G_{PRC}(S) = k_p + \sum_{h=1,3,..,29}^{n} \left(k_r \left(\frac{2\omega_c S}{S^2 + 2\omega_c S + (h\omega_0)^2} \right) \right) \tag{8.7}$$

where k_p, k_r, h, ω_c, and ω_0, represent the proportional gains, resonant gains, harmonics order, cut-off frequency, and angular frequency, respectively.

The optimum values that ensure stability during perturbations, as well as fast dynamic response with best phase margin, are selected as: $k_p = 0.9$, $k_r = 350$, and $\omega_c = 20$ rad/s [14].

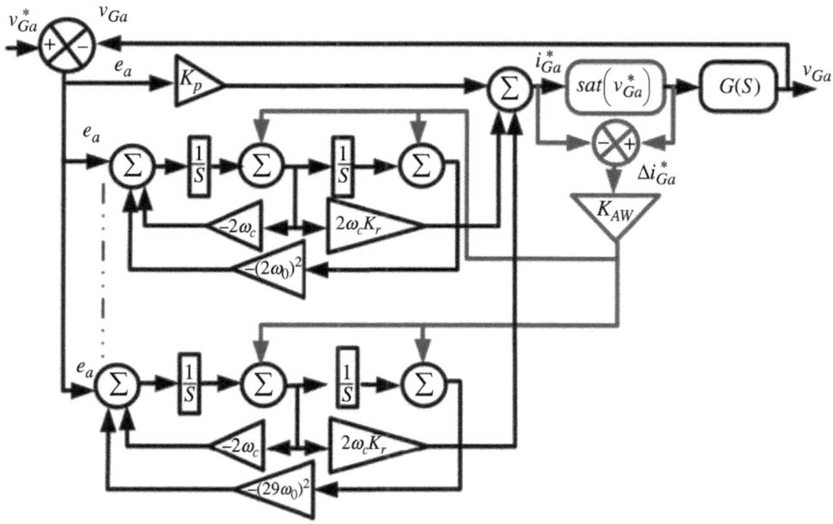

FIG. 8.5 Model of the AWPRC for outer control loop.

To avoid the saturation phenomena when the control variable reaches the actuator limit, the PRC controller is reinforced by an antiwindup strategy. However, when the output of the resonator exceeds the maximum or the minimum limits set by the saturation function expressed in Eq. (8.8), the difference between the output of the controller and the saturated output is fed back to the output of the integrators through a gain.

$$\begin{cases} i_{Gamax}\ if\ i^*_{Ga} \geq i^*_{Gamax} \\ i^*_{Ga}\ if\ i_{Gamin} < i^*_{Ga} < i_{Gamax} \\ i_{Gamin}\ if\ i^*_{Ga} \leq i_{Gamax} \end{cases} \tag{8.8}$$

where i_{camax} and i_{camin} are the maximum and minimum output limit of resonator.

As already mentioned and shown in Figs. 8.4 and 8.5, the outputs of the AWPRCs (u) are subtracted from the measured grid currents (iG) and the obtained signals, which represent the output reference currents (i_{sr}*) of the VSC1 added with the obtained harmonics currents. However, the obtained signals represent the series power converter current references (i_{sra}*,i_{srb}* and i_{src}*), which are expressed as follows:

$$\begin{cases} i^*_{sra} = u_a - i_{Ga} + i_{ha} \\ i^*_{srb} = u_b - i_{Gb} + i_{hb} \\ i^*_{src} = u_c - i_{Gc} + i_{hc} \end{cases} \tag{8.9}$$

The obtained series power converter current references are compared with the measured currents, and the errors are fed to AWPRCs, as is expressed in Eq. (8.10):

$$\begin{cases} d_{1a} = G_{PRC}(S)(i^*_{sra} - i_{sra}) \\ d_{1b} = G_{PRC}(S)(i^*_{srb} - i_{srb}) \\ d_{1c} = G_{PRC}(S)(i^*_{src} - i_{src}) \end{cases} \tag{8.10}$$

The outputs (d_{1a}, d_{1b}, and d_{1c}) are fed to pulse with modulation (PWM) to control the switches of the VSC1 (S_7 to S_{12}).

8.3.2 Control Strategy for the Shunt Power Converter (VSC2)

Fig. 8.6 shows the developed control strategy for the shunt power converter (VSC2). The detailed blocks used to estimate in-phase and quadrature units' templates, as well as to estimate the grid currents, are presented in Fig. 8.7. The Adaline NN-based learning LMS control algorithm is proposed to improve the power quality at PCC by compensating harmonics and balancing the grid currents. In addition, by controlling the DC-link voltage, one optimized the generated power from solar photovoltaic panels without using any MPPT method. In off-grid mode operation, the DC-DC buck/boost converter is controlled to achieve MPPT from solar photovoltaic panels and balance the power in the

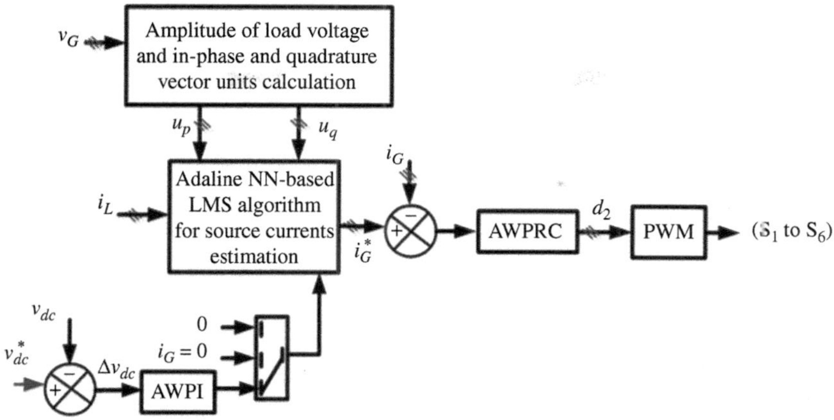

FIG. 8.6 Control strategy for the shunt power converter.

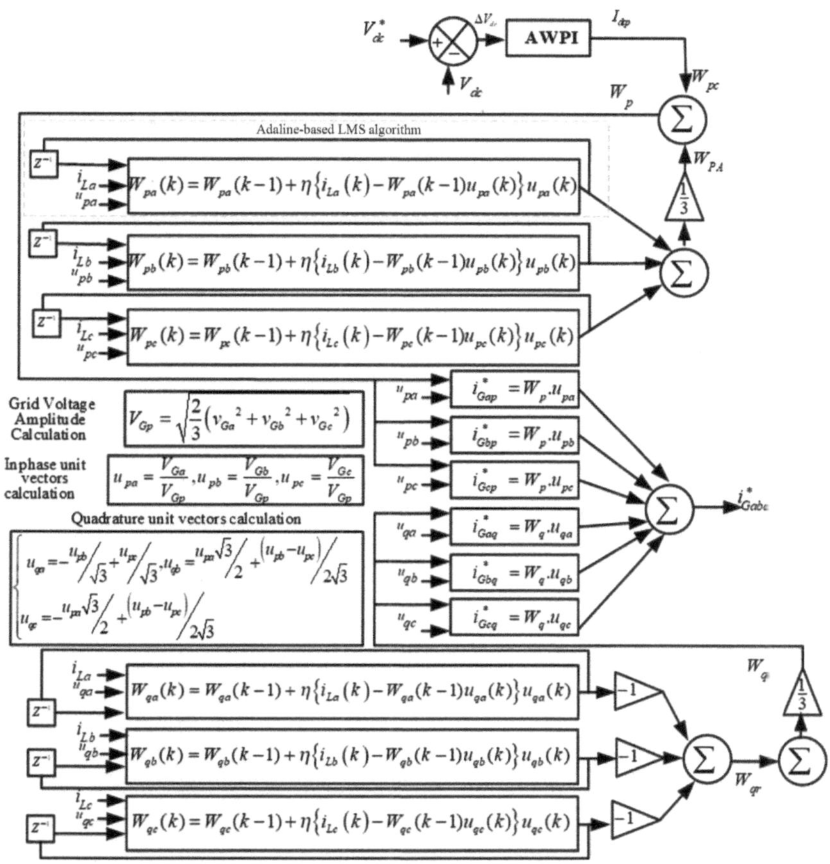

FIG. 8.7 Detailed structure of Adaline NN-based LMS algorithm for source currents estimation.

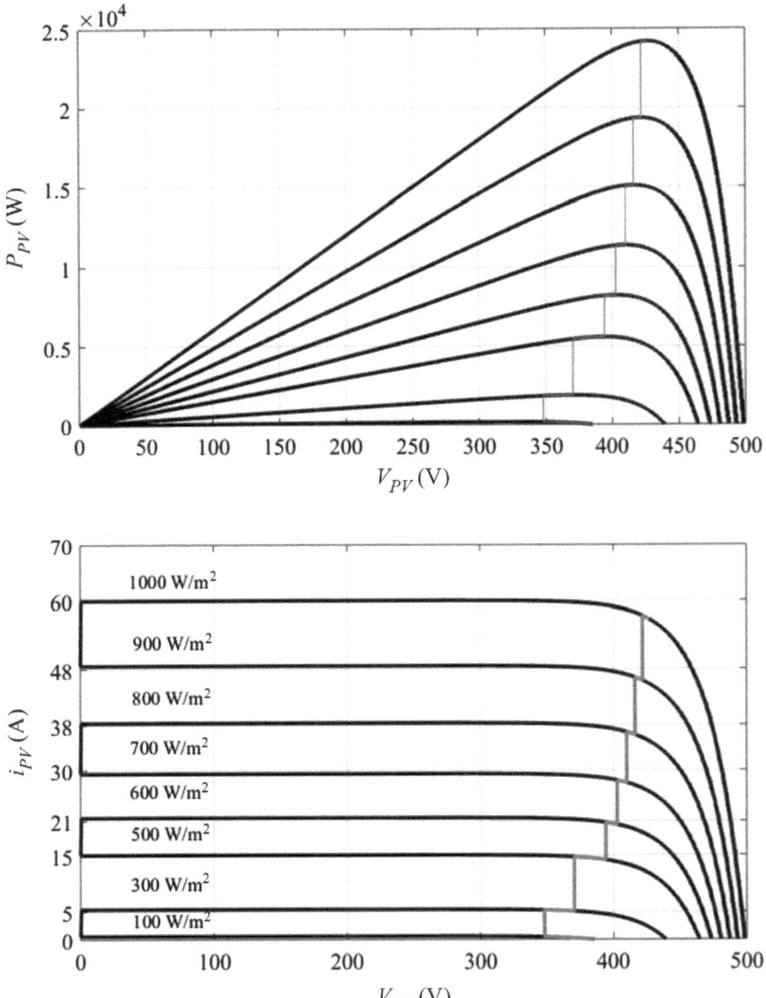

FIG. 8.8 $P_{PV}=f(V_{PV})$ and $i_{PV}=f(V_{PV})$ characteristics for fixed temperature and solar irradiation change.

system by charging and discharging the batteries. Regarding the approach used to achieve MPPT from the solar photovoltaic panels, as shown in Fig. 8.8, the near maximum power is extracted by controlling the DC link voltage. However, the output solar photovoltaic panels' voltage (V_{PV}) is equal to 400 V, which corresponds to the near maximum power. However, by selecting DC-link voltage equal to 400 V with a range of variation of ±10 V, one can easily extract the MPPT without using any MPPT method [15].

As already mentioned in Fig. 8.8, the grid current references are estimated as follows:

$$\begin{cases} i_{Ga}^* = i_{Gap}^* + i_{Gaq}^* \\ i_{Gb}^* = i_{Gbp}^* + i_{Gbq}^* \\ i_{Gc}^* = i_{Gcp}^* + i_{Gcq}^* \end{cases} \tag{8.10}$$

where p and q represent the active and reactive components, which are calculated as follows:

$$\begin{cases} i_{Gap}^* = W_p u_{pa} \\ i_{Gbp}^* = W_p u_{pb} \\ i_{Gcp}^* = W_p u_{pc} \end{cases} \tag{8.11}$$

where W_p and u_{pabc} denote the average weight of the fundamental active component of the estimated grid currents and the in-phase unit vectors of the grid voltage, respectively.

The reactive components of the estimated grid currents are expressed as follows:

$$\begin{cases} i_{Gaq}^* = W_q u_{qa} \\ i_{Gbq}^* = W_q u_{qb} \\ i_{Gcq}^* = W_q u_{qc} \end{cases} \tag{8.12}$$

where W_q and u_{qabc} denote the average weight of the fundamental reactive components of the estimated grid currents and the quadrature unit vectors of the grid voltage, respectively.

The quadrature unit vectors of the grid voltage are expressed as follows:

$$u_{qa} = -\frac{u_{pb}}{\sqrt{3}}, u_{qb} = \sqrt{3}\frac{u_{pa}}{2} + \frac{(u_{pb} - u_{pc})}{2\sqrt{3}}, u_{pc} = -\sqrt{3}\frac{u_{pa}}{2} + \frac{(u_{pb} - u_{pc})}{2\sqrt{3}} \tag{8.13}$$

The weight of the fundamental of active components of load currents is extracted using the least mean square (LMS) algorithm and its training through the Adaline NN control algorithm. However, the average weight of the fundamental of active and reactive ($W_p(k)$, $W_q(k)$) of estimated grid currents are expressed as follows:

$$\begin{cases} W_p(k) = \frac{1}{3}\left(W_{dc} - \left(W_{pa}(k) + W_{pb}(k) + W_{pc}(k)\right)\right) \\ W_q(k) = \frac{1}{3}\left(W_{qa}(k) + W_{qb}(k) + W_{qc}(k)\right) \end{cases} \tag{8.14}$$

where $W_{pabc}(k)$ and $W_{qabc}(k)$ denote the weight of the active and reactive components of load currents, which are calculated as

$$\begin{cases} W_{pa}(k) = W_{pa}(k-1) + \gamma\left\{i_{La}(k) - W_{pa}(k-1)u_{pa}\right\}u_{pa} \\ W_{pb}(k) = W_{pb}(k-1) + \gamma\left\{i_{Lb}(k) - W_{pb}(k-1)u_{pb}\right\}u_{pb} \\ W_{pc}(k) = W_{pc}(k-1) + \gamma\left\{i_{Lc}(k) - W_{pc}(k-1)u_{pc}\right\}u_{pc} \end{cases} \tag{8.15}$$

and

$$\begin{cases} W_{qa}(k) = W_{qa}(k-1) + \gamma\{i_{La}(k) - W_{qa}(k-1)u_{qa}\}u_{qa} \\ W_{qb}(k) = W_{qb}(k-1) + \gamma\{i_{Lb}(k) - W_{qb}(k-1)u_{qb}\}u_{qb} \\ W_{qc}(k) = W_{qc}(k-1) + \gamma\{i_{Lc}(k) - W_{qc}(k-1)u_{qc}\}u_{qc} \end{cases} \tag{8.16}$$

where i_{Labc} and γ represent the load currents and the convergence factor, which vary between 0.01 and 1 [15]. To achieve a fast response with high accuracy, γ is selected equal to 0.1.

The instantaneous compensation term (W_{dc}) is obtained at the output of the antiwindup PI DC-link voltage controller (AWPI). As is detailed in Fig. 8.8, a new PI controller scheme with antiwindup feedback to avoid the saturation issue during weather condition change is proposed. The proposed model of the AWPI is presented in Fig. 8.9 [16]. The constants of the antiwindup feedback, τ_i and τ_{c1}, and τ_{c2}, are selected equal to 2, 0.5, and 10, respectively.

The estimated grid currents, which are expressed in Eq. (8.10), are compared with the measured grid currents (i_{Gabc}), and the errors are fed to AWPRC controllers. The obtained output signals are fed to PWM to generate the pulses required by IGBTs of VSC1.

8.3.3 Control Strategy for the DC-DC Buck/Boost Converter

As already discussed, the proposed UPQC-S based microgrid configuration can operate during faults as a standalone system. However, in this operation mode, the DC-link voltage (v_{dc}) is regulated as is shown in Fig. 8.10 by controlling the DC-DC buck/boost converter.

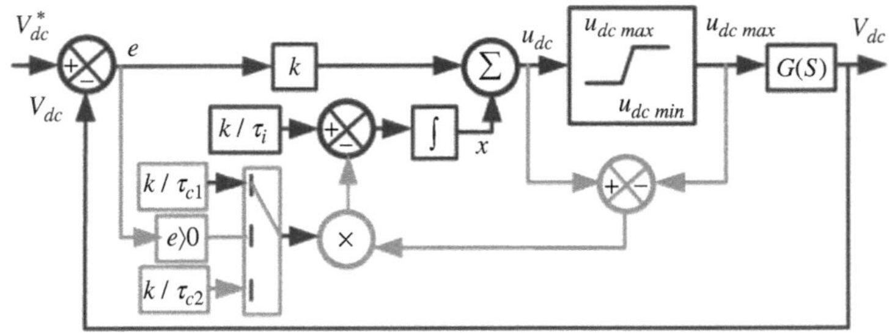

FIG. 8.9 Model of the AWPI controller.

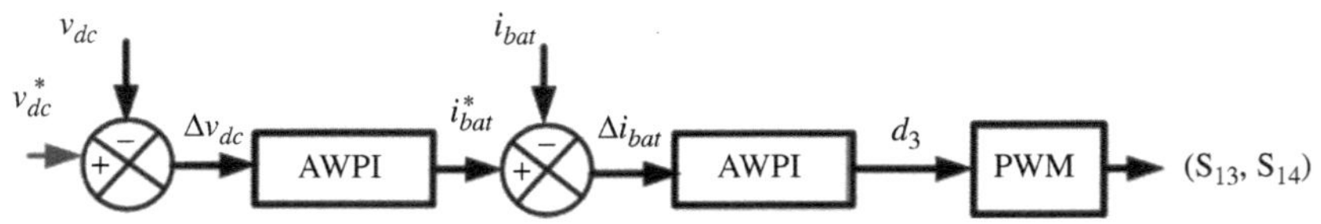

FIG. 8.10 Control strategy for DC-DC buck/boost converter.

By controlling the DC-link voltage, one can extract the MPPT from solar photovoltaic panels and balance the power in the system by charging and discharging the batteries. To avoid the saturation issue, AWPI controllers are used to control the DC-link voltage in the outer control loop and to control the battery current in the inner control loop.

However, the measured DC-link voltage is compared with its reference, which is equal to 400 V. The error of DC-link voltage (ΔV_{dc}) is fed to the AWPI controller. The output of the AWPI controller represents the reference battery current (i_{bat}*), which is compared afterwards with the sensed battery current (i_{bat}) and the error is fed to the AWPI controller. The obtained signal is fed afterwards to the PWM to control the switches (S_{13} and S_{14}).

8.4 Results and Discussion

The performance of the proposed UPQC-S based microgrid configuration with solar photovoltaic panels and its developed control strategies based on Adaline NN with LMS algorithm, as well as with AWPIs and AWPRCs, are validated using MATLAB/Simulink. The proposed system is supplied from the grid through step down transformer with three phase-supply voltages (110 V, 60 Hz) when it is operating in the grid-connected mode. In addition, the generated power from the solar photovoltaics system is supplied to the loads, and the rest of power is injected into the grid. When the system operates in the off-grid mode, loads are supplied from solar photovoltaic panels and battery storage through the VSC2. The performance of the complete system is tested with voltage sag and swell, as well as in the presence of linear and nonlinear load, and during absence of loads. In addition, the performances of both operation modes are tested under weather condition changes.

8.4.1 Performance Under Voltage Sag and Swell Conditions During Grid-Connected Mode

In Fig. 8.11, the waveforms of the grid (v_G), load (v_L), and series voltages (v_{sr}), are presented. The performances of the developed control strategy for the series power converter shown in Fig. 8.3 are tested under grid voltage decreasing from $t = 0.2$ s to $t = 0.4$ s and increasing from $t = 0.9$ s and $t = 1.2$ s. It can be observed that the series power converter injects the required compensating voltage during grid voltage variation. One can see clearly that instead of sudden increasing and decreasing of grid voltage, the load voltage is regulated at its value and is kept sinusoidal. One may also observe that during transition, the AWPRCs perform well. This confirms the robustness of the developed control strategy-based Adaline NN with LMS and AWPRC controller for series power converter.

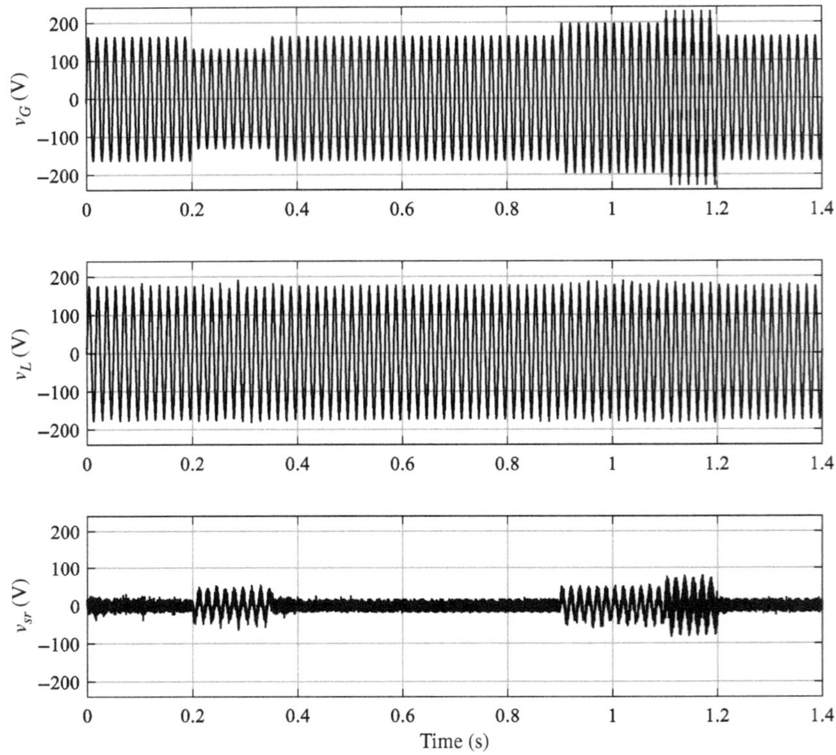

FIG. 8.11 Performance under voltage sag and swell conditions.

8.4.2 Performance Under Voltage Sag and Swell Conditions and Load Variation

In Fig. 8.12, the waveforms of phase "a" of the grid voltage (v_G) and current (i_G), load voltage (v_L), and current(i_L), shunt power converter current (i_{sh}), DC-link voltage (V_{dc}), and output current of solar photovoltaic panels (i_{PV}) are presented. The performance of the developed control strategy for shunt power converter (VSC2) to achieve high performance from solar photovoltaic panels without using the MPPT method, mitigate harmonics, balance the source current, and inject a clean power into the grid is tested under sudden decreasing and increasing of grid voltage, and sudden connecting and disconnecting of linear and nonlinear load, including RC and RL loads.

It can be observed that the grid voltage is decreased from $t=0$ s to $t=0.3$ s and is increased between $t=0.5$ s and $t=0.7$ s. Furthermore, the load is disconnected between $t=0.1$ s and $t=0.6$ s.

It can also be observed that the load voltage is regulated constant and sinusoidal during transition and when grid voltage is increased or decreased. One may observe that one load is disconnected, and the generated power from

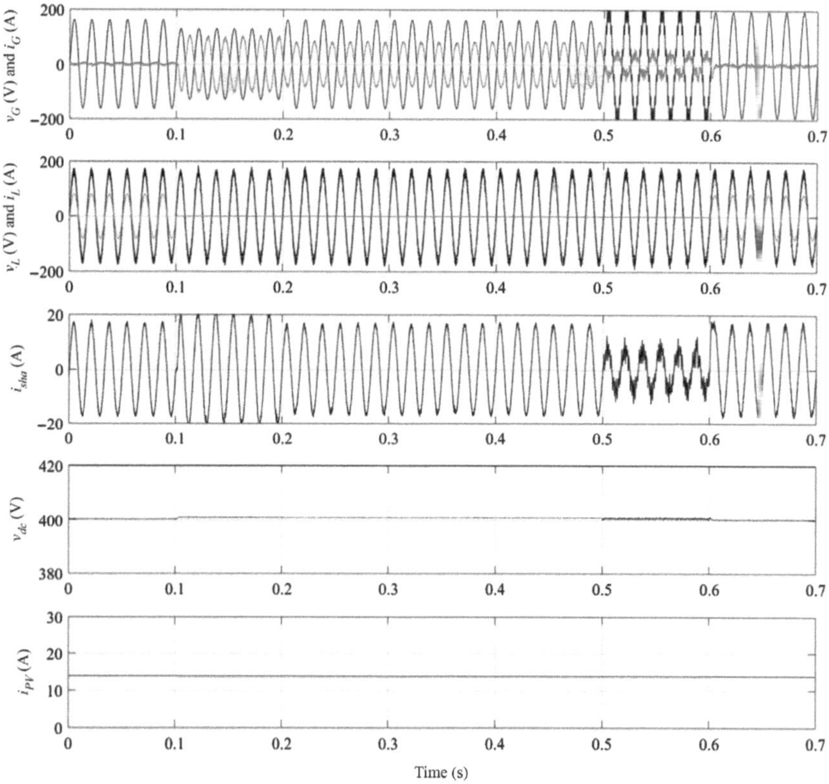

FIG. 8.12 Performance under linear load variation (switching off the phase "a" between $t=0.1$ s and 0.6 s) and fixed solar irradiation.

the solar photovoltaic panels is injected into the grid with high quality with less current and voltage harmonics distortions. The DC-link voltage is well regulated at its rated value, which is equal to 400 V. One can see clearly that by regulating the DC-link voltage, one can extract easily the maximum power from solar photovoltaic panels without using any MPPT method. However, the output current of the solar photovoltaic panels is precisely 15 A, which corresponds exactly to the maximum value of current obtained at 500 W/m^2, as shown in Fig. 8.8.

In Figs. 8.13 and 8.14, the waveforms of the grid voltage (v_G) and current (i_G), load current (i_L) and voltage (v_L), output current of the VSC2 (i_{sh}), DC-link voltage (V_{dc}), and the output current of the solar photovoltaic panels (i_{PV}) are presented. Nonlinear loads including RL and RC loads are connected between $t=0$ s and $t=0.3$ s, and disconnected from $t=0.3$ s to $t=0.7$ s.

In addition, the grid voltage is subjected to sudden increasing and decreasing between $t=0.1$ s and $t=0.2$ s and between $t=0.5$ s and $t=0.7$ s. In this test, the

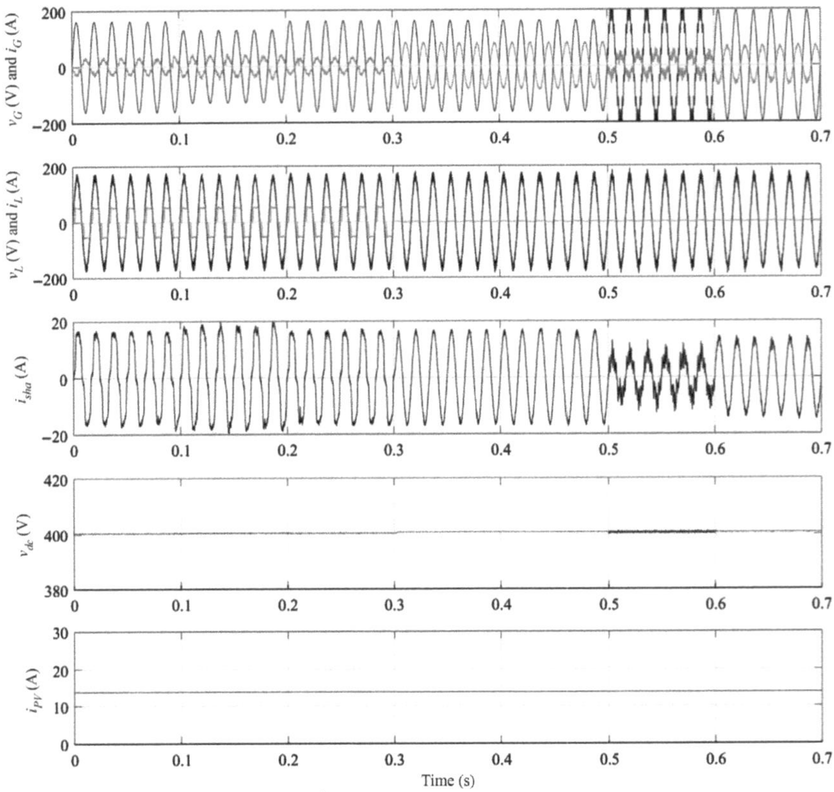

FIG. 8.13 Performance under nonlinear load type RL (switching off the phase "a" between $t=0.3$ s and 0.7 s) and fixed solar irradiation.

solar irradiation is kept equals to 500 W/m^2. It can be observed that the load voltage is kept constant and sinusoidal during voltage sag and swell condition. The grid current is balanced and sinusoidal when nonlinear loads are connected. It can also be observed that during connecting and disconnected of loads, system inject a clean power into grid with high power quality. One may observe that with regulating the DC-link voltage at its rated value, one achieves the MPPT from the solar photovoltaics without using any MPPT method. It may also be observed that the extract current at 500 W/s is equal to 15 A, which represents the maximum current that one extract as is shown in Fig. 8.8. It can be seen that the DC link-voltage and load voltage are well regulated during transition without any saturation issue and without oscillations, which confirms the robustness of the antiwindup feedbacks that integrated with PI and PRC controllers and their optimal gains. The obtained results show satisfactory performance, from the point of view of power quality and stability during load condition change and during grid voltage change, confirming the robustness

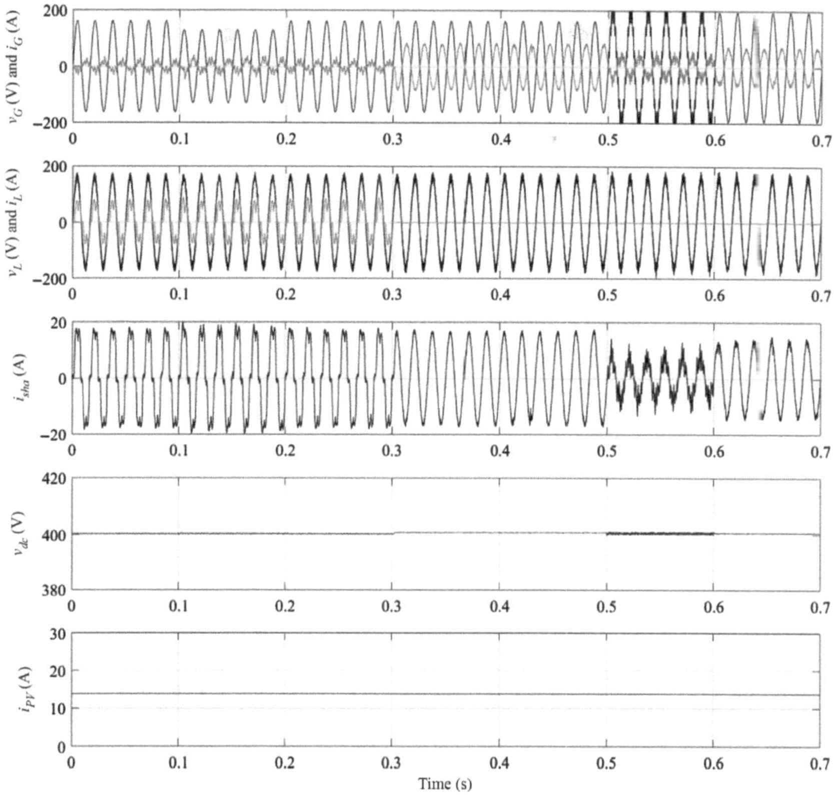

FIG. 8.14 Performance under nonlinear load type RC (switching off the phase "a" between $t=0.3$ s and 0.7 s) and fixed solar irradiation.

and capability of the proposed control approach based on Adaline NN with LMS algorithm to mitigate harmonics and balance the grid current, inject a clean power into the grid, and extract MPPT without using any MPPT method.

In Fig. 8.15, the waveforms and zooms between $t=0.6$ s and $t=0.8$ s of the grid voltage (v_G) and currents (i_G), load voltage (v_L) and current (i_L), output current of VSC2 (i_{sh}), DC-link voltage (V_{dc}), and output current of the photovoltaics panels (i_{PV}) are presented. To test the performance of the proposed control strategy for VSC2 to achieve high performance from solar photovoltaic panels without using any MPPT methods and to inject a clean power into grid, system is subjected to three different solar irradiations: 500 W/m^2, 600 W/m^2, and 800 W/m^2 at $t=0$ s, $t=0.2$ s, and at $t=0.7$ s. It is observed that the DC-link voltage varies slightly with these variations and the obtained currents corresponds to the maximum current, as shown in Fig. 8.8. One may observe that the grid and VSC2 currents vary with variation of the solar irradiation; it increases at $t=0.2$ s and increases more at $t=0.7$ s, which confirms that the system can extract maximum power from solar photovoltaic panels and inject this into the grid with high power quality.

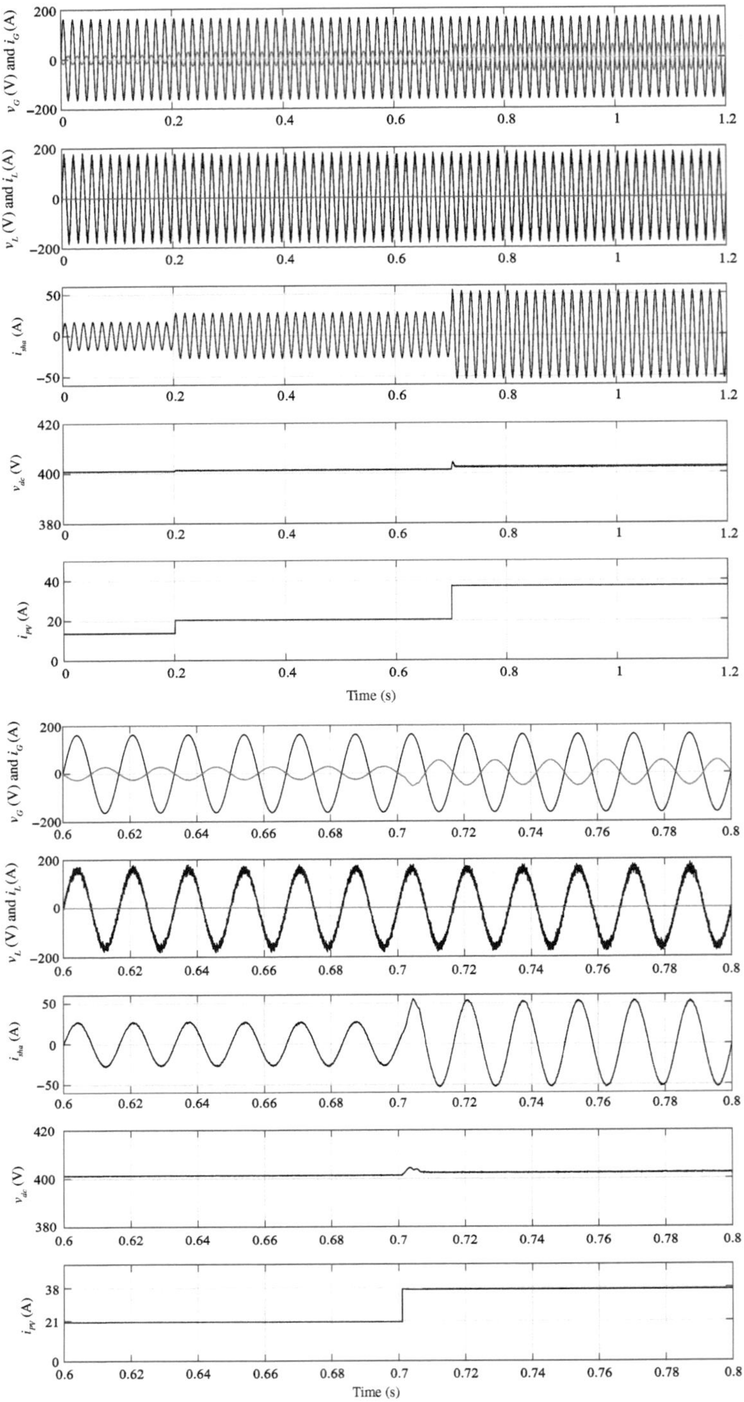

FIG. 8.15 Performance under no-load and with solar irradiation change and their zoom.

8.4.3 Performance During Transition Between Grid-Connected Mode to Off-Grid Mode

In Fig. 8.16, the waveforms and the zooms between $t=0.1$ s and $t=0.3$ s of the grid voltage (v_G) and current (i_G), load voltage (v_L) and current (i_L), VSC2 (i_{sh}), DC-link voltage (V_{dc}), and the output current of solar photovoltaic panels (i_{PV}) are presented. To test the performance of the proposed UPQC-S based microgrid configuration and its developed control strategies when it is connected to the grid, as well as when it operates as a standalone system, the grid is switched off at $t=0.2$ s. Furthermore, the system is subjected to solar irradiation change and is supplying a fixed nonlinear load type RL.

It can be observed that the output current from solar photovoltaic panels vary with variation of solar irradiations; it increases at $t=0.4$ s and increases more still at $t=0.4$ s. One can see clearly that the output current represents the maximum currents, as shown in Fig. 8.8. One can also observe that from $t=0$ s to $t=0.2$ s, the UPQC-S based microgrid is connected to the grid, which is why the grid is supplying the connected nonlinear load. In this period, VSC2 operates as a shunt active filter; it compensates harmonics and regulates the DC-link voltage at its rate value. At $t=0.2$ s, the grid is switched off and the UPQC-S based microgrid switches from the grid-connected mode to the off-grid mode, which is why the DC-DC buck/boost converter regulates the DC-link voltage instead of VSC2, in order to balance the power in the system. One can see clearly that the battery discharges from $t=0.2$ s to $t=0.4$ s and charges from $t=0.4$ s to $t=1.2$ s. The transition between the grid-connected mode and the off-grid mode is achieved quickly without disrupting the connected nonlinear load and without any saturations of the controllers. In addition, in the off-grid mode, the system frequency at PCC is regulated constant and the load voltage is kept stable and sinusoidal in the presence of nonlinear load. This confirms that the VSC2 supported by battery storage system is able to ensure regulation of the load voltage and frequency, mitigate harmonics, and extract the maximum of power from the solar photovoltaic panels.

8.5 Conclusion

In this chapter, the UPQC-S based microgrid configuration with solar photovoltaic panels for smart grid has been studied. The performance of the proposed configuration has been validated using MATLAB/Simulink and the obtained results show satisfactory performance when the UPQC-S based microgrid operates as a standalone system and when it is connected to the grid. The developed control strategies for the series and shunt power converters based on the Adaline NN with LMS algorithm demonstrate their capability to maintain the system stably during transition between both operations modes without any saturation issue, regulate the load voltage at a constant rate by compensating the voltage sag/swell, maintain the system frequency at a constant rate, compensate harmonics, and balance the grid currents, supplying the connected loads without

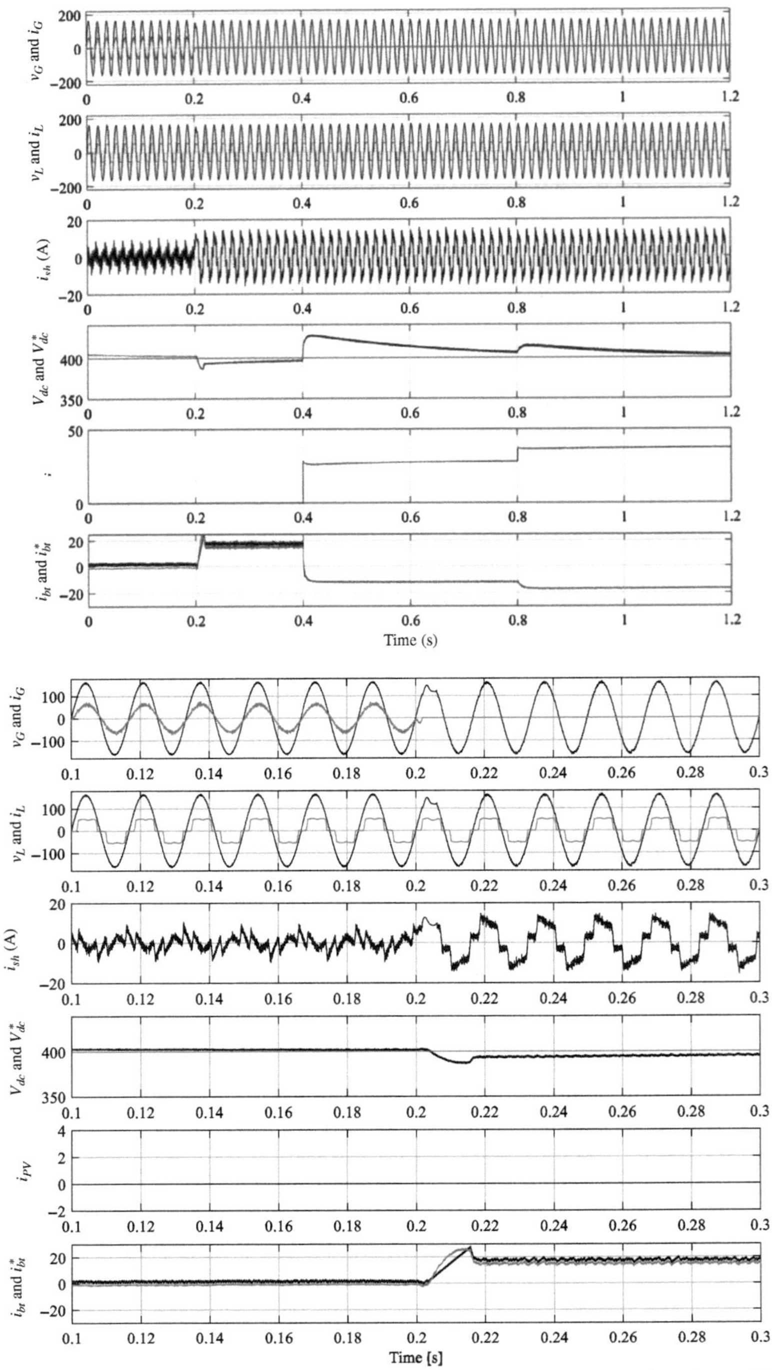

FIG. 8.16 Performance when system switch at $t = 0.2\,s$ from grid-connected mode to off-grid mode and their zoom.

interruption, injecting a clean power into the grid, and extracting the maximum of power from the solar photovoltaic panels without using any MPPT method.

References

[1] M. Rezkallah, A. Chandra, B. Singh, S. Singh, Microgrid: configurations, control and applications, IEEE Trans. Smart Grid (2017) 1. https://doi.org/10.1109/TSG.2017.2762349.

[2] R. Kappagantu, S. Arul Daniel, A. Yadav, Power quality analysis of smart grid pilot project, Puducherry, Proc. Technol. 21 (8) (2015) 560–568.

[3] IEEE Std 519-2014, IEEE recommended practices and requirements for harmonic control in electrical power systems, 2015.

[4] IEEE Std 1159-1995, IEEE recommended practice for monitoring electric power quality 1995.

[5] L. Morán, J. Dixon, M. Torres, Active power filters, in: Power Electronics Handbook, fourth ed., Elsevier, 2018, pp. 1341–1379.

[6] H.M.A. Antunes, S.M. Silva, D.I. Brandao, A.A.P. Machado, B. de Jesus Cardoso Filho, A new multifunctional converter based on a series compensator applied to AC microgrids, Int. J. Electr. Power Energy Syst. 102 (11) (2018) 160–170.

[7] F.H. Gandoman, A. Ahmadi, A.M. Sharaf, P. Siano, J. Pou, B. Hredzak, V.G. Agelidis, Review of FACTS technologies and applications for power quality in smart grids with renewable energy systems, Renew. Sustain. Energy Rev. 82 (2) (2018) 502–514.

[8] J. Ye, H.B. Gooi, F. Wu, Optimal design and control implementation of UPQC based on variable phase angle control method, IEEE Trans. Ind. Inform. (5) (2018) 1.

[9] V. Khadkikar, Enhancing electric power quality using UPQC: a comprehensive overview, IEEE Trans. Power Electron. 27 (5) (2012) 2284–2297.

[10] B. Singh, A. Chandra, K. Al-Haddad, Power Quality: Problems and Mitigation Techniques, John Wiley & Sons, 2014.

[11] P.G. Khorasani, M. Joorabian, S.G. Seifossadat, Smart grid realization with introducing unified power quality conditioner integrated with DC microgrid, Electr. Pow. Syst. Res. 151 (2017) 68–85.

[12] D. Kumar, F. Zare, A. Ghosh, DC microgrid technology: system architectures, AC Grid interfaces, grounding schemes, power quality, communication networks, applications, and standardizations aspects, IEEE Access 5 (2017) 12230–12256.

[13] M.A.A. Mohd Zainuri, M.A. Mohd Radzi, A. Che Soh, N. Mariun, N. Abd Rahim, S. Hajighorbani, Fundamental active current adaptive linear neural networks for photovoltaics shunt active power filters, Energies 9 (6) (2016) 397.

[14] M. Rezkallah, A. Chandra, M. Tremblay, H. Ibrahim, Experimental implementation of an APC with enhanced MPPT for standalone solar photovoltaic based water pumping station, IEEE Trans. Sustain. Energy 10 (1) (2018) 181–191.

[15] M. Rezkallah, A. Chandra, M. Saad, M. Tremblay, H. Ibrahim, High performance single-stage photovoltaic system interfacing grid without any MPPT algorithm, in: 43rd Annual Conference Industrial Electronics Society, IECON, 2017, pp. 7777–7782.

[16] M. Rezkallah, A. Hamadi, A. Chandra, B. Singh, Design and implementation of active power control with improved P&O method for wind-PV-battery-based standalone generation system, IEEE Trans. Ind. Electron. 65 (7) (2018) 5590–5600.

Further Reading

[17] M. Rezkallah, S. Sharma, A. Chandra, B. Singh, Hybrid standalone power generation system using hydro-PV-battery for residential green buildings, in: 41st Annual Conference of the IEEE Industrial Electronics Society, IECON, 2015, pp. 003708–003713.

Chapter 9

Smarter Solutions Based on Power Electronics

Man-Chung Wong, Chi-Seng Lam, Lei Wang and Wai-Hei Choi
University of Macau, Macau, China

Chapter Outline

9.1 Introduction

One of the pathways for a smarter power system involves using power electronic converters to deal with power system control and stability issues, such as control of active power and frequency, control of reactive power and voltage, subsynchronous torsional oscillations, voltage stability, frequency stability, small-signal stability, transient stability, renewable distributed generator energy injection and storage, power quality, security, resilience, and reliability etc. All this makes power system dynamic responses and optimization in real-time possible by energy Internet Of Things (IoTs) smart grid technologies. Fig. 9.1 shows development from the industrial revolution to today, when the cyber physical system connects everything together, such as power networks and human, through wireless communication and the internet forming

Pathways to a Smarter Power System. https://doi.org/10.1016/B978-0-08-102592-5.00009-0

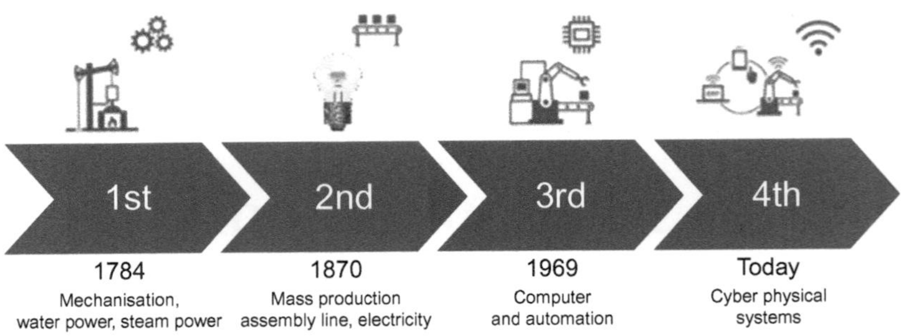

FIG. 9.1 Development from the industrial revolution to the present day.

a smart system with machine learning or artificial intelligence computing to make an optimal decision to control power electronic systems to regulate energy and power flow or storage in the systems in a precise way. Based on load demand response, power suppliers can optimize their resources into power generation, transmission, and distribution through fourth industrial revolution technologies. With the support of IoT and cyber physical system technologies, interactive collaboration among power consumers, suppliers, and computing machines, precision power generation, and transmission and distribution planning and operation can be possible. Power electronics is one of the keys to achieve a smarter power system development.

In this chapter, basic power electronic converters are reviewed in Section 9.2, such as DC/DC, AC/DC, AC/AC, and finally AC/AC converters. In Section 9.3, three-phase static inverters and their corresponding space vectors are discussed. In Section 9.4, flexible alternative current transmission systems (FACTS) and distributed FACTS devices are given. In this chapter, smarter solutions based on power electronics are introduced.

9.2 Power Electronics Converters [1–3]

The main purpose of power electronics is to control the flow of electric energy by processing the power electronics switches with storage elements to control its voltage and current to suit the user needs. Modern power electronics converters are involved in applications such as electrical-machine-motion-control, switched-mode power supplies, renewable energy and storage systems, distributed power generators, power quality compensators, and vehicle chargers. On top of power electronics converters, combining the IoTs connecting online power systems with data science technologies forms a smarter solution to optimize power and energy operation planning, management and control as micro-grids, and/or a smart grid making a superior energy network in a more effective and intelligent way.

The breakthrough and advancement of power semiconductor devices/switches forms the history of power electronics. The first power electronics device was a mercury arc rectifier developed in 1900. The second power electronics revolution began in 1958 with the development of the thyristor. From

1975 to 1995, more turn-off power-semiconductor elements were developed and implemented. After that, Bipolar Junction Transistors (BJTs), Metal-Oxide-Semiconductor Field-Effect Transistors (MOSFETs), Gate Turn-Off Thyristors (GTOs), Insulated-Gate Bipolar Transistors (IGBTs), Integrated Gate-Commutated Thyristors (IGCTs), and MOS Controlled Thyristor (MCTs) were developed. Correspondingly, different types of power semiconductor switches are employed into power conversion as power electronic converters. Accordingly, power electronic converters have the ability to convert shape, process, and control electrical power and energy in different applications.

Several examples of uses for power electronic systems are DC/DC converters (used in many mobile devices, such as cell phones), AC/DC converters (computer power supplies, battery chargers, etc.), AC/AC converters (motor drivers, railway traction motors, etc.), and DC/AC converters (electric vehicles, active power filters, distributed power supplies, etc.). Power scales can be small, medium, or large. Small-scale power electronic systems can be integrated circuits. Large-scale power electronics are used to control hundreds of megawatt of power flow across regions and countries.

Some of these converters are given below.

9.2.1 DC/DC Converters

Typically, DC/DC converters are used due to unstable DC input for having stable DC output or adjustment of specific dc level voltage for applications. DC to DC converters are used in portable electronic devices such as cellular phones and laptop computers, which are supplied with power mostly from batteries. In Figs. 9.2–9.5, a Buck (Step-Down) converter, a Boost (Step-Up) converter, a Buck-Boost converter and a Cuk converter are illustrated. In several applications, DC to DC converters are developed to maximize the energy harvest from photovoltaic systems, which are called power optimizers.

9.2.2 AC/DC Rectifiers

Rectifiers are power electronics converters to convert AC input power into DC output, which produces a DC current, and normally consist of pulses of current. In several applications of rectifiers, such as power supplies for television and

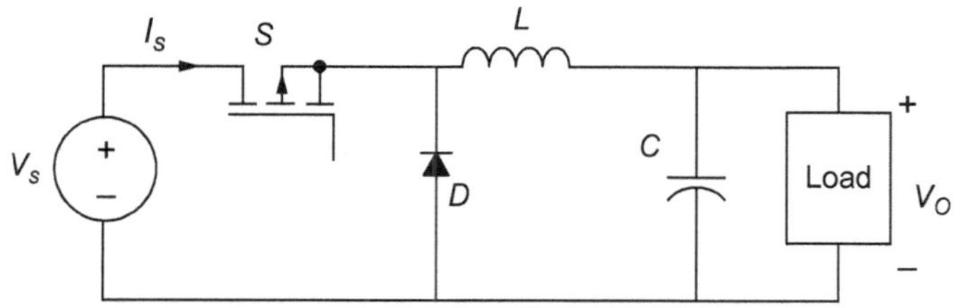

FIG. 9.2 Buck converter.

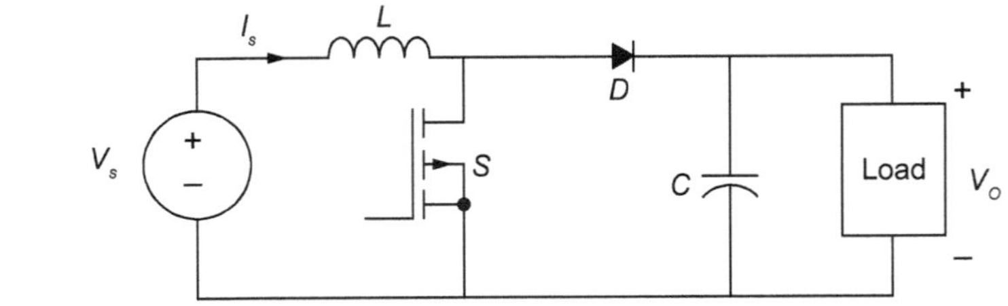

FIG. 9.3 Boost converter.

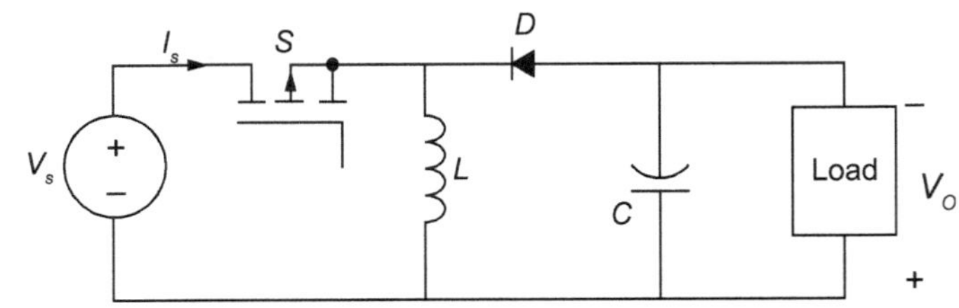

FIG. 9.4 Buck-Boost converter.

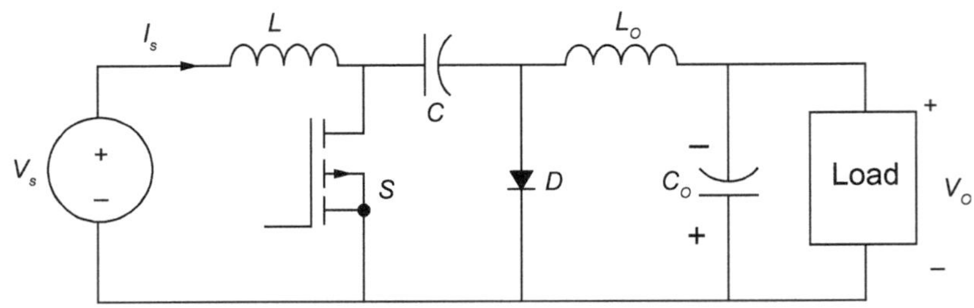

FIG. 9.5 Cuk converter.

computer facilities, they require a steady and stable DC current, in such cases smoothed by power quality compensators or filters, possibly followed by a voltage regulator to produce a steady current. Typical single-phase and three-phase rectifiers are given in Figs. 9.6 and 9.7, respectively. For very high-power rectifiers the 12-pulse bridge connection is used, which consists of two 6-pulse bridge circuits connected in series, as shown in Fig. 9.8, by connecting a transformer that produces a 30 degree phase shift between two bridges; in such a connection this cancels a lot of the characteristic harmonics that the six-pulse bridges produce. Fig. 9.8 shows a 30 degree phase shift achieved by a transformer with two sets of secondary windings, one in a star (wye) connection and another one in a delta connection. In summary, filtering, wye-delta angle-shifting transformers, and power factor correction circuits in rectifiers can reduce system harmonics and displacement power factor angle for increasing system performance.

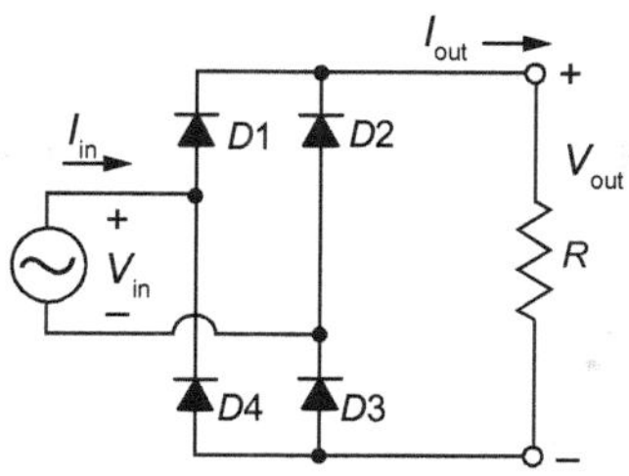

FIG. 9.6 A single-phase rectifier circuit.

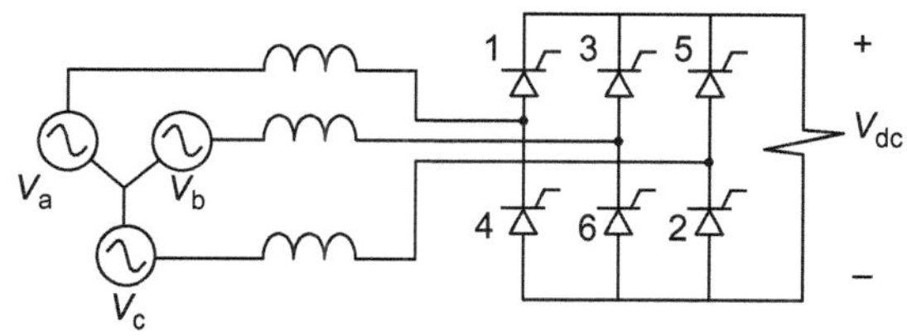

FIG. 9.7 A three-phase rectifier circuit.

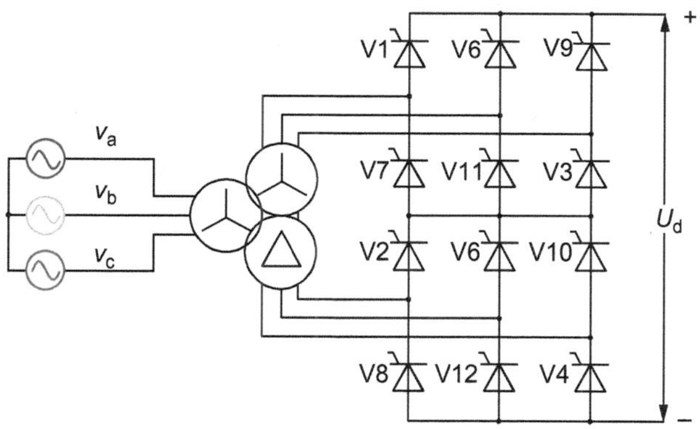

FIG. 9.8 A 12-pulse bridge three-phase rectifier.

9.2.3 AC/AC Converters

AC/AC converters are power electronic systems that can convert an AC waveform into another AC waveform by changing its voltage amplitude and frequency, which can be applied in electric locomotives, food processing, constant frequency power supplies, and high frequency applications such as induction heating, etc. Fig. 9.9 shows a simple AC/AC converter which can be controlled by on/off control and phase angle control. In such a case, it may generate high AC line distortion and low power factor. Switches are employed to represent power electronic switches in some of the given figures below. Fig. 9.10 shows a three-phase AC/AC converter, which is composed of a rectifier and an inverter together through a DC link voltage coupling by a

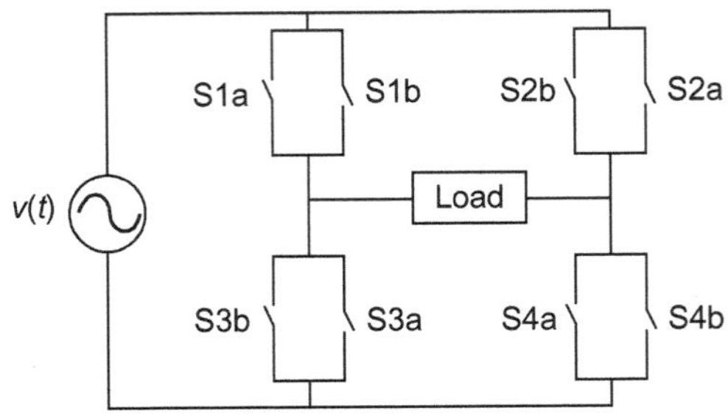

FIG. 9.9 Simple AC/AC converter.

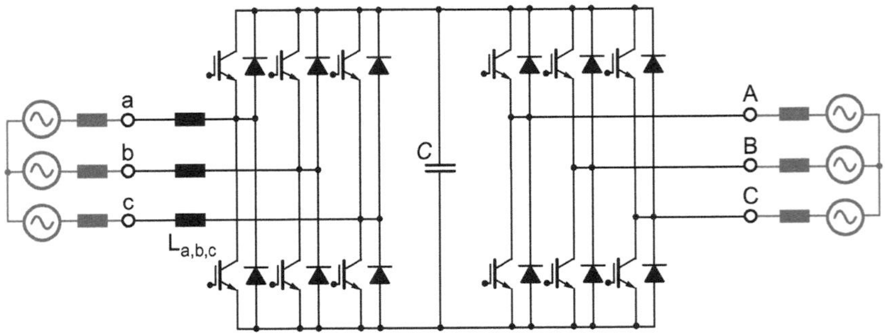

FIG. 9.10 An AC/AC converter in voltage source mode.

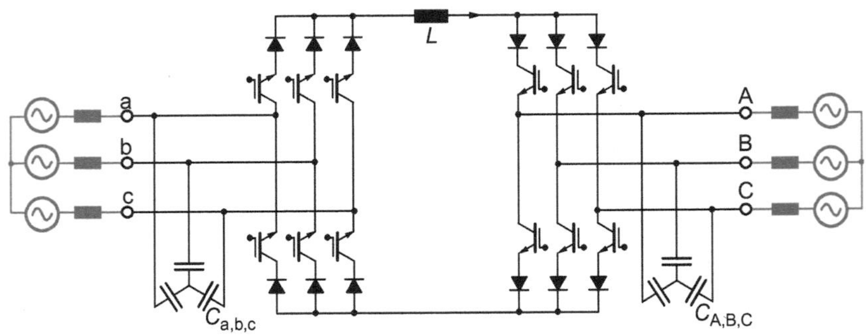

FIG. 9.11 An AC/AC converter in current source mode.

capacitor. However, its power flow can be bidirectional, which can be controlled by a corresponding Pulse Width Modulation. In Fig. 9.11, the DC link between two converters is replaced by an inductor, which supports current source system; meanwhile its switching components are replaced accordingly.

9.2.4 DC/AC Converters

DC/AC converters called inverters change DC input into AC output by power electronic systems through changing its amplitude and frequency. Static

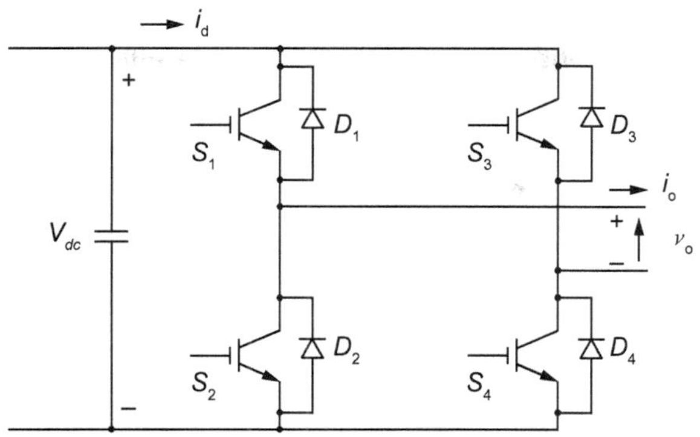

FIG. 9.12 Single-phase DC/AC converter.

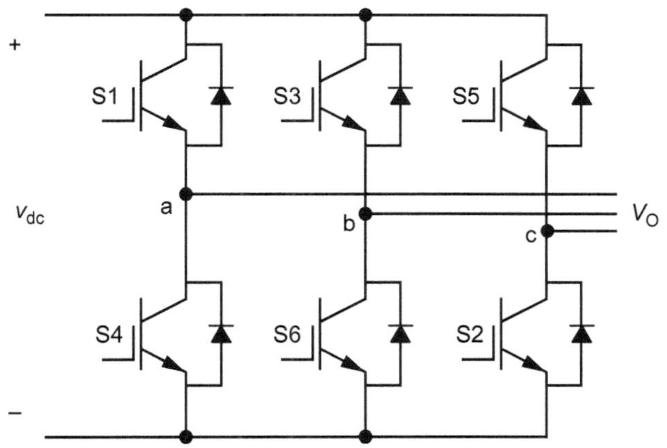

FIG. 9.13 Three-phase DC/AC converter.

inverters do not use mechanical moving parts in the conversion process. In applications, converters can be designed for electric vehicles, uninterruptible power supplies, refrigeration compressors, power quality compensators, renewable energy distributed generators, etc. Inverters are important for smarter power systems. More detailed explanation and information will be given below. However, several typical circuit topologies are given in Figs. 9.12–9.14. Figs. 9.12 and 9.13 show a typical DC source (vdc) with a full bridge converter in a single-phase AC (vo) output and three-phase DC/AC converter, respectively. Fig. 9.14 shows a DC/AC converter composed of two half-bridge circuits.

9.3 Three-Phase Static Inverters and Corresponding Space Vectors [4–9]

Due to the consideration that most of power electronic converters used for power network systems are inverters, as a result, three-phase static inverters

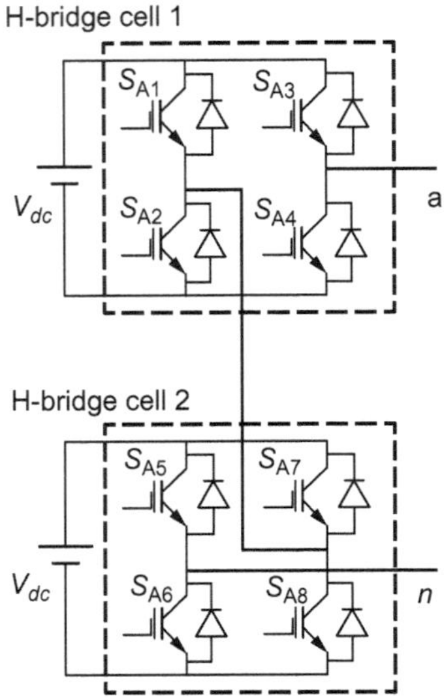

FIG. 9.14 H-bridge DC/AC converter.

are given in this section and their corresponding space vectors are discussed. Three-basic three-phase inverter topologies are analyzed and discussed in two level cases: for example, three-leg, three-leg center-split, and four-leg inverters in three-phase three-wire and three-phase four-wire systems.

9.3.1 Two-Level Three-Leg Converters

Based on Fig. 9.13, when power electronic devices are considered as equivalent switches, their equivalent model is given in Fig. 9.15, where Z is the load impedance, with a, b, and c representing phases a, b, and c, respectively. Point "0" represents the neutral point and point "n" means the reference. In Fig. 9.15, the DC link voltage can be expressed as Eq. (9.1). In an ideal case, $v_{dc1}=v_{dc2}=v_{dc}$. It should be noted that $v_{dc1}=v_{dcu}$ and $v_{dc2}=v_{dcL}$ are upper DC voltage source and lower DC voltage source, respectively, in the following sections.

$$v_{dc}=v_{dc1}+v_{dc2} \tag{9.1}$$

In a three-phase system, an instantaneous voltage vector, $\vec{v}$, can be represented by a linear combination of vector basis $B=\left\{\vec{n}_a, \vec{n}_b, \vec{n}_c\right\}$ as shown in Eq. (9.2), where v_a, v_b, and v_c are scalars. However, using the Clarke transformation [10–12], a three-phase voltage vector can be expressed in $\propto$-β-0 transformation, which shows that the voltage vector can span in a space. In a three-phase three-wire system, considering unbalance symmetric sequences, the zero sequence component is zero.

$$\vec{v}=v_a\vec{n}_a+v_b\vec{n}_b+v_c\vec{n}_c \tag{9.2}$$

Let $B'=\left\{\vec{n}_\alpha, \vec{n}_\beta, \vec{n}_0\right\}$ be a base; those two bases for a finite dimensional vector space have the same number of vectors, which can span in a space.

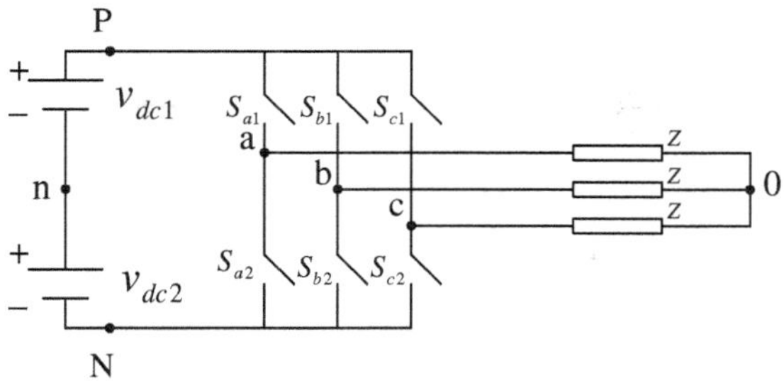

FIG. 9.15 DC/AC inverter equivalent model.

However, in a three-phase three-wire system, considering unbalance symmetric sequences, the zero sequence component is zero. As a result, the voltage vector can be expressed by Eq. (9.3) accordingly under the consideration that instantaneous power and reactive power computation are equal before and after transformations, so that the factor $\sqrt{2/3}$ is used instead of 2/3.

$$\begin{bmatrix} v_\alpha \\ v_\beta \end{bmatrix} = \sqrt{\frac{2}{3}} \begin{bmatrix} 1 & -\frac{1}{2} & -\frac{1}{2} \\ 0 & \frac{\sqrt{3}}{2} & -\frac{\sqrt{3}}{3} \end{bmatrix} \begin{bmatrix} v_a \\ v_b \\ v_c \end{bmatrix} \tag{9.3}$$

Fig. 9.16 shows vector bases in three dimensions, which demonstrates the relationship between $B = \left\{ \vec{n}_a, \vec{n}_b, \vec{n}_c \right\}$ and $B' = \left\{ \vec{n}_\alpha, \vec{n}_\beta, \vec{n}_0 \right\}$. The projection of $B = \left\{ \vec{n}_a, \vec{n}_b, \vec{n}_c \right\}$ on a two-dimensional plan is shown in Fig. 9.16, where $v_0 = 0$ in three-phase three-wire systems.

Considering Fig. 9.15, the converter output voltages in *a-b-c* coordinates can be expressed in Eq. (9.4). It is required to transform the voltages from *a-b-c* into *α-β-0* coordinates. Considering DC linked voltage, after taking the transformation from Eq. (9.4) into Eq. (9.5), it should be noted that the total DC linked voltage defined here is 2vdc between point *P* and *N* in Fig. 9.15.

$$\begin{bmatrix} v_{a0} \\ v_{b0} \\ v_{c0} \end{bmatrix} = \frac{1}{6}(2v_{dc}) \begin{bmatrix} 2 & -1 & -1 \\ -1 & 2 & -1 \\ -1 & -1 & 2 \end{bmatrix} \begin{bmatrix} S_a \\ S_b \\ S_c \end{bmatrix} \tag{9.4}$$

where $S_j = \{+1, -1\}$ and $j = \{a, b, c\}$.

Based on the Clarke transformation in Fig. 9.16, Eq. (9.5) is used for transformation.

$$\begin{aligned} P_{\alpha\beta 0} &= \sqrt{\frac{2}{3}} \begin{bmatrix} \cos 0^\circ & \cos(120^\circ) & \cos(240^\circ) \\ \sin 0^\circ & \sin(120^\circ) & \sin(240^\circ) \\ \sin(45^\circ) & \sin(45^\circ) & \sin(45^\circ) \end{bmatrix} \\ &= \sqrt{\frac{2}{3}} \begin{bmatrix} 1 & -1/2 & -1/2 \\ 0 & \sqrt{3}/2 & -\sqrt{3}/2 \\ 1/\sqrt{2} & 1/\sqrt{2} & 1/\sqrt{2} \end{bmatrix} \end{aligned} \tag{9.5}$$

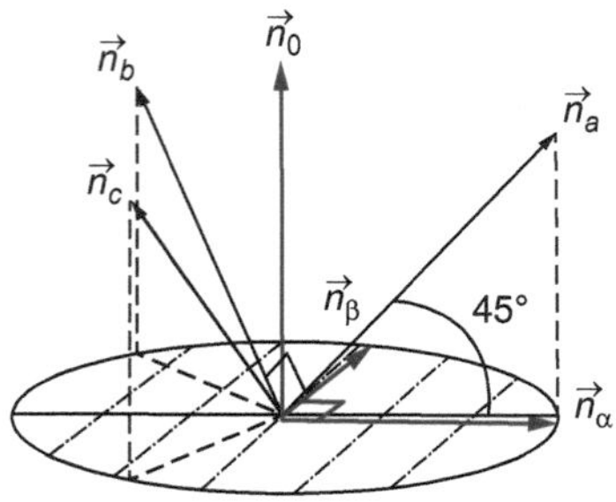

FIG. 9.16 Bases in 3D.

$$\begin{bmatrix} v_{\alpha 0} \\ v_{\beta 0} \\ v_{00} \end{bmatrix} = \frac{(2v_{dc})}{6}\left[P_{\alpha\beta 0}\right]\begin{bmatrix} 2 & -1 & -1 \\ -1 & 2 & -1 \\ -1 & -1 & 2 \end{bmatrix}\begin{bmatrix} S_a \\ S_b \\ S_c \end{bmatrix}$$

$$= v_{dc}\sqrt{\frac{2}{3}}\begin{bmatrix} 1 & -1/2 & -1/2 \\ 0 & \sqrt{3}/2 & -\sqrt{3}/2 \\ 0 & 0 & 0 \end{bmatrix}\begin{bmatrix} S_a \\ S_b \\ S_c \end{bmatrix} \tag{9.6}$$

Finally, considering the scalar values K_α, K_β, and K_0 by Eq. (9.7), an instantaneous converter output voltage vector of a three-leg converter can be expressed as Eq. (9.8) where $S_\alpha = S_a - \frac{1}{2}S_b - \frac{1}{2}S_c$ and $S_\beta = S_b - S_c$.

$$\vec{v} = K_a\vec{n}_a + K_b\vec{n}_b + K_c\vec{n}_c \tag{9.7}$$

$$\vec{v}_{\text{three-leg}} = v_{dc}\left[\sqrt{\frac{2}{3}}S_\alpha\cdot\vec{n}_\alpha + \frac{1}{\sqrt{2}}S_\beta\cdot\vec{n}_\beta\right] \tag{9.8}$$

so that the scalars K_α, K_β, and K_0 are $\sqrt{2/3}v_{dc}S_\alpha$, $1/\sqrt{2}v_{dc}S_\beta$, and 0, respectively. It should be noticed that v_{dc} is the half value of v_{PN}. $v_{dc1} = v_{dcu}$, $v_{dc2} = v_{dcL}$, and $v_{PN} = v_{dc1} + v_{dc2}$. There is no zero-component coincidentally, which changes 3D *a-b-c* coordinates on 2D *α-β* plane. Fig. 9.17 shows their corresponding voltage space vectors on *α-β* plane.

Referring to Fig. 9.18, related parameters are defined in 3D, then amplitudes and angles among different converters can be compared accordingly. Table 9.1 lists their corresponding voltage vector parameters defined in Eqs. (9.9)–(9.12) for two-level three-leg converters.

$$\rho = \sqrt{Z_\alpha^2 + Z_\beta^2 + Z_0^2} \tag{9.9}$$

$$r = \sqrt{Z_\alpha^2 + Z_\beta^2} \tag{9.10}$$

$$\theta = \tan^{-1}\left(\frac{Z_\beta}{Z_\alpha}\right) \tag{9.11}$$

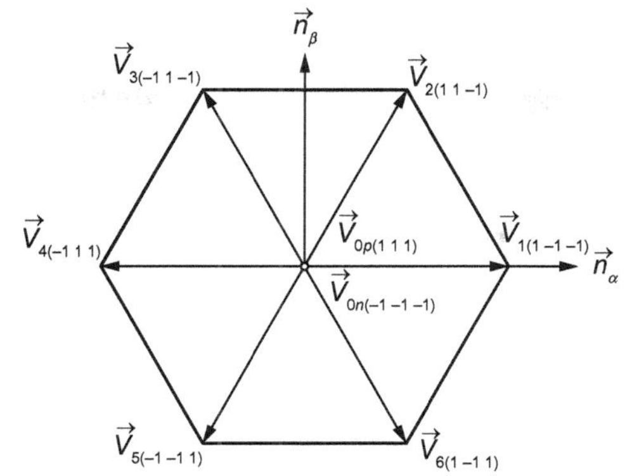

FIG. 9.17 Allocation of output voltage vectors in a two-level three-leg converter.

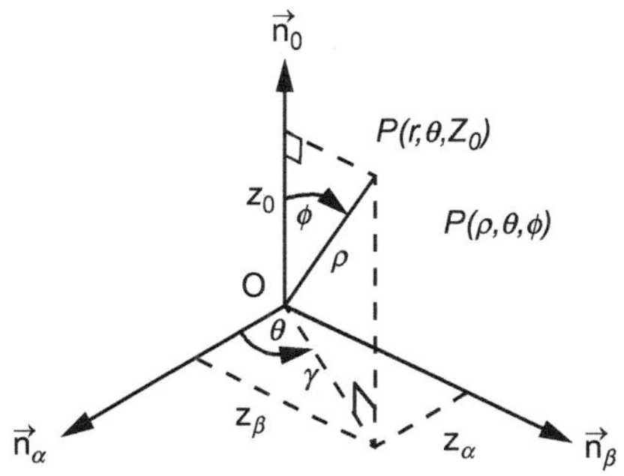

FIG. 9.18 3D Parameter definitions.

$$\varphi = \cos^{-1}\left(\frac{Z_0}{\rho}\right) = \sin^{-1}\left(\frac{r}{\rho}\right) \tag{9.12}$$

9.3.2 Two-Level Three-Leg Center-Split Converters

A two-level three-leg center-split converter is given in Fig. 9.19. It should be noted that two capacitors are used instead of one. In an idea case, vdcu and vdcL should be equal. However, in practical cases, vdcu should not be equal to vdcL due to neutral current passing through the point *n*. An equivalent circuit can be given according to Figs. 9.15 and 9.19, as shown in Fig. 9.20, where the three-phase load neutral point "o" is connected with DC linked voltage reference point "*n*."

According to the defined switching functions, the voltage vectors can be defined accordingly as given in Eq. (9.13) in *a-b-c* coordinates if vdc = vdcu = vdcL. It should be noted that the total dc linked voltage between terminals "*P*" and "*N*" is 2vdc; its half voltage is vdc = vdcu = vdcL.

TABLE 9.1 Two-Level Three-Leg Converter Voltage Vector Parameters

	S_a	S_b	S_c	S_α	S_β	S_0	r	θ	Z_0	ρ	φ
$\vec{V}_1$	1	−1	−1	2	0	0	1.633	0°	0	1.633	90°
$\vec{V}_2$	1	1	−1	1	2	0	1.633	60°	0	1.633	90°
$\vec{V}_3$	−1	1	−1	−1	2	0	1.633	120°	0	1.633	90°
$\vec{V}_4$	−1	1	1	−2	0	0	1.633	180°	0	1.633	90°
$\vec{V}_5$	−1	−1	1	−1	−2	0	1.633	240°	0	1.633	90°
$\vec{V}_6$	1	−1	1	1	−2	0	1.633	300°	0	1.633	90°
$\vec{V}_{0p}$	1	1	1	0	0	0	0	*	0	0	90°
$\vec{V}_{0n}$	−1	−1	−1	0	0	0	0	*	0	0	90°

**Unavailable.*

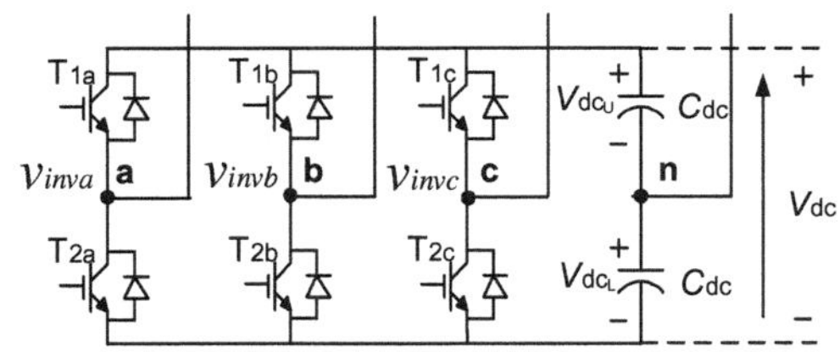

FIG. 9.19 A three-leg center-split voltage source inverter.

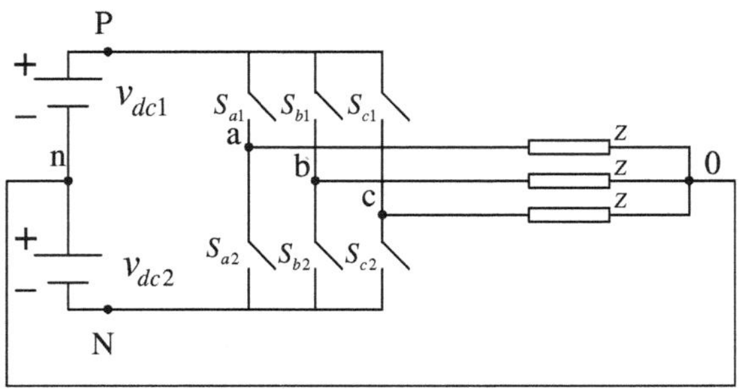

FIG. 9.20 Two-level three-leg equivalent center-split converter model.

$$\begin{bmatrix} v_{a0} \\ v_{b0} \\ v_{c0} \end{bmatrix} = v_{dc} \begin{bmatrix} 1 & 0 & 0 \\ 0 & 1 & 0 \\ 0 & 0 & 1 \end{bmatrix} \begin{bmatrix} S_a \\ S_b \\ S_c \end{bmatrix} \tag{9.13}$$

where $S_j = \{+1, -1\}$ and $j = \{a, b, c\}$.

The converter output voltages in α-β-0 coordinates are expressed in Eq. (9.14). An instantaneous converter output voltage vector of a three-leg center-split converter can be represented by Eq. (9.15).

$$\begin{bmatrix} v_{\alpha 0} \\ v_{\beta 0} \\ v_{00} \end{bmatrix} = v_{dc}\sqrt{\frac{2}{3}}\begin{bmatrix} 1 & -\frac{1}{2} & -\frac{1}{2} \\ 0 & \frac{\sqrt{3}}{2} & -\frac{\sqrt{3}}{2} \\ \frac{1}{\sqrt{2}} & \frac{1}{\sqrt{2}} & \frac{1}{\sqrt{2}} \end{bmatrix}\begin{bmatrix} 1 & 0 & 0 \\ 0 & 1 & 0 \\ 0 & 0 & 1 \end{bmatrix}\begin{bmatrix} S_a \\ S_b \\ S_c \end{bmatrix}$$

$$= v_{dc}\sqrt{\frac{2}{3}}\begin{bmatrix} 1 & -1/2 & -1/2 \\ 0 & \sqrt{3}/2 & -\sqrt{3}/2 \\ 1/\sqrt{2} & 1/\sqrt{2} & 1/\sqrt{2} \end{bmatrix}\begin{bmatrix} S_a \\ S_b \\ S_c \end{bmatrix} \tag{9.14}$$

In this case, scalars K_α, K_β, and K_0 are $\sqrt{2/3}v_{dc}S_\alpha$, $1/\sqrt{2}v_{dc}S_\beta$, and $1/\sqrt{3}v_{dc}S_0$, respectively.

$$\vec{v}_{center-split} = v_{dc}\left[\sqrt{\frac{2}{3}}S_\alpha \cdot \vec{n}_\alpha + \frac{1}{\sqrt{2}}S_\beta \cdot \vec{n}_\beta + \frac{1}{\sqrt{3}}S_0 \cdot \vec{n}_0\right] \tag{9.15}$$

where $S_\alpha = S_a - \frac{1}{2}S_b - \frac{1}{2}S_c$, $S_\beta = S_b - S_c$, and $S_0 = S_a + S_b + S_c$.

Table 9.2 lists corresponding voltage vectors and their parameter values accordingly. Comparing Tables 9.1 and 9.2, and especially voltage vectors

TABLE 9.2 Two-Level Three-Leg Center-Split Converter Voltage Vector Parameters

	S_a	S_b	S_c	S_α	S_β	S_0	r	θ	Z_0	ρ	φ
$\vec{V}_1$	1	−1	−1	2	0	−1	1.633	0°	−0.577	$\sqrt{3}$	109.46°
$\vec{V}_2$	1	1	−1	1	2	1	1.633	60°	0.577	$\sqrt{3}$	70.54°
$\vec{V}_3$	−1	1	−1	−1	2	−1	1.633	120°	−0.577	$\sqrt{3}$	109.46°
$\vec{V}_4$	−1	1	1	−2	0	1	1.633	180°	0.577	$\sqrt{3}$	70.54°
$\vec{V}_5$	−1	−1	1	−1	−2	−1	1.633	240°	−0.577	$\sqrt{3}$	109.46°
$\vec{V}_6$	1	−1	1	1	−2	1	1.633	300°	0.577	$\sqrt{3}$	70.54°
$\vec{V}_{0p}$	1	1	1	0	0	3	0	*	$\sqrt{3}$	$\sqrt{3}$	0°
$\vec{V}_{0n}$	−1	−1	−1	0	0	−3	0	*	$-\sqrt{3}$	$\sqrt{3}$	180°

*Unavailable.

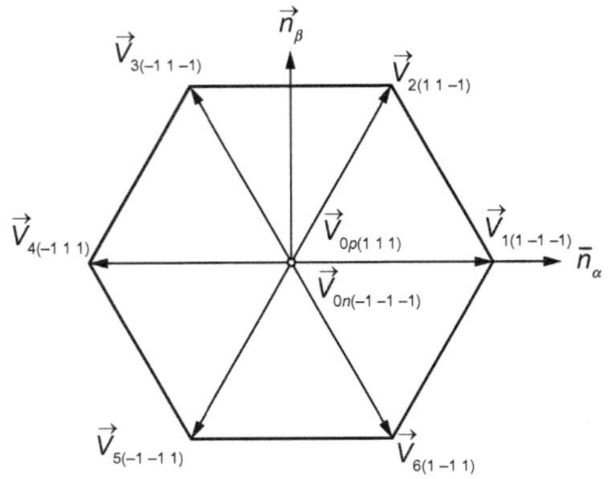

FIG. 9.21 Allocation of output voltage vectors in a two-level three-leg center-split converter.

on α-β plane are given in Fig. 9.21, amplitude r and angle θ are the same, corresponding to $\left\{\vec{V}_1, \vec{V}_2, \vec{V}_3, \vec{V}_4, \vec{V}_5, \vec{V}_6\right\}$. However, the vectors of the three-leg center-split converter have zero components that are not zero. In Table 9.2, $Z0$ can be +0.577, −0.577, $+\sqrt{3}$, and $-\sqrt{3}$. On α-β plane, voltage vectors generated by three-leg and three-leg center-split converters are equal.

In a two-level three-leg center-split converter, there are $2^3=8$ available voltage vectors with eight achievable switching states. The available vectors are projected from 3D a-b-c coordinates into 3D α-β-0 coordinates. Fig. 9.22 shows the allocation of available output voltage vectors of a two-level three-leg center-split converter in 3D.

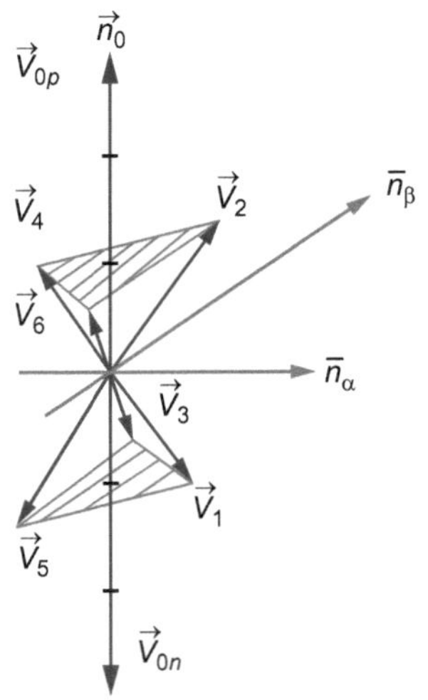

FIG. 9.22 Allocation of available output voltage vectors of a two-level three-leg center-split converter in 3D.

Comparing Figs. 9.17 and 9.21, we see that the number of available voltage vectors is equal to the number of available states in a two-level three-leg center-split converter, which is not the case in a two-level three-leg converter. In a two-level three-leg center-split converter, those two zero vectors can be employed for a neutral current compensation as well as optimizing the switching sequence to reduce switching loss.

In practical operations, $v_{dcu} \neq v_{dcL}$ is generally used in order to consider the influence of the unbalance DC upper and lower-leg voltage variations on 3D voltage space vectors of the three-phase converters. Switching functions are considered due to the unbalance between DC upper and lower capacitor voltages. V_{dc} is defined as Eq. (9.16). The upper-leg and lower-leg switching functions in a, b, or c phases can be defined as Eq. (9.17) accordingly, where S_j^N is defined as a switching function due to DC voltage unbalance, where $j = a, b$, or c for corresponding three-phase, respectively. In S_j^N, this switching function may not be 1; the value is changed due to voltage variation. When $v_{dcu} = v_{dcL} = V_{dc}$, $S_j^N = S_j$. However, when $v_{dcu} \neq v_{dcL} \neq V_{dc}$, $S_j^N \neq \pm 1$. Linking up the conventional switching function S_j and unbalance switching function S_j^N, their relationship can be found as given in Eq. (9.18).

$$V_{dc} = \frac{v_{dcu} + v_{dcL}}{2} \tag{9.16}$$

$$S_j^N = \begin{cases} \dfrac{v_{dcu}}{V_{dc}}, & upper\ leg \\ -\dfrac{v_{dcL}}{V_{dc}}, & lower\ \ leg \end{cases} \tag{9.17}$$

$$S_j^N = S_j + \Delta S \tag{9.18}$$

where $\Delta S = \dfrac{v_{dcu} - v_{dcL}}{2V_{dc}}$.

Therefore, the output voltage vector of a three-leg inverter can be recalculated by S_a^N,S_b^N, and S_c^N due to the DC voltage unbalance.

By checking Eq. (9.15), after α-β-0 transformation from a-b-c, original switching functions are as follows:

$$S_\alpha = S_a - \frac{1}{2}S_b - \frac{1}{2}S_c \tag{9.19}$$

$$S_\beta = S_b - S_c \tag{9.20}$$

$$S_0 = S_a + S_b + S_c \tag{9.21}$$

Under DC voltage variation, those DC unbalance switching functions are as follows:

$$S_\alpha^N = S_a^N - \frac{1}{2}S_b^N - \frac{1}{2}S_c^N \tag{9.22}$$

$$S_\beta^N = S_b^N - S_c^N \tag{9.23}$$

$$S_0^N = S_a^N + S_b^N + S_c^N \tag{9.24}$$

Finally, corresponding instantaneous voltage vector of a three-phase three-leg center-split converter can be expressed due to DC voltage variation by Eq. (9.25).

$$\vec{v}^{N}_{\text{center-split}} = V_{dc}\left[\sqrt{\frac{2}{3}}S^{N}_{\alpha}\cdot\vec{n}_{\alpha}+\frac{1}{\sqrt{2}}S^{N}_{\beta}\cdot\vec{n}_{\beta}+\frac{1}{\sqrt{3}}S^{N}_{0}\cdot\vec{n}_{0}\right] \tag{9.25}$$

Combining $\vec{v}_{\text{center-split}}$ and $\vec{v}^{N}_{\text{center-split}}$, the instantaneous voltage vector $\vec{v}^{N}_{\text{center-split}}$ can be expressed as Eq. (9.26), which shows that zero sequence voltage can be influenced due to DC voltage variation. As a result, voltage vector spans in 3D space can be varied, as shown in Fig. 9.23.

$$\vec{v}^{N}_{\text{center-split}} = \vec{v}_{\text{center-split}} + \vec{v}^{N}_{0} \tag{9.26}$$

where $\vec{v}^{N}_{0} = \sqrt{3}V_{dc}\Delta S\cdot\vec{n}_{0}$.

The voltage vectors may vary upward or downward along $\vec{n}_0$ axis due to the unbalance of DC voltage variation. However, the instantaneous voltage bounded volume can be either larger or smaller due to the DC voltage variation, V_{dc}, in which total DC linked voltage between terminals "*P*" and "*N*" is 2Vdc. It has 2Vdc = vdcu + vdcL.

9.3.3 Two-Level Four-Leg Converters

A two-level four-leg converter is given in Fig. 9.24. It should be noted that only one capacitor is used, and one more leg is connected with neutral point "*O*" through terminal "*f*." Its equivalent switch model is given in Fig. 9.25 in order to be compared with previous discussion, and its DC linked equivalent model is

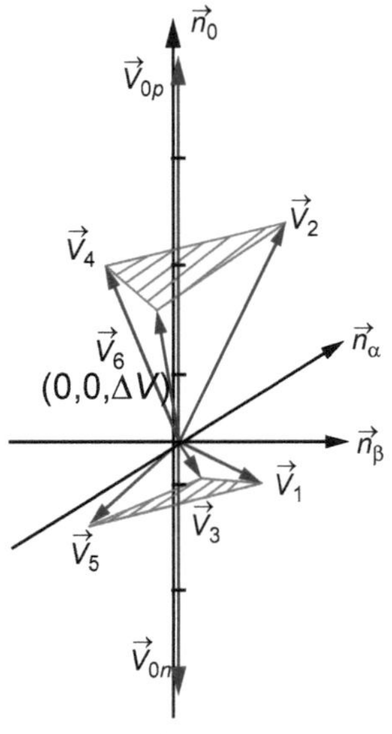

FIG. 9.23 Two-level 3D vector allocation due to DC voltage variation.

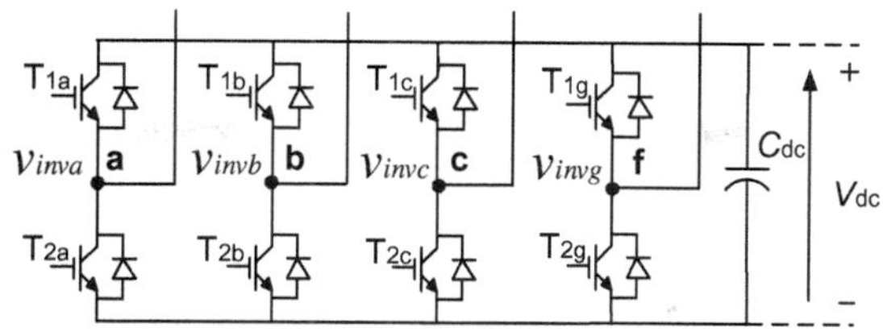

FIG. 9.24 A two-level four-leg converter configuration.

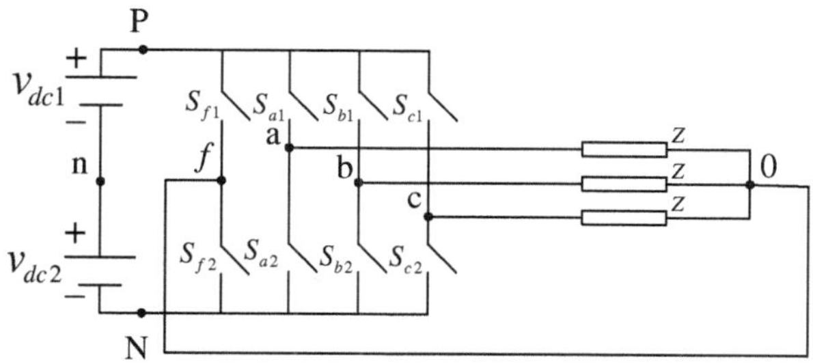

FIG. 9.25 Its equivalent two-level four-leg converter model.

decomposed into two parts: upper DC and lower DC parts with the expression vdc1 = vdcu and vdc2 = vdcL, respectively. Based on Fig. 9.25, when Sf1 is on, the fourth leg is connected to terminal "*P*." On the other hand, when Sf2 is on, the fourth leg is connected to terminal "*N*" correspondingly. Furthermore, the equivalent model can be expressed in two modes. When switching functions are only considered with +1 and −1, there are two operation modes for the four-leg converter, as shown in Figs. 9.26 and 9.27, respectively.

When S_f is +1, the node "*O*" is connected with node "*P*." In another mode, node "*O*" is connected with node "*N*" if S_f is −1.

By considering its switching possible modes and their related possible output voltage during switching, when the switches of the fourth leg are not turned "on" at all, the four-leg converter can be considered as a three-leg converter. However, when one of the switches of the fourth leg is on, the output voltage to the load is the difference between the DC voltage and another leg. It should be noted that vdc1 + vdc2 = 2vdc under the consideration of switching function difference between the fourth leg and another leg to generate the output voltage to the corresponding load. Then, its voltage vectors in $B = \left\{\vec{n}_a, \vec{n}_b, \vec{n}_c\right\}$ are given in Eq. (9.27).

$$\begin{bmatrix} v_{a0} \\ v_{b0} \\ v_{c0} \end{bmatrix} = v_{dc} \begin{bmatrix} 1 & 0 & 0 & -1 \\ 0 & 1 & 0 & -1 \\ 0 & 0 & 1 & -1 \end{bmatrix} \begin{bmatrix} S_a \\ S_b \\ S_c \\ S_f \end{bmatrix} \tag{9.27}$$

The four-leg inverter can produce three independent output voltages. Furthermore, the instantaneous converter output voltages in *α-β-0* coordinates from

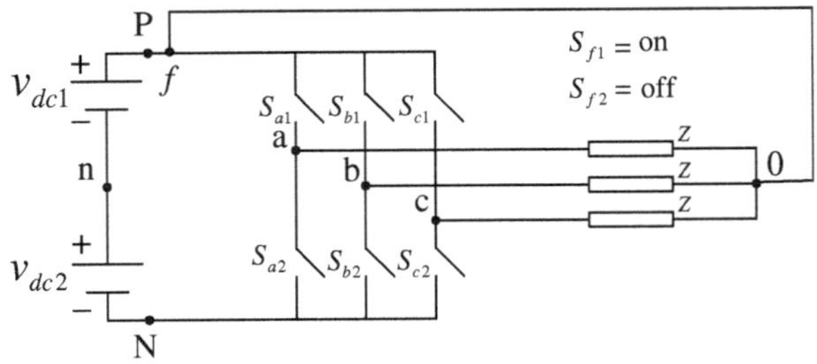

FIG. 9.26 Mode 1 when switching function of S_f is +1.

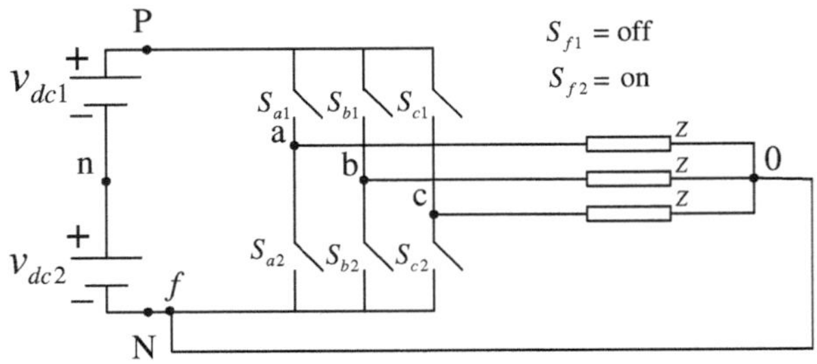

FIG. 9.27 Mode 2 when switching function of S_f is -1.

a-b-c coordinates are expressed in Eq. (9.28). An instantaneous converter output voltage vector of a four-leg converter can be represented by Eq. (9.29). The scalars K_α, K_β, and K_0 are $\sqrt{2/3}v_{dc}S_\alpha$, $1/\sqrt{2}v_{dc}S_\beta$, and $1/\sqrt{3}v_{dc}S_0$, respectively.

$$\begin{bmatrix} v_{\alpha 0} \\ v_{\beta 0} \\ v_{00} \end{bmatrix} = \sqrt{\frac{2}{3}}v_{dc} \begin{bmatrix} 1 & -\frac{1}{2} & -\frac{1}{2} \\ 0 & \frac{\sqrt{3}}{2} & -\frac{\sqrt{3}}{2} \\ \frac{1}{\sqrt{2}} & \frac{1}{\sqrt{2}} & \frac{1}{\sqrt{2}} \end{bmatrix} \begin{bmatrix} 1 & 0 & 0 & -1 \\ 0 & 1 & 0 & -1 \\ 0 & 0 & 1 & -1 \end{bmatrix} \begin{bmatrix} S_a \\ S_b \\ S_c \\ S_f \end{bmatrix}$$

$$= \sqrt{\frac{2}{3}}v_{dc} \begin{bmatrix} 1 & -1/2 & -1/2 & 0 \\ 0 & \sqrt{3}/2 & -\sqrt{3}/2 & 0 \\ 1/\sqrt{2} & 1/\sqrt{2} & 1/\sqrt{2} & -3/\sqrt{2} \end{bmatrix} \begin{bmatrix} S_a \\ S_b \\ S_c \\ S_f \end{bmatrix} \tag{9.28}$$

$$\vec{v}_{\text{four-leg}} = v_{dc}\left[\sqrt{\frac{2}{3}}S_\alpha \cdot \vec{n}_\alpha + \frac{1}{\sqrt{2}}S_\beta \cdot \vec{n}_\beta + \frac{1}{\sqrt{3}}S_0 \cdot \vec{n}_0\right] \tag{9.29}$$

where $S_\alpha = S_a - \frac{1}{2}S_b - \frac{1}{2}S_c$, $S_\beta = S_b - S_c$, and $S_0 = S_a + S_b + S_c - 3S_f$.

In a two-level four-leg converter, there are $2^4 = 16$ available vectors with 15 achievable switching states. The available voltage vectors are projected from 3D *a-b-c* coordinates into 3D *α-β-0* coordinates. Table 9.3 summarizes the

TABLE 9.3 Two-Level Four-Leg Converter Voltage Vector Parameters

	S_f	S_a	S_b	S_c	S_α	S_β	S_0	r	θ	Z_0	ρ	φ
$\vec{V}_{1p}$	−1	1	−1	−1	2	0	2	1.633	0°	1.155	2	54.73°
$\vec{V}_{2p}$	−1	1	1	−1	1	2	4	1.633	60°	2.309	2.828	35.27°
$\vec{V}_{3p}$	−1	−1	1	−1	−1	2	2	1.633	120°	1.155	2	54.73°
$\vec{V}_{4p}$	−1	−1	1	1	−2	0	4	1.633	180°	2.309	2.828	35.27°
$\vec{V}_{5p}$	−1	−1	−1	1	−1	−2	2	1.633	240°	1.155	2	54.73°
$\vec{V}_{6p}$	−1	1	−1	1	1	−2	4	1.633	300°	2.309	2.828	35.27°
$\vec{V}_{0p}$	−1	1	1	1	0	0	6	0	*	3.464	3.464	0°
$\vec{V}_{00n}$	−1	−1	−1	−1	0	0	0	0	*	0	0	90°
$\vec{V}_{1n}$	1	1	−1	−1	2	0	−4	1.633	0°	−2.309	2.828	144.73°
$\vec{V}_{2n}$	1	1	1	−1	1	2	−2	1.633	60°	−1.155	2	125.27°
$\vec{V}_{3n}$	1	−1	1	−1	−1	2	−4	1.633	120°	−2.309	2.828	144.73°
$\vec{V}_{4n}$	1	−1	1	1	−2	0	−2	1.633	180°	−1.155	2	125.27°
$\vec{V}_{5n}$	1	−1	−1	1	−1	−2	−4	1.633	240°	−2.309	2.828	144.73°
$\vec{V}_{6n}$	1	1	−1	1	1	−2	−2	1.633	300°	−1.155	2	125.27°
$\vec{V}_{0n}$	1	−1	−1	−1	0	0	−6	0	*	−3.464	−3.464	180°
$\vec{V}_{00p}$	1	1	1	1	0	0	0	0	*	0	0	90°

**Unavailable.*

voltage vector parameters for a two-level four-leg converter. Referring to amplitude r and phase angle θ, comparing with Tables 9.1 and 9.2, all shows that on an α-β plane, voltage vectors of the four-leg converter are the same as those of the original three-leg and the three-leg center-split converters; they are the projections of those vectors on the α-β plane from 3D space. Fig. 9.28 shows voltage vectors in 3D with $B' = \left\{\vec{n}_\alpha, \vec{n}_\beta, \vec{n}_0\right\}$ base.

Table 9.4 summarizes the above analysis results. Under the center-split capacitor approach, the three-phase converter becomes three single-phase half-bridge converters. Thus, it suffers a lower utilization of DC link voltage as compared to the three-leg and four-leg cases. Based on the maximum phase voltage that could be performed by the three-phase converters, reaching the

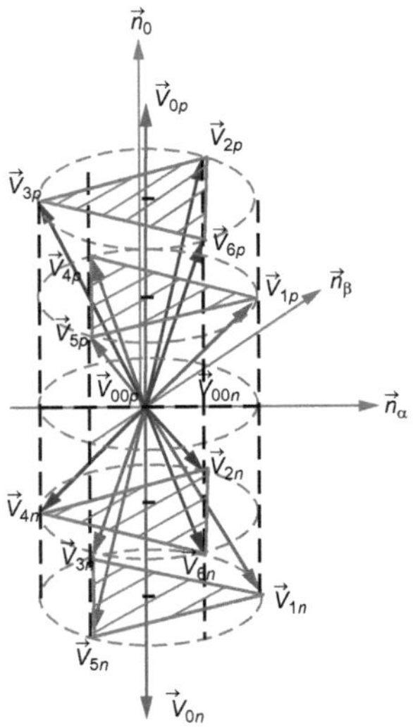

FIG. 9.28 Allocation of available output voltage vectors of a two-level four-leg converter in 3D.

TABLE 9.4 Comparisons Among Three Basic Three-Phase Converters

	Maximum Phase Voltage	Voltage Capacity on α-β Plane	Voltage Capacity on Zero Axis
Three-leg converters	$\frac{4}{3}V_{dc}$	Same	0
Three-leg center-split converters	V_{dc}	Same	Middle
Four-leg converters	$2V_{dc}$	Same	Highest

above conclusion was straightforward. However, all of the three-phase converters have the same amplitude, r, of the projected nonzero vectors on the α-β plane. It means that when the converters have the same DC linked voltage, they have the same α-β voltage capacity for the converters' operation. Comparing zero voltage injection capability, the four-leg converter has the highest. As a result, a three-leg center split and four-leg inverters have the same voltage capacity on the α-β plane when they have the same DC linked voltage.

Only the space voltage vectors of the three-leg center-split converter are affected due to the DC voltage unbalance. Conversely, the DC unbalance voltages can be controlled by shifting upward and downward of the reference voltage vector accordingly. However, the instantaneous voltage bounded volume can be either larger or smaller due to the DC voltage variation, V_{dc}.

9.4 Flexible AC Transmission System (FACTS) and Distributed FACTS Devices

9.4.1 FACTS and Principle of Power Transmission

The purpose of the transmission network is to connect power grids and loads; it is good to minimize the total power generation capacity. Transmission systems are taking advantage of diversity of loads, availability of sources, and fuel prices in order to supply electricity to the loads at minimum cost with the required reliability. From that perspective, transmission is often an alternative to a new generation resource. Less transmission capability means that more generation resources would be required, regardless of whether the system is made up of large or small power grids. In fact, small distributed generation becomes more economically viable with the developing of renewable energy. Nowadays, the studies of transmission systems are being pushed closer to their stability and thermal limits. However, the ability to transfer AC power is limited by factors such as:

- thermal limit;
- steady-state power transfer limit;
- voltage stability limit;
- transient stability limit;
- power system oscillation damping limit; and
- short-circuit current limit.

The above limits define maximum electric power transmitted efficiently through transmission systems without causing damage to electric equipment and transmission lines. A feasible solution without changing power system layout is FACTS, which can be directly controlled in the power system to change its system performance and push closer to its stability and thermal limits by varying:

- voltage amplitude;
- angle; and
- impedance.

With FACTS, the power flow within the power networks can be controllable. FACTS is a system that is composed of static equipment or devices to enhance the controllability and increase power transfer capability of the power network, and that is constructed mainly by power electronics. In IEEE, FACTS is defined as "A power electronic based system and other static equipment that provide control of one or more AC transmission system parameters to enhance controllability and increase power transfer capability" [13–16].

To model the power transmission operation, a transmission line can be considered by a reactance with the sending and receive terminal voltages. Fig. 9.29 shows the one-line diagram of the power flow between sending and receiving terminals, where E and V are sending and receive terminal voltages, X is transmission line impedance, and δ is the angle difference between two terminals.

Based on Fig. 9.29, the power flow equations can be deduced as Eqs. (9.30)–(9.33). The apparent power from the sending and receive terminals can be expressed as follows:

$$\begin{aligned} S_E &= \vec{E}\cdot\vec{I}^* = E\angle\delta\cdot\left(\frac{E\angle\delta - V\angle 0}{jX}\right)^* = E\angle\delta\cdot\left(\frac{E\cos(\delta) + jE\sin(\delta) - V}{jX}\right) \\ &= \frac{EV\sin(\delta)}{X} + j\frac{EV}{X}\cdot\left(\cos(\delta) - \frac{V}{E}\right) = P_E + jQ_E \end{aligned} \tag{9.30}$$

$$\begin{aligned} S_V &= V\cdot\vec{I}^* = V\cdot\left(\frac{E\angle\delta - V\angle 0}{jX}\right)^* = V\cdot\left(\frac{E\cos(\delta) + jE\sin(\delta) - V}{jX}\right)^* \\ &= \frac{EV\sin(\delta)}{X} + j\frac{EV}{X}\cdot\left(\frac{V}{E} - \cos(\delta)\right) = P_V + jQ_V \end{aligned} \tag{9.31}$$

Based on Eqs. (9.30) and (9.31), it can be seen that the active power is the same for the sending and receiving terminals, while the reactive power has the different direction.

$$P = P_E = P_V = \frac{EV\sin(\delta)}{X} \tag{9.32}$$

$$Q = -Q_E = Q_V = \frac{EV}{X}\cdot\left(\frac{V}{E} - \cos(\delta)\right) \tag{9.33}$$

Based on the deductions in Eqs. (9.32) and (9.33), the power flow amplitude in term of phase angle δ can be plotted as shown in Fig. 9.30.

As shown in this figure, the maximum active power flow will occur at phase angle with 90 degrees. The power flow between two terminals can be controlled

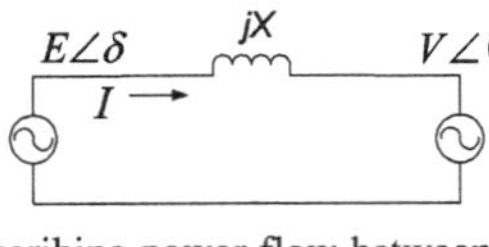

FIG. 9.29 One-line diagram describing power flow between two terminals.

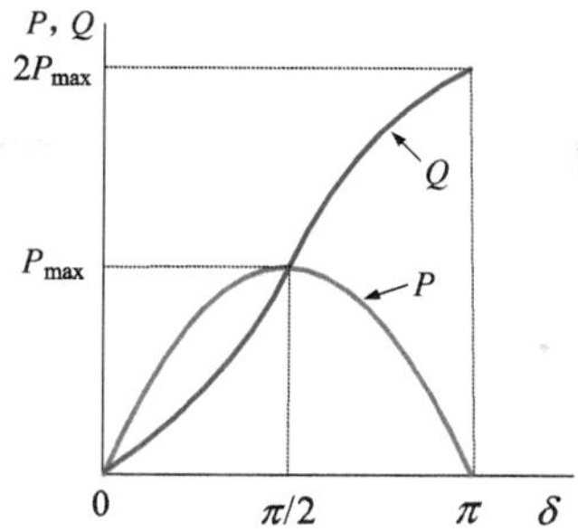

FIG. 9.30 Power angle curve before compensation.

by changing its voltage amplitudes, transmission line impedance, and phase angle up to its thermal limit. Moreover, Fig. 9.30 shows the reactive power flow, which can be regulated if the amplitude and phase angle δ of the line voltage are controlled. On the other hand, if the impedance X can be adjusted effectively by controlling the reactance of the transmission line, the consequence is the regulation of the active and reactive power flows in the line.

Conventional devices for enhancing power system control include series capacitors, switched shunt capacitors, transformers, phase shift transformers, synchronous condensers, etc. However, smarter solutions based on power electronics are FACTS-related equipment, such as Static VAR Compensators (SVCs), Static Synchronous Compensators (STATCOMs), Thyristor Controlled Series Compensators (TCSCs), Thyristor Controlled Phase Angle Regulars (TCPARs), Inter-line Power Flow Controller (IPFCs), Unified Power Flow Controllers (UPFCs), etc. The different FACTS can basically be classified as four types: series type, shunt type, combined series-series type, and combined series-shunt type.

By controlling voltage amplitude, angle, and impedance, the different types of FACTS devices can selectively affect active power transfer. Taking the series type and shunt type as examples, the active power equation can be illustrated in Figs. 9.31 and 9.32 with components selectively affected by FACTS.

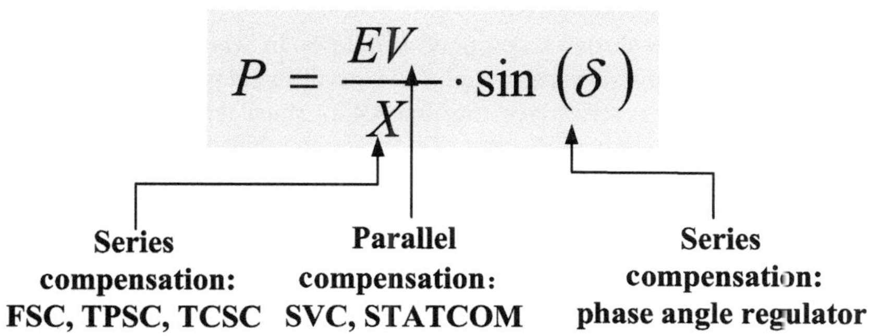

FIG. 9.31 The active power equation with transmission parameters selectively affected by FACTS.

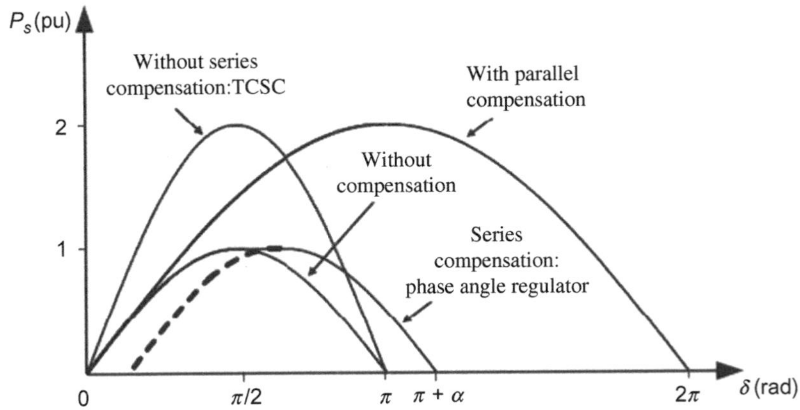

FIG. 9.32 The active power waveforms affected by FACTS.

Control of the line impedance X can provide a powerful means of active power by using series type compensation like Fixed Series Capacitors (FSCs), Thyristor Protected Series Capacitors (TPSCs), and TCSCs (discussed in Section 9.4.2).

Control of angle (with PARs, for example), which in turn controls the driving voltage, provides a powerful means of controlling the current flow and hence active power flow (discussed in Section 9.4.2).

When the angle is not large, controlling the magnitude of one or the other line voltages (e.g., with SVCs) can be a very cost-effective means for the control of reactive power flow through the interconnection (discussed in Section 9.4.3).

Injecting voltage in series with the line and with any phase angle with respect to the driving voltage can control the magnitude and the phase of the line current. This means that injecting a voltage phasor with variable phase angle can provide a powerful means of precisely controlling the active and reactive power flow. This requires injection of both active and reactive power in series (discussed in Section 9.4.4).

Combination of the line impedance control with a series compensation and voltage regulation with shunt compensation can also provide a cost-effective means to control both the active and reactive power flow between the two systems (discussed in Section 9.4.5).

The layout of this section is described as follows. In Section 9.4.1, the principle of power transmission is provided. Then, the different types of FACTS are introduced, including: series type (Section 9.4.2), shunt type (Section 9.4.3), combined series-series type (Section 9.4.4), and combined series-shunt type (Section 9.4.5). Finally, a summary is drawn in Section 9.4.6.

9.4.2 Series Type Compensation

Series type compensation devices have been developed from fixed or mechanically switched compensations to the Thyristor Controlled Series Compensators

(TCSCs) or even voltage source converter-based devices. The series type compensation devices can be variable impedance (such as TCSCs) or a power electronics-based variable (such as PARs).

The idea of the series type compensation is shown in Fig. 9.33. The representative series type compensators are introduced, which include FSCs, TPSCs, TCSCs, and PARs.

9.5.1.1 Fixed Series Capacitors and Thyristor-Protected Series Capacitors

The simplest and most cost-effective type of series compensation is the FSC, as shown in Fig. 9.34. FSCs comprise the capacitor banks and parallel arresters (Metal Oxide Varistors, MOVs) and a bypass switch. The surge arresters protect the capacitor from overvoltages during and after transmission system failures.

As a kind of dynamic short-circuit current limiting device in Fig. 9.35, the TPSCs, in line with a series reactor, can limit the short-circuit current within a few milliseconds by the fast increase of coupling reactance in response to a sudden short-circuit. As soon as a short-circuit occurs, the TPSCs are short-circuited by the thyristor and the reactor limits the short circuit current.

The basic idea behind both FSC and TPSC compensations is to decrease the overall effective series transmission impedance from the sending end to the receiving end. The model of the shunt capacitor compensation is illustrated in Fig. 9.36.

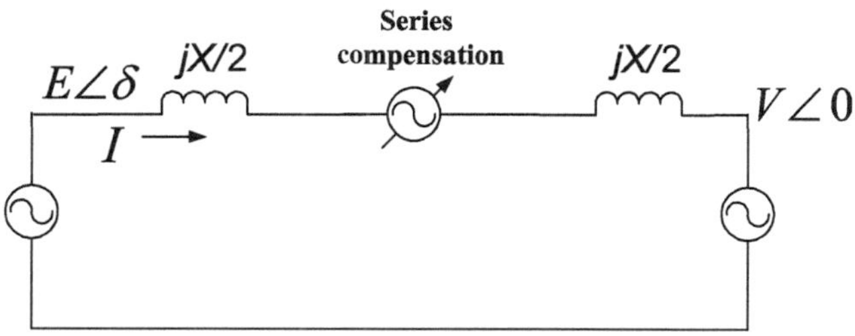

FIG. 9.33 Series type compensation.

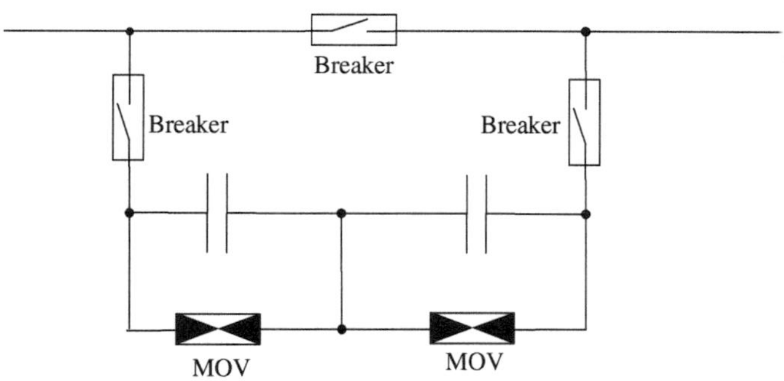

FIG. 9.34 Fixed Series Capacitor (FSC).

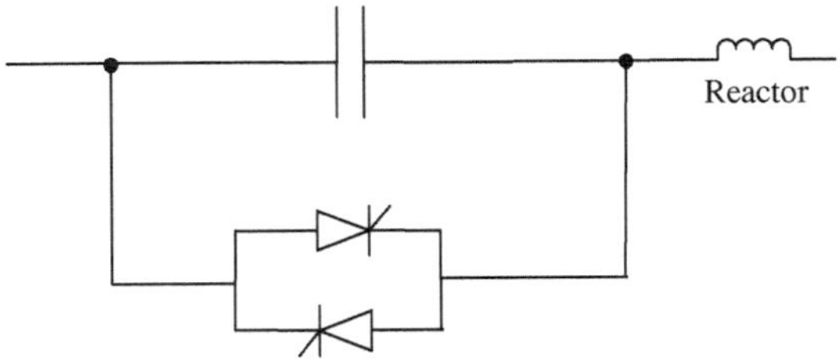

FIG. 9.35 Thyristor Protected Series Capacitor (TPSC).

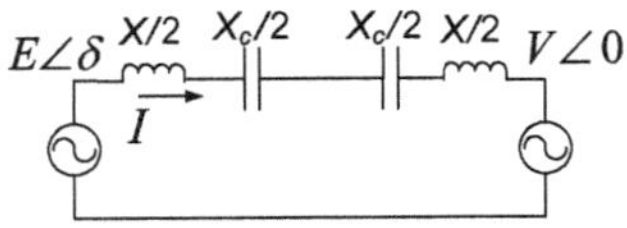

FIG. 9.36 The equivalent compensation model of FSCs and TPSCs.

Based on Fig. 9.36, the effective transmission impedance X_{eff} with the series capacitive compensation is given by the following:

$$X_{eff} = X - X_c = (1-k)\cdot X \tag{9.34}$$

where k is the degree of series compensation, which can be expressed as follows:

$$k = X_c/X \text{ and } 0 \le k \le 1 \tag{9.35}$$

Assuming $V=E$ in Fig. 9.36, the current and active power transmitted can be derived by following the deductions in Eqs. (9.30)–(9.35) as follows:

$$I = \frac{2V^2}{(1-k)\cdot X}\cdot \sin\frac{\delta}{2} \tag{9.36}$$

$$P = \frac{V^2}{(1-k)\cdot X}\cdot \sin\delta \tag{9.37}$$

The reactive power supplied by the series capacitor can be expressed as follows:

$$Q_c = I^2\cdot X_c = \frac{k}{(1-k)^2}\cdot\frac{2V^2}{X}(1-\cos\delta) \tag{9.38}$$

Based on Eqs. (9.37) and (9.38), the relationship among P, Q_c, and δ is shown plotted at different k in Fig. 9.37.

From Fig. 9.37, it can be observed that the active power increases with the degree of series compensation k. Similarly, the reactive power supplied by the series capacitor (FSC and TPSC) also increases with k and varies with angle δ.

9.4.2.1 *Thyristor Controlled Series Compensators (TCSCs)*

There are two main principles of the TCSC concept. The first one is to provide electromechanical damping between large electrical systems by changing the

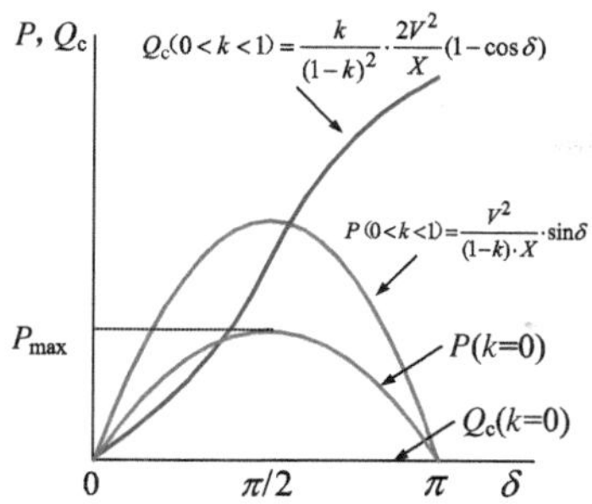

FIG. 9.37 Active power and series capacitor reactive power versus angle.

reactance of a specific interconnecting power line, that is, the TCSC will provide a variable capacitive reactance. The second one is to change its apparent impedance (as seen by the line current) for subsynchronous frequencies, so that a prospective subsynchronous resonance (SSR) can be avoided. Both objectives are achieved with the TCSC, using control algorithms that work concurrently. The TCSC's high-speed switching capability provides a mechanism for controlling line power flow, which permits increased loading of existing transmission lines, and allows for rapid readjustment of line power flow in response to various contingencies. The TCSC also can regulate steady-state power flow within its rating limits.

Fig. 9.38 shows the circuit configuration of a TCSC and its operational diagram. From a principal technology point of view, the TCSC resembles the conventional series capacitor. Thyristor valves are used to control the behavior of the main capacitor bank. Likewise, the control and protection is located on ground potential together with other auxiliary systems. The firing angle and the thermal limits of the thyristors determine the boundaries of the operational diagram.

The thyristor circuit is in parallel to the main capacitor bank so that controlled charges are added to the main capacitor, making it a variable capacitor

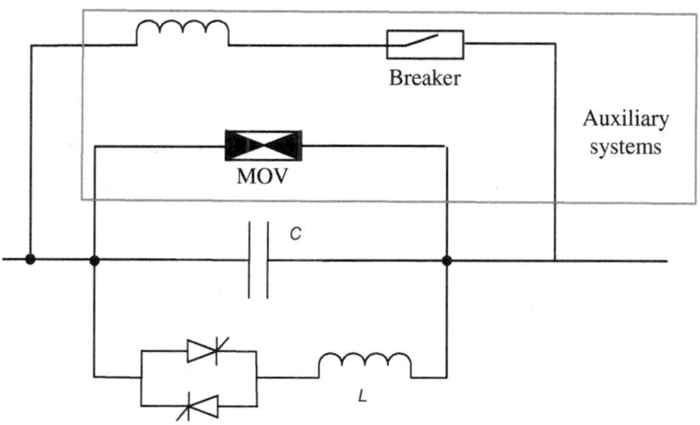

FIG. 9.38 Thyristor Controlled Series Compensator (TCSC).

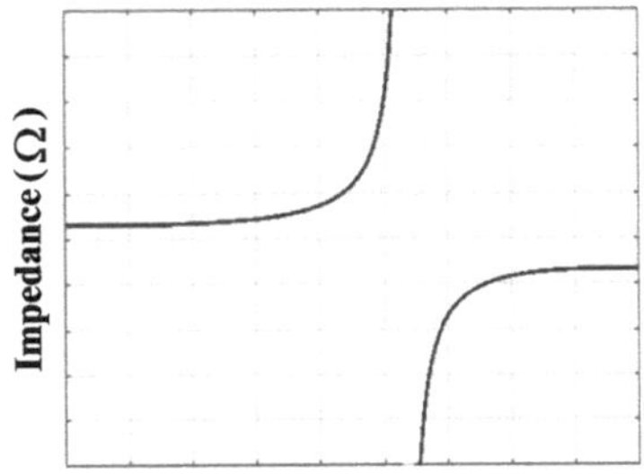

FIG. 9.39 The impedance of TCSCs with varying of firing angle.

at fundamental frequency but a "virtual inductor." The impedance of TCSC (X_{TCSC}) can be expressed as follows:

$$X_{\text{TCSC}}(\alpha) = \frac{X_C \cdot X_L(\alpha)}{X_L(\alpha) - X_C \cdot} \tag{9.39}$$

where the variable inductor impedance can be expressed as follows:

$$X_L(\alpha) = \frac{\pi}{2\pi - 2\alpha + \sin(2\alpha)} \cdot X_L \tag{9.40}$$

Based on Eqs. (9.39) and (9.39), the impedance of TCSC can be plotted as shown in Fig. 9.39.

The operation principle for the active and reactive power compensation by TCSC is similar with FSCs and TPSCs, as in Fig. 9.37.

9.4.2.2 Phase Angle Regulators (PARs)

Phase Angle Regulators have been in use since the 1930s for control of power flows in transmission lines in steady state. The PAR is a special case of the series compensator. It is inserted between the sending-end generator and the transmission line. This regulator is an AC voltage source with controllable amplitude and phase angle. A regulator controls the phase difference between the two AC systems and thereby can control the power exchanged between the two AC systems. The effective sending-end voltage is the sum of the sending-end voltage and the regulator voltage.

In Fig. 9.40, the power switches in converters are thyristors. The thyristor switches are connected, forming an on-load tap changer. Thyristors are connected in antiparallel, forming bidirectional naturally commutated switches. A thyristor tap-changing unit controls the voltage to the series transformer secondary.

Using phase control may control the magnitude of the series voltage. To avoid excessive harmonic generation, various taps are used. The tap changer can connect the excitation winding either completely or not at all. One

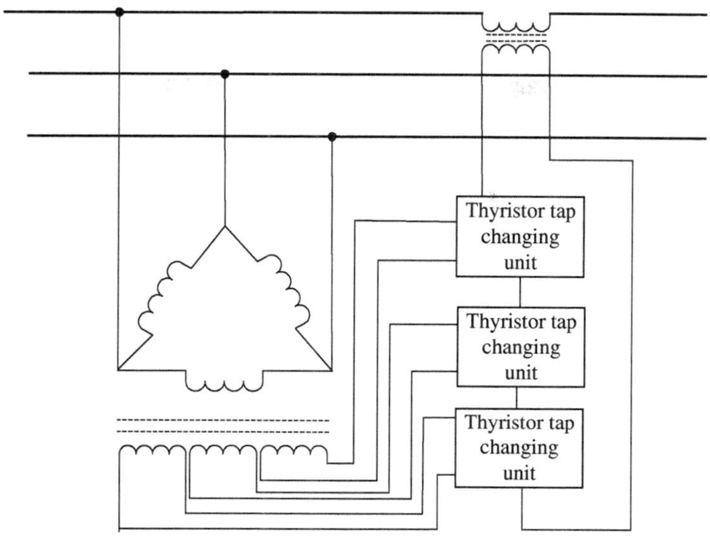

FIG. 9.40 Phase Angle Regulator.

important characteristic of the phase angle regulator is that the active power can only flow from the shunt to the series transformers. Thus reverse power flow is not possible.

The basic model of power flow control by PAR is illustrated in Fig. 9.41. The PAR is inserted between the sending-end bus and the transmission line. Theoretically, the PAR can be considered a sinusoidal ac voltage source with controllable amplitude and phase angle. Thus, the effective sending-end voltage V_{eff} becomes the sum of the prevailing sending end bus voltage E and the voltage V_{eff}, provided by the PAR. For an ideal PAR, the angle of phasor V_c relative to phasor E is stipulated to vary with σ so that the angular change does not result in a magnitude change, based on above discussions, the following expression can be given as follows:

$$\vec{V}_{eff} = \vec{E} + \vec{V}_\sigma \text{ and } |V_{eff}| = |E| = |V| \tag{9.41}$$

With the phase-angle control arrangement stipulated by Eq. (9.41) into Eqs. (9.30) and (9.31), the effective phase angle between the sending- and receiving-end voltages becomes (δ-σ), and with this the transmitted power P

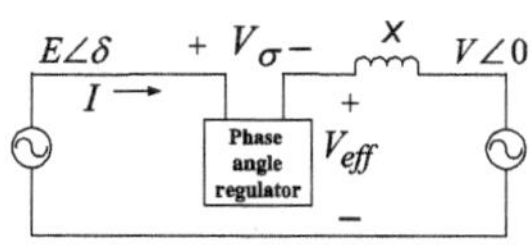

FIG. 9.41 The equivalent compensation model of PAR.

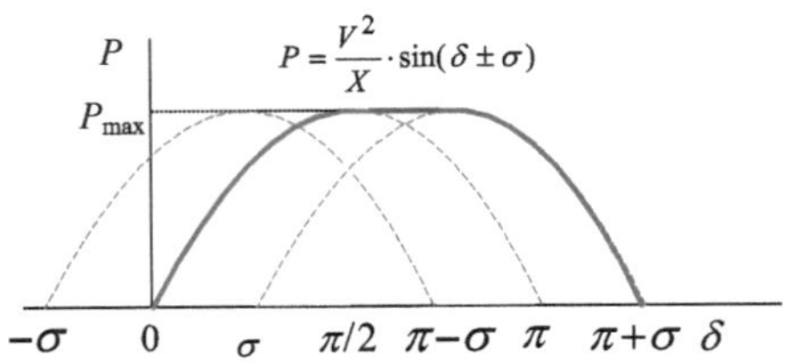

FIG. 9.42 The transmitted active power in terms of angle.

and the reactive power demands Q at the ends of the line can be expressed simply as follows:

$$P = \frac{V^2}{X}\sin(\delta - \sigma) \tag{9.42}$$

$$Q = \frac{V^2}{X}[1 - \cos(\delta - \sigma)] \tag{9.43}$$

The relationships between active power P, reactive power Q, and angles δ and σ are plotted in Fig. 9.42. It can be observed that although the PAR does not increase the transmittable active power, it makes it theoretically possible to keep the power at its maximum value at any angle δ in the range $\pi/2 < \delta < \pi/2 + \sigma$. Actually, the P curve in terms of δ curve can shift to both right and left sides, which depends on the polarity of the voltage of the angle regulator.

Similarly to the above FSCs, TPSCs, and TCSCs, the PAR allows control of power through the network and power sharing between parallel circuits. Unlike PAR, the FSCs, TPSCs, and TCLCs are more suitable for long-distance lines because they effectively reduce line reactance and hence reduce the reactive power and voltage control problems associated with long-distance transmission. PARs are more suitable for power flow control in compact high power density networks.

9.4.3 Shunt Type Compensation

Shunt type compensators can be variable impedance, variable source, or a combination of these. In principle, shunt type compensators inject current into the system at the point of connection. Even the variable shunt impedance connected to the line voltage causes a variable current flow and hence represents injection of current into the line. As long as the injected current is in phase quadrature with the line voltage, the parallel compensator only supplies or consumes variable reactive power. Any other phase relationship will involve handling of active power as well.

The ultimate objective of applying parallel compensators in a transmission system is to supply reactive power to increase the transmittable power and to make it more compatible with the prevailing load demand. Thus, the parallel compensators should be able to minimize the line overvoltage under light load conditions, and maintain voltage levels under load conditions. In the following parts, the representative are introduced, which include parallel compensators SVCs and STATCOMs.

9.4.3.1 Static VAR Compensators

In this part, SVCs are taken up for study. This compensator is a variable impedance device where the current through a reactor is controlled using back-to-back connected thyristor valves. The application of SVC was initially for load compensation of fast changing loads such as steel mills and arc furnaces. The objective is to provide dynamic power factor improvement and also balance the currents on the source side whenever required. The application of transmission line compensators commenced in the late 1970s.

SVCs are developed by replacing the mechanical switches in traditional shunt reactor and capacitor by thyristors. An SVC has no inertia compared to synchronous condensers and can be extremely fast in response (two or three cycles). This enables the fast control of reactive power in the control range. Typical SVCs include the Thyristor Controlled Reactor (TCR), Thyristor Switched Capacitor (TSC), or a combined SVC such as the Fixed Capacitor-Thyristor Switched Reactor (FC-TSR). The typical circuit schematics of an SVC are shown in Fig. 9.43.

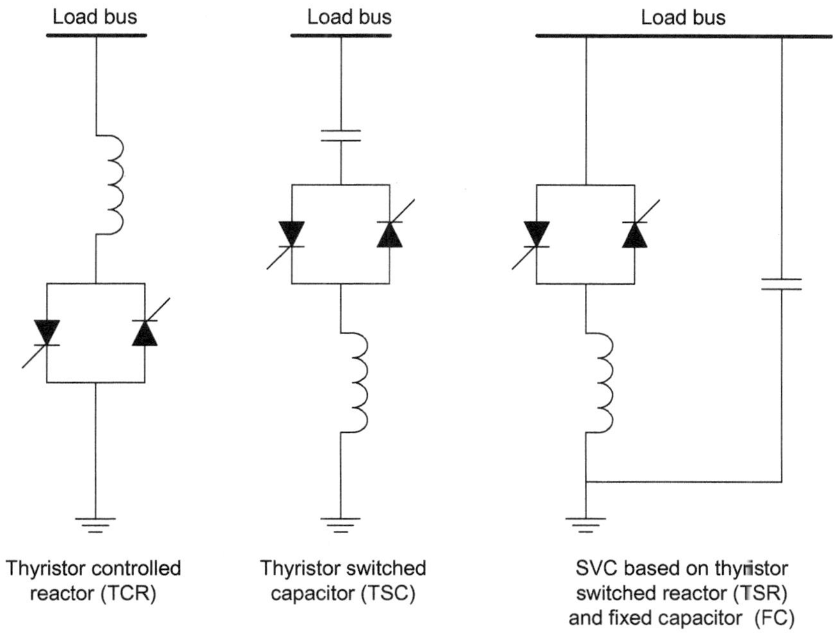

FIG. 9.43 Typical circuit schematics of the Static VAR Compensator (SVC).

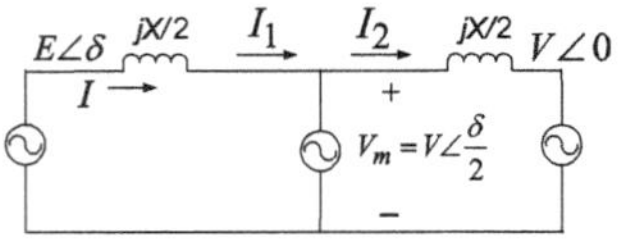

FIG. 9.44 The equivalent compensation model of SVC.

For example, when the load is capacitive (leading), the SVC is switched to reactor mode and absorbs reactive power from the system, thus lowering system voltage; on the other hand, when the load is inductive (lagging), the SVC is switched to capacitor mode and generates reactive power to the system, thus maintaining system voltage. However, the SVC cannot provide fast dynamic compensation performance and is not preferred when the load is dynamic and time varying. In addition, the compensation power of the SVC is relatively less compared to other advanced FACTS devices.

The equivalent model of the SVC compensation is provided in Fig. 9.44. In Fig. 9.44, the ideal Var compensator is shunt connected at the midpoint of the transmission line, and the line is represented by the series line inductance. The SVC is represented by a sinusoidal AC voltage source in-phase with the midpoint voltage V_m. Ideally, the amplitude of V_m is identical to the sending- and receiving-end voltages ($V_m = E = V$). The midpoint compensator in effect segments the transmission line into two independent parts with impedance of $X/2$.

For the lossless system assumed, the active power is the same at each terminal (sending-end, midpoint, and receiving-end) of the line, and it can be derived readily:

$$S_V = V \cdot \vec{I}^* = V \cdot \left(\frac{V\angle\delta/2 - V\angle 0}{jX/2}\right)^* = V \cdot \left(\frac{V\cos(\delta/2) + jV\sin(\delta/2) - V}{jX/2}\right)^*$$
$$= 2\frac{V^2}{X}\sin(\delta/2) + j\frac{4 \cdot V^2}{X} \cdot (1 - \cos(\delta/2)) = P + jQ \tag{9.44}$$

Based on deduction in Eq. (9.44), the active power and reactive can be expressed as follows:

$$P = \frac{2 \cdot V^2}{X}\sin(\delta/2) \text{ and } Q = j\frac{4 \cdot V^2}{X} \cdot (1 - \cos(\delta/2)) \tag{9.45}$$

The relationship between active power P, reactive power Q, and angle δ for the case of ideal shunt compensation is plotted in Fig. 9.45.

From Fig. 9.45, it can be observed that the midpoint shunt compensation can significantly increase the transmittable power (doubling its maximum value) at the expense of a rapidly increasing reactive power demand on the midpoint compensator (and also on the end-generators).

9.4.3.2 Static Synchronous Compensators

Although SVCs provide better compensation than conventional compensators, their dynamic performance is still far from satisfactory under dynamic loads.

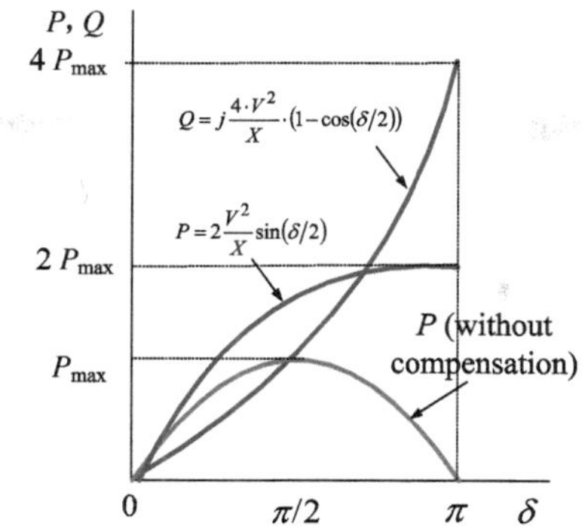

FIG. 9.45 The transmitted active and reactive power in terms of angle.

Controllable electronic switches such as IGBTs then contribute largely to the development of advanced compensators such as the STATCOM. This is sometimes known as an advanced version of the SVC. A typical circuit schematic is shown in Fig. 9.46.

The term "static" refers to a nonrotational device that is different from traditional compensation machine that may generate audible noise. The STATCOM is composed of a DC link and a voltage source inverter (VSI). The VSI is used to convert DC link power into AC so as to compensate harmonics, active power, or reactive power as desired.

As shown in Fig. 9.46, the converter produces a set of controllable three-phase output voltages from a DC input voltage source by the charged capacitor. Each output voltage is coupled to the corresponding AC system voltage via a relatively small (0.1–0.15 p.u.) reactance. By varying the amplitude of the output voltages, the reactive power exchange between the converter and the AC system can be controlled. Specifically, if the amplitude of the output voltage is increased larger than the AC system voltage, the current through the reactance from the converter to the AC system and the converter generates capacitive

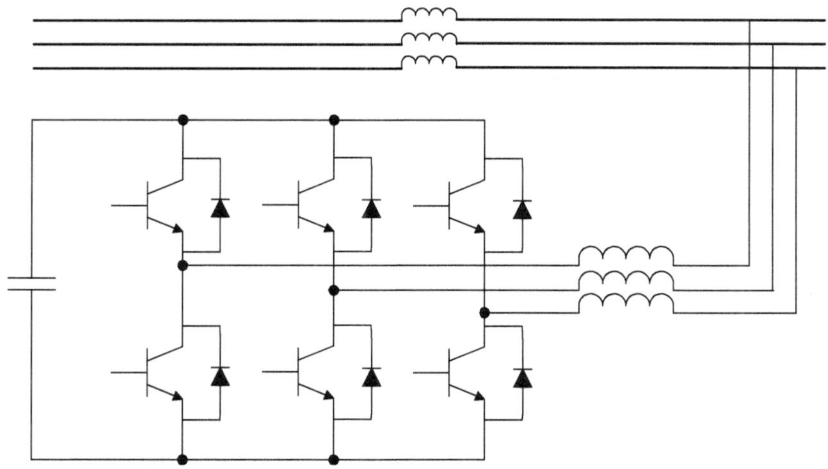

FIG. 9.46 Static synchronous compensator (STATCOM).

reactive power for the AC system. If the amplitude of the output voltage is decreased below that of the AC system, the reactive current flows from the AC system to the converter, and the converter absorbs inductive reactive power. In addition, if the amplitude of the output voltage is equal to that of the AC system voltage, the reactive power exchange is zero.

The model and the operation principle for the STATCOM are similar to that of the SVC, as shown in Fig. 9.45.

9.4.4 Combined Series-Series Type Compensation

A combination of separate series controllers are controlled in a coordinated manner in a multiline transmission system. Combined series-series type compensators provide independent series reactive compensation for each line but also transfer active power among the lines via the power link. The active power transfer capability of the combined series-series type compensator, referred to as Interline Power Flow Controllers (IPFCs), makes it possible to balance both the active and reactive power flow in the lines, and thereby maximize the utilization of the transmission system. In IPFCs, the DC terminals of all controller converters are all connected together for active power transfer.

In its general form, the IPFCs employ a number of DC-to-AC converters, thus providing series compensation for a different line. In other words, the IPFC comprises a number of Static Synchronous Series Compensators. However, within the general concept of the IPFC, the compensating converters are linked together at their DC terminals, as illustrated in Fig. 9.47. With this scheme, in addition to providing series reactive compensation, any converter can be controlled to supply active power to the common DC link from its own transmission line. Thus, an overall surplus power can be made available from the underutilized lines, which then can be used by other lines for active power compensation. In this way, some of the converters, compensating overloaded lines or lines with a heavy burden of reactive power flow, can be equipped with

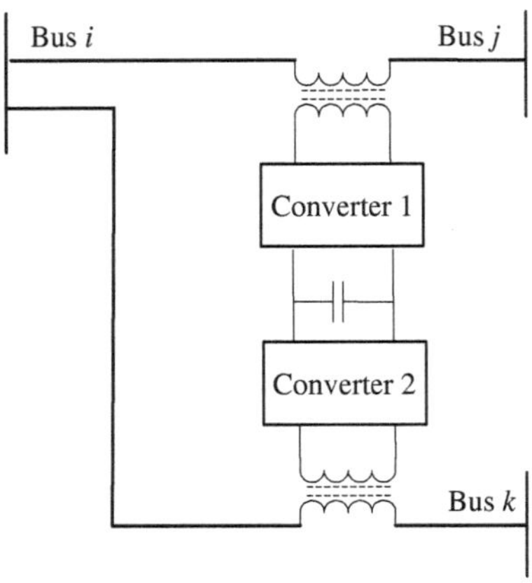

FIG. 9.47 Interline Power Flow Controllers (IPFCs).

full two-dimensional, reactive power, and active power control capability. Evidently, this arrangement mandates the rigorous maintenance of the overall power balance at the common DC terminal by appropriate control action, using the general principle that the underloaded lines are to provide help, in the form of appropriate active power transfer, for the overloaded lines.

The equivalent compensation model of IPFC is provided in Fig. 9.48. In the following analysis, the bus voltage can be expressed as follows:

$$\left|\vec{E}\right| = \left|\vec{E}_2\right| = \left|\vec{V}\right| = \left|\vec{V}_2\right| \tag{9.46}$$

where the E, E_2, V, and V_2 are the bus voltage. The active and reactive power in the first line can be given as follows:

$$\begin{aligned} S_V &= \vec{V} \cdot \vec{I}^* = V\angle 0 \cdot \left(\frac{E\angle\delta + V_c\angle\theta_1 - V\angle 0}{jX}\right)^* \\ & V\angle 0 \cdot \left(\frac{E\angle\delta - V\angle 0}{jX}\right)^* - V\angle 0 \cdot \left(\frac{V_c\angle\theta_1}{jX}\right)^* \\ &= \frac{V^2 \sin(\delta)}{X} + \frac{V \cdot V_c \sin(\delta+\theta_1)}{X} + j\frac{V^2}{X} \cdot (1 - \cos(\delta)) + j\frac{V \cdot V_c \cos(\delta+\theta_1)}{X} \\ &= P_V + jQ_V \end{aligned} \tag{9.47}$$

Based on Eq. (9.47), the power flow in line 1 can be expressed as follows:

$$P_V = \frac{V^2 \sin(\delta)}{X} + \frac{V \cdot V_c \sin(\delta+\theta_1)}{X} \tag{9.48}$$

$$Q_V = \frac{V^2}{X} \cdot [1 - \cos(\delta)] - \frac{V \cdot V_c \cos(\delta+\theta_1)}{X} \tag{9.49}$$

Moreover, the active power transfer from line 1 to line 2 can be expressed as follows:

$$P_{12} = \frac{V \cdot V_c}{X}[\sin(\delta+\theta_1) - \sin(\theta_1)] = \frac{2 \cdot V \cdot V_c}{X}\cos(\delta+\theta_1) \cdot \sin(\theta_1) \tag{9.50}$$

Based on Eq. (9.50), the transmitted active in terms of angle from line 1 to line 2 can be plotted as shown in Fig. 9.49.

As illustrated in Fig. 9.49, the active power can easily transfer from line 1 to line 2 with $P_{12}>0$ or vice versa $P_{12}<0$.

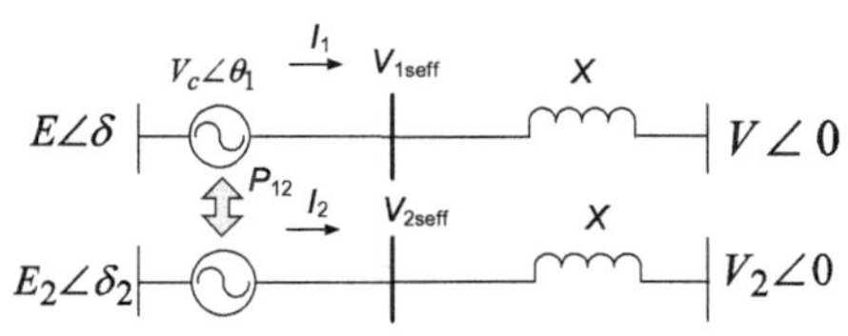

FIG. 9.48 The equivalent compensation model of IPFC.

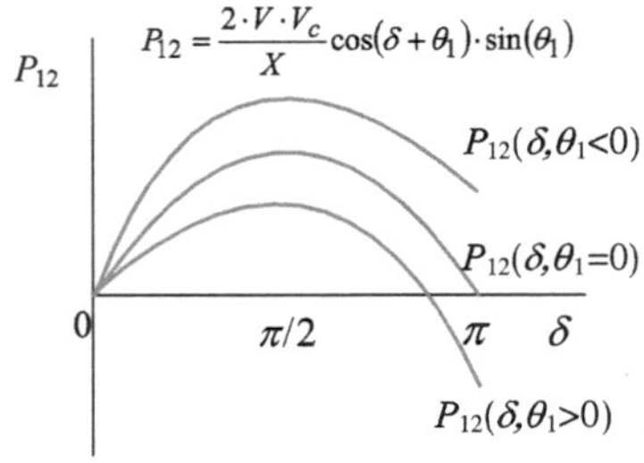

FIG. 9.49 The transmitted active in terms of angle from line 1 to 2.

9.4.5 Combined Series-Shunt Type Compensation

Combined series-shunt type compensators inject current into the system with the shunt part of the compensator and voltage in series in the line with the series part of the compensators. However, when the shunt and series compensators are unified, there can be active power exchange between the series and shunt compensators via the power link.

The Unified Power Flow Controller (UPFC) as one of the representative combined series-shunt type compensators was proposed by Gyugyi in 1991. The UPFC was devised for the real-time control and dynamic compensation of AC transmission systems, providing multifunctional flexibility required to solve many of the problems facing the power delivery industry. Differently to the above-mentioned compensators, the UPFC is able to control all the parameters affecting power flow in the transmission line (i.e., voltage, impedance, and phase angle). Alternatively, it can independently control both the active and reactive power flow in the line.

As illustrated in Fig. 9.50, the UPFC consists of two voltage source converters. These back-to-back converters, labeled "Voltage source converter 1" and "Voltage source converter 2" in the figure, are operated from a common DC link provided by a DC storage capacitor. As indicated before, this

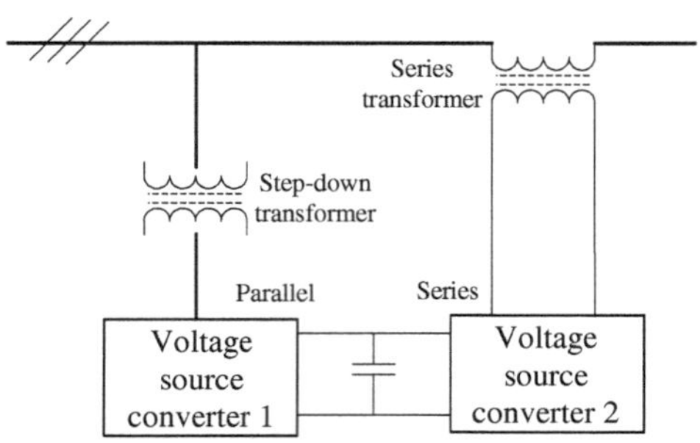

FIG. 9.50 Unified Power Flow Controllers (UPFCs).

arrangement functions as an ideal AC-to-AC power converter in which the active power can freely flow in either direction between the AC terminals of the two converters, and each converter can independently generate (or absorb) reactive power at its own AC output terminal.

The basic function of voltage source converter 1 is to supply or absorb the active power demanded by voltage source converter 2 at the common DC link to support the active power exchange resulting from the series voltage injection. This DC link power demand of converter 2 is converted back to AC by converter 1 and coupled to the transmission line bus via a parallel connected transformer. In addition to the active power need of converter 2, converter 1 can also generate or absorb controllable reactive power, if it is desired, and thereby provide independent shunt reactive compensation for the line. It is important to note that whereas there is a closed direct path for the active power negotiated by the action of series voltage injection through converters 1 and 2 back to the line, the corresponding reactive power exchanged is supplied or absorbed locally by converter 2 and therefore does not have to be transmitted by the line. Thus, converter 1 can be operated at a unity power factor or be controlled to have a reactive power exchange with the line independent of the reactive power exchanged by converter 2. Obviously, there can be no reactive power flow through the UPFC DC-link.

The equivalent compensation model of UPFC is provided in Fig. 9.51. From the conceptual viewpoint, the UPFC is a generalized synchronous voltage source (SVS), represented by voltage source, V_c, with controllable magnitude and angle, in series with the transmission line. Both exchanged reactive and active powers through the transmission system can be controlled by regulating voltage amplitude and phase angle.

The equivalent compensation model of IPFC is provided in Fig. 9.51. In the following analysis, the bus voltage can be expressed as follows:

$$\left|\vec{E}\right| = \left|\vec{V}_c\right| \tag{9.51}$$

where the E and V are the bus voltage. The active and reactive power in the first line can be given as follows:

$$\begin{aligned}
S_V &= \vec{V} \cdot \vec{I}^* = V\angle 0 \cdot \left(\frac{E\angle\delta + V_c\angle\theta - V\angle 0}{jX}\right)^* \\
&V\angle 0 \cdot \left(\frac{E\angle\delta - V\angle 0}{jX}\right)^* - V\angle 0 \cdot \left(\frac{V_c\angle\theta_1}{jX}\right)^* \\
&= \frac{V^2 \sin(\delta)}{X} + \frac{V \cdot V_c \sin(\delta+\theta)}{X} + j\frac{V^2}{X} \cdot (1 - \cos(\delta)) + j\frac{V \cdot V_c \cos(\delta+\theta)}{X} \\
&= P_V + jQ_V
\end{aligned} \tag{9.52}$$

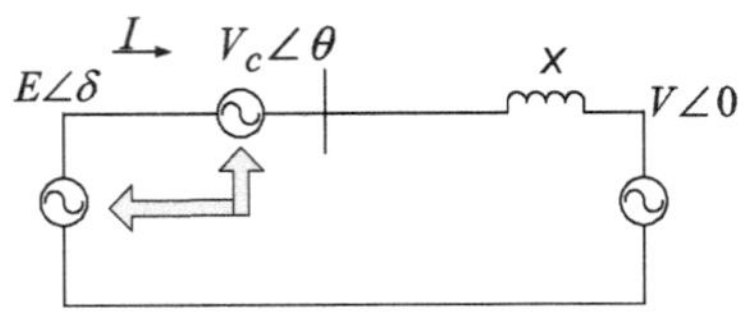

FIG. 9.51 The equivalent compensation model of UPFC.

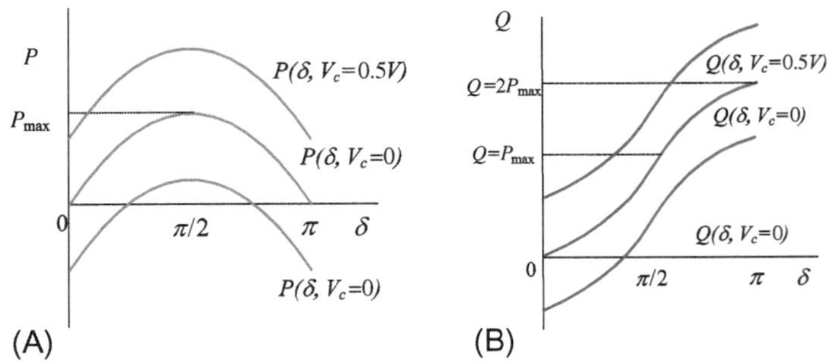

FIG. 9.52 The transmitted (A) active and (B) reactive power in terms of angle.

Based on Eq. (9.47), the power flow in line 1 can be expressed as follows:

$$P_V = \frac{V^2 \sin(\delta)}{X} + \frac{V \cdot V_c \sin(\delta+\theta)}{X} \tag{9.53}$$

$$Q_V = \frac{V^2}{X} \cdot [1 - \cos(\delta)] + \frac{V \cdot V_c \cos(\delta+\theta)}{X} \tag{9.54}$$

Based on Eqs. (9.53) and (9.54), the active and reactive power in terms of angle δ can be plotted as shown in Fig. 9.52.

Based on this figure, the transmitted active power P and the reactive power Q vary with the rotation of phasor V_c with angle θ. Therefore, the UPFC can control the magnitude and angular position of the injected voltage in real time so that to maintain or vary the real and reactive power flow in order to satisfy load demand and system operating conditions.

9.4.6 Distributed FACTS Devices

As new upgrade technologies, FACTS (Flexible AC Transmission System) and DFACTS (Distribution Flexible AC Transmission System) provide an effective approach to control the power flow and to improve the security, economy, and quality of electric power supply by rapidly adjusting the system parameters [17]. The concept of FACTS was introduced by N.G. Hingorani in 1986

[18], and aimed at reconstructing the traditional power system by introducing power electronics technologies to make it more flexible and reliable. Since then, the technology has been moving ahead at an increasing pace. Many kinds of FACTS controllers for different parameter regulating and in different configurations have been developed. Being the expanded use of FACTS, DFACTS brings the solution for "Custom Power" or power quality problem at the user-end. VSI-based dynamic power quality compensators are the main concern to solve the power quality problem. Compared with thyristor-switched capacitors and (or) reactors, these compensators have the features of fast response, continuous regulation, multiple functions, smaller physical size, etc. With the further evolution of electricity market and the progress of legislation in power quality, the reconstructed urban power network would rely on the application of new technology including DFACTS to improve its electric power supply. Further increasing the reliability, improving the power quality, and reducing the costs of power network are the main trends of the technology development of power system control and management. Table 9.5 lists most of the important DFACTS devices and their operational functions [19] in the distribution power network. According to their connection to the distribution power network, all the devices are classified three topologies: shunt connected, series connected, and shunt and series connected compensators.

Figs. 9.53–9.55 show the system configurations of three important power quality compensators: Active Power Filter (APF), Dynamic Voltage Restorer (DVR), and Unified Power Quality Conditioner (UPQC), respectively. The three compensators represent the shunt, the series, and the shunt and series connected compensator structures, respectively.

9.5 Summary

The amount of transfer power can be limited from sending-end to the receiving-end by controlling the operating parameters of the transmission line including the line impedance, phase angle between the sending and receiving voltages, and the magnitude of the voltages. The transferable power can be increased by one of four compensation methods: series, parallel, series-series, and series-parallel compensations. These methods are called FACTS, which are generally implemented by switching power electronics together with appropriate control strategy. For distributed sides, power quality, reactive power compensation, and renewable active power injections are important, so that DFACTS are combination of power electronic converters to allow power systems to operate in smarter ways to increase reliability, safety, and efficiency.

TABLE 9.5 DFACTS Devices

		Shunt Connected Compensators				Series connected Compensators		Shunt and Series Connected Compensators
		SVC	*APF*	*STATCOM*	*SMES*	*DVR*	*UPS*	*UPQC*
Current quality	Harmonics		**					*
	Unbalance		*					*
	Reactive	*	*	*				*
Voltage quality	Harmonics							*
	Unbalance					*		*
	Fluctuation/ stability	*		*	*	**	*	*
	Flicker					*		*
		S	N	S	S	N	S	N

Notes: *, the device has this function; **, this is the main function for this device; S, the device needs to generate sinusoidal waveform; N, the device needs nonsinusoidal waveform.
APF, active power filters; DVR, dynamic voltage restorers; SMES, superconducting magnetic energy storage; STATCOM, static synchronous compensators; SVC, static VAR compensators; UPS, uninterruptible power supply; UPQC, unified power quality conditioners.

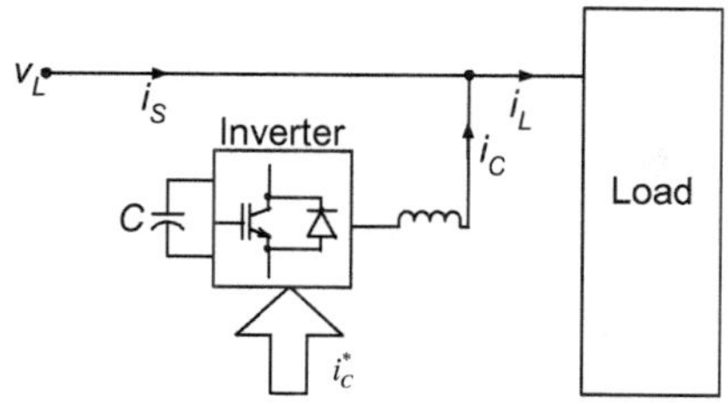

FIG. 9.53 Active Power Filter.

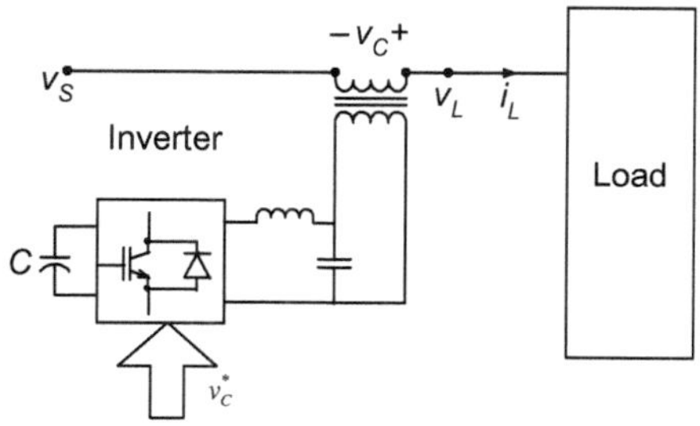

FIG. 9.54 Unified Power Quality Conditioner.

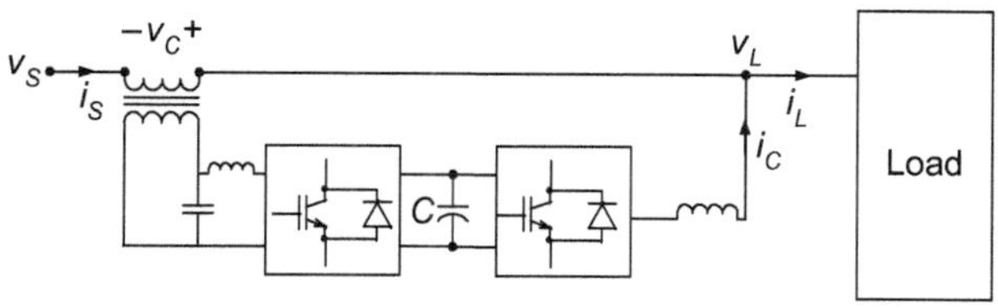

FIG. 9.55 Dynamic Voltage Restorer.

References

[1] B.K. Bose, Power electronics—a technology review, Proc. IEEE 80 (8) (1992) 1303–1334.

[2] M. H. Rashid, "Power Electronics: Circuits, Devices, and Applications", Third ed., 2004, Prentice-Hall International: Upper Saddle River, NJ.

[3] Mohan/Undeland/Robbins, "Power Electronics: Converters, Applications, and Design", Wiley, 2003

[4] M.C. Wong, Z.-Y. Zhao, Y.-D. Han, L.-B. Zhao, Three-dimensional pulse-width modulation technique in three-level power inverters for three-phase four-wired system, IEEE Trans. Power Electron. 16 (3) (2001) 418–427.

[5] M.C. Wong, J. Tang, Y.-D. Han, Cylindrical coordinate control of three-dimensional PWM technique in three-phase four-wired trilevel inverter, IEEE Trans. Power Electron. 18 (3) (2003) 208–219.

[6] N.Y. Dai, M.C. Wong, Y.D. Han, Three-dimensional space vector modulation with dc voltage variation control in three-leg center-split power quality compensators, IEE Proc. Electr. Power Appl. 151 (2) (2004) 198–204.

[7] N.Y. Dai, M.C. Wong, Y.D. Han, Application of a three-level NPC inverter as a three-phase four-wire power quality compensator by generalized 3DSVM, IEEE Trans. Power Electron. 27 (4) (2006) 1758–1772.

[8] N.Y. Dai, M.C. Wong, F. Ng, Y.D. Han, A FPGA-based generalized pulse width modulator for three-leg center-split and four-leg voltage source inverters, IEEE Trans. Power Electron. 23 (3) (2008) 1472–1484.

[9] M.J. Ryan, W. Doncker, R.D. Lorenz, Decoupled control of a four-leg inverter via a new 4X4 transformation matrix, IEEE Trans. Power Electron. 16 (5) (2001) 694–701.

[10] E.H. Watanabe, H. Akagi, M. Aredes, Instantaneous p-q power theory for compensating non-sinusoidal systems, in: International School on Nonsinusoidal Currents and Compensation (ISNCC), Lagow, Poland, 2008, pp. 1–10. June 10–13.

[11] H. Akagi, Y. Kanazawa, A. Nabae, Instantaneous reactive power compensators comprising switching devices without energy storage components, IEEE Trans. Ind. Appl. 20 (1984) 625–630.

[12] A. Nabae, T. Tanaka, A new definition of instantaneous active-reactive current and power based on instantaneous space vectors on polar coordinates in three-phase circuits, in: IEEE/PES Winter Meeting, Baltimore, Maryland, 1996, pp. 1238–1243. Jan. 21–25.

[13] N.G. Hingorani, L. Gyugyi, Understanding FACTS–Concepts and Technology of Flexible AC Transmission Systems, IEEE Press, New York, 1999.

[14] M.H. Rashid, Power Electronics, Circuits, Devices, and Applications, third ed., Prentice Hall, 2003.

[15] X.P. Zhang, C. Rehtanz, B. Pal, Flexible AC Transmission Systems: Modeling and Control, Springer, Berlin, 2012.

[16] Y.H. Song, A. Johns, Flexible AC Transmission Systems (FACTS), Institution of Electrical Engineers, London, 1999.

[17] H. Yingduo, C. Jianye, J. Qirong, W. Manchung, Study of FACTS and DFACTS in China, in: The Third International Power Electronics and Motion Control Conference Proceedings, PIEMC 2000, vol. 1, 2000.

[18] N.G. Hingorani, High power electronics and flexible AC transmission system, in: Joint APC/IEEE Luncheon Speech, April 1988 at the American Power Conference 50th. Annual Meeting in Chicago, IEE Power Engineering, 1988.

[19] M.-C. Wong, Y.-D. Han, L.-B. Zhao, Study of distribution system unified conditioner (DS-UniCon), in: IEEE Third International Power Electronics and Motion Control Conference, Beijing, 2000. Aug.

Chapter 10

A Survey of Recent Developments and Requirements for Modern Power System Control

Panayiotis Moutis, M. Hadi Amini, Irfan Ahmad Khan, Guannan He, Javad Mohammadi, Soummya Kar and Jay Whitacre
Carnegie Mellon University, Pittsburgh, United States

Chapter Outline

10.1 Introduction

The monopolies and oligopolies of energy generation and supply, owned and operated either by private companies or the state, have hardly ever been characterized as sustainable paradigms for the electricity sector. Energy [1–4] and financial [5–8] crises, and catastrophic events [9–12] affect the oligopolies gravely and extensively, while lack of proper regulation [13–16] has affected

Pathways to a Smarter Power System. https://doi.org/10.1016/B978-0-08-102592-5.00010-7

severely the price, reliability, and quality of electric power supplied to end customers, either residential, commercial, industrial, or in the public sector. System operators, regulators, and governing and policy bodies that introduce and oversee standardized codes, practices, and protocols regarding the operation and control of power systems have only been active since the 1980s [17, 18].

Additionally, since the early 1990s several concerns have been raised about the harmful, global effects of fossil fuel emissions to the environment and the climate [19–21]. A large portion of these emissions come from the conventional electricity generation sector [22–24], meaning that a shift to cleaner and more efficient sources and technologies is required. The shift was expressed by initiatives at the highest international level [25, 26], as well as regionally [27–30], and all of these initiatives have been steadily revisited since their inception, with the aim to update goals and assess progress. The initiatives focused on the promotion of generation from renewable energy sources (RES) [31, 32], as they are mostly emission-free ("mostly" due to the production lines of the said generators powered by conventional sources), and distributed generation (DG) in general [33], as it increases the efficiency of the system as a whole. Both RES and DG deployment was expected and has actually been realized by stakeholders outside the traditional utilities. Several subsidizing mechanisms [34–36] attracted participants as small as residential customers [37, 38] in the energy market. These latter mechanisms, although critical to the vision for the shift away from fossil fuels, have also introduced considerable irregularities in the electricity sector.

The aforementioned two factors have been the main drivers of what has been described as the deregulation of the energy markets [39]. Multiple stakeholders populate the market at various levels and with multiple roles including, but not limited to, generator owners, load customers of many types, energy providers/aggregators, services providers/aggregators, energy market regulators and operators, system operators, associations, and cooperatives [18, 40–42]. By the term "market" we refer to not only energy and power trading, but also the system as a whole. This last point brings into the discussion codes, standards, and other formalized protocols determining both the minimum requirements of the power system infrastructure, and also its operation within specified limits and at an adequate level of quality of service to customers [43–45].

The current state of the modern power systems, as has been briefly described above, implies that there is an ever increasing (considerably large) number of control points, dispersed over vast geographical areas, monitored and controlled by diverse entities with multiple aims, and maybe even competing interests. In this sense, what has been widely known and accepted as the "classical control approach of power systems" [46, 47] has to be revisited, redesigned, and formalized anew, while taking into account all of the aforementioned challenging, complex, and system-wide parameters.

The extensive interest in control theory and systems research has been influenced by the goal to shape ideas about the power system management by both academia and industry in the field. Going beyond the practical issues as these were outlined in the previous paragraphs, modern control technologies seek to address more abstract challenges in a way that cuts across multiple applications and problems by leveraging the positive features of the proposed innovations to enable efficiency at the system-wide level. This chapter also follows the same mind-set by organizing what is understood as modern control technologies from the perspective of the characteristics of each of these technologies. In other words, the aim of this chapter is to enable the reader to identify not application-specific power system challenges and how some control methods address them, but rather what should one expect and require as minimum or average features of any novel approach employed in power system control methods and strategies.

In Section 10.2, some of the most common classic control techniques and approaches for power systems are described, along the lines of their shortcomings, in the context of the modern operating paradigm. Next, in Section 10.3, the five main characteristics that modern control technologies are expected to have, in order to better serve their purpose in the current and future electric power system, are described in detail. In Section 10.4, four large application fields of power system control are identified and the corresponding advances in those fields are presented from the perspective of which characteristics of the control technologies are more crucial and are better expressed by the aforementioned advances.

10.2 Classic Control Applications in Power Systems

To have a better understanding of what the challenges are of employing control in modern power systems, one needs to assess the current status of the classic control requirements and functions in today's power systems and the very recent past. In this sense, a brief overview of the most common control problems is provided in the next subsections.

10.2.1 System Area Frequency Control and Voltage Regulation

Regulation of frequency and voltage at any given point of an operating area and at any given moment are the two most commonly understood operations in power systems stability and control [46]. The electric frequency of a system expresses the angular speed of the rotating electric currents in all electric machines, and must remain close to nominal value (50 or 60 Hz) for multiple reasons of infrastructure integrity and system protection. Voltage magnitude also needs to be kept as close to its rated values as possible to ensure efficient behavior of end-users' equipment and avoid collapse of any part of the system, which may lead to interrupts and blackouts. Both of these control functionalities need to be managed efficiently according to standardized grid codes, while taking into account their specific features [43, 47]. Frequency is common throughout an interconnected system and is an electromechanical phenomenon in the time frame of tens of seconds, while voltage regulation can only be handled by local resources and measures, and, as an electrical phenomenon, may require response times as fast as milliseconds.

Due to the aforementioned features, power systems operators have, traditionally, controlled frequency and voltage using hierarchical schemes. It is a type of scheme that is applied to achieve various control objectives in different spatio-temporal scales. In other words, the hierarchy of the control action may concern either the time-scale of operation (e.g., within milliseconds or up to minutes) or the area (e.g., starting closer to load centers rather than generating plants that can be involved in some following action) or some other perceived perspective/structure of the problem. Commonly, hierarchical control architectures comprise three main levels: primary, secondary, and tertiary, as shown in Fig. 10.1.

The primary control usually realizes control actions as a sort of automatic response to a change of a status variable. The aim of primary control is to limit excursions from nominal control points. Droop control is one of the most popular types of primary control that regulates the frequency by adjusting the output active power (commonly) of a controlled asset (usually a generator). Voltage control, although not explicitly, has traditionally been organized in primary and secondary levels, the primary being fast control action offered by automatic voltage regulators (AVRs) of generators or any dispersed assets, such as loads on demand response programs, properly equipped with controller topologies that monitor and drive accordingly the corresponding resources [48].

The secondary control follows the primary, resets the availability of primary reserves used in the previous stage, and aims to restore frequency and voltage to nominal values, if that has not been achieved earlier. As primary control is, commonly, a local proportional type of control, it may not be able to achieve the reference values accurately and some offset may still exist, thus requiring

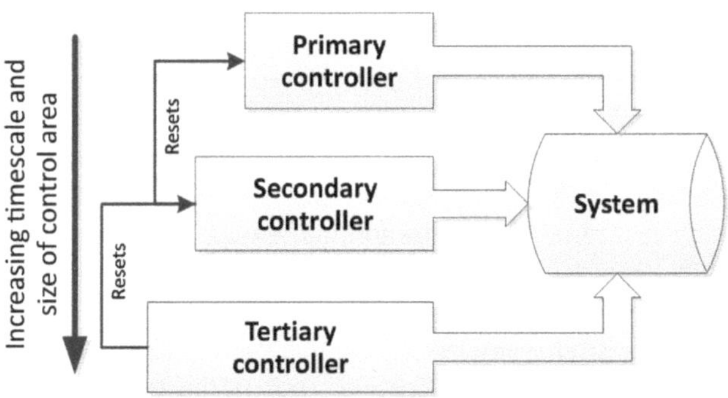

FIG. 10.1 Hierarchical control concept of power systems.

an integral type of control. Since it follows primary control, secondary control is implemented for longer time horizons (around minutes) and is a regional type of control dealing with a zone of an electrical system. In other words, an operator is responsible for scheduling the contribution of the various assets in a control area. To regulate the frequency and voltage, secondary control procures reserves of generation capacity, usually reimbursed as ancillary services and scheduled properly in the day-ahead markets of the regional operators [49]. In this sense, the control also becomes a market problem with optimal formulation characteristics. Hence, at this stage one may identify the secondary voltage control and load-frequency control (LFC)—otherwise known as automatic generation control (AGC). AGC is relevant to another crucial matter of power system control, the area control error and the interarea exchanges of active power and their effects on the stability of neighboring systems [46].

The third level of control, the tertiary control, is driven mostly by cost criteria of sorts, that is, optimal economic dispatching, minimization of losses, operation scheduling, and other optimization problems of similar nature. It is a system level, multiple time-steps ahead control technique, which takes into account the whole system (consisting of many, loosely connected regions) over a longer time horizon (to the length of hours, days, etc.). It is, essentially, the operators' framework according to which all reserves for primary and secondary control are allocated based on minimum cost considerations. Popular examples of tertiary control are economic dispatch (ED) and unit commitment. The distinguishing characteristics of primary, secondary, and tertiary control schemes are delineated in Table 10.1.

In modern power systems with high penetrations of DG, it is critical to involve in frequency and voltage control as many of such units as possible,

TABLE 10.1 Comparative Characteristics of Primary, Secondary, and Tertiary Control

Primary Control	Secondary Control	Tertiary Control
Reduces system state excursions from nominal values	Restores system state to nominal values	Restores reserves of primary and secondary control
Realized in less than 1 min	Realized in less than an hour	Realized/scheduled in multi-time-steps ahead horizon
Automatic proportional	Automatic integral	Allocation by operator
Strictly local (not aware of total system state)	Regional (aware of total system state)	Regional and interareal

due to the displacement of conventional sources they cause, which severely affects the grid stability [50, 51]. Despite the strong motivation, there are numerous challenges. For example, when employing DG in frequency regulation, due to their unpredictability and operational constraints such as those imposed by power extraction strategies, the performance of the respective control has to be monitored continuously, while some sources might not be able to contribute with reserves at all, due to the said strategies, which do not make sense otherwise [52, 53]. For the case of voltage regulation by DG units equipped with an inverter (a particularly common topology for most renewables), their output will be actively controlled in response to some voltage sensing, depending also, though, on the network characteristics [54]; obviously, the inverters need to be able to adapt and ensure proper control performance in all cases.

10.2.2 Dispatching the System Resources

The steady-state dispatch problem of power system is usually formulated as an optimization problem to find the control set-point of all assets in the system with respect to an objective function (usually the system operational cost) subject to a set of operating constraints (power balance, voltage limit, line flow limit, etc.) and given parameters (load, equipment status, etc.) [55]. The control models of power system take different mathematical forms. The ED model neglects transmission constraints and only implements steady-state power balance constraints. On top of ED, the direct-current (DC) optimal power flow (OPF) model adds transmission constraints, but still ignores the reactive power and assumes a flat voltage magnitude of all buses to keep a linear model. Alternating-current (AC) OPF further calculates reactive power and its impact on bus voltage and considers voltage constraints. As AC OPF is nonlinear and nonconvex, its computational complexity is significantly higher than ED and DCOPF. There is a trade-off between convergence and accuracy. Since its first formulation [56], ACOPF and its approximate models have attracted intensive research [57–59]. In the conventional power system operation, the independent system operator (ISO) solves the ED problem of the available generation units on an hourly basis by scheduling the on/off cycles of generation units by solving the day-ahead unit commitment (UC). There is a rich literature on different solutions to both ED and UC problems, with certain objectives and constraints [60–62]. While these studies solve the formulated problem corresponding to power system operation with acceptable accuracy, they mostly require expensive computation platforms due to the centralized nature of their algorithms [62a, 62b].

The increasing complexity and large scale of power systems motivates the need for novel algorithms to enable efficient computations of power system operation and control. This is, however, a challenging task owing to various sources of uncertainty [52, 53], the physical constraints of the network [63], and the novel technologies introduced by the smart grid (e.g., RES and demand response programs) [64, 65, 65a]. For instance, unpredictable renewable energy is being integrated into the generation side, which not only increases the uncertainty but also injects harmonics to the power systems, which may cause instability and unexpected voltage deviation. After deregulation of power systems, the possibility of having mismatch between supply and demand is threatening grid stability. Further, the deregulation of the energy markets complemented by the aforementioned smart grid paradigms has been causing a continuous increase in the number of participants at both the generation and demand sides of power systems, leading to more decision variables, more constraints, and consequently more complicated optimization problems for ISO and other entities [66–68]. This leads to a major concern regarding the scalability of the control methods in power systems.

10.3 Characteristics to Best Serve Modern Power System Control Applications

10.3.1 Scalability

The matter of addressing the control of large numbers of actors and for multiple purposes is evident and urgent following the deregulation of the energy markets [39] and the wide promotion of the highest penetrations of DG units [25–30]. Units of sizes as small as a few kilowatts are dispersed throughout the power system, and especially at the distribution level where no direct control of the assets has traditionally been employed or practically explored. To this end, all control methods and paradigms need to be rethought to address this evolving status.

Recently, decentralized/distributed and hierarchical control and optimization techniques have been widely deployed in the power system control to answer the challenges by the scalability requirements. They introduce potential advantages, as compared with centralized methods, including but not limited to the following [69–71, 71a–71e]:

- Each controlled asset needs to broadcast limited information with some neighbors.
- Limited information exchange requirement ensures more privacy among the assets, as well as reduced costs related to communication infrastructure.
- The algorithms are more robust in terms of communication link failure and certainly against single-point failures.
- Due to the distributed nature of computation, the run-time reduces considerably.

These techniques include three major classes: (i) distributed optimization with agents that only communicate with their neighbors without a need to centralized controller [69, 72, 73]; (ii) hierarchical approaches with agents who communicate with the higher level agents at each iteration [74]; and (iii) decentralized methods with agents that solve their problem in a purely local fashion and require a centralized controller for the coordination purpose at each iteration [75–77].

10.3.2 Adaptive Control and Topologies

Traditionally, a controller is adaptive if it is able to update its parameters (for the sake of performance or stability) in response to changes occurring to the controlled asset/equipment [78]. This would translate to control methods or systems, which either estimate the state of the asset parameters online and adapt accordingly or adapt once they identify changes to the asset according to an exact (reference) model representation of the said asset. Fig. 10.2 offers a schematic abstraction of adaptive control.

Although adaptive control methods are not new in the field [79–86], the power system and the electricity markets have become more prone to showing changing behaviors, due to the involvement of multiple market players, the operation of numerous DG units throughout the system, and the unprecedented interactions among them all. "Changing behavior," in the context of adaptive methods, should not be confused with "variability," since the latter implies that there are inherent characteristics that could be described through some probabilistic models (see more in Section 10.3.4 about robustness).

With the growing concerns about cyber-physical security issues [87, 87a, 87b], adaptive control will be put center stage as the only viable counter measure to attacks of that kind. In this sense, not only the control techniques but also the corresponding topologies should be made adaptive. Put in practical terms, a control topology is adaptive if the overlying control method can be executed successfully (or with little loss of performance efficiency) by some remaining

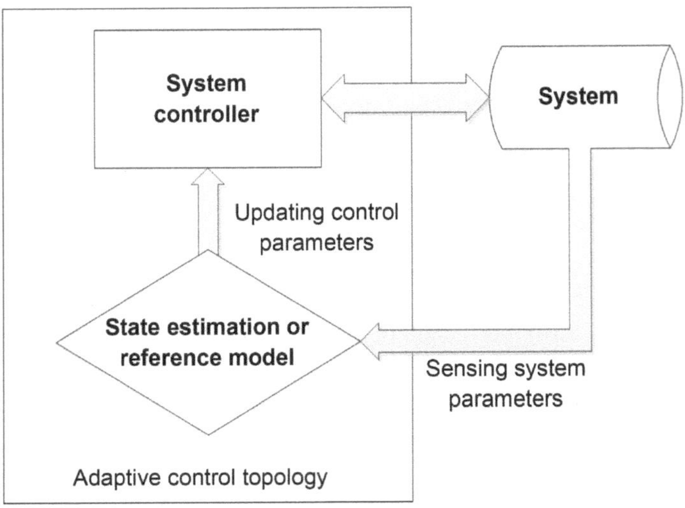

FIG. 10.2 General representation of adaptive control approaches.

part of the topology following an attack or a failure. Some typical examples here are the various decentralized and distributed methods [69–76], which do not depend on any single point of their architectures.

From the above, it may be noted that novel, adaptive control techniques and topologies are also expected to be characterized by resiliency [88]. Resiliency pertains to showing adequate control performance under failures of the control architecture or changes to the controlled asset that would otherwise affect the performance. Further to this, if the control performance remains the same (in all or most of its measures), the adaptive control and its topology may also be described as reliable.

10.3.3 Flexibility

Flexibility is a control property that plays a crucial role for the transition of traditional power systems, many of them based on fossil fuels, toward power systems that can efficiently accommodate high shares of variable distributed energy generations. Flexibility pertains to control implementations that can handle a problem in several time scales, regions, or other structured organization blocks that have a minimum affect among them and may thus be handled separately. Flexibility is best expressed in power systems by employing hierarchical schemes of control actions.

As hierarchical control schemes have multiple levels of timescales and control areas, they are proven very effective, because they offer a high degree of practical flexibility for power systems. This is especially so for the very timely case of widely deployed DG units based on RES that differ from conventional power plants—prominently due to their high intermittency, efficient management of their operation is essential. As discussed in Section 10.2.1, control hierarchy is not a new concept in power systems. However, the diversity of the characteristics of DG and RES implies that flexible paradigms and controls must be conceptualized for the case of better handling such characteristics. A very common and obvious such concept is putting together hybrids of intermittent resources and storage systems [89], or more abstractly and at a greater extent combining stochastic and "deterministic" generators in microgrids [90] or virtual power plants [91]. Based on such set-ups, [92] introduces hierarchical control to integrate RES and energy-storage devices in the existing power system, while [93] reviews advanced control techniques for microgrids, including hierarchical control scheme as one of the proposed paradigms for integration of RES with the main grid.

An analogy of how flexibility should be understood for the case of modern power systems is described in Fig. 10.3. Any events and phenomena occurring at the smallest scale (one unit or a set of units behind a point of common coupling/control) can be handled through hybrids of energy sources that ensure a consistent behavior with regard to power quality standards. At the next level,

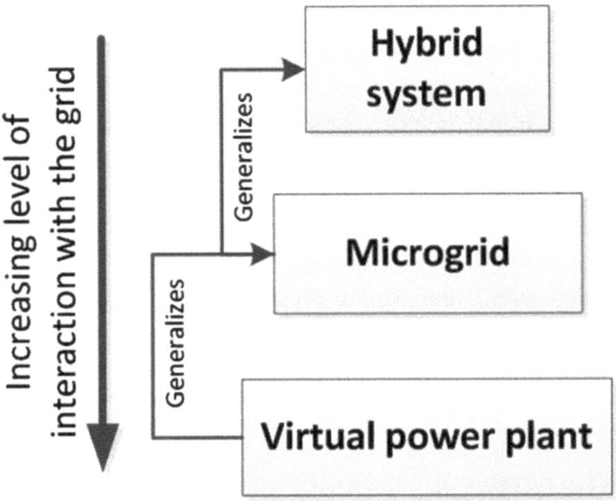

FIG. 10.3 The hierarchy abstraction for modern power system control concepts.

microgrids are paradigms that can offer services to the grid and control the encapsulated resources in a fully vertical manner. At the highest level, virtual power plants can execute control actions and respond to operator requests by accounting for the diversity of the resources they comprise, while not relying on any given status of the power system (or monitor it), but rather ensuring that the control is efficient under a variety of circumstances.

10.3.4 Advanced Hardware Platforms and Relevant Enablers

With the increasing numbers and diversity of power system assets, the classic centralized control centers [94] will be required either to handle unprecedented amounts of data and control actions or to follow the characteristics of the assets and adapt accordingly. The Internet of Things (IoT) has been lately introduced in power systems [95] with the deployment of smart meters [96], which have been gradually yet widely replacing the classic end-users' energy meters throughout the grid. The IoT poses firstly the challenge of handling huge amounts of data that are required (online or in later times) [97] for planning and control actions on behalf of system operators. In this sense, it is imperative to promote methods that can handle data in a meaningful way, retain the value of information while reducing its size, and ensure that control centers are able to process them [98, 99]. This challenge has been identified clearly in the early 2000s and led to the adoption of multiagent approaches [100]. Agents may act as autonomous monitoring, control, and planning software entities operating on microprocessors, thus relieving central control centers from the burden described earlier. Due to this, the hardware that is hosting agents need not be of high processing capacities, but is required to enable the secure and reliable exchange of data with other hardware units hosting agents throughout the grid or any lesser part of it [100].

This latter requirement describes the challenge of standardizing the respective hardware platforms and their interactions with the software entities that they will be hosting. These platforms will be equipped with sensors gathering systems readings, controllers driving power system assets and resources, and the integrated software for the operation of this equipment. The interface of the integrated software with the agent may or may not follow standardized protocols [101], thus accounting for added complexity to the control problems at hand. Several start-up companies have also been entering the market of power system control hardware and software [102], offering a wide range of options in addition to what the traditional electric equipment companies have also been developing and deploying [103, 104]. This implies that the interactions among all of these options will appear as an eminent challenge in the very near future.

Lastly, the direct control of power system assets, especially those that mainly expressed the path toward the deregulation of the energy markets, that is, power electronics and automation of DG units, has been required to adapt to increased power system requirements [105]. Implementing ancillary services and offering access to system operators for monitoring and control means that the control hardware needs to be properly designed and purposed in the updated operating frameworks.

10.3.5 Robustness—Handling Uncertainty

In power systems, "the only certainty is uncertainty," and addressing this is necessary and crucial to the stability and economics of power system control. The power system uncertainties root from both supply and demand. On the supply side, the penetration of intermittent RES brings significant uncertainties to the system, since the wind speed and the solar irradiation are hard to estimate accurately even in the near future, for example, half a day from now. On the demand side, it is also challenging to make accurate predictions on the random energy consumption of each household.

The power system uncertainties are usually modeled with various stochastic methods, according to the control scenarios and the characteristics of the uncertainty. The scenario-based approach is one of the most popular probabilistic methods, in which different scenarios with different levels of wind or solar generation or load are generated based on the probability distribution of each of them [106]. Many scenario generation and reduction techniques have been developed to reduce the number of scenarios considered, while ensuring that they properly represent the studied uncertainty [107, 108]. Robust optimization, which models the uncertainties with confidence intervals as an interval-based method, has been popular in recent decades [109]. This methodology has the merits of lower requirements for forecasting data and computational burden compared to other stochastic approaches, while still yielding practical results for the power system [110]. In some cases, multiscenario and robust methods can be incorporated into the same model to account for numerous uncertainties with different characteristics [111]. Other common stochastic methods include chance-constrained optimization [112] and risk-based optimization [113, 113a].

The other challenge is how to determine the optimal control policy given appropriate uncertainty modeling. Usually, power system control aims to optimize the objective function, for example, the total system cost subject to a series of constraints, such as power balance, start-up time, plant ramping capability, etc. If the constraints are temporally coupled, intertemporal programming is required to model the control problem. Model predictive control (MPC), also known as "receding horizon" control or "rolling horizon planning," is a powerful approach to address intertemporal decisions with uncertainty, as an approximate form of dynamic programming (DP). MPC typically determines the discrete control policy for the next T steps in a rolling manner given the prediction of the future states and outcomes [114]. This manner enables the utilization of the most updated forecasting information, and its performance can be guaranteed under certain conditions.

Uncertainty has been and will continue to be one of the major concerns in power system control, as we move forward to a low-carbon energy sector. Without meeting this requirement, the stability, economics, and resilience will all be compromised.

10.4 The Path Forward in Power System Control

Many cases of novel power system control approaches are discussed in the following subsections. Research by academia and the industry in the field has been multilevel and has spanned many applications. Nevertheless, all of the proposed ideas bear the five main characteristics analyzed in the previous section, in order to address the control challenges in the new framework, and also what it is expected of the power systems in the coming decades.

10.4.1 Control of Distributed Resources

10.4.1.1 Decentralized and Distributed Optimization Methods

The different characteristics and features of employing distributed and decentralized methods on the voltage regulation problem are described as follows. Let it be noted here that for this problem, a decentralized approach is, practically, equivalent to employing local controllers with very infrequent (if none at all) updates of their parameters with regard to some central entity recalculating some relevant objectives.

Starting with the decentralized techniques, according to [71] a solution to optimal voltage control can be obtained by the unconstrained optimization problem of minimizing the weighted sum of voltage deviations as well as power losses. This further highlights a trade-off between the local voltage deviation and the local reactive power measurements, referred to as droop control, more details of which can be found in [115, 116]. Further studies on the topic considered various cost functions and implemented local gradient-based approaches, for example, dual gradient ascent method for reactive power loss minimization [117], projected gradient method for power transfer maximization [118], and local proximal gradient schemes for voltage deviation minimization [119].

In order to enable fully decentralized control, projected (sub)gradient methods were used in [76, 120]. The proposed approaches perform as effectively as the droop-based method in IEEE standard 1547.8 in terms of convergence, while outperforming this standard in terms of transient performance. Especially for [76], an incremental control algorithm has been employed based on some preliminary elaboration of a class of nonincremental voltage control strategies. According to this work, droop-based techniques using nonincremental voltage control may lead to unacceptable oscillatory patterns. In order to deal with this issue, [76] proposed to use the (sub)gradient approach to achieve an incremental control method. The reason for referring to this approach as "incremental control" is the gradual update of reactive power based on the previous time steps. Although authors mention that their proposed approach is distributed in nature, as the reactive power values are updated only based on local variables and measurement, we would emphasize that "distributed control" techniques involve higher level of peer-to-peer communication.

Although the mentioned fully decentralized approaches provide efficient performance for voltage control, according to [121, 122] some of them may fail, because local controllers will not be able to keep the voltage in the preferred interval, due to the limited reactive power injection by the compensators. To address this matter, one should enable peer-to-peer communication to allow wider coordination. The framework, in which each agent (compensator) performs local computation while coordinating its decision based on information exchange with some other agent, is referred to as the distributed algorithm [72]. There are two major categories of distributed algorithms for voltage regulation in power distribution systems. The first category, relevant to the majority of the aforementioned works, relies on using linearized power flow equations around the operation point [123, 124]. The other category is based mainly on convexification of the problem using second-order cone programming [125] or semidefinite programming (SDP) relaxation [126].

Distributed algorithms provide feasible solutions to the reactive power compensation problem [127], as well as simultaneous voltage regulation and loss minimization [128]. In [129], an average consensus on reactive power injection ratios is used for optimizing a reactive power sharing objective. There are also several methods based on decomposing the original voltage control problem into subproblems, including online distributed optimization [130], SDP [125], ADMM, and leader-follower methods [131].

In [72], first a linearized network model is introduced to enable the design of distributed voltage control strategy. Then an analytical scheme is proposed to verify the convergence of distributed method to local minimum in the nonconvex branch-flow model. The error during iterations is also characterized. Minimizing deviations of voltage from rated values and the reactive power injections (as a form of inverter operation cost) comprise a weighted objective function for the problem. In order to enable the distributed computation, two communication schemes are considered using a tree communication network and complete communication network, with the latter case showing improved performance and efficiency in achieving the voltage regulations goals.

10.4.1.2 Novel Model Predictive Control Understanding

MPC has been intensively studied for power system applications in recent years. It became popular because of its effectiveness in addressing uncertainty issues that are temporally coupled.

The researchers in [132] proposed an MPC-based model that controls the aggregation of heterogeneous thermostatically controlled loads (TCL) to provide frequency regulation service. The model aims to decide which TCL in control to switch on/off to track the frequency regulation signal while keeping the temperatures within a certain range. Similarly, [133] proposes an MPC model that controls residential air conditioners to track solar power fluctuations. The TCL dynamics and regulation signal/solar power are both temporally coupled in this application.

Energy storage is a key solution that provides flexibility to address the intermittency of RES production and fast ramping capability to enhance power system stability. The control of energy storage is subject to the physical power constraint and energy constraints, while the latter is temporally coupled; the accumulated energy has to be kept within a range. Therefore, many studies have applied MPC in storage control [134–137].

The degradation issue of electrochemical energy storage (EES) can also be addressed with MPC incorporated in an intertemporal framework [138]. Typically, the capacity and efficiency of EES both degrade as more energy is processed and time goes on. The end of life of EES is usually defined as when its capacity decreases to a certain level, which limits the total energy that can be processed during the EES life cycle. Because of this limit, the degradation incurred by usage at present will affect the functionality and profitability of EES all through the remaining EES life, which presents a long-term effect. The intertemporal framework proposed in [138] addresses the long-term effect by defining a metric that constrains the usage/degradation or sets the degradation cost of EES in the short-term MPC horizon. This metric, named the marginal benefit of usage, brings long-term information into MPC-based short-term control decisions to guarantee that when the finite-horizon MPC decisions are sequentially made, the long-term/life-cycle objective of EES is optimized.

Intrinsically, electric vehicle (EV) charging is EES control. In this case, the arrival time of future EV is usually random and needs prediction, and the consumer usually requires its EV to be fully charged within a certain amount of time, which brings temporally coupled constraints. In [139], MPC was applied to determine the near optimal real-time EV charging profiles in a charging station. [140] analyzed the performance of MPC in EV charging scheduling and concluded that MPC performs very closely to the finite-horizon continuous DP and could reduce computational complexity.

From the whole system perspective, MPC can be applied to control each agent in the system including demand response, energy storage, FACTS, fast-responding generation, etc., and consequently determine the power flow. [141] proposed an MPC-based DC-power-flow model that alleviates line-temperature overloads to prevent cascading failures. [142] further expanded the MPC-based dispatch model by accounting for voltage magnitudes and reactive power.

MPC can be incorporated in a distributed/decentralized manner, which can enhance the robustness of the system against communication failure. In the distributed control scheme, there is typically no centralized coordinator, and each agent in the system communicates with its neighbors to determine the optimal actions. [143] proposed a Consensus + Innovations scheme for microgrid energy management, using MPC to develop a multi-time-step control model. [144] proposed a noniterative distributed MPC model for a class of spatially interconnected systems with communication constraints, in which a cost function switching strategy is developed to guarantee the stability of the overall system.

10.4.2 Frequency Control and Active Power Reserves

10.4.2.1 Extending Frequency Control Implementation and Application

A state-of-the-art review of LFC schemes was offered in [145]. The work in [146] proposed a robust LFC using genetic algorithms and linear matrix inequalities. Similarly, a fuzzy logic controller was proposed in [147] to control the load frequency in two areas of power systems.

Frequency of power systems can be controlled using varying controllable loads, energy storage devices, and generation units. In this sense, [148] presented a frequency control scheme by controlling Hybrid Electric Vehicles, controllable loads, and a generation unit whereas real-time pricing control technique was utilized to regulate frequency of powers in [149].

At a higher level of realizing this type of regulation and due to the recent expansion and growing complexity of power systems, the requirement to involve DG resources in tasks as the LFC requires the accurate and timely transmission of enormous amount of data. To deal with this challenge, a singular value decomposition (SVD) based real-time LFC can be realized [150]. According to the proposed strategy, the measured data in each control area is decomposed using SVD before being transmitted through the communication network, and only the most valuable information is transmitted to the control center.

10.4.2.2 Active Power Reserves Beyond the Load-Frequency Control Problem

Catering for closer control of active power became crucial as RES-based DG grew rapidly and vastly, due to its high efficiency compared to conventional generation [151] and its limited environmental impact. RES, however, depends heavily on weather and the surrounding terrains, thus accounting for the intermittent and stochastic nature of the relevant DG units. Operators and research in the field focused on proactive and online strategies to deal with the above problems, especially beyond the limits of traditional LFC, as was discussed earlier. The employed control approaches are essentially driving the operating set-points of dispatchable assets; hence, they are less of control in the classic sense and more of planning tools.

Proactively and based on RES generation forecasts, large power reserves may be procured [152, 153]. In [152] the size of the reserves scheduled were determined according to reliability indexes in light of the increased penetration of wind power. A somehow symmetrical approach was employed in [153], where the reliability of the intermittent RES (a photovoltaic system) was assessed, so as to adjust reserves accordingly. As is evident, the fact that both these methods leverage forecasting tools characterizes them as robust or adaptive, depending on whether the forecasts are incorporated as probability functions or otherwise lead to frequent updates of the set-points of the controlled assets.

Stochastic methods have been another approach to handle RES variance by accounting for their output through probability density functions incorporated as cost factors and inequality constraints in security-aware dispatching set-ups [154]. Applying the stochastic methods in top-down approaches starting from the perspective of transmission system operators (the grid of a whole control area) and all the way to that of aggregators (trading energy at distribution feeders) adds much flexibility to their already very robust profile.

In [91, 155] machine learning was used (binary decision trees, BDTs) to prepare dispatching set-points for handling multiple levels of excess or deficit of DG power. The BTDs were trained with actual cost and AC power flow data for either cases of DG variance. The resulting control was shown to favor RES and properly accounted for their stochastic behavior. This was achieved thanks to the structure of output of the BTD tool, which allowed for a control interval rather than a specific set-point of active/reactive power. This control method is a definitive example of a robust approach, while its application in an actual test-bed, as discussed in [156], highlights the requirements for versatile and low-cost hardware platforms with dispersed processing power.

10.4.3 Voltage Regulation

Voltage regulation has become a major control concern in the past few years, especially at the level of distribution and for a twofold reason. Firstly, network reinforcements to overcome voltage drops, due to the increase of loads, can be postponed thanks to DG deployment, but this cannot be achieved automatically; DG units need to be coordinated either among them or additionally with the actions of local operators. Secondly, at valley loading times, increased DG power outputs can lead to opposite effects of increased voltages. In most of the control following approaches, the effect of locality in the voltage regulation problem will attest to the requirement for flexible solutions.

In [157], a decentralized control action over a few network assets is employed, based on optimal formulations. It is worth noting that the "hunting behavior" [158] identified by the authors of [157] is a persistent issue even in the context of such modern control methods. The decentralized approach enhances with scalability the flexibility of this technique.

Many control strategies based on sensitivity analysis, that is, assessment of the voltage dependent derivatives of the inverse Jacobian matrix of the Newton-Raphson power flow method, have also been proposed frequently. One should note the flexibility of [159], which handled sensitivities over loosely coupled control areas, and also of [160], which went all the way down to the detail of controlling every single node. Combining sensitivity analysis with classic rule-based control, as in [160], adds some adaptive behavior to these control approaches.

In [161], a particularly flexible online dispatching tool was proposed for the voltage regulation of radial (not necessarily, though) distribution networks. The distribution network is viewed as a dimensionless body with the injections (from DG units) and absorptions (of loads) of power as point weights, the centers of which need to be counterbalanced to reduce the voltage drops. The method iterates for all lesser parts of a distribution line to ensure the improvement of the overall voltage profile. Its online operation, which may handle DG variance, also offers some robustness characteristics to this control approach.

Focusing back on classic control, the relations between active power and frequency, and reactive power and voltage magnitude were used in [162] for designing generic active/reactive power controllers for DG units operating at the distribution level. However, as reminded by [54], the ratio of resistance to reactance of distribution lines may well affect the validity of the aforementioned relations. Hence, these flexible controllers, residing on the processors of commercially available power electronic components, should be further developed to be more adaptive to the parameters of the grid, for example, based on network impedance identification methods [163].

Particle swarm optimization was proposed to minimize voltage deviation in electric power systems using reactive power control in [164]. Similar work was presented in [165] to minimize power loss and voltage deviation using optimal reactive power control. Voltage of a synchronous generator was regulated by designing and implementing an AVR in [166] with the aim to focus on hardware implementations of novel control approaches.

Reactive power control of wind farms was proposed in [167], where a control strategy was developed for wind farms made up with double-fed induction generators to regulate the voltage of the electrical grid to which farms are connected. Likewise, reactive power control for synchronous machines was developed in [168] to minimize voltage deviation in systems with synchronous machine-based DGs.

In [169], the authors proposed optimal reactive power control to minimize power loss, voltage deviation, and reactive power generation cost simultaneously. The reason to consider the opportunity cost of reactive power generation is that it affects the frequency control capability of the generator to some degree. This paper proposed a distributed nonlinear control-based algorithm to achieve the optimal reactive power generation for multiple generators in a power grid. The reactive power control setting update for each generator only requires local measurement and information exchange with its neighboring buses. It was demonstrated that the proposed algorithm could reduce the nonconvex objective function monotonically until convergence and achieve comparable solutions to the centralized technique: particle swarm optimization with faster convergence speed.

10.4.4 Advances on Automation and Control Equipment

A very recent initiative, implemented through the IEEE standard 2660.10 working group, has identified the need for a protocol on how a software agent should efficiently interact with automation devices for the monitoring and control, among others, of electric power and energy systems [170]. The working group has been surveying existing topologies, mostly found at the research stage and less at the wide industry application level. The multitude of interactions identified to this point implies that such a standard is timely and will facilitate the effective deployment of IoT in power systems.

As for DG equipped with power electronics, the updated IEEE standard 1547 sets multiple connection and operating criteria for the DG resources [171]. Voltage regulation through active and reactive power, low-voltage ride-thru capabilities, automatic reconnection to the grid, and considerations about interactions with microgrids are a few of the criteria that will require the control hardware to monitor the grid extensively and ensure that the DG units controlled perform as determined in the said document.

10.5 Conclusions

The ever-increasing complexity of power systems, as large-scale dynamical systems, originates from integration of novel smart grid technologies (e.g., RES, demand response, and distributed energy storage), constraints imposed by limitation of physical infrastructure, and growth of electric consumption. These resources and technologies, on the one hand, can potentially improve the reliability of power systems by reducing dependency on bulk generation units, using more distributed and environmentally friendly alternatives, and enabling intelligent demand-side management. On the other hand, they introduce more uncertainty that makes the control and operation of power systems a challenging task. These uncertainties may also lead to unexpected operation difficulties, such as voltage/frequency deviations caused by mismatch between demand and supply. Further, deregulation of power systems has increased the number of stakeholders and entities, which results in a major concern regarding the scalability of the existing power systems control approaches.

To cope with the increasing complexity and uncertainties of modern power systems, in this chapter we have reviewed emerging control methods and technologies. After a brief review of classic power system control strategies and practices, we have identified five major characteristics that can improve modern power system control: scalability, adaptive control, flexibility, hardware advances, and robustness. Finally, we have reviewed recent studies and advances for control in modern power systems considering the above-mentioned characteristics. From the above discussion, it is natural to focus future efforts in modern power systems control on enhancing these five characteristics by means of integrated design of efficient algorithms and software, such as distributed algorithms and parallel computing platforms, as well as deploying modern hardware-based technologies. As understood by the examples presented, most of the ongoing research in the field attempts to address as many of these characteristics as possible to offer readily applicable solutions to the ongoing challenges of the deregulated energy markets and the highly interacting control areas and regions throughout the grids worldwide.

References

[1] F. Venn, The oil crisis, Routledge, 2016.

[2] G. Verbong, F. Geels, The ongoing energy transition: lessons from a socio-technical, multi-level analysis of the Dutch electricity system (1960–2004), Energy Policy 35 (2007) 1025–1037.

[3] A.K.N. Reddy, A strategy for resolving India's oil crisis, Curr. Sci. 50 (1981) 50–53.

[4] J.D. Hamilton, This is what happened to the oil price-macroeconomy relationship, J. Monetary Econ. 38 (1996) 215–220.

[5] B. Declercq, E. Delarue, W. D'haeseleer, Impact of the economic recession on the European power sector's CO_2 emissions, Energy Policy 39 (2011) 1677–1686.

[6] J.W. Markham, A Financial History of the United States: From Enron-era Scandals to the Subprime Crisis (2004–2006); From the Subprime Crisis to the Great Recession (2006–2009), Routledge, 2015.

[7] M.D. Hurd, S. Rohwedder, Effects of the financial crisis and great recession on American households, Nat. Bureau Econ. Res. (2010)

[8] D. Haerer, L. Pratson, Employment trends in the US Electricity Sector, 2008–2012, Energy Policy 82 (2015) 85–98.
[9] A.A. Rafindadi, I. Ozturk, Effects of financial development, economic growth and trade on electricity consumption: evidence from post-Fukushima Japan, Renew. Sustain. Energy Rev. 54 (2016) 1073–1084.
[10] E. Park, J.Y. Ohm, Factors influencing the public intention to use renewable energy technologies in South Korea: effects of the Fukushima nuclear accident, Energy Policy 65 (2014) 198–211.
[11] M.A. Fields, V. Janjigian, The effect of Chernobyl on electric-utility stock prices, J. Bus. Res. 18 (1989) 81–87.
[12] R. Kalra, G.V. Henderson Jr., G.A. Raines, Effects of the Chernobyl nuclear accident on utility share prices, Quar. J. Bus. Econ. (1993) 52–77.
[13] Y. Li, The case analysis of the scandal of Enron, Int. J. Bus. Manag. 5 (2010) 37.
[14] S. Carley, State renewable energy electricity policies: an empirical evaluation of effectiveness, Energy Policy 37 (2009) 3071–3081.
[15] H. Nagayama, Effects of regulatory reforms in the electricity supply industry on electricity prices in developing countries, Energy Policy 35 (2007) 3440–3462.
[16] J. Arrillaga, M.H. Bollen, N.R. Watson, Power quality following deregulation, Proc. IEEE 88 (2000) 246–261.
[17] A. Foley, B.Ó. Gallachóir, J. Hur, R. Baldick, E. McKeogh, A strategic review of electricity systems models, Energy 35 (2010) 4522–4530.
[18] T. Jamasb, M. Pollitt, Electricity market reform in the European Union: review of progress toward liberalization & integration, The Energy J. (2005) 11–41.
[19] R.J. Andres, G. Marland, I. Fung, E. Matthews, A 1×1 distribution of carbon dioxide emissions from fossil fuel consumption and cement manufacture, 1950–1990, Global Biogeochem. Cycles 10 (1996) 419–429.
[20] D.A. Lashof, D.R. Ahuja, Relative contributions of greenhouse gas emissions to global warming, Nature 344 (1990) 529.
[21] A.A. Lacis, D.J. Wuebbles, J.A. Logan, Radiative forcing of climate by changes in the vertical distribution of ozone, J. Geophys. Res. Atmos. 95 (1990) 9971–9981.
[22] A. Ito, J.E. Penner, Historical emissions of carbonaceous aerosols from biomass and fossil fuel burning for the period 1870–2000, Global Biogeochem. Cycles 19 (2005).
[23] Y. Zhang, J.J. Schauer, Y. Zhang, L. Zeng, Y. Wei, Y. Liu, et al., Characteristics of particulate carbon emissions from real-world Chinese coal combustion, Environ. Sci. Technol. 42 (2008) 5068–5073.
[24] C. Hendriks, Carbon dioxide removal from coal-fired power plants, vol. 1, Springer Science & Business Media, 2012.
[25] K. Protocol, United Nations framework convention on climate change, in: Kyoto Protocol, Kyoto, vol. 19, 1997.
[26] P. Agreement, United nations framework convention on climate change, *Paris, France.* (2015).
[27] M.A. Delmas, M.J. Montes-Sancho, US state policies for renewable energy: Context and effectiveness, Energy Policy 39 (2011) 2273–2288.
[28] J.E. Payne, On the dynamics of energy consumption and output in the US, Appl. Energy 86 (2009) 575–577.
[29] E. Union, Directive 2009/28/EC of the European Parliament and of the Council of 23 April 2009 on the promotion of the use of energy from renewable sources and amending and subsequently repealing Directives 2001/77/EC and 2003/30/EC, Off. J. Eur. Union 5 (2009) 2009.

[30] S. Mittal, H. Dai, S. Fujimori, T. Masui, Bridging greenhouse gas emissions and renewable energy deployment target: comparative assessment of China and India, Appl. Energy 166 (2016) 301–313.

[31] J. Twidell, T. Weir, Renewable Energy Resources, Routledge, 2015.

[32] R. Margolis, C. Coggeshall, J. Zuboy, SunShot vision study, in: US Dept. of Energy, vol. 2, 2012.

[33] N.D. Hatziargyriou, E.I. Zountouridou, A. Vassilakis, P. Moutis, C.N. Papadimitriou, A.G. Anastasiadis, Overview of distributed energy resources, in: Smart Grid Handbook, 2016, pp. 1–44.

[34] A. Campoccia, L. Dusonchet, E. Telaretti, G. Zizzo, Comparative analysis of different supporting measures for the production of electrical energy by solar PV and Wind systems: four representative European cases, Solar Energy 83 (2009) 287–297.

[35] T. Couture, Y. Gagnon, An analysis of feed-in tariff remuneration models: implications for renewable energy investment, Energy Policy 38 (2010) 955–965.

[36] L. Butler, K. Neuhoff, Comparison of feed-in tariff, quota and auction mechanisms to support wind power development, Renew. Energy 33 (2008) 1854–1867.

[37] M. Green, Recent developments in photovoltaics, Solar energy 76 (2004) 3–8.

[38] L. Ledo, P. Kosasih, P. Cooper, Roof mounting site analysis for micro-wind turbines, Renew. Energy 36 (2011) 1379–1391.

[39] G. Rothwell, T. Gomez, Electricity economics, in: IEEE Series on Power Engineering, 2003.

[40] B. Andersson, L. Bergman, Market structure and the price of electricity: an ex ante analysis of the deregulated Swedish electricity market, Energy J. (1995) 97–109.

[41] M. Ventosa, A. Baíllo, A. Ramos, M. Rivier, Electricity market modeling trends, Energy Policy 33 (2005) 897–913.

[42] S. Borenstein, J.B. Bushnell, F.A. Wolak, Measuring market inefficiencies in California's restructured wholesale electricity market, Am. Econ. Rev. 92 (2002) 1376–1405.

[43] C. E. O. H. ENTSO-E, P1-Policy 1: Load-Frequency Control and Performance, Tech. Rep. 2000-130-003, May 2000. Available: http://www.eurelectric.org. (2009).

[44] R.C. Dugan, M.F. McGranaghan, H.W. Beaty, S. Santoso, Electrical Power Systems Quality, vol. 2, Mcgraw-Hill, New York, 1996.

[45] E. Standard, Voltage Characteristics of Electricity Supplied by Public Distribution Systems, BS EN, vol. 50, (1995) 160.

[46] P. Kundur, N.J. Balu, M.G. Lauby, Power System Stability and Control, vol. 7, McGraw-Hill, New York, 1994.

[47] T. Van Cutsem, C. Vournas, Voltage Stability of Electric Power Systems, Springer Science & Business Media, 2007.

[48] N. Martins, S. Corsi, G. Taranto, Coordinated voltage control in transmission networks, in: *CIGRE Technical Brochure 310 Task Force C4. 602*, 2007.

[49] S. Stoft, Power system economics, J. Energy Lit. 8 (2002) 94–99.

[50] M. Tsili, S. Papathanassiou, A review of grid code technical requirements for wind farms, IET Renew. Power Gener. 3 (2009) 308–332.

[51] G. Lalor, J. Ritchie, S. Rourke, D. Flynn, M.J. O'Malley, Dynamic frequency control with increasing wind generation, in: *Power Engineering Society General Meeting*, 2004. IEEE, 2004, pp. 1715–1720.

[52] T. Ackermann, Wind Power in Power Systems, John Wiley & Sons, 2005.

[53] A. Luque, S. Hegedus, Handbook of Photovoltaic Science and Engineering, John Wiley & Sons, 2011.

[54] A. Engler, N. Soultanis, Droop control in LV-grids, in: *Future Power Systems, 2005 International Conference on*, 2005, p. 6.
[55] P.P. Varaiya, F.F. Wu, J.W. Bialek, Smart operation of smart grid: risk-limiting dispatch, Proc. IEEE 99 (2011) 40–57.
[56] J. Carpentier, Contribution to the economic dispatch problem, Bulletin de la Societe Francoise des Electriciens 3 (1962) 431–447.
[57] M. Huneault, F.D. Galiana, A survey of the optimal power flow literature, IEEE Trans. Power Syst. 6 (1991) 762–770.
[58] J.A. Momoh, R. Adapa, M.E. El-Hawary, A review of selected optimal power flow literature to 1993. I. Nonlinear and quadratic programming approaches, IEEE Trans. Power Syst. 14 (1999) 96–104.
[59] J.A. Momoh, M.E. El-Hawary, R. Adapa, A review of selected optimal power flow literature to 1993. II. Newton, linear programming and interior point methods, IEEE Trans. Power Syst. 14 (1999) 105–111.
[60] Y. Fu, M. Shahidehpour, Z. Li, Security-constrained unit commitment with AC constraints, IEEE Trans. Power Syst. 20 (2005) 1001–1013.
[61] S.H. Low, Convex relaxation of optimal power flow—Part I: formulations and equivalence, IEEE Trans. Control Netw. Syst. 1 (2014) 15–27.
[62] J. Zhu, Optimization of Power System operation, vol. 47, John Wiley & Sons, 2015.
[62a] J. Wang, N. Elia, A control perspective for centralized and distributed convex optimization, in: 50th IEEE Conference on Decision and Control and European Control Conference, IEEE, (2011).
[62b] J. Mohammadi, G. Hug, S. Kar, Agent-based distributed security constrained optimal power flow, IEEE Trans. Smart Grid 9 (2) (2018) 1118–1130.
[63] A. Kumar, S. Srivastava, S. Singh, Congestion management in competitive power market: A bibliographical survey, Electr. Pow. SYST. RES. 76 (2005) 153–164.
[64] P. Palensky, D. Dietrich, Demand side management: Demand response, intelligent energy systems, and smart loads, IEEE transactions on industrial informatics 7 (2011) 381–388.
[65] O. Erdinc, Optimization in Renewable Energy Systems: Recent Perspectives, Butterworth-Heinemann, 2017.
[65a] S. Bahrami, et al., A decentralized renewable generation management and demand response in power distribution networks, IEEE Trans. Sustain. Energy 9 (4) (2018) 1783–1797.
[66] H. Farhangi, The path of the smart grid, IEEE Power Energ. Mag. 8 (2010).
[67] V.C. Gungor, D. Sahin, T. Kocak, S. Ergut, C. Buccella, C. Cecati, et al., Smart grid technologies: Communication technologies and standards, IEEE Trans. Ind. Inf. 7 (2011) 529–539.
[68] X. Fang, S. Misra, G. Xue, D. Yang, Smart grid—the new and improved power grid: a survey, IEEE Commun. Surv. Tutorials 14 (2012) 944–980.
[69] S. Kar, G. Hug, J. Mohammadi, J.M. Moura, Distributed State Estimation and Energy Management in Smart Grids: A Consensus + Innovations Approach, IEEE J. Sel. Top. Sign. Proces. 8 (2014) 1022–1038.
[70] A. Kargarian, J. Mohammadi, J. Guo, S. Chakrabarti, M. Barati, G. Hug, et al., Toward distributed/decentralized DC optimal power flow implementation in future electric power systems, IEEE Trans. Smart Grid (2016).
[71] D.K. Molzahn, F. Dörfler, H. Sandberg, S.H. Low, S. Chakrabarti, R. Baldick, et al., A survey of distributed optimization and control algorithms for electric power systems, IEEE Trans. Smart Grid 8 (2017) 2941–2962.

[71a] S. Kar, J.M.F. Moura, Distributed consensus algorithms in sensor networks: quantized data and random link failures, IEEE Trans. Sig. Proc. 58 (3) (2010) 1383–1400.

[71b] J. Mohammadi, G. Hug, S. Kar, Role of communication on the convergence rate of fully distributed DC optimal power flow, in: 2014 IEEE International Conference on Smart Grid Communications (SmartGridComm), IEEE, (2014).

[71c] J. Mohammadi, G. Hug, S. Kar, Fully distributed DC-OPF approach for power flow control, in: 2015 IEEE Power & Energy Society General Meeting, IEEE, (2015).

[71d] M.H. Amini, et al., Distributed security constrained economic dispatch, in: 2015 IEEE Innovative Smart Grid Technologies-Asia (ISGT ASIA), IEEE, (2015).

[71e] M.H. Amini, et al., Decomposition methods for distributed optimal power flow: panorama and case studies of the DC model, in: Classical and Recent Aspects of Power System Optimization, 2018, pp. 137–155.

[72] C. Wu, G. Hug, S. Kar, Distributed voltage regulation in distribution power grids: utilizing the photovoltaics inverters, in: *American Control Conference (ACC), 2017*, 2017, pp. 2725–2731.

[73] K. Baker, J. Guo, G. Hug, X. Li, Distributed MPC for efficient coordination of storage and renewable energy sources across control areas, IEEE Trans. Smart Grid 7 (2016) 992–1001.

[74] K.T. Chaturvedi, M. Pandit, L. Srivastava, Self-organizing hierarchical particle swarm optimization for nonconvex economic dispatch, IEEE Trans. Power Syst. 23 (2008) 1079–1087.

[75] M. Farivar, L. Chen, S. Low, Equilibrium and dynamics of local voltage control in distribution systems, in: *Decision and Control (CDC), 2013 IEEE 52nd Annual Conference on*, 2013, pp. 4329–4334.

[76] M. Farivar, X. Zhou, L. Chen, Local voltage control in distribution systems: An incremental control algorithm, in: *SmartGridComm*, 2015, pp. 732–737.

[77] I. Kamwa, R. Grondin, Y. Hébert, Wide-area measurement based stabilizing control of large power systems-a decentralized/hierarchical approach, IEEE Trans. Power Syst. 16 (2001) 136–153.

[78] K.J. Åström, B. Wittenmark, Adaptive Control, Courier Corporation, 2013.

[79] C. Ross, Error adaptive control computer for interconnected power systems, IEEE Trans. Power Syst. (1966) 742–749.

[80] T.E.D. Liacco, The adaptive reliability control system, IEEE Trans. Power Syst. (1967) 517–531.

[81] J. Kanniah, O. Malik, G. Hope, Excitation control of synchronous generators using adaptive regulators part I-Theory and simulation results, IEEE Trans. Power Syst. (1984) 897–903.

[82] A. Ghosh, G. Ledwich, O. Malik, G. Hope, Power system stabilizer based on adaptive control techniques, IEEE Trans. Power Syst. (1984) 1983–1989.

[83] P. Anderson, M. Mirheydar, An adaptive method for setting underfrequency load shedding relays, IEEE Trans. Power Syst. 7 (1992) 647–655.

[84] M. Abido, Y. Abdel-Magid, Adaptive tuning of power system stabilizers using radial basis function networks, Electr. Pow. Syst. Res. 49 (1999) 21–29.

[85] A.A. Ghandadly, Adaptive enhancement of synchronous generator stabilizer performance by including the external utility system dynamics, Electr. Pow. Syst. Res. 2 (1998) 111–117.

[86] F.-J. Lin, R.-J. Wai, R.-H. Kuo, D.-C. Liu, A comparative study of sliding mode and model reference adaptive speed observers for induction motor drive, Electr. Pow. Syst. Res. 44 (1998) 163–174.

[87] S. Sridhar, A. Hahn, M. Govindarasu, Cyber-Physical System Security for the Electric Power Grid, Proc. IEEE 100 (2012) 210–224.

[87a] A.N. Samudrala, M.H. Amini, S. Kar, R.S. Blum, Optimal sensor placement for topology identification in smart power grids, in: Conference on Information Sciences and Systems, March 20-22, 2019, Baltimore, MD.

[87b] A. Gholami, et al., Toward a consensus on the definition and taxonomy of power system resilience, IEEE Access 6 (2018) 32035–32053.

[88] M. Panteli, P. Mancarella, The grid: Stronger, bigger, smarter?: Presenting a conceptual framework of power system resilience, IEEE Power Energ. Mag. 13 (2015) 58–66.

[89] S.V. Papaefthymiou, E.G. Karamanou, S.A. Papathanassiou, M.P. Papadopoulos, A wind-hydro-pumped storage station leading to high RES penetration in the autonomous island system of Ikaria, IEEE Trans. Sustainable Energy 1 (2010) 163–172.

[90] N. Hatziargyriou, H. Asano, R. Iravani, C. Marnay, Microgrids, IEEE Power Energ. Mag. 5 (2007) 78–94.

[91] P. Moutis, N.D. Hatziargyriou, Decision trees-aided active power reduction of a virtual power plant for power system over-frequency mitigation, IEEE Trans. Ind. Inf. 11 (2015) 251–261.

[92] J.M. Guerrero, J.C. Vasquez, J. Matas, L.G. De Vicuña, M. Castilla, Hierarchical control of droop-controlled AC and DC microgrids—A general approach toward standardization, IEEE Trans. Ind. Electron. 58 (2011) 158–172.

[93] J.M. Guerrero, M. Chandorkar, T.-L. Lee, P.C. Loh, Advanced control architectures for intelligent microgrids-Part I: Decentralized and hierarchical control, IEEE Trans. Ind. Electron. 60 (2013) 1254–1262.

[94] K. Stouffer, J. Falco, Guide to Supervisory Control and Data Acquisition (SCADA) and Industrial Control Systems Security, National institute of standards and technology, 2006.

[95] M. Yun, B. Yuxin, Research on the architecture and key technology of Internet of Things (IoT) applied on smart grid, in: *Advances in Energy Engineering (ICAEE), 2010 International Conference on*, 2010, pp. 69–72.

[96] S.S.S.R. Depuru, L. Wang, V. Devabhaktuni, Smart meters for power grid: Challenges, issues, advantages and status, Renew. Sustain. Energy Rev. 15 (2011) 2736–2742.

[97] B. Babcock, S. Babu, M. Datar, R. Motwani, J. Widom, Models and issues in data stream systems, in: *Proceedings of the Twenty-First ACM SIGMOD-SIGACT-SIGART Symposium on Principles of Database Systems*, 2002, pp. 1–16.

[98] Y. Song, G. Zhou, Y. Zhu, Present status and challenges of big data processing in smart grid, Power System Technol. 37 (2013) 927–935.

[99] P. Moutis, N.D. Hatziargyriou, Using wavelet synopsis techniques on electric power system measurements, in: *Innovative Smart Grid Technologies (ISGT Europe), 2011 2nd IEEE PES International Conference and Exhibition on*, 2011, pp. 1–7.

[100] S.D. McArthur, E.M. Davidson, V.M. Catterson, A.L. Dimeas, N.D. Hatziargyriou, F. Ponci, et al., Multi-agent systems for power engineering applications—Part I: Concepts, approaches, and technical challenges, IEEE Trans. Power Syst. 22 (2007) 1743–1752.

[101] I. Modbus, Modbus application protocol specification v1. 1a, *North Grafton, Massachusetts*. www.modbus.org/specs.php(2004).

[102] http://zaphiro.ch.

[103] T. Taylor, H. Kazemzadeh, Integrated SCADA/DMS/OMS: Increasing distribution operations efficiency, Electr. Energy T&D Mag. 9 (2009) 32–34.

[104] Z. Hameed, S. Ahn, Y. Cho, Practical aspects of a condition monitoring system for a wind turbine with emphasis on its design, system architecture, testing and installation, Renew. Energy 35 (2010) 879–894.

[105] T.S. Basso, R. DeBlasio, IEEE 1547 series of standards: interconnection issues, IEEE Trans. Power Electron. 19 (2004) 1159–1162.

[106] Q.P. Zheng, J. Wang, A.L. Liu, Stochastic Optimization for Unit Commitment—a Review, IEEE Trans. Power Syst. 30 (2015) 1913–1924.

[107] J. Dupačová, N. Gröwe-Kuska, W. Römisch, Scenario reduction in stochastic programming, Math. Program. 95 (2003) 493–511, 2003/03/01.

[108] H. Heitsch, W. Römisch, Scenario reduction algorithms in stochastic programming Comput. Optim. Appl. 24 (2003) 187–206, 2003/02/01.

[109] D. Bertsimas, E. Litvinov, X.A. Sun, J. Zhao, T. Zheng, Adaptive robust optimization for the security constrained unit commitment problem, IEEE Trans. Power Syst. 28 (2013) 52–63.

[110] L. Baringo, A.J. Conejo, Offering strategy via robust optimization, IEEE Trans. Power Syst. 26 (2011) 1418–1425.

[111] G. He, Q. Chen, C. Kang, Q. Xia, Optimal offering strategy for concentrating solar power plants in joint energy, reserve and regulation markets, IEEE Trans. Sustainable Energy 7 (2016) 1245–1254.

[112] D. Pozo, J. Contreras, A chance-constrained unit commitment with an $n-k$ security criterion and significant wind generation, IEEE Trans. Power Syst. 28 (2013) 2842–2851.

[113] N. Zhang, C. Kang, Q. Xia, Y. Ding, Y. Huang, R. Sun, et al., A convex model of risk-based unit commitment for day-ahead market clearing considering wind power uncertainty, IEEE Trans. Power Syst. 30 (2015) 1582–1592.

[113a] S. Bahrami, et al., A decentralized trading algorithm for an electricity market with generation uncertainty, Appl. Energy 218 (2018) 520–532.

[114] E.F. Camacho, C.B. Alba, Model Predictive Control, Springer, 2013.

[115] S. Kundu, S. Backhaus, I.A. Hiskens, Distributed control of reactive power from photovoltaic inverters, in: *Circuits and Systems (ISCAS), 2013 IEEE International Symposium on*, 2013, pp. 249–252.

[116] J.W. Simpson-Porco, F. Dörfler, F. Bullo, Voltage stabilization in microgrids via quadratic droop control, IEEE Trans. Autom. Control 62 (3) (2017) 1239–1253.

[117] N. Li, G. Qu, M. Dahleh, Real-time decentralized voltage control in distribution networks, in: *Communication, Control, and Computing (Allerton), 2014 52nd Annual Allerton Conference on*, 2014, pp. 582–588.

[118] P. Nahata, S. Mastellone, F. Döorfler, Decentralized optimal projected control of PV inverters in residential microgrids, IFAC-PapersOnLine 50 (2017) 6624–6629.

[119] V. Kekatos, L. Zhang, G.B. Giannakis, R. Baldick, Voltage regulation algorithms for multiphase power distribution grids, IEEE Trans. Power Syst. 31 (2016) 3913–3923.

[120] H. Zhu, H.J. Liu, Fast local voltage control under limited reactive power: optimality and stability analysis, IEEE Trans. Power Syst. 31 (2016) 3794–3803.

[121] F. Olivier, P. Aristidou, D. Ernst, T. Van Cutsem, Active management of low-voltage networks for mitigating overvoltages due to photovoltaic units, IEEE Trans. Smart Grid 7 (2016) 926–936.

[122] G. Cavraro, S. Bolognani, R. Carli, S. Zampieri, The value of communication in the voltage regulation problem, in: *Decision and Control (CDC),* 2016 IEEE *55th Conference on*, 2016, pp. 5781–5786.

[123] K. Turitsyn, P. Šulc, S. Backhaus, M. Chertkov, Distributed control of reactive power flow in a radial distribution circuit with high photovoltaic penetration, in: *Power and Energy Society General Meeting, 2010 IEEE*, 2010, pp. 1–6.

[124] O. Alsac, J. Bright, M. Prais, B. Stott, Further developments in LP-based optimal power flow, IEEE Trans. Power Syst. 5 (1990) 697–711.

[125] B. Zhang, A.Y. Lam, A.D. Domínguez-García, D. Tse, An optimal and distributed method for voltage regulation in power distribution systems, IEEE Trans. Power Syst. 30 (2015) 1714–1726.

[126] J. Lavaei, S.H. Low, Zero duality gap in optimal power flow problem, IEEE Trans. Power Syst. 27 (2012) 92.

[127] S. Bolognani, S. Zampieri, A distributed control strategy for reactive power compensation in smart microgrids, IEEE Trans. Autom. Control 58 (2013) 2818–2833.

[128] S. Bolognani, R. Carli, G. Cavraro, S. Zampieri, Distributed reactive power feedback control for voltage regulation and loss minimization, IEEE Trans. Autom. Control 60 (2015) 966–981.

[129] J. Schiffer, T. Seel, J. Raisch, T. Sezi, Voltage stability and reactive power sharing in inverter-based microgrids with consensus-based distributed voltage control, IEEE Trans. Contr. Sys. Techn. 24 (2016) 96–109.

[130] M. Todescato, J.W. Simpson-Porco, F. Dörfler, R. Carli, F. Bullo, Online distributed voltage stress minimization by optimal feedback reactive power control, IEEE Trans. Control Netw. Syst. (2017)

[131] A. Bernstein, L. Reyes-Chamorro, J.-Y. Le Boudec, M. Paolone, A composable method for real-time control of active distribution networks with explicit power setpoints. Part I: Framework, Electr. Pow. Syst. Res. 125 (2015) 254–264.

[132] M. Liu, Y. Shi, Model predictive control of aggregated heterogeneous second-order thermostatically controlled loads for ancillary services, IEEE Trans. Power Syst. 31 (2016) 1963–1971.

[133] N. Mahdavi, J.H. Braslavsky, M.M. Seron, S.R. West, Model predictive control of distributed air-conditioning loads to compensate fluctuations in solar power, IEEE Trans. Smart Grid 8 (2017) 3055–3065.

[134] R. Kumar, M.J. Wenzel, M.J. Ellis, M.N. ElBsat, K.H. Drees, V.M. Zavala, A stochastic model predictive control framework for stationary battery systems, IEEE Trans. Power Syst. 33 (2018) 4397–4406.

[135] K. Meng, Z.Y. Dong, Z. Xu, S.R. Weller, Cooperation-driven distributed model predictive control for energy storage systems, IEEE Trans. Smart Grid 6 (2015) 2583–2585.

[136] T. Wang, H. Kamath, S. Willard, Control and optimization of grid-tied photovoltaic storage systems using model predictive control, IEEE Trans. Smart Grid 5 (2014) 1010–1017.

[137] D. Zhu, G. Hug, Decomposed stochastic model predictive control for optimal dispatch of storage and generation, IEEE Trans. Smart Grid 5 (2014) 2044–2053.

[138] G. He, Q. Chen, P. Moutis, S. Kar, J.F. Whitacre, An intertemporal decision framework for electrochemical energy storage management, Nature Energy 3 (2018) 404–412, 2018/04/23.

[139] R. Wang, G. Xiao, P. Wang, Hybrid centralized-decentralized (HCD) charging control of electric vehicles, IEEE Trans. Veh. Technol. 66 (2017) 6728–6741.

[140] W. Tang, Y.J. Zhang, A model predictive control approach for low-complexity electric vehicle charging scheduling: optimality and scalability, IEEE Trans. Power Syst. 32 (2017) 1050–1063.

[141] M.R. Almassalkhi, I.A. Hiskens, Model-predictive cascade mitigation in electric power systems with storage and renewables—Part I: theory and implementation, IEEE Trans. Power Syst. 30 (2015) 67–77.

[142] J.A. Martin, I.A. Hiskens, Corrective model-predictive control in large electric power systems, IEEE Trans. Power Syst. 32 (2017) 1651–1662.

[143] G. Hug, S. Kar, C. Wu, Consensus + innovations approach for distributed multiagent coordination in a microgrid, IEEE Trans. Smart Grid 6 (2015) 1893–1903.

[144] P. Liu, U. Ozguner, Distributed model predictive control of spatially interconnected systems using switched cost functions, IEEE Trans. Autom. Control 63 (2018) 2161–2167.

[145] H. Shayeghi, H. Shayanfar, A. Jalili, Load frequency control strategies: a state-of-the-art survey for the researcher, Energ. Conver. Manage. 50 (2009) 344–353.

[146] D. Rerkpreedapong, A. Hasanovic, A. Feliachi, Robust load frequency control using genetic algorithms and linear matrix inequalities, IEEE Trans. Power Syst. 18 (2003) 855–861.

[147] E. Çam, I. Kocaarslan, Load frequency control in two area power systems using fuzzy logic controller, Energ. Conver. Manage. 46 (2005) 233–243.

[148] M.D. Galus, S. Koch, G. Andersson, Provision of load frequency control by PHEVs, controllable loads, and a cogeneration unit, IEEE Trans. Ind. Electron. 58 (2011) 4568–4582.

[149] A.W. Berger, F.C. Schweppe, Real time pricing to assist in load frequency control, IEEE Trans. Power Syst. 4 (3) (1989) 920–926.

[150] I. Khan, Y. Xu, S. Kar, H. Sun, Compressive sensing-based optimal reactive power control of a multi-area power system, IEEE Access 5 (2017) 23576–23588.

[151] J.P. Lopes, N. Hatziargyriou, J. Mutale, P. Djapic, N. Jenkins, Integrating distributed generation into electric power systems: A review of drivers, challenges and opportunities, Electr. Pow. Syst. Res. 77 (2007) 1189–1203.

[152] W. Wangdee, R. Billinton, Reliability assessment of bulk electric systems containing large wind farms, Int. J. Electr. Power Energy Syst. 29 (2007) 759–766.

[153] P. Zhang, Y. Wang, W. Xiao, W. Li, Reliability evaluation of grid-connected photovoltaic power systems, IEEE Trans. Sustainable Energy 3 (2012) 379–389.

[154] F. Bouffard, F.D. Galiana, Stochastic security for operations planning with significant wind power generation, in: *Power and energy society general meeting-conversion and delivery of electrical energy in the 21st century, 2008 IEEE*, 2008, pp. 1–11.

[155] P. Moutis, N.D. Hatziargyriou, Decision trees aided scheduling for firm power capacity provision by virtual power plants, Int. J. Electr. Power Energy Syst. 63 (2014) 730–739.

[156] P. Moutis, S. Skarvelis-Kazakos, M. Brucoli, Decision tree aided planning and energy balancing of planned community microgrids, Appl. Energy 161 (2016) 197–205.

[157] A.R. Di Fazio, G. Fusco, M. Russo, Decentralized control of distributed generation for voltage profile optimization in smart feeders, IEEE Trans. Smart Grid 4 (2013) 1586–1596.

[158] P.N. Vovos, A.E. Kiprakis, A.R. Wallace, G.P. Harrison, Centralized and distributed voltage control: impact on distributed generation penetration, IEEE Trans. Power Syst. 22 (2007) 476–483.

[159] L. Yu, D. Czarkowski, F. De León, Optimal distributed voltage regulation for secondary networks with DGs, IEEE Trans. Smart Grid 3 (2012) 959–967.

[160] M. Brenna, E. De Berardinis, L.D. Carpini, F. Foiadelli, P. Paulon, P. Petroni, et al., Automatic distributed voltage control algorithm in smart grids applications, IEEE Trans. Smart Grid 4 (2013) 877–885.

[161] P. Moutis, P.S. Georgilakis, N.D. Hatziargyriou, Voltage regulation support along a distribution line by a virtual power plant based on a center of mass load modeling, IEEE Trans Smart Grid 9 (4) (2018) 3029–3038.

[162] H. Bevrani, S. Shokoohi, An intelligent droop control for simultaneous voltage and frequency regulation in islanded microgrids, IEEE Trans. Smart Grid 4 (2013) 1505–1513.

[163] M. Céspedes, J. Sun, Online grid impedance identification for adaptive control of grid-connected inverters, in: *Energy Conversion Congress and Exposition (ECCE), 2012 IEEE*, 2012, pp. 914–921.

[164] Y. Fukuyama, H. Yoshida, A particle swarm optimization for reactive power and voltage control in electric power systems, in: *Evolutionary Computation, 2001. Proceedings of the 2001 Congress on*, 2001, pp. 87–93.

[165] I. Khan, Y. Xu, H. Sun, V. Bhattacharjee, Distributed optimal reactive power control of power systems, IEEE Access 6 (2018) 7100–7111.

[166] I.A. Khan, Y. Xu, B. Tahir, Design and manufacturing of digital MOSFET based-AVR for synchronous generator, in: *Cyber Technology in Automation, Control, and Intelligent Systems (CYBER), 2015 IEEE International Conference on*, 2015, pp. 217–222.

[167] A. Tapia, G. Tapia, J. Ostolaza, Reactive power control of wind farms for voltage control applications, Renew. Energy 29 (2004) 377–392.

[168] F.A. Viawan, D. Karlsson, Voltage and reactive power control in systems with synchronous machine-based distributed generation, IEEE Trans. Power Delivery 23 (2008) 1079–1087.

[169] I. Khan, Z. Li, Y. Xu, W. Gu, Distributed control algorithm for optimal reactive power control in power grids, Int. J. Electr. Power Energy Syst. 83 (2016) 505–513.

[170] P. Leitão, S. Karnouskos, L. Ribeiro, P. Moutis, J. Barbosa, T.I. Strasser, Common practices for integrating industrial agents and low level automation functions, in: *Industrial Electronics Society, IECON 2017—43rd Annual Conference of the IEEE*, 2017, pp. 6665–6670.

[171] Distributed Generation Photovoltaics and Energy Storage, IEEE Standard for Interconnection and Interoperability of Distributed Energy Resources with Associated Electric Power Systems Interfaces, in IEEE Std 1547-2018 (Revision of IEEE Std 1547-2003), pp.1–138, 6 April 2018.

Chapter 11

New Protection Schemes in Smarter Power Grids With Higher Penetration of Renewable Energy Systems

Payman Dehghanian*, Bo Wang* and Mohammad Tasdighi†
*The George Washington University, Washington, DC, United States, †S&C Electric, Chicago, IL, United States

Chapter Outline

11.1 Introduction

Natural resources are mainly of two types: renewables, which are replenished at a rate higher than that of exploitation; and nonrenewables, which are replenished at a rate slower than that of exploitation. Thorough investigations on recent global practices in the deployment of many types of environmentally

Pathways to a Smarter Power System. https://doi.org/10.1016/B978-0-08-102592-5.00011-9

friendly renewable sources of energy (e.g., solar energy, wind power, hydropower, geothermal, and biofuel) reveal that solar energy, wind energy, and biofuels are leading in this race as the favorite sources of energy for the future [1].

Under the overall umbrella of high-level policy commitments on the global deployment and utilization of renewable energy resources, a number of renewable energy targets, for example, climate change targets, were set. Even though climate change is typically a driving factor in shaping and characterizing such targets, it is commonly envisaged as an advantage of renewable energy integration. The convening power of international climate change negotiations offers a strategic setting and guiding mechanisms for countries to envision long-term renewable energy targets that encapsulate national renewable energy strategies and political commitments [2]. The District of Columbia (District) Department of Energy Environment (DOEE) provides a climate and energy plan, entitled Clean Energy DC (Plan). The Plan is DOEE's proposal to reduce greenhouse gas (GHG) emissions by 50% below 2006 levels by 2032, while increasing renewable energy use and reducing energy consumption, as directed by the District's sustainability plan, sustainable DC. Achieving the 50% reduction by 2032 will put the District on track to reduce GHG emissions by 80% by 2050. Therefore, the plan enforces an increase in deployment of renewable energy, to make up 50% of the District's energy supply by 2032 [3]. Renewable energy sources (RES) are clean, environmentally friendly, and will not run out. The deployment of renewable and distributed energy resources (DER) has been observed to be significantly on the rise in the United States and the world. The existing policies for reduction of GHG emissions are expected to ensure—and even expedite—a sustainable growth in RES penetration in the electric industry. To put a figure on this, 37 United States states have already implemented renewable portfolios that mandate meeting 10%–33% of energy consumption. Likewise, the energy policies of the European Union (EU) have legislated targets of 20% renewable energy penetration by 2020 [1, 2].

As such, additional variability and uncertainty in the net-load will be realized in future, contributing to the advent of new operational challenges in bulk power grids, ranging from operation limit violations, frequent start-ups and shut-downs of dispatchable generating units, reduction in available lead time, and increased ramping and reserve requirements [3, 4]. Among such technical challenges, the power grid of the future requires a suite of modified and complemented protection schemes compatible and capable to cope with the rushing arrival of uncertainties from RES. For instance, such schemes are essential in the case of an inverter-based photovoltaic (PV) RES, as the short-circuit currents of these resources are extremely poor, and protection devices may fail to discriminate between the normal loading and fault conditions due to the narrow margin between them [5].

The green energy transition increases the risk of power failures, if not accompanied by a grid technology transition. Traditional large coal and gas power plants, with the high inertia of their heavy, rotating generators secure network stability and ensure a minimal deviation from the desired 50 Hz or 60 Hz output frequency even under a relatively large, temporary imbalance between mechanical power supplied and electrical power demanded. Smaller generators, such as wind turbines, are less stable and may need to be disconnected from the grid promptly after occurrence of a fault, as they may affect system stability, which in turn might result in a cascading outage effect, ultimately leading to a blackout. In present practice, these small power generating resources are disconnected during fault periods in order to protect them from the reverse fault current from the main system [6].

Faults will always happen within a power grid, and there is no single best solution to deal with them. The challenge is to keep these faults localized, to minimize the number of consumers affected and to ensure a speedy and timely recovery. Whereas the original, 100-year-old power grid design allowed for hard-wired mechanical solutions following simple rules, the ever-increasing complexity of the current grid demands smart(er) solutions. Protection relays need to monitor the relevant parameters—such as voltage, frequency, and current—at the proper sampling frequency. The use of communication-assisted protection schemes and adaptive protection systems can be suggested, the performance of which are closely driven and dependent on reliable real-time communication facilities. Algorithms are only given a short time interval (several milliseconds up to tens of milliseconds, depending on the system design and architecture) to make the correct decision autonomously, based on the received and/or conceived data [7]. This chapter discusses the merits and demerits of various state-of-the-art protection schemes with the integration of RESs in power systems. The trends and possible solutions for protection schemes of future smart power systems are discussed.

This chapter is organized as follows. Protection issues in RES-integrated transmission systems are presented in Section 11.2; protection issues in RES-integrated distribution systems are discussed in Section 11.3; and microgrid protection schemes are introduced in Section 11.4. Finally, we present the conclusions and discuss the future scope of research on massively RES-integrated power systems in Section 11.5.

11.2 Protection Challenges and Solutions in RES-Integrated Bulk Transmission Systems

Bulk windfarm-based renewables have penetrated the transmission network at the subtransmission and transmission levels, and continue to do so. In addition to windfarms, bulk penetration of solar power has seen a tremendous increase in Megawatt size. This intensified variability and the stochastic nature of RES affects the existing protection philosophy in transmission systems. Special attention should be paid to the design of protection schemes for transmission systems integrated with large-scale renewables. Particularly, the design of protection schemes for compensated transmission lines under these conditions becomes more complicated [8].

11.2.1 Main Protection Challenges in Transmission Grid

11.2.1.1 RES Integration

When the bulk Megawatt renewables are integrated at the transmission level, the reach of the distance relays may be affected, possibly resulting in under reach. For a large number of integrated renewables, it is not practically feasible to set the reach of each distance relay for different renewable infeeds. Adaptive distance and intelligent technique based differential relaying protection schemes are needed, which can adapt the relay protection settings as per the network operational requirements [9].

11.2.1.2 Flexible AC Transmission system (FACTS) Compensated High-Voltage Transmission Lines

In order to effectively integrate the RES in the transmission grid and fulfill the supply requirements of increasing demand for electricity, electric power utilities today are continuously seeking opportunities for optimal utilization of the existing transmission infrastructure and built-in flexibilities [10–16]. However, the existing constraints in interconnected power systems, such as voltage and transient stability, restrict the full and effective utilization of the existing transmission corridors. In order to alleviate such constraints, and for the main purpose of efficient utilization and control of the existing transmission network, the utility industry has focused on deployment and commissioning of advanced power-electronic-based FACTS in the grid [17]. These deployments, in turn, mandate a significant modification in the protection requirements of the transmission lines reinforced by FACTS. The thyristor-controlled series capacitor and the unified power flow controller based high-voltage transmission lines require modern signal processing and computational intelligence techniques, since the fault type and operating mode of the FACTS-based distance relays suffer from either underreaching or overreaching concerns and challenges [18].

11.2.2 Main Protection Schemes for RES-Integrated Transmission Lines

11.2.2.1 PMU Based Protection Scheme

Over the past 35 years, deployment of Phasor Measurement Units (PMUs) and PMU-based Intelligent Electronic Devices (IEDs) has seen a tremendous record, with revolutionized synchrophasor-based applications adopted that, in turn, have facilitated a better understanding of modern power systems through high-resolution and high-precision grid observations. Several critical applications, for example, State Estimation, Fault Detection/Location, Remedial Actions, and Wide Area Monitoring Systems and Protection, are backboned with PMUs and PMU measurements. Recently, detection of incipient voltage instabilities has also been viable thanks to a number of PMUs located at strategic points across the grid and a central assessment unit that calculates the power transfer stability across a specific transmission line based on the voltage and current phasor measurements, as shown in Fig. 11.1. GPS time-stamped signals from PMUs located at both ends of the transmission lines help to provide synchronized voltage and current phasors for grid-scale assessments [19].

In [20], an adaptive protection coordination scheme for active power networks is presented, which transfers the desired relay settings to overcurrent (OC) relays and distance relays, as per the operational network topology (TP). Whenever there is a change in the network fault level, either due to change in line parameters or incoming of new RES, new relay group settings are communicated to corresponding relays.

Furthermore, while some research efforts have been made [21–23], additional studies should be carried out to investigate the impact of RES integrations on PMU measurements that will eventually affect the synchrophasor-based protection schemes. Specifically, different PMU algorithms manifest different performance accuracy under changing network conditions, which may compromise the trustworthiness of the end-user protection application, especially under massive penetration of renewables. An example of such studies is [21], where the impact of errors in the PMU response on synchrophasor-based fault location algorithms is investigated. The impact of PMU errors on a synchrophasor-based fault location algorithm requiring synchronized phasor measurements at both ends of a transmission line is demonstrated in Fig. 11.2. It is revealed from this

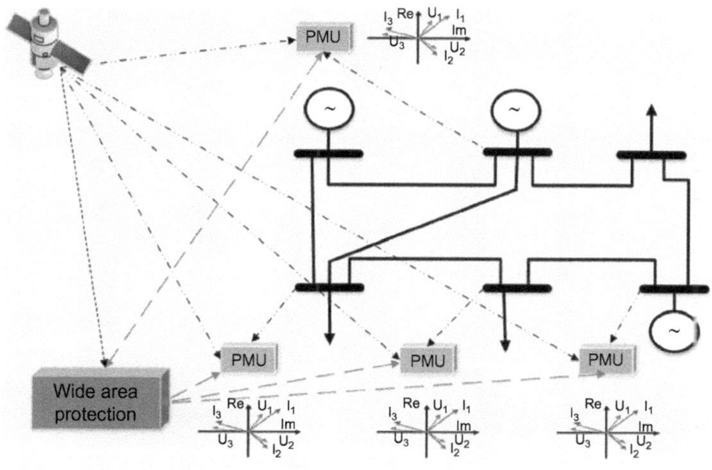

FIG. 11.1 Overview of the PMU-installed transmission system [19].

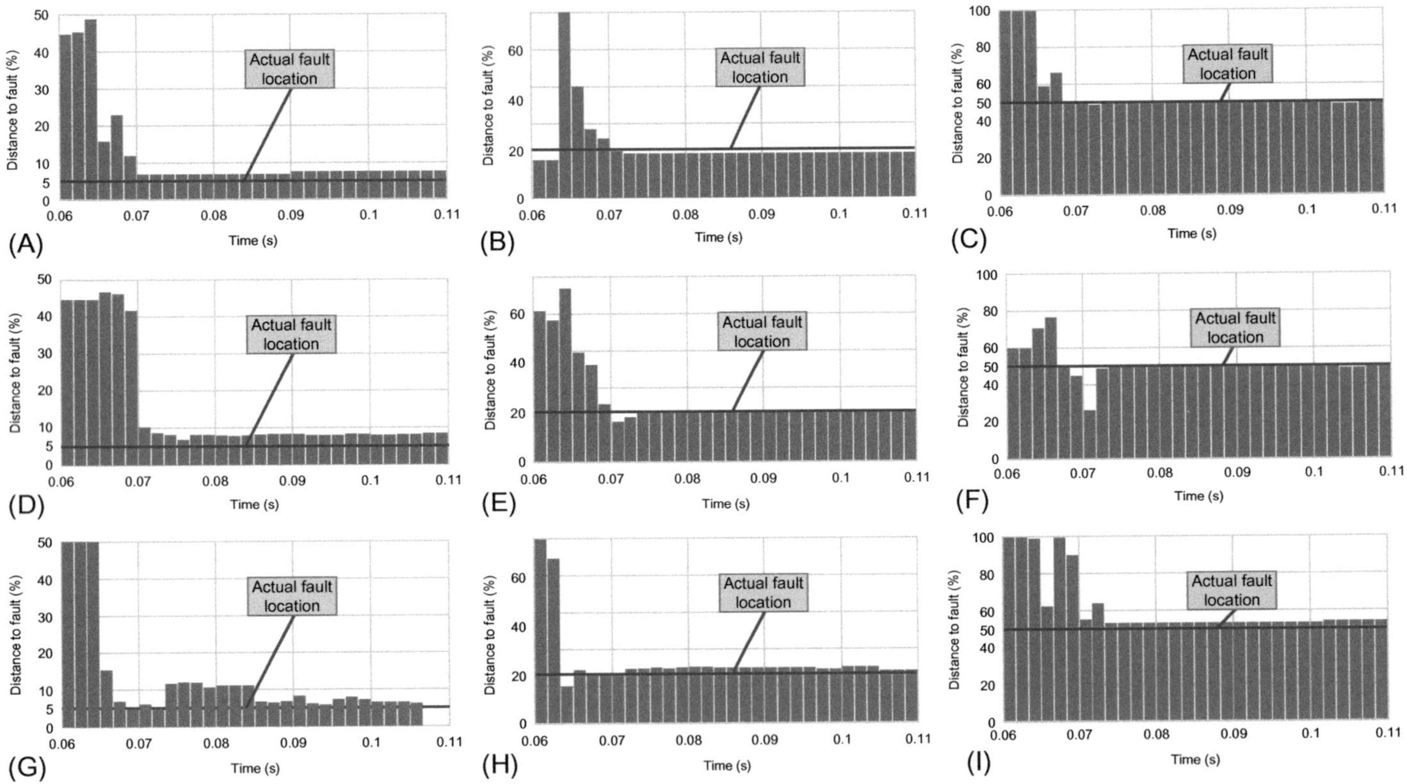

FIG. 11.2 Synchrophasor-based fault location algorithm performance: calculated fault locations for different fault scenarios (single line to ground fault was simulated with the three values of fault resistance: the bolted fault use case scenario (0 ohm) is illustrated in figures (A)–(C); the 20 ohm fault resistance use case scenario is depicted in figures (D)–(F); and the high-resistance 100 ohm fault scenario is illustrated in figures (G)–(I), each containing fault locations at the 5%, 20%, and 50% pf the line length).

figure that: (i) during the higher resistance fault scenarios, the time taken to locate the fault is comparatively longer; (ii) the minimum error is corresponding to the scenario when the fault has happened in the middle of the line; (iii) the error is at its maximum when the fault occurs at the begging of the line Errors introduced by the communication network and phasor data concentrators (PDCs) should also be taken into account in future research.

11.2.2.2 Integrated Wide Area Protection and Control Scheme

This concept is largely founded on the integration of protection and control principles, particularly in a wide-area, promising a number of advantages for the next-generation protection and control systems in modern power grids. With the real-time synchronized data collected and mined in the wide-area information platform, such technology enables merging and integrating the traditional lines of defense systems and the online self-healing paradigms, helping in prevention and mitigation of cascading outages with a higher level of ensured sensitivity, reliability, and fault tolerance. The data received from the wide area information platform (e.g., static, dynamic, transient measurements and status of circuit breakers, etc.) is assigned to various focused computation algorithms in the platform responsible for performing and implementing various advanced protection and control functions at substation or regional levels through a distributed cloud system. Such functions include wide area fault location, power quality monitoring, fault line selection, protection settings, etc. The cloud computing platform equipped with advanced big data analytics significantly reduces the computational burden on the secondary terminal equipment [24].

In a holistic wide-area backup protection (WABP) scheme, shown in Fig. 11.3, there exist six different layers that mainly contribute to the total delay: (i) data acquisition, (ii) wide-area network (WAN) for data transmission, (iii) PDCs, (iv) application solution packages and algorithms, (v) WAN for command transmission, and (vi) circuit breakers [25]. A detailed formulation centered on network regionalization is suggested in [25] that provides solutions to constraint the total latency of the WABP schemes. The suggested solution is a 3-stage linear model, which optimizes the number and location of measurement devices (MDs), the number and boundary of protection regions, location of protection rooms, and the total number of communication links between MDs and protection rooms. With such characteristics, the model can be applied, with minimum adjustments, to any latency-critical WABP scheme.

11.2.2.3 Hierarchically Coordinated Protection (HCP) Scheme

Power system control methods are primarily focused in response to the classification of power system operating states for mitigating the prevailing conditions in a power grid (voltage, transient, frequency, and small-signal instability) and maintaining them within a secure operating state. Future

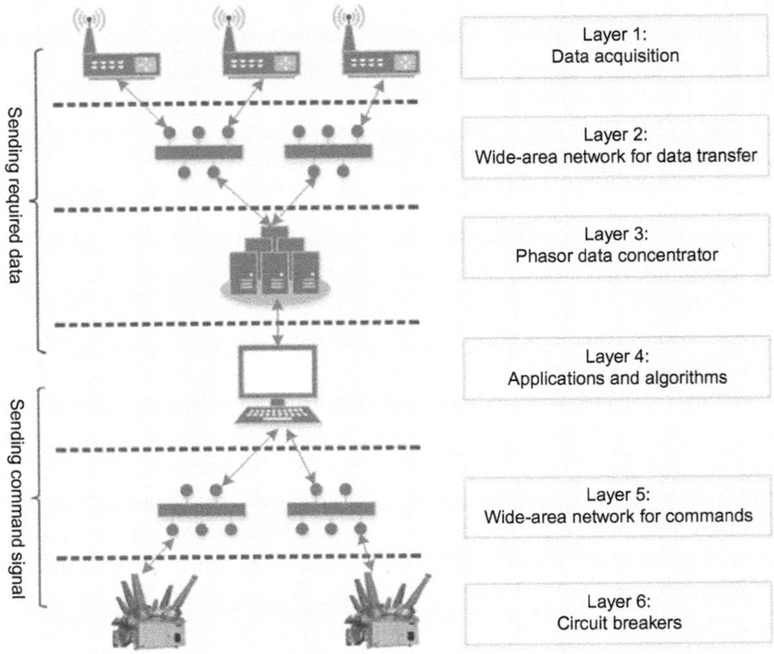

FIG. 11.3 Layers of the WABP Scheme [25].

hierarchical protection and control systems will be expected to benefit from look-ahead features, enabling predictive protection [26]. The look-ahead features may include:

i. *Analytical computer programs* that may employ many new and timelier system measurements to recognize the system's actual and predicted challenges, and to take immediate action, automatically, to prevent or mitigate problems. Where appropriate, and when time allows, manual operator intervention options and solutions are provided to the system operators.

ii. *Probabilistic risk analytics* that can detect system threats under projected normal operating conditions, as well as during single-order failures, double-order failures, and out-of-service maintenance intervals.

iii. *Load forecasting* models that are significantly enhanced and generally encapsulate several time horizons:

short-term—minutes, hours, and days, in support of system real-time, hour-ahead, and day-ahead operational decisions;

mid-term—monthly, quarterly, and annually, to support the operations and maintenance (O&M) plans; and long-term—multiyear horizons, in support of long-term investment and reinforcement decisions.

iv. *Fast simulation and modeling* with look-ahead features and capabilities that enable accurate prediction of system disturbances ahead of time while continuously ensuring an optimized grid performance.

An HCP approach, proposed in [27], relies on three protection layers: (i) predictive protection, (ii) adaptive/settingless protection, and (iii) corrective protection under unintended tripping and other conditions. All three layers are integrated and analytically correlated in order to achieve the full potential of the envisioned ideology [27]. While characterized and equipped with high-accuracy modules, the tool may suffer from an expensive computational burden that may impose an unacceptable operational latency, specifically for real-time applications and decision making. Hence, this module should be triggered in coordination with a legacy protection scheme to monitor and, if needed, rectify the original relay performance. At the substation level, fast and accurate fault location and event tree algorithms based on synchronized sampling are employed to detect the false line tripping and relay misoperations, respectively. Particularly, the fault location algorithm will be immediately triggered upon a line tripping incident in order to validate the relay operation. If the fault condition remains unconfirmed, the tripped line will be swiftly restored.

Based on the developments mentioned above, several efficient criteria should be devised and utilized to evaluate the protection schemes of the future. The design of the future protection schemes should follow the steps illustrated in Fig. 11.4. The predictive protection should detect the prospective power system operating states and leave ample time for a protection system to adjust the relay and algorithm settings if needed. This could be achieved by coordinating with the Energy Management Systems and enabling the big data analytics and

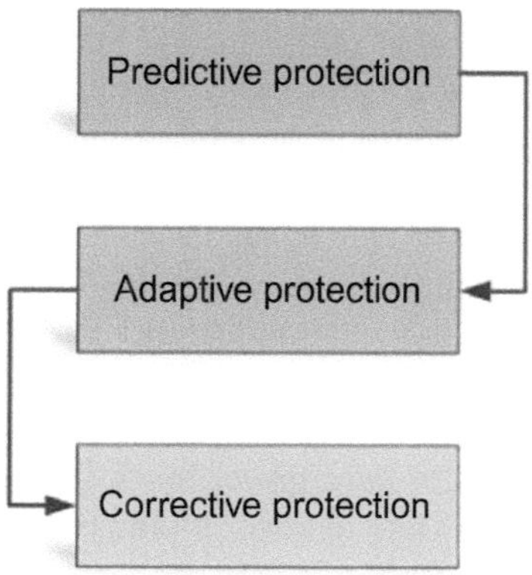

FIG. 11.4 The procedure to design a new protection scheme in the transmission system.

mining of the data being acquired from wide-area communication systems. The adaptive protection algorithm should make the right decision to trip the relay and react in a short lead time. Regionalized and distributed algorithms should be adopted to achieve a fast and accurate response, yet with fewer communication requirements. The corrective protection at the third layer should consider the incorrect relay tripping, measurement errors, and communication system failures/delays caused by either equipment degradation or cyberattacks. Such a forward design would allow each layer to take into account and sequentially carry the impact of the layers forward.

The effectiveness of the protection schemes should be evaluated backwards. Maintaining the system operation in the normal operating state and avoiding the cascading failure of system components that may result in partial or system-wide blackouts is critical to system operators. Thus, corrective protection should be validated first. The adaptive protection and predictive protection can be evaluated next, to achieve better and more accurate protection results and to minimize economic losses.

An effective HCP scheme can also be suggested to verify different approaches to corrective, adaptive, and predictive protection strategies aimed at improving power system resilience. Relay misoperation is known to be a major contributing factor in 75% of major disturbances in North America. During abnormal conditions, the backup relays may not be able to differentiate faults from no-fault conditions (balance between dependability and security is inadequate), such as when overload and large power swings occur. It has been observed that while redundancy (providing backup protection and improving dependability) reduces the probability of a dependability-based protection system failure, it may increase the probability of a security-based protection system failure. As a result, maintaining the balance between dependability and security of protective relay operation remains a challenge. An improved protection system design must provide inherently dependable and secure operation, where an HCP aims at achieving that goal [28].

To clarify better the impact of protection reliability on the power grid's resilience, one may compare the legacy (reliability-based) protection perspective with what the requirements for resilience ask for today. While the legacy protection on transmission may be able to handle an $N-1$ contingency case without causing any misoperation, it may not be able to cope with $N-m$ contingency cases (high-impact, low-probability events) effectively. In other words, the static predetermined balance between dependability and security might not necessarily be able to meet the system's resilience criteria. It may not be reasonable to expect to have protection solutions, for example, proper relay settings, developed off-line to cover all the possible $N-m$ contingency cases. Instead, a dynamic trade-off between security and dependability is needed as the system goes through various events and operating conditions. Fig. 11.5 illustrates an HCP that supervises traditional distance relay functions for transmission lines. The basic idea is to maintain a dynamic balance between

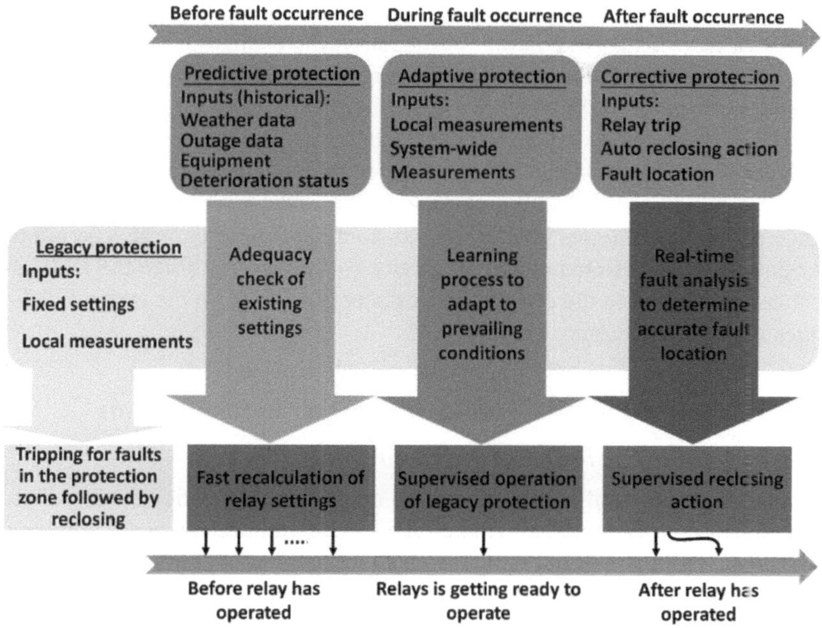

FIG. 11.5 Traditional distance protection supervised by HCP.

dependability and security, which is achieved through predictive, adaptive, and corrective relaying actions. This provides flexibility in the protection scheme's behavior to handle the uncertainties associated with the protection operation.

11.3 Protection Challenges and Solutions in RES-Integrated Distribution Systems

Power distribution systems are primarily protected via a number of current-sensing devices, such as OC relays, re-closers, and fuses. These devices monitor the current flow in the system through the protected element and generate trip signals to the circuit breakers if the fault current flow is greater than a prespecified value. The protection philosophy in power distribution systems is generally designed based on the assumption that the systems are radial in nature and power flow is always unidirectional from the source to the consumers [29].

In the case of a multigenerator/multiloop distribution system, the power flows are not unidirectional, and the fault currents will flow in either direction based on its location in the network. Directional overcurrent relays (DOCRs) are the most promising solution to avoid sympathetic trippings in multiloop distribution systems. In the case of a large penetration of RES in the distribution grid, however, network fault levels will change dynamically with respect to the RES penetration and the corresponding stochastic behavior. DOCR with fixed time dial setting and plug multiplier setting will not protect the feeder when there is a large RES penetration. Particularly, the change in the fault current primarily depends on the type, rating (penetration), and location of the RES integration in the grid [30].

The phase-ground or phase-phase faults in conventional distribution systems normally result in an OC. The basic precondition of the traditional OC protection paradigms in distribution grids recognizes the fault current to be in the range of five to six times greater than its nominal (rated) current. The limited current rating of powered electronic devices, however, can force the fault current corresponding to the converter-interfaced RES to be in the range of 1.1–2 times the nominal current [31].

11.3.1 Main Protection Challenges in Distribution Systems

11.3.1.1 Fault-Current Contributions

Compared to the traditional grid with no distributed generation (DG) connection, higher penetration of distributed RES in distribution networks will result in a drastic reduction in the fault currents measured by the protective relays at the front end of the distribution feeder, which, in turn, may lead to delayed fault detection and relay operation. This is mainly because of the fault-current contributions of the distributed RES, where the fraction of the fault currents measured by the OC relays decreases [32]. In the worst-case scenario, where the fault is missed or not instantly detected, high voltage violations (while the fault current is low) and unsafe conditions in the field would be the consequences. Furthermore, the fault can spread throughout the entire distribution network, resulting in further severe damage if it remains undetected for a long time.

11.3.1.2 Reduction in Reach of Impedance Relays

The maximum fault distance (corresponding to the maximum fault impedance or minimum fault current) that triggers the relay in a certain impedance zone, or in a certain time instant (depending on the configuration), characterizes the reach of an impedance relay [33]. When a fault is detected in the downstream section of the bus to which the distributed RES is connected, the upstream relay sees and measures a higher impedance than the actual fault impedance. Thus, the increased voltage due to the additional infeed at the common bus will cause an amplified fault distance, triggering the relay with a faster time response [34].

11.3.1.3 Auto Re-Closure

Re-closers are devices that are designed to respond swiftly to temporary faults in distribution systems, isolate the faulty segments, and allow timely self-clearance of the fault. DG and massive deployment of renewables have brought about significant challenges to the safe and secure operation of auto re-closers (e.g., out-of-phase re-closing). As a solution, DG must be entirely detached before the re-closure process gets prompted [35]. Continuous operation of a DG unit under a single fault scenario, if the automatic re-closure is prompted after a short interruption, may result in two main concerns: (i) the arc may have been fed by the DG, it may not be entirely cleared, and this does not allow the instantaneous re-closure to succeed; and (ii) due to a change in frequency of the islanded segments, resulting from the active power unbalance during faults, the switch reclosure attempt will fail, as it couples two asynchronously operating systems with active sources on both sides of the re-closer [36].

11.3.2 Main Protection Schemes for RES-Integrated Distribution Systems

11.3.2.1 Distance Protection

Compared with the OC relays, distance relays are less influenced by changes in network topology. To overcome the false tripping of OC relays in RES-penetrated distribution systems, high set instantaneous (50H or 50NH) OC elements were replaced with instantaneous elements with quadrilateral characteristics; low set instantaneous (50L or 50NL) elements were substituted with instantaneous mho elements; and time-overcurrent (51P) elements are replaced with mho elements [37]. Distance relay with mho characteristics is an inherently directional relay; thus, it can discriminate the forward and reverse faults. Fault resistance is ignored in [37], but it is also an important factor for distance relays in distribution systems, as distribution feeders are short in length and most faults are of high impedance characteristics. For higher RES penetration levels, distance relays may also face underreach or overreach problems in distribution systems.

11.3.2.2 Limiting the RES Capacity

The protection coordination index (PCI) evaluated in [30] is defined as the relationship between changes in the RES penetration power, with respect to the change in the coordination time interval. Attempts are made to penetrate the RES in the distribution system without compromising the network existing protection setting. This is primarily achieved by limiting the size of the RES penetration. The suggested PCI index in [30] provides insightful information about which locations are safe for RES integration or where protection coordination margins are less affected or compromised with RES. The fall in coordination time interval is evaluated with respect to percentage penetration of the RES at the node with the highest PCI value. If the RES are customer-owned, then the size of the RES is optimized in such a way that does not compromise the existing relay settings in the distribution network. The application of PCI is also extended to multiple RES penetrations in a utility network [38]. This index is also helpful for utilities to decide on the optimal location of the customer-owned RES where the customer can inject their RES power into the distribution system without negatively influencing the existing operation and control of the system.

11.3.2.3 Fault Current Limiters

As the principle philosophy behind this approach, the fault current can be blocked from the RES during fault intervals so that the relay setting, originally decided at the planning stage without RES, remains valid and intact under scenarios with variable penetration of RES. This is achieved by connecting series devices with the RES, which offers low impedance during normal operation and high impedance during faults to block the fault currents. In order to restore the fault current levels to those original values without DER, fault current limiters (FCLs) are suggested, in [39], to be connected in series with the DERs and utility interconnection point. A nonlinear programming optimization problem is formulated for FCL allocation in distribution systems that minimize the DER-caused changes in network fault current levels. Solid state fault current limiters (SSFCL) with fast switching time characteristics are also suggested as a promising and cost-effective solution in distribution systems to quickly block DER-caused fault currents and minimize the protection-related side-effects of DER penetration in the grid. A genetic algorithm approach is suggested and employed in [40] for optimal allocation (sizing and siting) of SSFCLs in the distribution network that is needed to keep the DER fault consequences blocked.

11.3.2.4 Differential Protection

The difference between the incoming and outgoing currents of a protection zone is the basis for the differential protection principles; as soon as this exceeds a predefined value, the protection scheme will be triggered. A differential protection scheme is known with a very fast response time of ~5 ms. In the case of active distribution systems with high penetration of renewables and DGs, the DG at the faulty zone is signaled to be disconnected by the corresponding protection relay. Discrete Fourier transform is utilized in [41] on the fault currents and voltage signals to extract the differential features, followed by decision tree data mining for final decision making.

11.3.2.5 Wavelet Transform-Based Protection

A fault detection algorithm for PV arrays is suggested in [42], in which multiresolution signal decomposition by wavelet transform as well as a two-stage support vector machine classifier are utilized for feature extraction and decision making, respectively. The boundary discrete wavelet transform is deployed within a fast and simple OC protection scheme in [43], where its coefficient energy helps achieve an accurate, fast, and real-time fault inception time detection.

Both differential protection and wavelet transform-based protection rely on the communication infrastructure. This imposes a significant level of vulnerability to communication failures, in case of which, the entire protection scheme may be dysfunctional.

11.3.2.6 Adaptive Protection

A networked protection scheme in power distribution systems is introduced in [44], where the goal is accurate and fast fault detection centered on utilizing a centralized fault detector (CFD). In order to detect the fault-induced disturbances, the CFD simultaneously processes the 3-phase voltage and frequency quantities from the synchrophasor datasets received from the main point of common coupling (PCC) and the local PCCs corresponding to DG systems. Upon the synchrophasor-based data-driven detection of faults, persistent disturbances, or a significant prevailing condition by the CFD, the relevant protective relays, IEDs, and DG systems will be commanded via signals transferred through communication links.

A hardware-in-the-loop (HIL) adaptive protection scheme is introduced in [45], which ensembles a variety of real-time simulation tools—for example, real-time digital simulators (RTDS), the programmable logic controller (PLC), multifunction digital relays, etc.—enabling a multifunction protection scheme with centralized control and accurate protection settings. Distribution networks are simulated in RTDS with digital relays as protective devices for distribution feeders and the PLC as the centralized controller. See Fig. 11.6 for more details on the developed architecture. This adaptive protection scheme first identifies the optimal relay setting groups, following by an online self-adjustment process, enabling a dynamic, continuously tuned, and adaptive protection scheme responsive to the changing system operating conditions. The PLC encapsulates the status of circuit breakers, and monitors the accurate relay transitions to

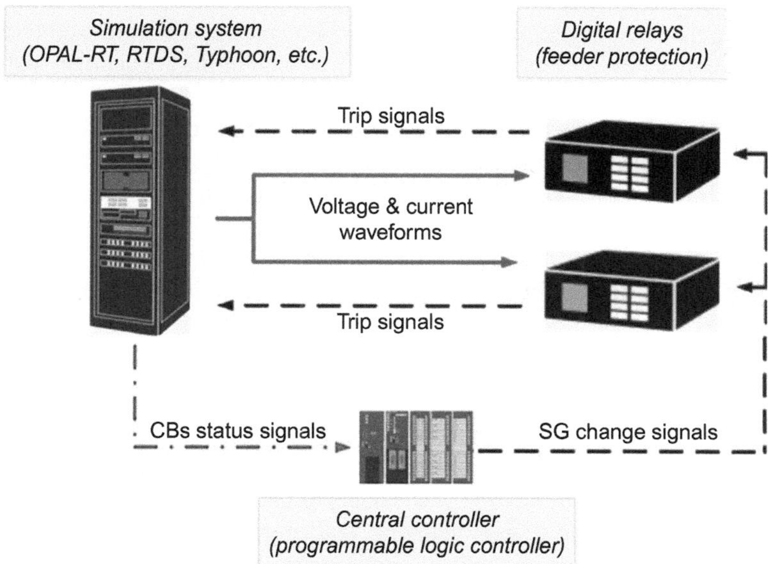

FIG. 11.6 Illustration of the suggested HIL testbed [45].

effective setting groups in the face of a major change in the system operating condition, for example, DG connection/disconnection, or network reconfiguration. This robust approach is characterized by a significantly high rate of convergence, making the real-time evaluation of the optimal relay setting groups feasible.

A multiagent system-based protection and control scheme is suggested in [46], which is a centrally controlled adaptive protection and control scheme for DGs. A basic three-level control scheme—proposed for DGs, relays, and load agents—includes several functions, such as information collection, decision making, and execution of commands, as well as communication to remote agents. This scheme assumes that a strong and reliable communication network exists in the distribution system. If any network changes are originated from system reconfigurations or DG integration (connection or disconnection), the DG agents and other agents adapt their feasible control modes accordingly, and relay agents will modify the relay settings corresponding to the changes in the system operating conditions. In this scheme, a strong communication network is required to make it possible for each of the element agents in the network to communicate. This multiagent oriented protection and control scheme offers several advantages in power distribution systems: (i) it can prevent relay misoperations; (ii) it can harmonize and coordinate the distributed relays and converter controllers, realizing a resourceful information sharing platform.

While providing decent flexibility in decision making, adaptive relays and their wide-scale implementation anticipates several main hurdles:

(i) reinforcing the grid and supplanting (substituting) all the existing protective relays in the distribution network with upgraded adaptive schemes is extremely costly;
(ii) dependable, secure, accurate, and timely operation of adaptive protection schemes heavily rely on reliable communication infrastructures; and
(iii) their implementation is burdensome, especially in large distribution systems with massive penetration of DERs, since the history and prior knowledge on a significant number of DER possible configurations should be documented and accustomed based on the adaptive adjusting and tuning rules [47].

11.4 Protection Challenges and Solutions in Microgrids

A microgrid is typically formed of several interconnected loads and DER units, with clearly defined electrical boundaries, representing a single controllable entity in the grid. Microgrids can electrically connect and disconnect from the grid, enabling participation in both grid-connected and islanded operating modes [48]. Most importantly, a microgrid can operate in isolation from the grid in order to mitigate sustained power outages as they unfold.

Although microgrid design, planning, and operation bring about potentials for improved network reliability, proliferated integration of renewables, enhancing power quality, and investment cost optimization, they may also raise a number of concerns and challenges, among them, the need for modified protection and control strategies can be highlighted. In particular, one of the major challenges is to design and implement protection schemes in microgrids that can respond dependably and securely to faults in both grid-connected and islanded modes.

11.4.1 Main Protection Challenges in Microgrids

11.4.1.1 Fault Current Levels

Fault levels in microgrid networks are higher when they operate in a grid-connected mode as compared to an islanded mode of operation [49]. In an islanded operation mode, the DER units are the only fault current feeding sources, and have limited short circuit capabilities. During these operational modes, discrimination of fault levels from loading levels may lead to nuisance tripping of the protective relays.

11.4.1.2 Coordination and Selection of Protection Devices

Selection of the protection device depends on the operating speed, voltage levels, and fault currents. The voltage level of a DER integrated terminal bus is highly variable due to the intermittency imposed by the stochastic DERs [50]. The topology (grid-connected/islanded mode) of the microgrid certainly affects the magnitude and directions of the fault currents in the microgrid network. Reliable coordination of the protective relays, fuses, and re-closers in a microgrid is a complicated protection challenge. The microgrid may also suffer from other protection issues similar to those in traditional distribution systems listed in the previous section.

11.4.1.3 Communication Systems

For a microgrid with varying architectural designs and configurations, comprehensive protection studies should be performed to identify proper relay setting groups that ensure system protection against various fault scenarios. On the other hand, advanced protection strategies heavily rely on communication systems and, thus, their implementation is overshadowed by economic considerations. More importantly, in the case of communication network failure or delays, the protection reliability is significantly compromised, which calls for designing and implementing a backup protection system capable of handling such communication-related circumstances [51].

11.4.1.4 Hybrid AC/DC Microgrids

In recent years, a combination of both AC and DC microgrids—so-called hybrid microgrids—interconnected by power electronic interfaces has seen enormous attention in the electric industry. The idea behind designing and implementing hybrid microgrids is to conjoin the advantages and maximum benefits of both

AC and DC microgrids, while shortening the investment requirements on the number of power electronic converters needed for integrating DER units, energy storage systems, and loads to network AC or DC buses. The interfaced power electronic converters facilitate the reactive power flow to the AC subgrid and are responsible for importing/exporting the required active power to/from the DC subgrid. Hybrid microgrids structurally integrate two independent subgrids, one AC and the other DC (with minimized power transfer for loss reduction), each representing its own portfolio of DER units, energy storage systems, and loads. New adaptive, yet secure and dependable, protection devices and schemes should be developed in hybrid AC/DC microgrids capable of handling the protection requirements of such complex architectures [52].

11.4.2 Main Protection Schemes for RES-Integrated Microgrid Systems

Most of the current projects for microgrid protection are founded based on AC microgrids. Many protection schemes in distribution systems could be utilized, with some adjustments, in microgrid systems. Examples are distance protection, differential protection, wavelet transform-based protection, and adaptive protection. Other protection schemes commonly employed in microgrid systems are as follows.

11.4.2.1 OC and Overload Protection Schemes

Due to limited fault current contribution from weak DER units, the time inverse OC protective relay fails to maintain a coordinated operation in the microgrid. A voltage function is added to the time-inverse OC relay logic to enhance the operating time of the microgrid relays [53]. During the fault periods, the voltages at the DER terminals fall below their ratings, and there will be a small increase in the fault currents. The operating region of the voltage–current-based time-inverse protective relays is demonstrated in Fig. 11.7. The operating time of the voltage–current-based time-inverse OC relays can be enhanced if effectively designed and coordinated in microgrid systems.

Two voltage control protection schemes for OC and overload protection of microgrids with a high proliferation of DER units are suggested and fully addressed in [54]. The first protection scheme enables a sequence of (i) detecting the faults; (ii) constraining the output currents of the DER units; and (iii) restoring the microgrid back to its normal operating state following a successful fault clearance process. The second scheme, which focuses on overload protection, (i) detects the overload conditions in the microgrid using the voltage measurements, (ii) adjusts and constrains the DER output power through the interface terminals of the DER voltage source controllers. Worthy of note is that the voltage controlled protection scheme is faster than the conventional OC relay protection scheme.

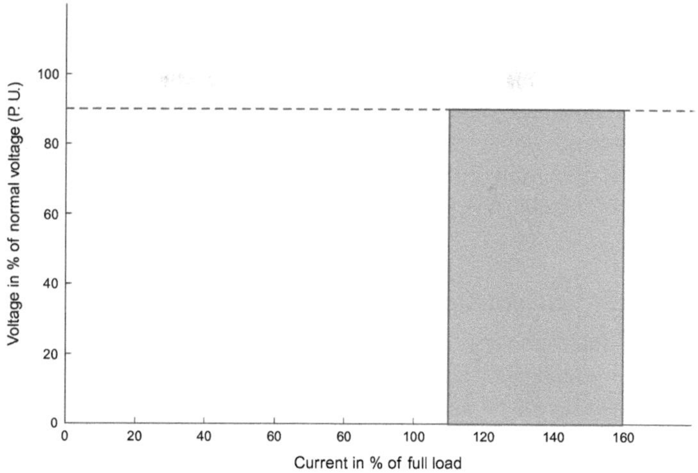

FIG. 11.7 Operating region of voltage–current based time-inverse protective relay [9].

11.4.2.2 Harmonics Content-Based Schemes

Microgrid protection is approached in [55] through a developed scheme, centered on the inverter output voltage and particularly on its harmonic contents. The voltage signals at the inverter terminals are commonly low-distorted when the microgrid is connected to the grid (i.e., the grid-connected operation mode) since, in such circumstances, the distribution network manifests itself as a low-impedance voltage source. In the face of fault incidents or other prevailing conditions in the network, however, the microgrid islanded operation will result in an increase in the impedance at the inverter terminals. This is realized as the low-impedance main grid gets disconnected, and hence, existing harmonics of the inverter output current increases the harmonic magnitudes at the terminal voltage signal. In order to tackle this challenge, a protective relay is proposed in [55] to evaluate and monitor continuously the total harmonic distortion (THD) associated with the terminal voltage signal where each inverter-based DER unit is located. Upon violating a prespecified THD value, the relay signals the corresponding local circuit breakers to trip, as the fault is found within the relay's operation zone.

Different from the previous approach, and centered solely on low harmonic components, a protection scheme is suggested in [56] where a certain proportion of the 5th harmonics to the fault current is injected, through which the digital protective relay can detect the fault. In this way, limitations of the conventional OC-based protection schemes are effectively tackled, as this approach and its performance do not depend on huge fault currents.

While offering several distinguished advantages as stated above, there are a number of critical concerns and challenges known for such harmonics-oriented protection schemes:

(i) they are not able to effectively detect high-impedance faults;
(ii) the algorithm computations and filterings in such protection schemes create quite noticeable latencies; and
(iii) some of such harmonic-based protection schemes do not account for the transformer connections in distribution systems—this is particularly critical as the fault current and voltage waveforms are significantly affected by a delta-wye transformer connection.

11.4.2.3 Fault-Current Compensation

Since the DER fault-current level changes depending on the microgrid mode of operation (grid-connected and islanded), and particularly in the case of massively distributed inverter-interfaced sources of generation, designing a secure and dependable protection scheme compatible with both modes of operation is a major challenge. An extra fault-current source (FCS) can be deemed as a solution that not only enables compatibility of the OC protection scheme with the traditional schemes but also can compensate and adjust the fault-current levels in various operation modes. In so doing, synchronous condensers or storage systems—for example, flywheels, batteries, and ultra-capacitors—can be regarded as the possible FCS options in the microgrid to inject high current levels during faults [57]. Storage-enabled FCS systems are actually systems of systems (SoS) and are typically composed of a storage element, a power electronic converter, a triggering circuit, and a charging module. The FCS will be switched on to inject the requisite levels of current needed for voltage restoration when the fault detection is confirmed, and will be switched off upon fault clearance. Integration of FCS for fault current compensation in microgrids with massive penetration of DERs calls for a significant investment, and this remains a major challenge for the wide adoption and implementation of this approach in practice.

A resilient, adaptive protection algorithm in microgrids should be able to effectively handle the communication failures and delays with optimal cost and minimum damage to hardware assets, as such failures are deemed to be high-priced risks imposing inevitable consequences. Energy storage devices are suggested and employed in [58] to realize an adaptive, resilient protection scheme in microgrids. Relying on the existing energy storage devices in the network with minimum additional cost, this framework enhances the microgrid protection scheme resilience to communication failures—for example, cybersecurity attacks, loss of or delays in communication—by injecting the required fault current to activate and trip the responsible relays. The main challenge is when there concurrently exist both a heavy-pulsed load and a fault in the network. Under such worst-case scenarios, an existing supercapacitor in the designed system could activate the relays by successfully increasing the fault currents.

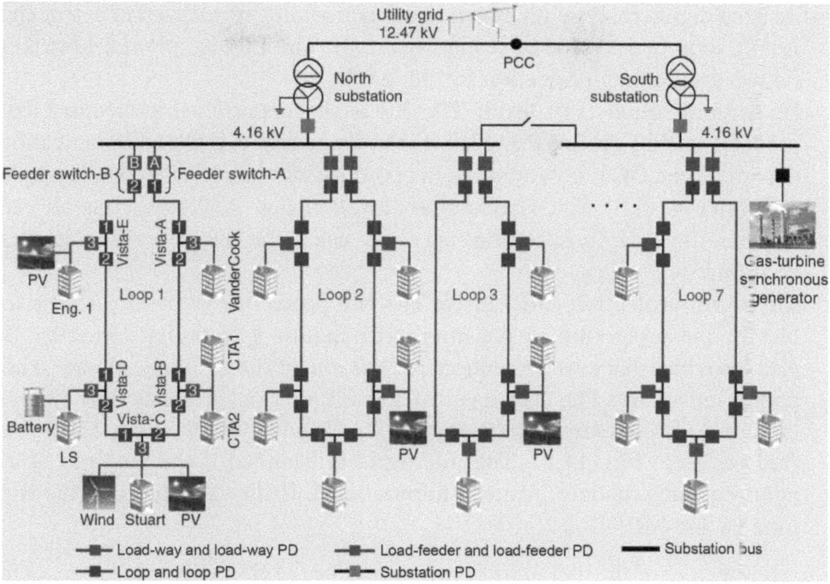

FIG. 11.8 The IIT microgrid protection system [59].

11.4.2.4 Hierarchical Protection

The Illinois Institute of Technology (IIT) microgrid, with a peak load of 12 MW, embeds a variety of DERs, including 8 MW natural gas turbine, 300 kW PV system, 8 kW wind generation, 500 kWh flow battery, and 4 MW backup generation. Fig. 11.8 depicts a schematic diagram of the IIT microgrid, including the location and type of the DERs. As illustrated, the microgrid is fed through two substations (i.e., north substation and south substation) to ensure seamless operation of the system if one of the feeders fails. Both substations are supplied by 12.47 kV/4.16 kV transformers equipped with proper protective devices. Fig. 11.8 also indicates the hierarchical protection scheme applied and implemented in the IIT microgrid, primarily centered on differential protection in four coordinated levels. The protection levels will be described in the following [59]:

- Load-way protection (LWP) level: The LWP (green boxes in Fig. 11.8) consists of directional OC relays that are responsible for load-level faults. If the LWP breakers fail to operate, backup trip signals will be sent to the adjacent breakers on the same switch. The under-/overvoltage and under-/overfrequency protections can also be implemented on the LWP level to enable load shedding and/or other control schemes.
- Loop protection (LP) level: Differential protection is used to clear faults impacting the line sections between two switches. Thus, communication-

assisted digital relays with directional functionality are utilized to detect and isolate loop faults (blue boxes in Fig. 11.8). The LP also provides breaker failure and backup protection for the LWP.

- Feeder protection (FP) level: The FP is the upper-level protection that provides backup for LWP and LP levels, for example, if the communication network fails. OC relays are used in coordination with LWP and LP (purple boxes in Fig. 11.8). Since fault current levels change in different operational modes, the adaptive setting may be used for FP relays to ensure coordination.
- Microgrid protection (MP) level: The MP protection is mainly responsible for the protection of the microgrid against grid faults. However, it also provides backup protection for the other three levels in the grid-connected mode. The MP level of the IIT microgrid protection system includes OC, under-/overfrequency, and under-/overvoltage functions (red boxes in Fig. 11.8). The microgrid is islanded if the network voltage/frequency deviates from its normal level. Utility code defines the trip time of the MP.

The architecture of the hierarchical protection for the IIT microgrid is shown in Fig. 11.9. Compared with the hierarchical protection which combines predictive protection, adaptive protection, and corrective protection altogether for the transmission system, the different layers of hierarchical protection in micro-grids deal with the different fault levels/types, and the upper layer is the backup protection for the lower layer against communication failure.

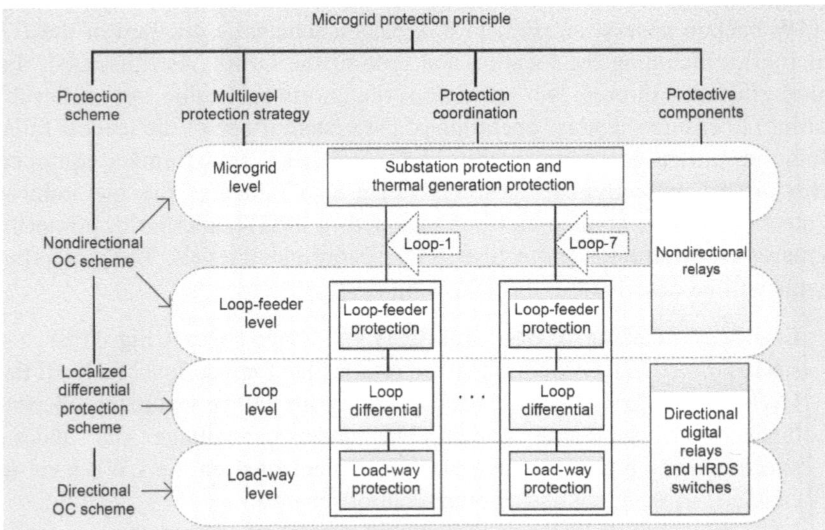

FIG. 11.9 The hierarchical protection principle of the IIT microgrid [59].

11.5 Conclusion

The deployment of renewable and DERs has been observed to be significantly on the rise in the United States and the world. As such, additional variability and uncertainty in the net-load will be realized in future, contributing to the advent of new operational challenges in bulk power grids, ranging from operation limit violations, frequent start-ups and shut-downs of dispatchable generating units, reduction in available lead time, and increased ramping and reserve requirements. Among such technical challenges, the future power system requires a modified, adaptive, robust, and further compatible protection schemes to cope with the uncertainties driven by the rushing arrival of RES.

This chapter discussed the merits and demerits of various state-of-the-art protection schemes with the integration of RESs in power systems. In particular, the protection challenges in RES-integrated bulk transmission systems, distributions systems, and microgrids were comprehensively covered. The trends and possible solutions for protection schemes of future smart power systems were discussed. The use of the communication-assisted protection schemes and adaptive protection systems was suggested, the performance of which are closely driven and dependent on reliable real-time communication facilities. Nevertheless, future research is still needed to address the protection challenges in the face of massive RES penetrations as well as cyberattacks, and to develop new adaptive techniques, transformative mindsets, and advanced technologies for secure, affordable, and dependable protection systems in the electricity grids of the future.

References

[1] K. Jaiswal, A Brief Comparison of Renewable and Non-Renewable Resources of Energy, Department of Mechanical Engineering, Tezpur University, 2013. [Online] Available at: https://www.researchgate.net/publication/254559905_A_brief_comparison_of_renewable_and_non-renewable_resources_of_energy.

[2] Renewable Energy Target Setting, The International Renewable Energy Agency, 2015. [Online] Available at: http://www.irena.org/publications/2015/Jun/Renewable-Energy-Target-Setting.

[3] The Department of Energy Environment, Clean Energy DC, Washington, DC, 2016. [Online] Available at: https://doee.dc.gov/sites/default/files/dc/sites/ddoe/publication/attachments/Clean_Energy_DC_2016_final_print_single_pages_102616_print.pdf.

[4] D. Pravitasari, S. Nisworo, New and renewable energy: a review and perspectives. in: International Conference on Sustainable Information Engineering and Technology (SIET), Malang, 2017, 2017, pp. 284–287, https://doi.org/10.1109/SIET.2017.8304149.

[5] A. Darwish, A.S. Abdel-Khalik, A. Elserougi, S. Ahmed, A. Massoud, Fault current contribution scenarios for grid-connected voltage source inverter-based distributed generation with an LCL filter, Electr. Power Syst. Res. 104 (2013) 93–103.

[6] M. Islam, H. Gabbar, Study of microgrid safety & protection strategies with control system infrastructures, Smart Grid and Renewable Energy 3 (1) (2012) 1–9.

[7] TUDELFT, Protection of Future Power Systems, Technical University of Delft, 2018. [Online] Available at: https://www.tudelft.nl/en/2018/ewi/protection-of-future-power-systems.
[8] R. Dubey, S.R. Samantaray, B.K. Panigrahi, G.V. Venkoparao, Adaptive distance relay setting for parallel transmission network connecting wind farms and UPFC, Int. J. Electr. Power Energy Syst. 65 (2015) 113–123.
[9] V. Telukunta, J. Pradhan, A. Agrawal, M. Singh, S.G. Srivani, Protection challenges under bulk penetration of renewable energy resources in power systems: a review, CSEE J. Power Energy Syst. 3 (4) (2017) 365–379.
[10] P. Dehghanian, B. Zhang, T. Dokic, M. Kezunovic, Predictive risk analytics for weather-resilient operation of electric power systems, IEEE Trans. Sustainable Energy 10 (1) (2019) 3–15.
[11] P. Dehghanian, Quantifying power system resilience improvement through network reconfiguration in cases of extreme emergencies, (M.Sc. thesis). Texas State University, 2017.
[12] P. Dehghanian, Y. Wang, G. Gurrala, E. Moreno-Centeno, M. Kezunovic, Flexible implementation of power system corrective topology control, Electr. Power Syst. Res. 128 (2015) 79–89.
[13] P. Dehghanian, M. Kezunovic, Impact assessment of power system topology control on system reliability, in: IEEE Conference on Intelligent System Applications to Power Systems (ISAP), 11–16 Sep. 2015, Porto, Portugal, 2015.
[14] P. Dehghanian, M. Kezunovic, Probabilistic impact of transmission line switching on power system operating states, in: IEEE Power and Energy Society (PES) Transmission and Distribution (T&D) Conference and Exposition, May 2–5, 2016, Dallas, USA, 2016.
[15] P. Dehghanian, S. Aslan, P. Dehghanian, Maintaining electric system safety through an enhanced network resilience, IEEE Trans. Ind. Appl. 54 (5) (2018) 4927–4937.
[16] P. Dehghanian, S. Aslan, P. Dehghanian, Quantifying power system resiliency improvement using network reconfiguration, in: IEEE 60th International Midwest Symposium on Circuits and Systems (MWSCAS), Boston, MA, 2017, pp. 1364–1367.
[17] S. Biswas, P.K. Nayak, State-of-the-art on the protection of FACTS compensated high-voltage transmission lines: a review, High Voltage 3 (1) (2018) 21–30.
[18] Z. Moravej, M. Pazoki, M. Khederzadeh, Impact of UPFC on power swing characteristic and distance relay behavior, IEEE Trans. Power Delivery 29 (1) (2014) 261–268.
[19] V. Lohmann, New strategies for substation control, protection and access to information, in: 16th International Conference and Exhibition on Electricity Distribution (CIRED), Amsterdam, Netherlands, vol. 3, 2001, p. 5.
[20] M. Singh, T. Vishnuvardhan, S.G. Srivani, Adaptive protection coordination scheme for power networks under penetration of distributed energy resources, IET Gener. Transm. Distrib. 10 (15) (2016) 3919–3929.
[21] T. Becejac, P. Dehghanian, M. Kezunovic, Impact of PMU errors on the synchrophasor-based fault location algorithms, in: 48th North American Power Symposium (NAPS), Sep. 2016, Denver, Colorado, USA, 2016.
[22] T. Becejac, P. Dehghanian, M. Kezunovic, Probabilistic assessment of PMU integrity for planning of periodic maintenance and testing, in: International Conference on Probabilistic Methods Applied to Power Systems (PMAPS), Oct. 2016, Beijing, China, 2016.
[23] T. Becejac, P. Dehghanian, M. Kezunovic, Analysis of PMU algorithm errors during fault transients and out-of-step disturbances, in: IEEE Power and Energy Society (PES) Transmission & Distribution (T&D) Conference and Exposition Latin America, Sep. 2016, Morelia, Mexico, 2016.
[24] Z.Q. Bo, X.N. Lin, Q.P. Wang, et al., Developments power system protection and control. Protection and Control of Modern Power Systems (2016) 1–8, https://doi.org/10.1186/s41601-016-0012-2. [Online] Available at:.

[25] J. Zare, F. Aminifar, M. Sanaye-Pasand, Communication-constrained regionalization of power systems for synchrophasor-based wide-area backup protection scheme, IEEE Trans. Smart Grid 6 (3) (2015) 1530–1538.

[26] P.I. Obi, G.C. Chidolue, I.I. Okonkwo, Protection and control of power systems—a review, Int. J. Adv. Res. Technol. 3 (5) (2014) 158–166.

[27] M. Kezunovic, B.M. Cuka, Hierarchical Coordinated Protection With High Penetration of Smart Grid Renewable Resources, Power Systems Engineering Research Center (PSERC) Future Grid Initiative, 2013.

[28] M. Tasdighi, M. Kezunovic, Hierarchically coordinated protection: a key element in improving power system resilience, in: IEEE International Energy Conference (ENERGYCON), Limassol, 2018, 2018, pp. 1–6.

[29] IEEE Report, Impact of distributed resources on distribution relay protection, in: Line Protection Subcommittee of the Power System Relay Committee of the IEEE Power Engineering Society, Working Group D3 Report, 2004. Aug. [Online] Available at: http://www.pes-psrc.org/Reports/wgD3ImpactDR.pdf.

[30] H.H. Zeineldin, Y.A.R. Mohamed, V. Khadkikar, V. Ravikumar Pandi, A protection coordination index for evaluating distributed generation impacts on protection for meshed distribution systems, IEEE Trans. Smart Grid 4 (3) (2013) 1523–1532.

[31] J. Keller, B. Kroposki, Understanding fault characteristics of inverter-based distributed energy resources, Technical report REL/TP-550–46698. National Renewable Energy Laboratory (NREL), Golden, CO, 2010.

[32] D.Q. Hung, N. Mithulananthan, Multiple distributed generators placement in primary distribution networkers for loss reduction, IEEE Trans. Ind. Electron. 60 (4) (2013) 1700–1708.

[33] M.A. Kashem, G. Ledwich, Impact of distributed generation on protection of single wire earth return lines, Electr. Power Syst. Res. 62 (1) (2002) 67–80.

[34] B. Hussain, S.M. Sharkh, S. Hussain, M.A. Abusara, Integration of distributed generation into the grid: protection challenges and solutions, in: 10th IET International Conference on Developments in Power System Protection (DPSP 2010), Managing the Change, Manchester, 2010, pp. 1–5.

[35] L. Kumpulainen, K. Kauhaniemi, Analysis of the impact of distributed generation on automatic reclosing, in: IEEE Power and Energy Society (PES) Power Systems Conference and Exposition (PSCE), New York, NY, vol. 1, 2004, pp. 603–608.

[36] M. Geidl, Protection of power systems with distributed generation-state of the art, PhD dissertation. Power Systems Laboratory, Swiss Federal Institute of Technology (ETH), Zurich, 2005.

[37] A. Sinclair, D. Finney, D. Martin, P. Sharma, Distance protection in distribution systems: how it assists with integrating distributed resources, IEEE Trans. Ind. Appl. 50 (3) (2014) 2186–2196.

[38] L. Huchel, H.H. Zeineldin, E.F. El-Saadany, Protection coordination index enhancement considering multiple DG locations using FCL, IEEE Trans. Power Delivery 32 (1) (2017) 344–350.

[39] H.-C. Jo, S.-K. Joo, K. Lee, Optimal placements of superconducting fault current limiters (SFCLs) for protection of electric power system with distributed generations (DGs), IEEE Trans. Supercond. 3 (3) (2013) 5600–5604.

[40] S.-H. Lim, J.-S. Kim, M.-H. Kim, J.-C. Kim, Improvement of protection coordination of protective devices through application of a SFCL in a power distribution system with a dispersed generation, IEEE Trans. Appl. Supercond. 22 (3) (2012) 5601001–5601004.

[41] S. Kar, S.R. Samantaray, M.D. Zadeh, Data-mining model based intelligent differential microgrid protection scheme, IEEE Syst. J. 11 (2) (2017) 1161–1169.

[42] Z. Yi, A.H. Etemadi, Line-to-line fault detection for photovoltaic arrays based on multiresolution signal decomposition and two-stage support vector machine, IEEE Trans. Ind. Electron. 64 (11) (2017) 8546–8556.

[43] F.B. Costa, A. Monti, S.C. Paiva, Overcurrent protection in distribution systems with distributed generation based on the real-time boundary wavelet transform, IEEE Trans. Power Delivery 32 (1) (2017) 462–473.

[44] Y. Seyedi, H. Karimi, Design of networked protection systems for smart distribution grids: a data-driven approach, in: IEEE Power & Energy Society General Meeting, Chicago, IL, 2017, 2017, pp. 1–5.

[45] V.A. Papaspiliotopoulos, G.N. Korres, V.A. Kleftakis, N.D. Hatziargyriou, Hardware-in-the-loop design and optimal setting of adaptive protection schemes for distribution systems with distributed generation, IEEE Trans. Power Delivery 32 (1) (2017) 393–400.

[46] Z. Liu, C. Su, H.K. Hoidalen, Z. Chen, A multi-agent system-based protection and control scheme for distribution system with distributed-generation integration, IEEE Trans. Power Delivery 32 (1) (2017) 536–545.

[47] F. Blaabjerg, Y. Yang, D. Yang, X. Wang, Distributed power-generation systems and protection, Proc. IEEE 105 (7) (2017) 1311–1331.

[48] Energy, Microgrid Exchange Group, DOE Microgrid Workshop Report, Aug. 2011. [Online] Available at: https://energy.gov

[49] W.K.A. Najy, H.H. Zeineldin, W.L. Woon, Optimal protection coordination for microgrids with grid-connected and islanded capability, IEEE Trans. Ind. Electron. 60 (4) (2013) 1668–1677.

[50] A.A. Memon, K. Kauhaniemi, A critical review of AC microgrid protection issues and available solutions, Electr. Power Syst. Res. 129 (2015) 23–31.

[51] J. Shiles, et al., Microgrid protection: an overview of protection strategies in North American microgrid projects, in: IEEE Power & Energy Society General Meeting, Chicago, IL, 2017, pp. 1–5.

[52] S. Mirsaeidi, X. Dong, S. Shi, D. Tzelepis, Challenges, advances and future directions in protection of hybrid AC/DC microgrids, IET Renew. Power Gener. 11 (12) (2017) 1495–1502.

[53] A. Agrawal, M. Singh, M.V. Tejeswini, Voltage current based time inverse relay coordination for PV feed distribution systems, in: National Power Systems Conference (NPSC), Bhubaneswar, Dec. 19–21, 2016, 2016, pp. 1–6.

[54] A.H. Etemadi, R. Iravani, Overcurrent and overload protection of directly voltage-controlled distributed resources in a microgrid, IEEE Trans. Ind. Electron. 60 (12) (2013) 5629–5638.

[55] H. Al-Nasseri, M.A. Redfern, Harmonics content based protection scheme for micro-grids dominated by solid state converters, in: 12th International Middle-East Power System Conference, Aswan, Egypt, 2008, pp. 50–56.

[56] Z. Chen, X. Pei, L. Peng, Harmonic components based protection strategy for inverter-interfaced AC microgrids, in: IEEE Energy Conversion Congress and Exposition (ECCE), Milwaukee, WI, 2016, pp. 1–6.

[57] B. Kroposki, K. Johanson, Y. Zhang, V. Gevorgian, P. Denholm, B.-M. Hodge, B. Hannegan, Achieving a 100% renewable grid: operating electric power systems with extremely high levels of variable renewable energy, IEEE Power Energ. Mag. 15 (2) (2017) 61–73.

[58] H.F. Habib, C.R. Lashway, O.A. Mohammed, A review of communication failure impacts on adaptive microgrid protection schemes and the use of energy storage as a contingency, IEEE Trans. Ind. Appl. 54 (2) (2018) 1194–1207.

[59] L. Che, M.E. Khodayar, M. Shahidehpour, Adaptive protection system for microgrids: protection practices of a functional microgrid system, IEEE Electrific. Mag. 2 (1) (2014) 66–80.

Chapter 12

ICT Requirements and Recent Developments

Nils Dorsch, Stefan Böcker and Christian Wietfeld
TU Dortmund University, Dortmund, Germany

Chapter Outline

12.1 Introduction

Today's power systems are challenged by the transition from conventional towards renewable energy generation [1]. In addition, since the liberalization of energy markets, increasing amounts of energy are traded and shifted. Thus, in future, not only power systems' loads, but also feed-in, will be extremely volatile. Moreover, decentralized generation leads to bidirectionality of power flows. In combination, these developments cause the power system to be operated close to its limits. Established (N-1) criteria cannot be fulfilled at all

Pathways to a Smarter Power System. https://doi.org/10.1016/B978-0-08-102592-5.00012-0

times anymore. Meanwhile, new concepts and devices, such as demand side management and charging management of electric vehicles, are introduced.

Altogether, these developments require intense monitoring and control of the power system on all levels in order to maintain grid stability. In turn, monitoring and control rely on adequate Information and Communication Technology (ICT) infrastructures, fulfilling latency, reliability, and availability requirements; all of this at the lowest costs possible [2, 3]. Consequently, this chapter is dedicated to analyzing Smart Grid communications requirements and deriving corresponding solution approaches. Fig. 12.1 provides an overview of the different power levels, corresponding applications, and ICT infrastructures. It is complemented by an illustration of the most significant communication demands.

The remainder of this chapter is structured as follows. First, a more in-depth introduction to the different requirements towards communications in future energy systems is provided, including a matching with the promises of future 5G mobile network infrastructures (Section 12.2). Next, to address the mentioned challenges, solution approaches based on 5G are presented and discussed (Section 12.3). This analysis is supplemented by various case studies, giving examples of possible technical realizations of such concepts (Section 12.4). Finally, a summary, a conclusion, and an outlook on further research issues are provided in Section 12.5.

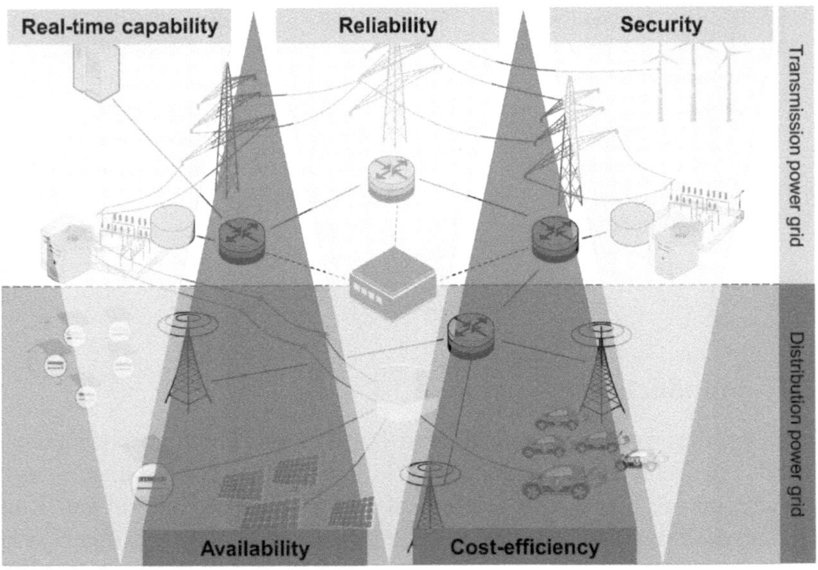

FIG. 12.1 Overview of power system and corresponding ICT levels, highlighting major research challenges, where the width of the *triangle* indicates the significance of respective requirements.

12.2 Smart Grid Communications Requirements

In general, Smart Grid communication requirements vary among the different voltage levels of power systems. Hence, requirements can be classified into those of the distribution power grid on the one hand and those of the transmission power grid on the other.

12.2.1 Requirements in the Context of 5G Mobile Networks

Currently, there are ongoing developments aiming to evolve existing Long Term Evolution (LTE) technology towards a new standard of 5th generation mobile networks. Examples of these efforts are given in terms of whitepapers [4–6] and standardization [7–9]. 5G aims to improve the performance of mobile networks with regard to nearly all relevant criteria, for example, data rate, mobility, and spectral efficiency. An overview of these enhancements is given in Fig. 12.2. At the same time, 5G also involves major changes to the underlying core network/transport infrastructure. In contrast to previous standards, 5G embraces all kinds of applications, including, but not being limited to, M2M (Machine-to-Machine) communications. Hence, it can be considered a major enabler for future Smart Grid communications. The main goals of 5G and corresponding use cases are illustrated in Fig. 12.3. Respecting the different requirements of smart distribution and transmission grid, these applications are added to the overview in Figs. 12.2 and 12.3. On the basis of this matching, 5G-driven solution approaches can be derived, which will be detailed in the following section.

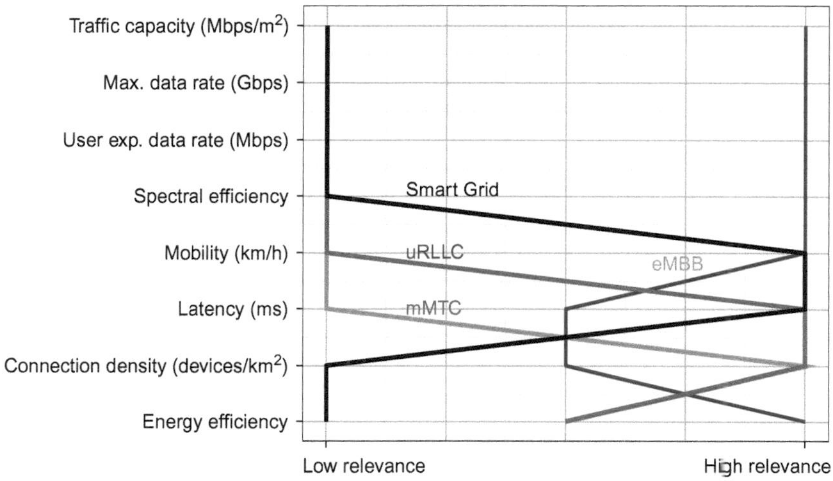

FIG. 12.2 Requirements of intelligent energy systems in the context of 5G infrastructures. *((Deduced from ITU-R International Telecommunication Union, IMT Vision—Framework and overall objectives of the future development of IMT for 2020 and beyond, 2015. https://www.itu.int/dms_pubrec/itu-r/rec/m/R-REC-M.2083-0-201509-I!!PDF-E.pdf.))*

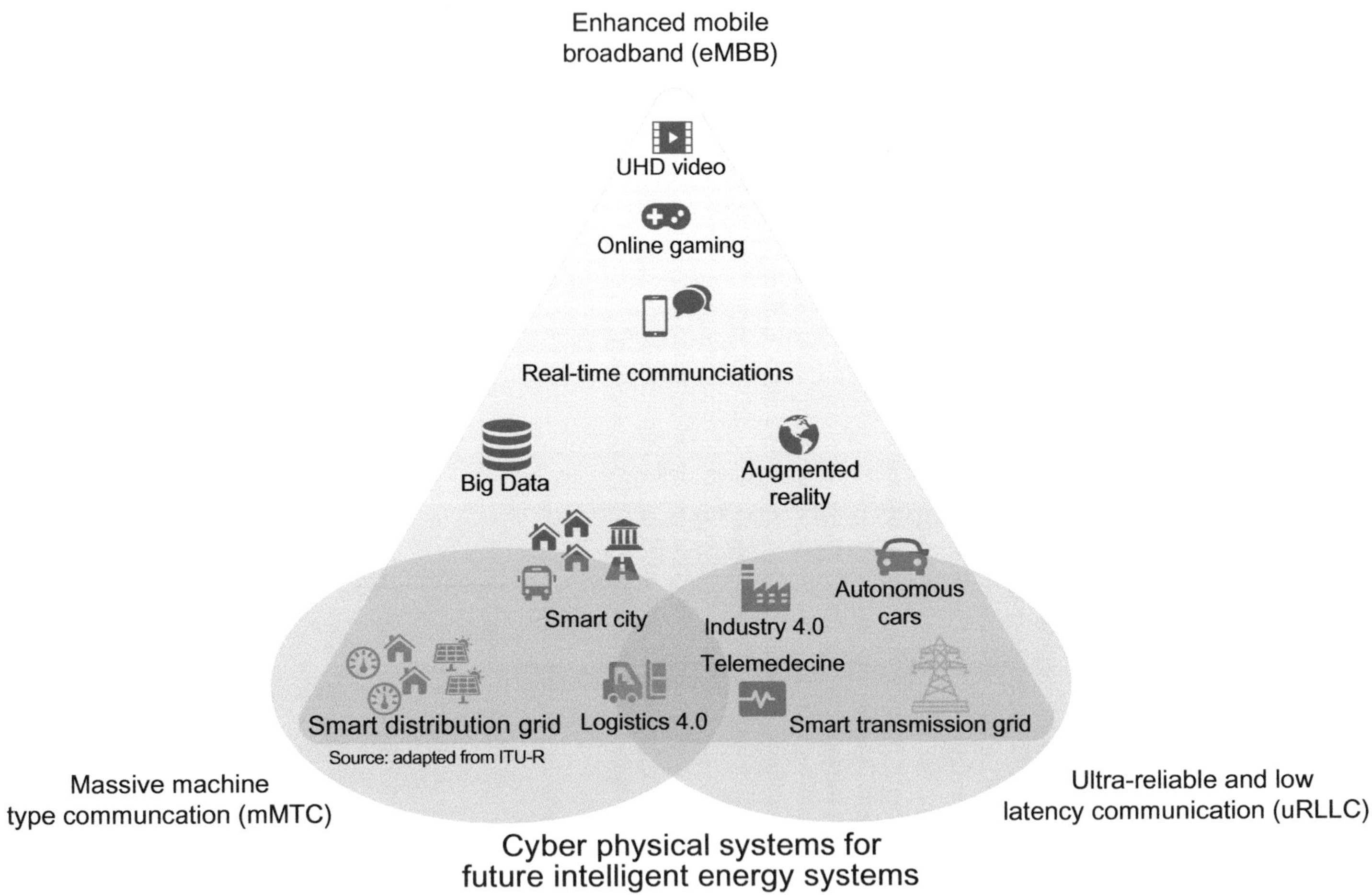

FIG. 12.3 Mapping of 5G communication requirements to applications. *(Deduced from ITU-R International Telecommunication Union, IMT Vision—Framework and overall objectives of the future development of IMT for 2020 and beyond, 2015. https://www.itu.int/dms_pubrec/itu-r/rec/m/R-REC-M.2083-0-201509-I!!PDF-E.pdf.)*

12.2.2 Distribution Power Grid Requirements

An increased level of volatile renewable distributed energy resources, such as photovoltaic and shiftable loads (e.g., electric vehicles) creates the need for continuous monitoring and active power coordination to ensure the short- and long-term stability of future distribution grids. These Smart Grid functionalities present lots of new challenges, but at the same time facilitate opportunities for Smart Market ancillary services.

A progressive smart meter rollout and an increased installation of intelligent substations provide means for distribution grid-wide monitoring, as well as energy flow control management [10]. All of these applications are characterized by a very high device density with limited transmission requests and data volumes per device, resulting in a high comprehensive capacity requirement for all devices covered by a communication network cell. The requirement profile for latency and data rate is strongly dependent on the underlying applications and varies from seconds to minutes for latency and from bps to kbps for data rates [11]. In contrast, the ITU-R has defined an intended density of 1,000,000 devices per square kilometer [12]—probably the most challenging demand for future applications in the distribution network. An associated essential requirement is the challenging spatial availability even in difficult communication environments (e.g., smart meters in basements).

In the case of mobile radio-based technologies, it is desired to establish a connection directly to metering locations without any structural measures. However, in the context of achieving basement coverage, building characteristics need to be considered. The building attenuation is frequency-dependent and varies greatly due to different designs and materials. Within the ITU-R, a model for the prediction of building entry loss (BEL) is defined based on data from extensive measurement campaigns carried out for different frequency ranges and building variants [13]. This model is valid for frequency ranges from 80 MHz to 10 GHz and recommended for comparative studies [14]. However, this model does not take specific locations of smart meters into account. For this reason, results of measurement campaigns [15] carried out at specific locations of smart meters and extensive simulations [16] propose BEL values with specific reference to metering applications.

Consequently, this chapter illustrates technology solutions to meet these requirements with focus on future 5G IoT (Internet of Things) technologies.

12.2.3 Transmission Power Grid Requirements

On the transmission grid level, power is transmitted over long distances, up to several hundreds of kilometers. From an operational point-of-view, the main challenges arise from maintaining grid stability in terms of voltage and frequency levels. As the transition towards renewable energy resources involves increasingly fluctuating feed-in, grids need to be operated close to their limits, complicating the realization of this goal. Hence, protection and control measures need to be applied more frequently. In addition, tight monitoring of the system is required to obtain a full view of the grid's current status. Failures in receiving corresponding messages or delayed transmission may have severe consequences, such as cascading outages or even complete blackouts. Subsequently, communication demands on the transmission grid level are focused on high reliability and real-time capability, with latency requirements as challenging as 5 ms for local protection [17]. An important basis for transmission grid communications is provided by the standard IEC 61850. Its high relevance is associated with its broad applicability and comprehensiveness.

12.2.3.1 IEC 61850 for Power System Communications

IEC 61850 was originally specified by the International Electrotechnical Committee (IEC) for substation automation purposes (in 1995) [18]. Since then it has been extended significantly to cover diverse use cases of the Smart Grid [19–21]. The main contribution of IEC 61850 lies in its comprehensive data model, which allows for a holistic description of energy system functionalities and devices. In addition, it includes abstract specifications of communication protocols, as well as proposals for their implementation. The following section details this important aspect of communication services.

Communication Services

IEC 61850 comprises several communication services, designed for different use cases. In the following, it is further elaborated on the most relevant of these.

The sampled value (SV) service utilizes lightweight messages for transmitting measurement data [22]. It applies directly to layer 2 of the ISO/OSI reference stack (Ethernet), avoiding additional, unnecessary packet overhead. It was originally considered for intra-substation communication only. However, subsequent extensions enabled its use on wide-area networks, employing the term routable SV [20]. SV messages are transmitted regularly, in intervals as small as 74 μs (256 samples per power cycle).

Generic object oriented substation events (GOOSE) describe lightweight packets, applied for transmitting status information and switching commands. Their structure is similar to SV, yet message transmission is only partly periodic. GOOSE make use of a heartbeat mechanism (typical interval: 1 s), which provides regular updates on devices' states. Yet, it also includes event-driven communication. In this case, a message is issued in response to a random event and repeated eight times in increasing intervals until the heartbeat frequency is retained. In this way, transmission reliability is ensured, protocol-wise.

Manufacturing message specification (MMS) offers a client/server-based communication service, using the standard TCP/IP protocol stack. It is employed for multiple purposes, such as configuration, status updates, and measurement reports. Message types include requests, responses, and reports.

Besides these main services, SNTP is included by IEC 61850 for synchronization purposes.

Latency Requirements

Additionally, IEC 61850 includes information on the latency requirements of different Smart Grid applications [17]. This is summarized in Fig. 12.4 and grouped according to different levels of the communication infrastructure. It can be obtained that local protection functions involve the most challenging latency demands, whereas longer transmission times are tolerated for wide-area applications. In contrast to end consumer services—for example, for voice communication—delay requirements are not only higher, but also stricter. This can be attributed to the fact that every single latency violation may endanger grid stability, no matter how improbable the event. In the case of end consumer applications, exceeding defined delay bounds is deemed acceptable for some rare events.

12.3 5G Based Solution Approaches for Smart Grids

This section picks up some important technological developments in the context of 5G, which may provide relevant benefits for Smart Grid communications.

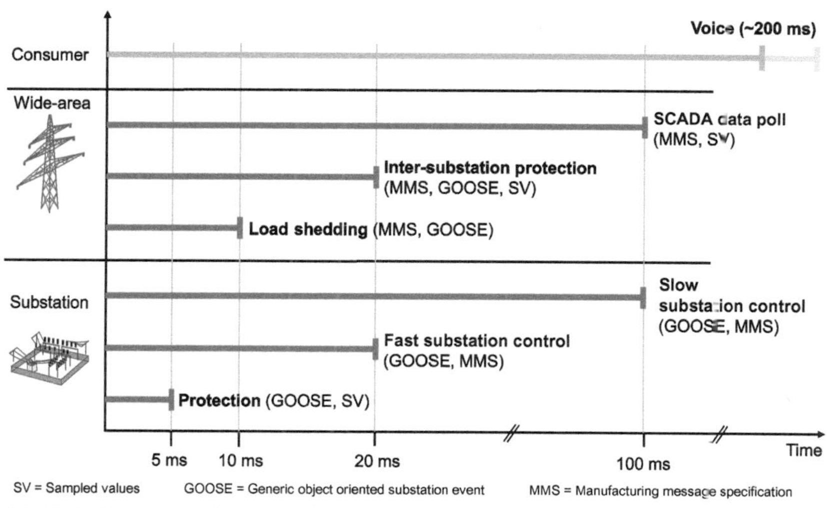

FIG. 12.4 Latency requirements of typical Smart Grid functions based on IEC 61850 [17].

12.3.1 IoT-Technologies for Enhanced Coverage in Distribution Grids

In order to cope with all arising challenges and capabilities of distribution grids, the implementation of holistic Information and Communication Technologies (ICT) is an essential requirement for monitoring and control in the distribution grid domain [23]. Underlying heterogeneous ICT infrastructures offer a great diversity of possible technology and networking options, as well as many approaches for planning and operation of Smart Grids. Currently two major groups of technologies, providing functionalities to solve defined 5G mMTC (massive Machine Type Communication) requirements, are discussed [24]. On the one hand, low power wide area networks (LPWAN) in unlicensed frequency bands enable simple, cost-effective network operation independent of licensed operators. On the other hand, cellular IoT technologies, standardized by 3GPP in licensed frequency bands, can provide QoS (quiality of service) guarantees and do not require any expertise from the user's point of view. In the following, a brief overview is given on current mobile radio technologies, for both unlicensed as well as licensed communication technologies, which contribute to the fulfillment of 5G requirements for mMTC.

12.3.1.1 IoT-Technologies in Unlicensed Frequencies Bands

The development of 5G mobile networks in licensed frequency bands is intended to provide connection density of 1,000,000 devices per square kilometer [12] in the area of massive Machine Type Communication (mMTC). However, it faces a growing number of technologies in unlicensed frequency bands. Such technologies are limited by decentralized congestion control mechanisms in order to cope with the challenge of an unpredictable cochannel interference. In the following, relevant IoT technologies grouped under the term Low Power Wide Area Networks are briefly discussed. LoRaWAN [25] is a specification of the Low Power WAN (LPWAN) network protocol family defined by the LoRa Alliance. The specification is freely available and based on the LoRa modulation specified by Semtech. LoRaWAN is mainly operated in the short-range device (SRD) band, available in Europe at around 868 MHz and at around 915 MHz in the US. Depending on the environment, long ranges of up to 11 km and good basement penetration can be achieved. Typical system data rates vary, depending on the system parameters, from 0.25 to 11 kbit/s, without consideration of cochannel interference, resulting in a transmission delay of more than one 1 s [26, 27]. Taking into account the regulatory duty cycle limitation of 1% (LoRaWAN uplink), the peak data rates are reduced to a significantly low average throughput of 1.5 to 97 bit/s, due to multiple inactivity time frames (time off) in relation to the transmission time per packet (time on air) [28]. For data transmissions with large data volumes (e.g., firmware upgrade/update), this might involve an intolerable overall delay. Nevertheless, LoRaWAN is attractive for distribution grid applications, due to a very high communication range of multiple kilometers, enabling high coverage even with a small number of cells. With a limited but still suitable scalability of multiple thousands of devices per network cell [29], LoRaWAN provides high potential, especially for non-time-critical sensor applications.

The IEEE 802.11ah standard [30] proposes a WLAN version specially designed for IoT requirements. The lower operating frequency in the 868 MHz SRD band supports significantly better propagation characteristics compared to the 2.4 and 5 GHz ISM band, while maintaining sufficiently higher typical system data rates. However, due to the implementation in the unlicensed, globally shared SRD band, the interference situation cannot be controlled. Studies have shown that a transmission delay of up to 1.5 s is to be expected in high load situations (>2000 participants) [31]. IEEE802.11ah takes advantage of the more sophisticated channel access approach CSMA/CA, which leads to a higher number of users being able to achieve a good data rate and low latency. However, IEEE 802.11ah only achieves ranges of less than one 1 km and is therefore applicable for campus networks in particular.

12.3.1.2 IoT-Technologies in Licensed Frequencies Bands

On the way to future 5G mobile radio networks, the licensed LTE extensions (Rel. 13) enhanced Machine-Type Communications (eMTC), and Narrowband-IoT (NB-IoT) offer possible solutions to cover distribution grid specific IoT requirements. A major advantage of these cellular IoT technology solutions is the fast and straightforward expandability of existing LTE infrastructures, so that area-wide availability can be implemented with little effort. For both technologies, a significant increase in robustness can be achieved, based on the optimized spectral power density enabled by redundant data packet transmissions. A gain of up to 23 dB (for an NB-IoT Maximum Coupling Loss (MCL) of 164 dB, see Fig. 12.5) results in a significantly increased range or an improved potential for penetration of difficult communication environments (e.g., smart meter with basement location), compared to typical LTE infrastructures for the end consumer market (MCL: 140, 7 dB).

In addition, the LTE extensions NB-IoT and eMTC introduce new energy saving mechanisms. After data transmission, LTE terminals switch to DRX (discontinuous reception) mode, only listening to the downlink channel for paging messages at short, defined intervals. With eDRX (extended discontinuous reception), NB-IoT and eMTC enable significantly longer time intervals between paging scans, so that the end devices can remain in standby mode for an extended period of time. After the eDRX timer expires, devices can switch to power saving mode by disabling the LTE modem and associated peripherals, further reducing power consumption without losing network registration. It can be shown that the battery lifetime for eMTC and NB-IoT devices is increased significantly and reaches up to 10 and more years, depending on its activity levels [32].

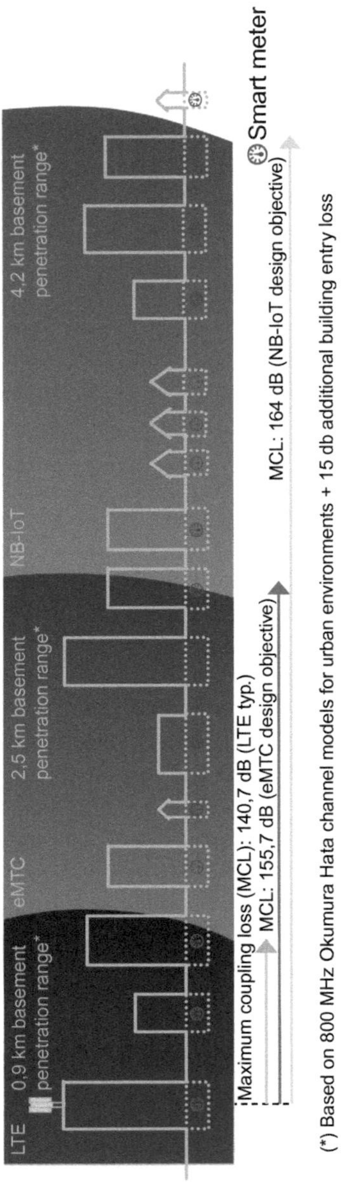

FIG. 12.5 Improved range and building penetration of LTE extensions eMTC and NB-IoT (LTE Rel. 13).

The development of the mobile radio standards NB-IoT and eMTC, developed for the IoT, will be continued in future issues of 3GPP Releases. In addition to improvements in data rate for the uplink and downlink, a multicast option will be specified that allows data to be transmitted simultaneously to multiple devices, further increasing spectral efficiency. Concluding, dependent on the application-driven requirements, NB-IoT and eMTC are suitable technology solutions for the distribution grid with limited performance demands [11, 32].

12.3.2 Software-Defined Networking for Enabling Hard Service Guarantees

Software-Defined Networking (SDN) is considered one of the major enablers of future 5G core infrastructures [33, 34], providing the basis for network slicing [35]. Its flexibility in network configuration is regarded as a major benefit for future energy systems as well.

12.3.2.1 Main Concepts of Software-Defined Networking

SDN was first proposed at Stanford University in 2008 [36]. It is based on the idea of separating data and control plane functionalities. By concentrating network intelligence at a central controller instance, a global network view is obtained. Moreover, the controller is designed as a programmable unit, enabling flexibility and dynamic adaption of network configurations. As shown in Fig. 12.6, SDN provides several interfaces for interconnecting the different

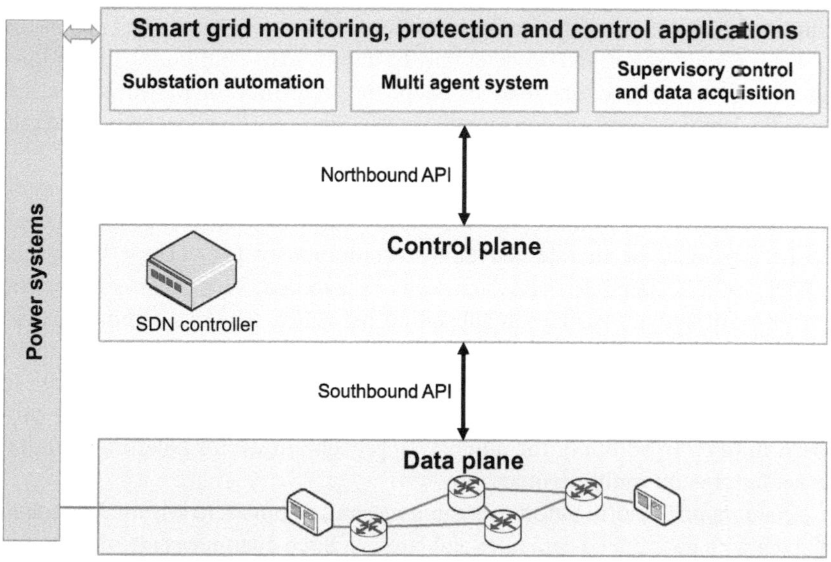

FIG. 12.6 Three-layered SDN architecture, extended for serving specific requirements of Smart Grid communications [51].

planes of the architecture. The southbound interface is utilized by the controller to manage and configure the switches on the data plane. Hence, switches are reduced to their forwarding functionality, whereas higher functionalities are relayed to the controller. The most prominent protocol for this interface is OpenFlow (OF) [37]. In the opposite direction—between control and application plane—the northbound interface is situated. This interface is particularly relevant for integrating the concept of SDN with Smart Grid communications, as it allows applications to inform the controller of their specific requirements, such as required data rate, latency, or priority. Consequently, network configurations can be adapted according to this information. Meanwhile, applications do not require any knowledge of the underlying communication infrastructure. The use of redundant SDN controllers and their coordination is handled via the eastbound interface, whereas the westbound interface enables interaction with legacy networks.

12.3.2.2 Application of Software-Defined Networking for Smart Grid Communications

The flexibility and programmability introduced by the SDN approach can be leveraged to establish hard service guarantees for transmission system communications. In this way, QoS levels comparable to multiprotocol label switching (MPLS) based systems can be achieved, while administration overhead is reduced [38].

In particular, SDN can be applied to enable automated substation configuration and management. In [39] such an approach was proposed and verified, using an actual intelligent electronic device.

Network utilization and latencies may be improved by configuring OpenFlow-based rules at switches in order to adapt the frequency of transmissions, for example, from phasor measurement units [40]. Similarly, load balancing and rate limiting for QoS provisions in substation networks were analyzed in [41].

Fast recovery methods can be established, enabling the handling of network failures within a few milliseconds [42–45]. Meanwhile, optimal routes through the network can be maintained almost continuously [42]. Thus, hazardous impacts on associated power systems can be avoided. To further enhance the reliability of active distribution substation networks, wireless communication links may be utilized as back-up paths by SDN-enabled recovery mechanisms [46]. As multicast is of utter importance for distributing measurement data to multiple sources within the power grid, concepts for reliable multicast were proposed in [47]. In addition, the authors' approach allows for handling simultaneous failures on multiple links.

Queuing and prioritization mechanisms can be applied to ensure data transmission with guaranteed data rates and latency. Such guarantees are of foremost concern for time-critical services like IEC 61850 GOOSE. Exploiting the interconnection between SDN controller and Smart Grid applications via the northbound interface, corresponding network configurations can be adapted dynamically. This may be extremely relevant to react to changing communication requirements of applications, for example, under overload situations in the power system [48]. In such cases, the SDN controller—triggered via the northbound interface—may temporarily reallocate resources to match altered communication demands.

With regard to fulfilling latency requirements, close monitoring of the communication system is of utmost importance. For this purpose, the global network view of the SDN controller is regarded as a major enabler. This reactive supervision can even be surpassed by integrating analytical modeling techniques such as network calculus (NC) [49, 50] into the controller. By applying such concepts for online delay supervision, potential latency violations can be identified ahead of time and dealt with accordingly [51].

Furthermore, SDN is regarded as an important enabler for enhancing the security of Smart Grid communications. This includes the integration of network-based intrusion detection into next generation Supervisory Control and Data Acquisition (SCADA) systems, gathering and evaluating statistics via the SDN controller [52, 53]. In addition, controller-based communication network verification is proposed in [53] in order to achieve self-healing network management in microgrid infrastructures. The authors of [54] aim at mitigating link insecurities with the help of SDN-driven, double constrained QoS-aware routing. For this purpose, link vulnerabilities, for example, potential man-in-the-middle-attacks, are considered as a further constraint of the routing problem. Thus, QoS guarantees for Smart Grid communications can be ensured, while the risk of cyber-attacks is minimized.

12.3.3 Network Slicing for Isolated Network Resources on Shared Infrastructures

Building upon SDN and network function virtualization [55–57], the concept of network slicing introduces full network virtualization, in the sense of isolated resources, which can be allocated dynamically. Network slicing is considered a fundamental part of 5G architectures, both in core and access networks [4, 5]. Using this concept, application-specific end-to-end service guarantees can be obtained over common, shared infrastructures.

12.3.3.1 Main Concepts of Network Slicing

Besides SDN, which is introduced in the previous section, NFV is regarded as a major basis of network slicing [33–35, 55–57]. Virtualizing network functions provides an alternative to the utilization of dedicated networking hardware. In traditional infrastructures, where specific hardware is deployed, the introduction of new functions is a lengthy and complicated process. In contrast, virtualization allows for dynamic deployment and replacement of network services.

Due to the separation of virtual network functions from the underlying physical devices, they can be run flexibly and cost efficiently on standard server hardware.

Network slicing proposes the arrangement of the infrastructure into virtual networks, so-called slices, each tailored to the requirements of a specific application. For this purpose, network resources are allocated dynamically, including, for example, link transmission capacities, queues, ports, or radio spectrum. Such slices can be rented by third parties and used in the same manner as physical networks. The network operator ensures that there is no interference between different slices, enabling hard service guarantees for each slice. Thus, critical infrastructure communications can be transmitted over the same physical network as end consumer traffic, even in situations of high load, without the risk of violating service requirements.

Creation and administration of network slices is performed using a centralized architecture, referred to as management and orchestration (MANO). The so-called MANO controller enables the network operator to dynamically configure slices and assign resources. In addition, for each slice, VNFs can be deployed automatically and combined in complex topologies. This process is described as network service chaining.

Fig. 12.7 illustrates the main ideas of network slicing on three different example applications.

A popular implementation of an SDN-based Network Slicing approach is called FlowVisor [58]; it acts as proxy between OF-enabled switches and

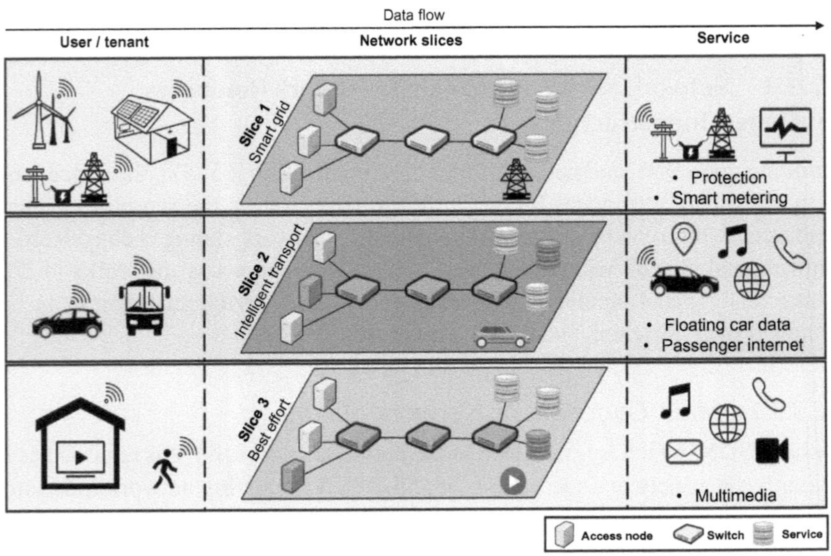

FIG. 12.7 Mapping between application-specific, virtual network slices and underlying physical infrastructure [60, 61].

multiple SDN controllers. Several studies base their work on this concept and provide extensions, such as general QoS schemes using flow redirection [59]. However, the original FlowVisor approach does not include resource allocations and service provisions, but is rather focused on providing transparent network access.

12.3.3.2 Application of Network Slicing for Smart Grid Communications

Network Slicing can be regarded as key enabler for utilizing public communication infrastructures to transmit critical Smart Grid traffic. In this way, hard service provisions are upheld even in the presence of interfering end consumer broadband traffic during peak hours [60, 61]. Such service guarantees may refer to minimum data rates as well as maximum latencies or packet loss. Shared use of infrastructures allows for significant cost savings as the set-up and operation of multiple, individual systems is avoided [62]. At the same time, the power grid operator maintains full control over its respective slice and can configure it internally according to its specific needs. For example, an own SDN controller could be deployed for this purpose. Thus, critical switching commands, that is, IEC 61850 GOOSE messages, may be given higher priority—slice-internally—compared to grid configuration messages. In [63], SDN is applied to create and adapt virtual network slices, meeting Smart Grid communication requirements in a secure, dynamic, and cost-efficient manner. The feasibility of the concept is tested using SDN-enabled Ethernet switches. In comparison to a standard solution using IEEE 802.1Q, significant advantages of the SDN-enabled approach are highlighted, in particular with regard to flexibility. In addition, FlowVisor-based Network Slicing can be deployed to ensure confidentiality, authentication, and authorization of Smart Grid communications [64].

12.3.4 Edge Computing for Real-Time Capability

Edge computing is considered a central architectural feature of 5G infrastructures [4, 5]. It describes decentralized processing of information at the edge of the network, which is intended to reduce communication load and latencies. In this way, edge computing perfectly fits real-time requirements of Smart Grid applications.

12.3.4.1 Main Concepts of Edge Computing

Edge computing can be seen as contrary to cloud computing, which transfers all data processing to high performance servers in the cloud [65]. It is focused on the idea of hosting as many services as possible at the network edge in order to avoid data transmissions deep into the core network. Thus, latencies can be minimized, as shorter transmission distances incur lower propagation delays. In addition, the number of traversed nodes is reduced, further decreasing delays. Moreover, edge computing is of major interest from a network operation point-of-view. Reduced communications over long distances and into the core network lead to decreasing overall load. Meanwhile, edge computing avoids the use of dedicated hardware at the location of data generation by running virtualized functions on standard server hardware at the edge of the network.

The concept of edge computing is often utilized in the context of content distribution and caching, for example, for orchestrating webpage platforms [66].

12.3.4.2 Application of Edge Computing for Smart Grid Communications

In the context of Smart Grids, edge computing is of particular interest for time critical applications of the transmission power grid. For example, the processing of measurement data close to its source enables ultra-low latencies when merging protection functions at edge cloud servers [67]. The approach also fits quite well with decentralized control approaches, using, for example, distributed multiagent systems (MAS). In [68], edge computing is proposed for data processing in intelligent substations, as well as for online monitoring of transmission line status. Edge servers can be utilized for applying deep learning algorithms to monitoring data, enabling advanced real-time control of Smart Grids [69]. This approach was validated with the help of real-world experiments and simulations, indicating reduced delays in detecting threats. A latency-aware scheduling and resource provisioning algorithm enables time-critical Smart Grid control tasks to be executed within their given latency requirements, while sharing the same edge computing platform with other applications [70]. Using a hierarchical cloud-fog model, cloud and edge cloud computing resources can be provided to different services for resource management in Smart Grid [71].

12.4 Case Studies

In the following section of this chapter, case studies are presented, providing a more detailed view on specific aspects of Smart Grid communications.

12.4.1 Software-Defined Networking for Smart Grid Communications

As pointed out previously, SDN can be applied to achieve reliability in the sense of fast recovery from link failures. In addition, it can be utilized to provide an interface to Smart Grid applications, allowing them to specify their communication requirements. These requirements can be mapped to corresponding network configurations by the SDN controller, enabling hard service guarantees based on prioritization and queuing. The global view of the SDN controller further facilitates the integration of analytical techniques, such as NC for supervising delay bound compliance. This case study is divided according to the three topics listed above: fast failure recovery, dynamic network configuration, and NC-based delay supervision.

12.4.1.1 Fast Recovery From Link Failures

Resilience against failures of the underlying communication infrastructure is of major importance for future Smart Grids, in particular considering time-critical applications of the transmission power grid. SDN allows for straightforward integration and control of decentralized link failure detection mechanisms such as bidirectional forwarding detection (BFD) [45]. BFD establishes periodic message exchange between neighboring switches, applying a defined time-interval to identify failed links [72]. A predetermined back-up path is stored at the switches, relying on OpenFlow Fast Failover Groups [37]. Thus, nearly instantaneous recovery (<5 ms) to an arbitrary back-up path is achieved. SDN-enabled recovery shows significantly better performance than standard approaches, as illustrated in Fig. 12.8. The figure depicts end-to-end recovery delays for different fast failover approaches. Recovery delays are found to be in the same range as those of combining MPLS with BFD. In order to restore optimal network configuration as fast as possible, the SDN controller ought to be notified of the failure as well. Thereby, traffic can be redirected to new optimal paths, considering the current network state, which allows upholding service guarantees even in the face of network failures [42]. Besides reduced administration efforts, such concepts are regarded a major benefit of SDN-based compared to MPLS-enabled fast recovery [38].

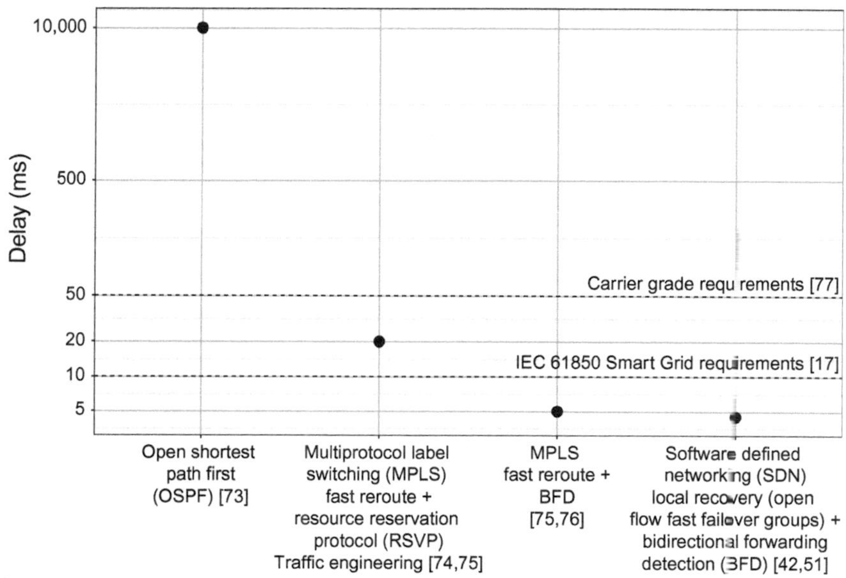

FIG. 12.8 Comparison of SDN-enabled and traditional approaches regarding recovery delay [42, 51].

12.4.1.2 Dynamic Network Configuration

The implementation of the SDN northbound interface enables Smart Grid applications to inform the SDN controller of their specific communication requirements. This approach has been evaluated in [48, 51], using an experimental setup, which includes a total of nine SDN-enabled virtualized and hardware switches. These were utilized to create several evaluation scenarios based on common reference energy systems (Nordic 32 reference system [78]). Fig. 12.9 shows the example of a multiagent system for distributed power grid control, interacting with the controller. The figure is structured in two parts. The lower part displays end-to-end latencies of MAS messages, whereas the upper part indicates corresponding data rate requirements. On the x-axis, different columns indicate different situations and configurations of the communication network. As a first step, the data rate requirement and corresponding latencies of MAS traffic are determined on an empty network, visualized by the leftmost column in Fig. 12.9. Confronted with overlapping IEC 61850 traffic—under full load conditions—MAS traffic would be delayed heavily (second column). Such situations can be avoided if the MAS communicates its data rate and latency requirements to the controller beforehand; thus, the controller is able to allocate sufficient resources by assigning the traffic to a matching queue (third column). Here, queue capacity and link utilization by IEC 61850 traffic is added to the upper part of Fig. 12.9. Moreover, the northbound interface allows dynamic adaptation of network configurations according to changing requirements, for example, in emergency situations. Such reconfiguration is

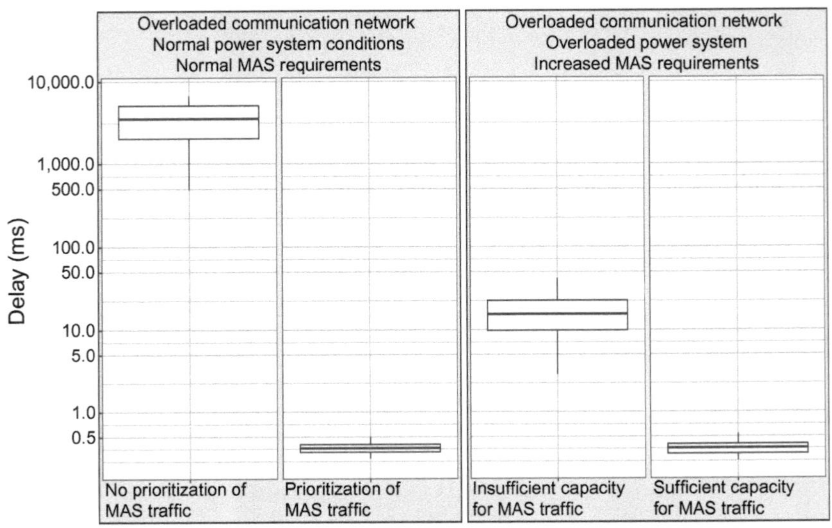

FIG. 12.9 Successive steps of handling multiagent system traffic in response to changing network conditions, employing northbound interface requests for dynamic priority/queue assignment [48, 51].

displayed by the transition between the forth and the last column in Fig. 12.9. By switching traffic to higher priority queues, potential latency violations, caused by overlapping cross traffic flows, can be resolved.

12.4.1.3 Network Calculus Enabled Delay Supervision

Real-time capability can be ensured by supervising traffic flows and identifying potential latency violations ahead of time. For this purpose, the analytical modeling concept of NC can be applied and integrated into the SDN controller [51]. In its deterministic version, NC perfectly matches the needs of Smart Grid communications, as it can be applied to arbitrary types of traffic (i.e., arrival distributions) and follows a worst-case approach [79]. To this end, NC provides upper bounds on network performance. In contrast, stochastic models are inapplicable for critical infrastructures, as latency violations—even with extremely low probability of occurrence—may incur fatal consequences, for example, cascading outages or blackouts. NC-enabled delay supervision acts on any kind of network reconfiguration, such as the introduction of new traffic flows, failure recovery, or in response to northbound interface requests. Delay bounds of traffic flows, affected by these reconfigurations, are recalculated and compared to their respective latency requirements. In the case of a potential violation being identified, further reconfiguration is triggered to resolve the issue—measures include rerouting, queue adaptation, or even dropping of low priority flows if network resources are insufficient. The principles of this concept are shown in Fig. 12.10. The illustration also includes applying NC for delay-aware routing. It is completed by an overview of the corresponding experimental set-up, used for validation.

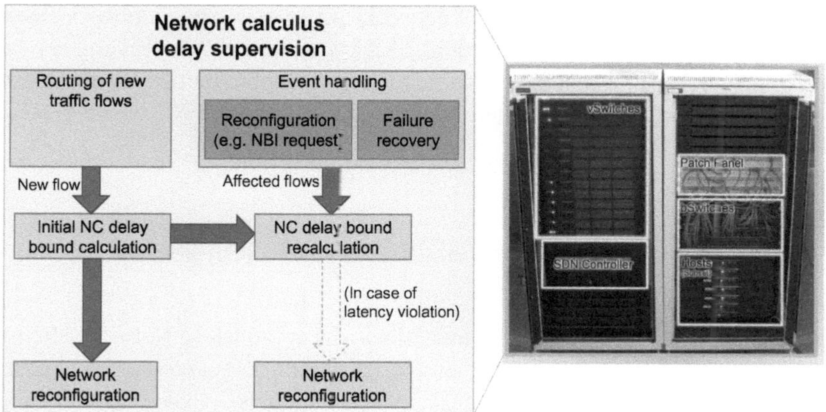

FIG. 12.10 Concept for Network Calculus integration into the SDN controller [51].

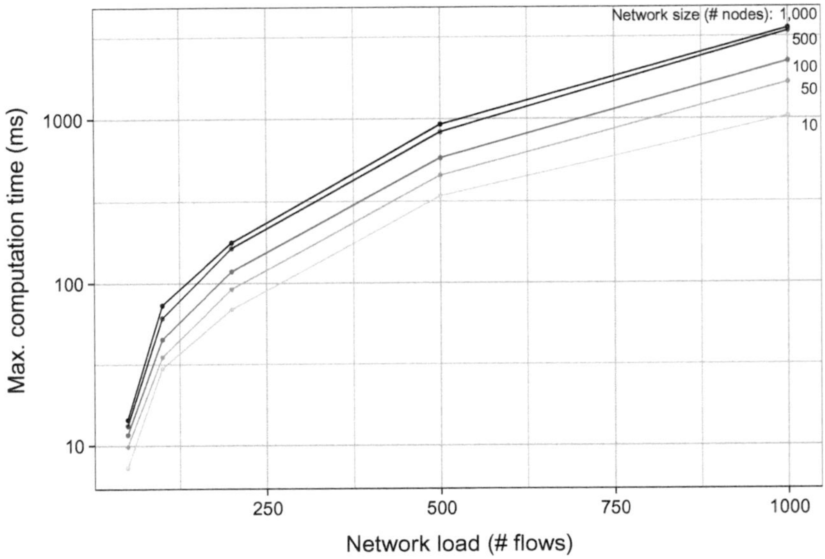

FIG. 12.11 Scalability of Network Calculus algorithms, integrated into the SDN controller, with regard to computation times, with varying network sizes and numbers of flows [51].

Computation times of the corresponding algorithms can be described as a major criterion for the online deployment of NC in a real-time system. Subsequently, Fig. 12.11 depicts computation times of NC-enabled routing and delay supervision. The two columns (x-axis) indicate different computation objects, which map to the respective applications of NC logic in the SDN controller. Further, a proposed standard algorithm and an optimized concept are distinguished (y-axis). For each sector of the plot, computation times are determined for different combinations of network size (x-axis) and number of flows (set of curves). In this way, limitations of the proposed NC integration can be derived. Vice versa, it becomes possible to derive network partitions, which can be managed by a single controller, integrating the described NC-enabled delay supervision.

12.4.2 Network Slicing for Reliable, Cost-Efficient Shared Infrastructures

In the context of smart grid communications, the approach of Network Slicing can be regarded as a major enabler for hard service guarantees over cost-efficient, shared infrastructures. Hence, this case study is divided into analyzing a possible technical implementation of Network Slicing on the one hand and associated economic benefits on the other hand.

12.4.2.1 Technical Realization of Network Slicing

Network Slicing is typically based upon the concepts of SDN and NFV. For this particular realization of slicing, SDN is utilized to configure queues at the switches of the network. Such queues can be dimensioned and adapted according to application requirements, providing resource isolation of individual slices. Considering virtualized SDN switches based on Open vSwitch, the hierarchical token bucket is applied as a queuing strategy.

The central component of the proposed slicing architecture is the MANO controller. It creates one main bridge per switch, which includes all physical ports of the device. This bridge acts as a link between the virtualized network and the underlying physical hardware. For this purpose, slice specific bridges are connected to the main bridge. These slice bridges are handled by the MANO controller as well. It is able to set up, dynamically reconfigure, and delete the bridges. Traffic newly entering the network is forwarded to the MANO controller for identification. It then assigns the traffic to the corresponding slice bridge based on packet header information. For actual traffic routing, slice internal prioritization and other functionalities, the packet is further forwarded to the appropriate slice controller. This controller establishes slice-specific end-to-end routes, which are mapped to corresponding queues and ports of the main bridge.

This general concept has been evaluated using an experimental testing environment in [60], consisting of three SDN-enabled, virtualized switches, six traffic generating servers, and four SDN controllers. Thus, it can be shown that the data rate, available to best effort traffic, is reduced to match the requirements of incoming high priority Smart Grid communications.

12.4.2.2 Economic Impact of Network Slicing

The techno-economic analysis performed in [62] highlights that shared communication infrastructures on basis of Network Slicing are highly profitable for all involved parties, that is, telecommunication network operators (telcos) and power grid operators (utilities).

Fig. 12.12 shows the total cost of ownership of different options for realizing Smart Grid communication infrastructures from the utility's point of view. It can be observed that expenses for dedicated utility networks (first column) are significantly higher compared to shared solutions based on public infrastructures (second and third columns). Outsourcing distribution power grid traffic to the telco's network (second column) helps to achieve cost savings of about one billion euros for the given scenario. The scenario covers the whole area of New England, providing an extension of the well-known New England Test System (NETS)/IEEE 39 bus system [80]. Transferring transmission power grid traffic to the public infrastructure as well, enables comparable results (third column). In all cases, traffic isolation on basis of Network Slicing is considered a fundamental precondition for maintaining the required hard service guarantees of critical Smart Grid communications.

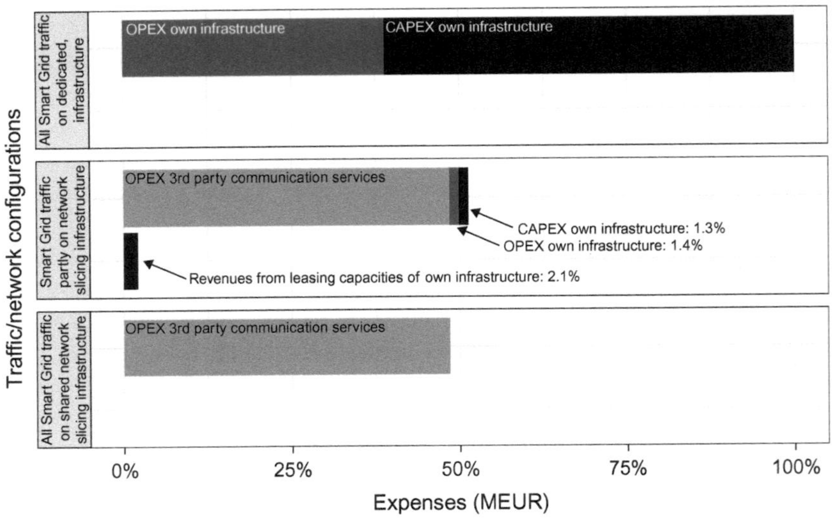

FIG. 12.12 Expenses for different combinations of dedicated, private infrastructures and purchased communication services over shared, public networks from the perspective of the utility [62].

12.5 Conclusion

Within this chapter, major challenges of communications in future Smart Grids were identified and associated to the different levels of the power grid. This involves, in particular, the requirements of availability, real-time capability, reliability, security, and cost-efficiency. Subsequently, these demands were matched with the opportunities offered by 5G communication infrastructures. SDN, considered in 5G specifications to create a more flexible core network, enables dynamic network configuration according to the requirements of Smart Grid applications. This includes queue/priority adaption as well as fast recovery to new optimal network path after failures. In addition, the central view of the SDN controller allows for advanced network monitoring and the prediction of bottlenecks or threats, in particular with regard to real-time capability and cyber security.

End-to-end network slicing as a fundamental feature of 5G infrastructures is esteemed for providing hard service guarantees to critical Smart Grid services, based on strict resource isolation. Moreover, it allows the shared use of common physical infrastructures, offering considerable economic benefits for the implementation of diverse power system applications.

Handling of time-critical Smart Grid functions may be improved using edge computing, which moves processing to servers at the edge of the network. Thus, communication latencies for transmission into the core of the infrastructure can be avoided.

Further promising advances of Smart Grid ICT infrastructures may involve the integration of artificial intelligence, applying machine learning to predict critical situations within the communication and the energy system. In this way, the risk of outages and complete blackouts in the power grid can be further reduced.

References

[1] X. Fang, S. Misra, G. Xue, D. Yang, Smart grid—the new and improved power grid: a survey, IEEE Commun. Surv. Tutorials 14 (4) (2012) 944–980.

[2] Y. Yan, Y. Qian, H. Sharif, D. Tipper, A survey on smart grid communication infrastructures: motivations, requirements and challenges, IEEE Commun. Surv. Tutorials 15 (1) (2013) 5–20.

[3] V.C. Gungor, D. Sahin, T. Kocak, S. Ergut, C. Buccella, C. Cecati, G.P. Hancke, A survey on smart grid potential applications and communication requirements, IEEE Trans. Ind. Inf. 9 (1) (2013) 28–42.

[4] Next Generation Mobile Networks (NGMN) Alliance, NGMN 5G White Paper, (2015). https://www.ngmn.org/uploads/media/NGMN_5G_White_Paper_V1_0.pdf.

[5] International Telecommunication Union, IMT Vision—Framework and overall objectives of the future development of IMT for 2020 and beyond, (2015). https://www.itu.int/dms_pubrec/itu-r/rec/m/R-REC-M.2083-0-201509-I!!PDF-E.pdf.

[6] The 5G Infrastructure Public Private Partnership (5G PPP), 5G PPP 5G Architecture White Paper Revision 2.0, (2017). https://5g-ppp.eu/wp-content/uploads/2018/01/5G-PPP-5G-Architecture-White-Paper-Jan-2018-v2.0.pdf.

[7] 3rd Generation Partnership Project (3GPP), Technical Specification Group Radio Access Network, Study on new radio access technology: Radio access architecture and interfaces (Release 14), (2017).

[8] 3rd Generation Partnership Project (3GPP), Technical Specification Group Radio Access Network, NR, Use Equipment (UE) radio transmission and reception, Part 2: Range 2 Standalone (Release 15), (2018).

[9] 3rd Generation Partnership Project (3GPP), Technical Specification Group Radio Access Network, Study on channel model for frequencies from 0.5 to 100 GHz (Release 15), (2018).

[10] S. Böcker, P. Jörke, C. Wietfeld, Performance Evaluation of an IEEE 802.11 Mesh-based Smart Market and Smart Grid Communication System, in: In International Conference on Smart Grid Communications (SmartGridComm), IEEE Dresden Oct 17, IEEE, 2017, pp. 183–188.

[11] Y. Li, X. Cheng, Y. Cao, D. Wang, L. Yang, Smart choice for the smart grid: narrowband internet of things (NB-IoT), IEEE Internet Things J. 5 (3) (2018) 1505–1515.

[12] International Telecommunication Union, Radiocommunication Sector (ITU-R), Report M.2410–0: Minimum requirements related to technical performance for IMT-2020 radio Interface(s), (2017).

[13] International Telecommunication Union, Radiocommunication Sector (ITU-R), Report P. 2346: Compilation of measurement data relating to building entry loss, (2017).

[14] International Telecommunication Union, Radiocommunication Sector (ITU-R), Report P. 2109–0: prediction of Building entry loss, (2017).

[15] S. Monhof, S. Böcker, J. Tiemann, C. Wietfeld, Cellular network coverage analysis and optimization in challenging smart grid environments, in: In International Conference on Smart Grid Communications (SmartGridComm), IEEE Aalborg Oct 18, IEEE, 2018.

[16] C. Hägerling, C. Ide, C. Wietfeld, Coverage and Capacity Analysis of Wireless M2M Technologies for Smart Distribution Grid Services, in: In International Conference on Smart Grid Communications (SmartGridComm), IEEE Venice Nov. 14, IEEE, 2014, pp. 368–373.
[17] International Electrotechnical Commission (IEC) Technical Committee (TC) 57, IEC 61850 Communication Networks and Systems for Power Utility Automation—Part 5: Communication Requirements for Functions and Devices Models, 2003.
[18] International Electrotechnical Commission (IEC) Technical Committee (TC) 57, IEC 61850 Communication Networks and Systems for Power Utility Automation, 2003.
[19] International Electrotechnical Commission (IEC) Technical Committee (TC) 57, IEC 61850 Communication Networks and Systems for Power Utility Automation—Part 90–2: Use of IEC 61850 for the Communication Between Substations and Control centers, 2013.
[20] International Electrotechnical Commission (IEC) Technical Committee (TC) 57, IEC 61850 Communication Networks and Systems for Power Utility Automation—Part 90–1: Use of IEC 61850 for the Communication Between Substations, 2009.
[21] International Electrotechnical Commission (IEC) Technical Committee (TC) 57, IEC 61850 Communication Networks and Systems for Power Utility Automation—Part 90–5: Use of IEC61850 to Transmit Synchrophasor Information According to IEEE C37.118, 2010.
[22] International Electrotechnical Commission (IEC) Technical Committee (TC) 57, IEC 61850 Communication Networks and Systems for Power Utility Automation—Part 9–2: Specific Communication Service Mapping (SCSM)—Sampled Values Over ISO/IEC 8802-3, 2005.
[23] Hägerling C, Kurtz F.M, Olsen R.L, Wietfeld C. Communication Architecture for Monitoring and Control of Power Distribution Grids Over Heterogeneous ICT Networks. In EnergyCon, IEEE Cavtat May 14 2014 (pp. 838–845). IEEE.
[24] C. Bockelmann, et al., Towards massive connectivity support for scalable mMTC communications in 5G networks, IEEE Access 6 (2018) 28969–28992.
[25] LoRaTM Alliance, Specification: LoRaWAN TM Specification, (2016).
[26] F. Feltrin, C. Buratti, E. Vinciarelli, R. De Bonis, R. Verdone, LoRaWAN: evaluation of link- and system-level performance, IEEE Internet Things J. 5 (3) (2018) 2249–2258.
[27] P. Jörke, C. Falkenberg, C. Wietfeld, Power Consumption Analysis of NB-IoT and eMTC in Challenging Smart City Environments, in: In Global Communications Conference (Globecom) Workshops, IEEE Abu Dhabi Dec 18, IEEE, 2018, pp. 1–6.
[28] K. Mikhaylov, J. Petäjäjärvi, T. Hänninen, Analysis of Capacity and Scalability of the LoRa Low Power Wide Area Network Technology, in: In European Wireless, VDE Oulu May 16, VDE, 2016, pp. 119–124.
[29] F. Adelantado, et al., Understanding the limits of LoRaWAN, IEEE Commun. Mag. 55 (9) (2017) 34–40.
[30] Institute of Electrical and Electronics Engineers (IEEE), Standard IEEE 802.1ah: IEEE Standard for Information Technology—Telecommunications and Information Exchange Between Systems Local and Metropolitan Area Networks—Specific Requirements Part 11: Sub 1 GHz, 2016.
[31] A. Šljivo, et al., Performance evaluation of IEEE 802.11 ah networks with high-throughput bidirectional traffic, Sensors 18 (2) (2018).
[32] P. Jörke, S. Böcker, D. Liedmann, C. Wietfeld, Urban Channel Models for Smart City IoT-Networks Based on Empirical Measurements of LoRa-links at 433 and 868 MHz, in: In International Symposium on Personal, Indoor and Mobile Radio Communications (PIMRC) Workshops, 2017 IEEE Montreal Oct, IEEE, 2017, pp. 1–6.

[33] J. Ordonez-Lucena, P. Ameigeiras, D. Lopez, J.J. Ramos-Munoz, J. Lorca, J. Folgueira, Network slicing for 5G with SDN/NFV: concepts, architectures, and challenges, IEEE Commun. Mag. 55 (5) (2017) 80–87.
[34] H. Zhang, N. Liu, X. Chu, K. Long, A.H. Aghvami, V.C.M. Leung, Network slicing based 5G and future Mobile networks: mobility, resource management, and challenges, IEEE Commun. Mag. 55 (8) (2017) 138–145.
[35] I. Afolabi, T. Taleb, K. Samdanis, A. Ksentini, H. Flinck, Network slicing and Softwarization: A survey on principles, enabling technologies, and solutions, IEEE Commun. Surv Tutorials 20 (3) (2018) 2429–2453.
[36] N. McKeown, T. Anderson, H. Balakrishnan, G. Parulkar, L. Peterson, J. Rexford, S. Shenker, J. Turner, OpenFlow: enabling innovation in campus networks, ACM SIGCOMM Computer Communication Review 38 (2) (2008) 69–74.
[37] Open Networking Foundation, Open Flow Switch Specification Version 1.3.0, (2012). https://www.opennetworking.org/images/stories/downloads/sdn-resources/onf-specifications/openflow/openflow-spec-v1.3.0.pdf.
[38] A. Sydney, J. Nutaro, C. Scoglio, D. Gruenbacher, N. Schulz, Simulative comparison of multi-protocol label switching and OpenFlow network technologies for transmission operations, IEEE Trans. Smart Grid 4 (2) (2013) 763–770.
[39] A. Cahn, J. Hoyos, M. Hulse, E. Keller, Software-Defined Energy Communication Networks: From Substation Automation to Future Smart Grids, in: In International Conference on Smart Grid Communications (SmartGridComm), IEEE Vancouver 2013 Dec 13, IEEE, 2013, , pp. 558–563.
[40] A. Goodney, S. Kumar, A. Ravi, Y. Cho, Efficient PMU networking with software defined networks, in: In International Conference on Smart Grid Communications (SmartGridComm), IEEE Vancouver 2013 Dec 13, IEEE, 2013, pp. 378–383.
[41] E. Molina, E. Jacob, J. Matias, N. Moreira, A. Astarloa, Using software defined networking to manage and control IEC 61850-based systems, Comput. Electr. Eng. 43 (2015) 142–154.
[42] N. Dorsch, F. Kurtz, F. Girke, C. Wietfeld, Enhanced fast failover for software-defined smart grid communication networks, in: In Global Communications Conference (Globecom), IEEE Washington 2016 Feb 17, IEEE, 2016, pp. 1–6.
[43] J. Kempf, E. Bellagamba, A. Kern, D. Jocha, A. Takacs, P. Skoldstrom, Scalable fault management for OpenFlow, in: In International Conference on Communications (ICC), IEEE Ottawa 2012 Nov 12, IEEE, 2012, pp. 6606–6610.
[44] S. Sharma, D. Staessens, D. Colle, M. Pickavet, P. Demeester, Fast failure recovery for in-band OpenFlow networks, in: In Design of Reliable Communication Networks (DRCN), IEEE Budapest 2013 Jun 13, IEEE, 2013, pp. 52–59.
[45] N.L.M. van Adrichem, B.J. van Asten, F.A. Kuipers, Fast recovery in software-defined networks, in: In European Workshop On Software Defined Networks, IEEE London 2014 Dec 14, IEEE, 2014, pp. 61–66.
[46] A. Aydeger, K. Akkaya, M.H. Cintuglu, A. Selcuk Uluagac, O. Mohammed, Software defined networking for resilient communications in Smart Grid active distribution networks, in: In International Conference on Communications (ICC), Kuala Lumpur 2016 Jul 16, IEEE, 2016, pp. 1–6.
[47] T. Pfeiffenberger, J.L. Du, P. Bittencourt Arruda, A. Anzaloni, Reliable and flexible communications for power systems: Fault-tolerant multicast with SDN/OpenFlow, in: In International Conference on New Technologies, Mobility & Security (NTMS), IEEE Paris 2015 Sep 15, IEEE, 2015, pp. 1–6.

[48] N. Dorsch, F. Kurtz, S. Dalhues, L. Robitzky, U. Häger, C. Wietfeld, Intertwined: Software-defined communication networks for multi-agent system-based smart grid control, in: In International Conference on Smart Grid Communications (SmartGridComm), IEEE Sydney 2016 Dec 16, IEEE, 2016, pp. 254–259.

[49] R.L. Cruz, A Calculus for network delay I. network elements in isolation, IEEE Trans. Inf. Theory 37 (1) (1991) 114–131.

[50] J.Y. Le Boudec, P. Thiran, Network Calculus: A Theory of Deterministic Queueing Systems for the Internet, Springer, Berlin, Heidelberg, 2001.

[51] N. Dorsch, F. Kurtz, C. Wietfeld, Enabling hard services guarantees in software-defined smart grid infrastructures, Comput. Netw. 147 (2018) 112–131.

[52] E.G. Da Silva, A.S. da Silva, J.A. Wickboldt, P. Smith, L. Zambenedetti Granville, A. Schaeffer-Filho, A one-class NIDS for SDN-based SCADA systems, in: In International Conference on Computers, Software and Applications (COMPSAC), IEEE Atlanta 2016 Aug 16, IEEE, 2016, , pp. 303–312.

[53] D. Jin, Z. Li, C. Hannon, C. Chen, J. Wang, M. Shahidehpour, C.H. Lee, Toward a cyber resilient and secure microgrid using software-defined networking, IEEE Trans. Smart Grid 8 (5) (2017) 2494–2504.

[54] E. Hammad, J. Zhao, A. Farraj, D. Kundur, Mitigating Link Insecurities in Smart Grids via QoS Multi-Constraint Routing, in: In International Conference on Communications (ICC) Workshops, IEEE Kuala Lumpur 2016 Jul 16, IEEE, 2016, , pp. 380–386.

[55] X. Foukas, G. Patounas, A. Elmokashfi, M.K. Marina, Network slicing in 5G: Survey and challenges, IEEE Commun. Mag. 55 (5) (2017) 94–100.

[56] P. Rost, C. Mannweiler, D.S. Michalopoulos, C. Sartori, V. Sciancalepore, N. Sastry, O. Holland, S. Tayade, B. Han, D. Bega, D. Aziz, H. Bakker, Network slicing to enable scalability and flexibility in 5G Mobile networks, IEEE Commun. Mag. 55 (5) (2017) 72–79.

[57] B. Chatras, U.S. Tsang Kwong, N. Bihannic, NFV enabling network slicing for 5G, in: In Innovation in Clouds, Internet and Networks (ICIN), IEEE Paris 2017 Apr 17, IEEE, 2017, , pp. 219–225.

[58] R. Sherwood, G. Glibb, K.K. Yap, G. Appenzeller, N. McKeown, G. Parulkar, M. Casado (Ed.), FlowVisor: A Network Virtualization Layer, 2009https://www.gta.ufrj.br/ensino/cpe717-2011/openflow-tr-2009-1-flowvisor.pdf.

[59] C.W. Tseng, Y.K. Huang, Y.T. Yang, C.C. Liu, D.Y. Huang, L.D. Chou, A network traffic shunt system in SDN network, in: In International Conference on Computer, Information and Telecommunications Systems (CITS), IEEE Dalian 2017 Sep 17, IEEE, 2017, , pp. 195–199.

[60] F. Kurtz, C. Bektas, N. Dorsch, C. Wietfeld, Network Slicing for Critical Communications in Shared 5G Infrastructures—An Empirical Evaluation, in: In Conference on Network Softwarization (NetSoft), IEEE Montreal 2018 Sep 18, IEEE, 2018, , pp. 393–399.

[61] C. Bektas, S. Monhof, F. Kurtz, C. Wietfeld, Towards 5G: An empirical evaluation of software-defined end-to-end network slicing, in: In Global Communications Conference (Globecom) Workshops, IEEE Abu Dhabi 2018 Jan 19, IEEE, 2018, , pp. 1–6.

[62] N. Dorsch, F. Kurtz, C. Wietfeld, On the economic benefits of software-defined networking and network slicing for smart grid communications, Netnomics Econ. Res. Electron. Netw. (2018) 1–30.

[63] Y.J. Kim, K. He, M. Thottan, J.G. Deshpanda, Virtualized and self-configurable utility communications enabled by software-defined networks, in: In International Conference on Smart Grid Communications (SmartGridComm), IEEE Venice 2014 Jan 15, IEEE, 2014, , pp. 416–421.

[64] A. Irfan, N. Taj, S.A. Mahmud, A novel secure SDN/LTE based architecture for smart grid security, in: In Computer and Information Technology; Ubiquitous Computing and

Communications; Dependable, Autonomic and Secure Computing; Pervasive Intelligence and Computing, IEEE Liverpool 2015 Dec 15, IEEE, 2015, pp. 762–769.

[65] T. Taleb, K. Samdanis, B. Mada, H. Flinck, S. Dutta, D. Sabella, On multi-access edge computing: A survey of the emerging 5G network edge cloud architecture and orchestration, IEEE Commun. Surv. Tutorials 19 (3) (2017) 1657–1681.

[66] A. Lertsinsrubtavee, A. Ali, C. Molina-Jimenez, A. Sathiaseelan, J. Crowcroft, PiCasso: A lightweight edge computing platform, in: In International Conference on Cloud Networking (CloudNet), IEEE Prague 2017 Oct 17, IEEE, 2017, pp. 1–7.

[67] F. Kurtz, I. Laukhin, C. Bektas, C. Wietfeld, Evaluating software-defined networking-driven edge clouds for 5G critical communications, in: In International Conference on ICT Convergence (ICTC), 2018 IEEE Jeju Island 2018 Nov 18, IEEE, 2018, pp. 405–410.

[68] Y. Zhang, K. Liang, S. Zhang, Y. He, Applications of edge computing in PIoT, in: In Conference on Energy Internet and Energy System Integration (EI2), IEEE Beijing 2017 Jan 18, IEEE, 2017, pp. 1–4.

[69] Y. Huang, Y. Lu, F. Wang, X. Fan, J. Liu, V.C.M. Leung, An edge computing framework for real-time monitoring in smart grid, in: In International Conference on Industrial Internet (ICII), IEEE Seattle 2018 Nov 18, IEEE, 2018, pp. 99–108.

[70] J. Xu, B. Palanisamy, H. Ludwig, Q. Wang, Zenith: Utility-aware Resource Allocation for Edge Computing, in: In International Conference on Edge Computing (EDGE), IEEE Honolulu 2017 Sep 17, IEEE, 2017, pp. 47–54.

[71] S. Zahoor, N. Javaid, A. Khan, B. Ruqia, F.J. Muhammad, M. Zahid, A cloud-fog-based smart grid model for efficient resource utilization, in: In International Wireless Communications and Mobile Computing Conference (IWCMC), IEEE Limassol 2018 Aug 18, IEEE, 2018, pp. 1154–1160.

[72] D. Katz, D. Ward, Bidirectional Forwarding Detection (BFD), Internet Engineering Task Force (IETF) RFC 5880, (2010)www.ietf.org/rfc/rfc5880.txt.

[73] J. Moy, OSPF Version 2, Internet Engineering Task Force,(IETF) RFC 2328, www.ietf.org/rfc/rfc2328.txt, 1998.

[74] O. Komolafe, J. Sventek, Analysis of RSVP-TE Graceful Restart, in: In IEEE International Conference on Communications (ICC), IEEE Glasgow 2007, Jun 17, IEEE, 2007, pp. 2324–2329.

[75] Y. Lei, C.H. Lung, A. Srinivasan, A cost-effective protection and restoration mechanism for Ethernet based networks: an experiment report, in: In Workshop on High Performance Switching and Routing, IEEE Phoenix 2004, Aug 04, IEEE, 2004, pp. 350–354.

[76] M. Tacca, K. Wu, A. Fumagalli, J.P. Vasseur, Local Detection and Recovery from Multi-Failure Patterns in MPLS-TE Networks, in: IEEE International Conference on Communications (ICC), IEEE Istanbul 2006, Jun. 06, IEEE, 2006, pp. 658–663.

[77] B. Niven-Jenkins, D. Brugard, M. Betts, N. Sprecher, S. Ueno, Requirements of an MPLS Transport Profile. Internet Engineering Task Force (IETF) RFC 5654, (2009). www.ietf.org/rfc/rfc5654.txt.

[78] IEEE PES Power System Dynamic Performance Committee, TEST systems for voltage stability analysis and security assessment, Technical Report PES-TR19, (2015).

[79] N. Dorsch, H. Georg, C. Wietfeld, Analysing the Real-Time Capability of Wide Area Communication in Smart Grids, in: In International Conference on Computer Communications (Infocom) Workshops, IEEE Toronto 2014 Jul 14, IEEE, 2014, pp. 682–687.

[80] T. Athay, R. Podmore, S. Virmani, A practical method for the direct analysis of transient stability, IEEE Trans. Power Syst. 98 (2) (1979) 573–584.

Chapter 13

Data Security in the Smart Grid Environment

Ashok Kumar Das* and Sherali Zeadally[†]
Center for Security, Theory and Algorithmic Research, International Institute of Information Technology, Hyderabad, India, [†]College of Communication and Information, University of Kentucky, Lexington, KY, United States

Chapter Outline

13.1 Introduction

Smart grid (SG) technology is regarded as the next generation of the power grid, which makes use of the two-way flows of electricity, as well as information,

Pathways to a Smarter Power System. https://doi.org/10.1016/B978-0-08-102592-5.00013-2

in order to build a broadly distributed automated energy delivery network. The Smart grid is a critical infrastructure that plays a critical role in the daily lives of people. Both the Smart Grid and its data must be protected from cyberattacks at all times [1, 2]. In SG, four components are present, namely sensing, control, communication, and actuation systems. The most important component of SG is the smart meter (SM), which consists of sensing and communication modules. In addition, there are service systems from the service providers (SPs), which consist of several modules for control, communication, and actuation. SMs are broadly used in homes for monitoring energy consumption in real time. Moreover, power pricing information to consumers is also provided by the SMs [3–6]. A detailed survey on SG including its applications can be found in [7–14].

In the following subsections, we now discuss the smart grid framework developed by the National Institute of Standards and Technology (NIST) and its taxonomy, based on domains and targeted research areas.

13.1.1 Taxonomy of Smart Grid Domains Based on NIST's Framework

Fig. 13.1 shows the smart grid framework proposed by NIST, which includes seven domains, namely: (1) customers, (2) markets, (3) service providers, (4) operations, (5) bulk generation, (6) transmission, and (7) distribution [15]. The taxonomy of this framework is shown in Fig. 13.2, and the domains and actors in the SG are presented in Table 13.1 [15, 16].

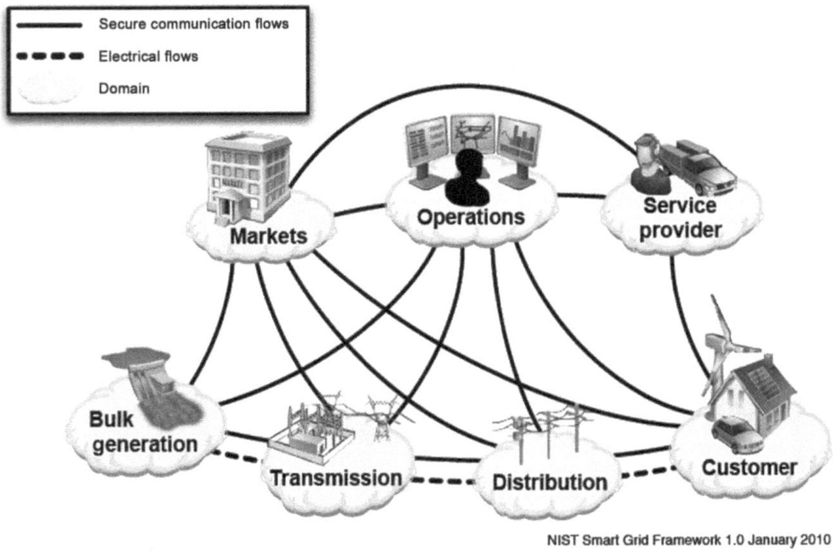

FIG. 13.1 NIST smart grid framework. *((Based on G. Locke, and P. D. Gallagher. NIST Framework and Roadmap for Smart Grid Interoperability Standards, Release 1.0, 2010. Avialble at: http://www.nist.gov/sites/default/files/documents/public affairs/releases/smartgrid interoperability final.pdf. Accessed January 2018, 2010.))*

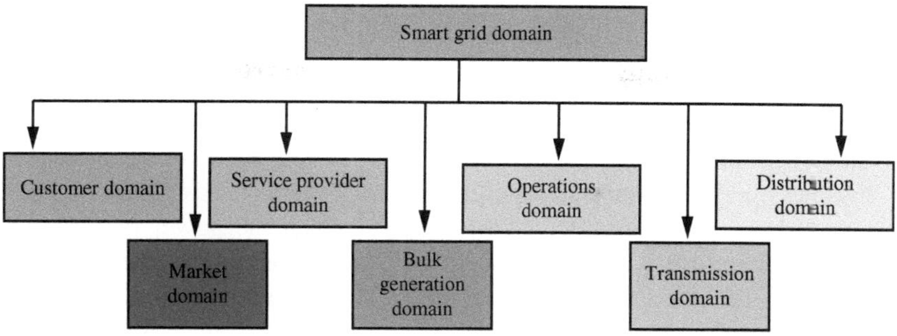

FIG. 13.2 Taxonomy of smart grid domains based on the NIST framework for the smart grid. *(Adapted from M. H. Cintuglu, O. A. Mohammed, K. Akkaya, A. S. Uluagac. A survey on smart grid cyber-physical system testbeds.* IEEE Commun. Surv. Tutor. *19(1) (2017) 446–464.)*

TABLE 13.1 Domains and Actors in the SG Framework [15, 16]

Domain	Actors in the Domain
Customers	The actors are the end users of electricity. They can generate and store, as well as manage the use of power. Three types of customers are included: residential, commercial, and industrial.
Markets	The actors are considered as the operators and participants in electricity markets. The typical market domain applications include market management, retailing, aggregation, market operations, trading, and ancillary operations.
Service providers	The actors are mainly the organizations that provide services to electrical customers and utilities. The typical applications in this domain include customer management, smart building management, and smart device installment.
Operations	The actors in this domain include the managers of the movement of electricity. The typical applications of this domain include extensive power system operations (e.g., monitor, control, protection, and analysis).
Bulk generation	The generators of electricity in bulk quantities are considered as actors in this domain. These may store energy for later distribution. The typical bulk generation units are considered as traditional large scale generation units (e.g., nuclear, hydro plants, wind farms, thermal, and large scale solar generation).
Transmission	The actors in this domain are the carriers of bulk electricity over long distances. These may store as well as generate electricity.
Distribution	The actors included in this domain are the distributors of electricity to and from customers. These may store as well as generate electricity. The typical components of the distribution domain include metering points, loads, and micro-grids.

13.1.2 Taxonomy Based on Targeted Research Areas

In Fig. 13.3, we identify various targeted research areas [16].

- *Demand response and consumer energy efficiency*: The demand response in deregulated electricity markets yields a method and also incentives for business, utilities, and industrial as well as residential customers to lower energy use at the time of peak demand or when the power reliability is at risk. Therefore, it is essential to have a reliable operation of the electricity system which can achieve a perfect balance between demand and supply in - real-time [17].
- *Distributed energy resources*: The primary research area in this category includes utility-independent generation units (nonbulk) and energy storage behind the prosumer energy meter.
- *Energy storage:* This research area includes the conversion of electrical energy from a power network (smart grid) into a form of energy that can be stored and converted back to electrical energy [18].
- *Wide-area situational awareness:* To avoid disasters, such as the one that occurred on August 14, 2003, in North America Eastern Interconnection, where there was a wide-scale power loss for millions of consumers [19], it is essential to monitor and display the components related to the power system across the interconnection over large geographical areas in real-time. Therefore, for a safe, reliable and economical electric power system, we need advanced wide-area monitoring and protection, as well as control capabilities with applications.
- *Advanced metering infrastructure*: The primary research goal in this area is to find a way for integrating various technologies that can support an intelligent interface between consumers and system operators [20].
- *Electric transportation*: The research in electrical transportation includes plug-in electric vehicle battery banks, large-scale grid integration, and also wired-wireless charging stations [21].
- *Distributed grid management*: The focus of this research area is on active distribution operations, such as voltage reduction, system reliability and power quality improvements, and outage management [22].

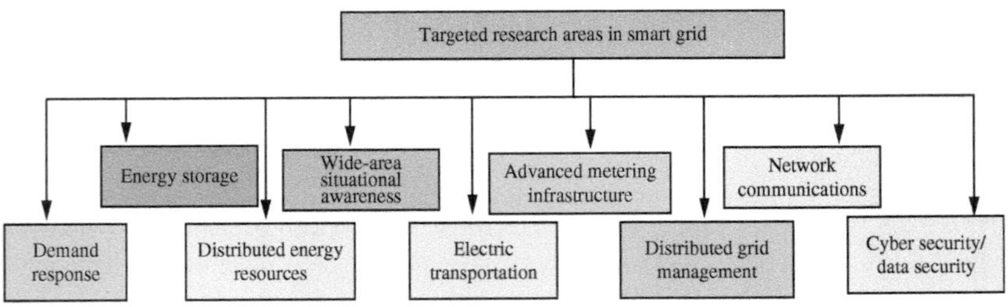

FIG. 13.3 Taxonomy based on targeted research areas. *((Adapted from M. H. Cintuglu, O. A. Mohammed, K. Akkaya, A. S. Uluagac. A survey on smart grid cyber-physical system testbeds.* IEEE Commun. Surv. Tutor. *19(1) (2017) 446–464.))*

- *Network communications*: A variety of public as well as private communication networks, based on both wired and wireless communication, are deployed in the SG environment [23]. In this environment, wireless communication is a key aspect for realizing the SG vision. The wireless technologies include IEEE 802.16-based WiMAX, IEEE 802.11-based wireless local area network (LAN), long-term evolution (LTE), 3G/4G cellular, IEEE 802.20-based MobileFi, and ZigBee based on IEEE 802.15 [24, 25].
- *Cyber security/data security*: The power system communication and information infrastructures are heavily needed for the development of SG solutions [26]. Existing cyber security solutions are not enough or efficient for SG cyber-physical system security solutions. Therefore, we need domain specific mechanisms and solutions to enforce cyber security [27, 28]. Some interesting research areas on cyber security for the SG are integrity, authentication, availability, confidentiality, and vulnerability assessment [29, 30].

  The pervasive use of wireless communications in SG opens up several threats where a malicious adversary can intercept, modify, delete, and interrupt the transmitted messages between the communicating entities, such as between SMs and SPs. SMs are typically deployed in close proximity to homes and are protected by physical locks. Then, there will be a possibility to compromise the SMs physically by an adversary when he/she breaks the physical lock. If secret credentials are stored in the compromised SMs, the adversary can perform smart meter impersonation attacks on behalf of the compromised SMs [31]. The energy usage data of a particular customer can be stored at the SM, and then the behaviors and habits (e.g., activities like watching television have detectable power consumption signatures) can be easily revealed to the adversary [32]. Hence, the SMs are typically the most attractive targets for the adversary because the vulnerabilities can be easily exploited [31, 33]. The rapid exponential growth in smart grid technology in recent years has fueled data security research in this area. Research efforts include the design of security protocols, such as key distribution, authentication, access control, trust computing, and intrusion detection (as discussed in Section 13.2).

The rest of this chapter is organized as follows. Section 13.2 presents a taxonomy of various security protocols in the SG. We then discuss various security issues in the SG in Section 13.3. We discuss various existing security solutions proposed for SG security in Section 13.4. Next, we discuss various deployment and implementation issues related to security solutions in the SG in Section 13.5. Finally, we conclude the chapter in Section 13.6.

13.2 Taxonomy of Security Protocols in Smart Grid

In this section, we discuss various security protocols that have been proposed in the smart grid environment to provide data security. Fig. 13.4 presents a taxonomy of security protocols in the SG environment. Next, we discuss the principles of various security mechanisms in the SG.

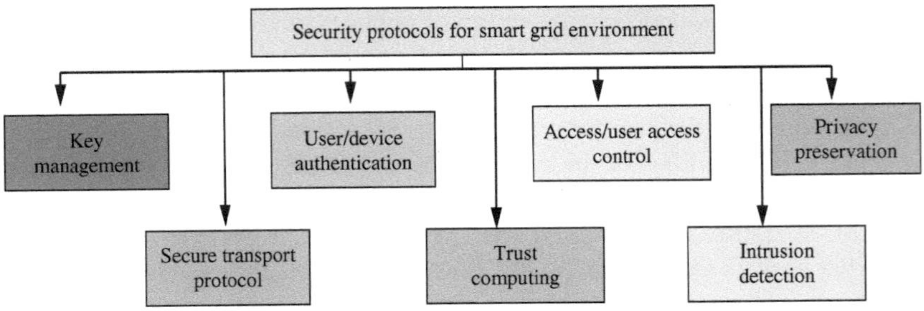

FIG. 13.4 Taxonomy of security protocols in the smart grid environment.

13.2.1 Key Management

The key management mechanism establishes a secret key between two communicating entities in the SG environment. Two types of key management can be used in the SG environment, namely public key-based key management and symmetric key-based key management. In a public key-based key management scheme, a secure public key cryptographic technique (e.g., the station-to-station key agreement protocol of the Diffie–Hellman key exchange protocol [34]) helps to establish a session key between two participants (e.g., between two SMs, or between SM and SP) over a public channel. The station-to-station key agreement protocol applies the digital signatures with the public key certificates to establish a session key between two participating entities. Afterwards, the entities can encrypt or decrypt the secret messages using the symmetric key cryptographic techniques (e.g., advanced encryption standard (AES) [35]) for their secure communication.

In contrast, in symmetric key-based key management, prior to installing the SMs and deploying the SPs, some secret credentials are preloaded in their memory in the offline mode by a trusted party. After their deployment, using the preloaded credentials they will establish a symmetric secret key for their secure communication in the future. In symmetric key-based key management, it is also essential to support dynamic SMs addition phase after initial deployment, because some SMs can be physically compromised by an adversary (as explained in the threat model in Section 13.3.1) or they may fail due to battery usage. The phases related to a symmetric key-based key management scheme can be similar to that used in wireless sensor networks (WSNs) [36–42]. Thus, a typical key management scheme in the SG environment should have the following phases:

- *Key predistribution phase*: This phase is executed in offline mode by the trusted party in the SG. The trusted party generates the secret credentials (known as the keying materials) for each SM, and also for the SP in the SG environment prior to their deployment.
- *Shared key discovery phase*: This phase is executed immediately after the nodes (SMs and SPs) are deployed in the SG environment. Using the

preloaded keying materials, any two neighbor nodes, such as two neighbor SMs (or an SM and an SP) can establish a symmetric key between them.
- *Path key establishment phase*: This phase is applicable if necessary. Once it is executed, it enables secure network connectivity. Here, secure network connectivity is defined as the probability of establishing a secret key between two neighbor nodes. Assume that two neighbor nodes, say SM_i and SM_j, fail to come up with a direct pairwise secret key between them during the shared key discovery phase. On discovering the secure path between SM_i and SM_j, a new random pairwise symmetric key is transmitted securely along that secure path so that both SM_i and SM_j will now have the same pairwise key.

Additionally, it is also necessary to deploy new nodes in the network because some nodes (smart meters) may be physically captured by an adversary or some smart meters may be exhausted due to a power failure (if the smart meters are battery powered). In this case, the key management mechanism should be able to establish secret keys between the newly deployed node and the existing neighbor nodes in the network.

Let the secure network connectivity be denoted by P_{con}. If $0 < P_{con} < 1$, we call the key management scheme probabilistic (random); otherwise, if $P_{con} = 1$, we call the key management scheme deterministic. Examples of some probabilistic key management schemes are included in [37–42], whereas some deterministic key management schemes are included in [36, 43–46].

13.2.2 User Authentication/Device Authentication

In the SG real-time applications, it is essential to access the real-time data directly from the desired smart meters (SMs) by an external party (called a user). The data access is only allowed if the mutual authentication between an accessed SM and a user is successful. After successful mutual authentication, they will then establish a secret session key for their future secure communication.

A user authentication scheme in the SG environment will have the following phases [47]:

- *System setup phase*: The system parameters are selected by some trusted authority (*TA*) in the offline mode.
- *Predeployment phase*: In this phase, each smart meter SM_i and service provider SP_j are registered with the *TA*, and the *TA* loads the essential credentials in their memories prior to their deployment in the SG environment.
- *User registration phase*: To access real-time information from designated smart devices (SM_i), a user U_i needs to register with the *TA*. For this purpose, U_i first chooses his/her credentials (e.g., identity, password, and personal biometrics) to the *TA* via a secure channel (e.g., submitting the credentials in person). After receiving those secret credentials, the *TA* will issue a smart card or mobile device securely to the registered user U_i.

- *Login phase*: In this phase, U_i inputs his/her secret credentials, which are then authenticated by his/her smart card (mobile device). After successful verification of the credentials, a login request message is created, which is sent to the service provider SP_j through a public channel.
- *Authentication and key agreement phase*: This phase is executed after receiving the login request message. Once the login request message is received, the SP_j verifies it and, if the verification is successful, then only the SP_j creates an authentication request message and sends it to the accessed smart meter SM_i via a public channel. SM_i then validates the received message and dispatches the authentication reply to the user U_i. Finally, the user U_i verifies the received authentication reply message from SM_j. If mutual authentication between U_i and SM_j is successful, both U_i and SM_j establish a session key, say SK_{ij}, between them. In the future, the session key SK_{ij} is used for future secure communication between U_i and SM_j.
- *Password and biometric update phase*: For security reasons, it is good practice to update/change the password and personal biometrics (if needed) by a legitimate registered user U_i locally using his/her smart card or mobile device without further contacting the *TA*.
- *Mobile device/smart card revocation phase*: Suppose the smart card (mobile device) of a legitimate user is lost or stolen by an adversary. In this case, the revocation phase to issue a new smart card (mobile device) is necessary so that a new set of credentials can be stored in it.
- *Dynamic node addition phase*: This phase is essential in a scenario where some nodes (smart meters) are physically captured by an adversary or some smart meters are exhausted due to a power failure (if the smart meters are battery powered).

Based on the number of factors used in a user authentication, it is called a single-factor (if only the smart card (mobile device) or password is used), two-factor (if both the smart card (mobile device) and password are used), three-factor (if all the smart card (mobile device), password, and biometrics such as fingerprint, are used), and so on. It is worth noting that in the case of multi-factor user authentication, the number of factors is higher and, hence, it is expected to have a better security level.

Device authentication [48] in the SG environment is another kind of authentication method when two nodes, such as two smart meters or a smart meter and an SP_j need to authenticate each other for secure communications between them.

13.2.3 Access Control/User Access Control

To sustain the lifetime of the smart meters, it is necessary to deploy new smart meters in the SG environment. This is because the nodes may stop functioning due to battery power or because of a physical node capture by an adversary in the network. A deployed smart meter is not necessarily a legitimate node, because it may be a fake node that has been deployed by an adversary in the network. Hence, it is important to separate a malicious node from genuine nodes in the SG environment. This can be achieved by designing a secure access control scheme in order to prevent malicious nodes entering in the network. In an access control mechanism, the following two tasks are performed:

- *Node authentication*: The newly deployed node must authenticate itself to its neighbor nodes in order to prove that it is a legitimate node for accessing the information from the other nodes in the network.
- *Key establishment*: A newly deployed node should be able to establish shared secret keys with its existing neighbor nodes to assure secure communication during the transmission of information.

As with the user authentication mechanism, the addition of a new node (e.g., a smart meter) phase is essential for access control mechanisms in the SG environment. Access control methods can be classified into two categories based on their authentication type: (1) certificate-less and (2) certificate-based. In a certificate-based access control mechanism, each deployed node is loaded with a digital certificate (e.g., X.509 certificate [49]) issued by the *TA*. The certificate is then used to prove the identity of a node to its neighbor node. In a certificate-less access control mechanism, typically a hash-chain-based scheme is applied.

To provide access rights only to registered legitimate users for different services, information, and resources available in the SG environment, a user access control is another important security mechanism.

13.2.4 Secure Transport Protocol

Large amounts of data will be generated by different measuring devices, such as advanced meters, intelligent sensors, and electric vehicle charging stations that are embedded in the power grid of the smart grid networks. Therefore, it is extremely important to send the huge volume of generated data from the measuring devices to the utility control centers for wide-area monitoring and control, and also to determine the overall grid status accurately and with minimal delays. To improve power stability, the gathered data is also applied for encouraging consumer participation. This motivates the use of secure transport protocols for such periodic collection of grid measurement data [50].

13.2.5 Privacy-Preservation

The SG integrates several intelligent sensing devices and communication networks into the existing power grid to collect data from the grid for operational intelligence. Smart meters—critical components in such an infrastructure—frequently transmit readings to the electric utility (e.g., a reading every 15 min) [51]. Such reading streams benefit the utilities (e.g., load balancing) and the energy consumers (e.g., optimizing electricity usage). However, such features may lead to serious breaches of consumers' privacy [51]. For example, smart meter readings could likely disclose the consumers' personal daily behavior or habits (e.g., cooking time (by the stove or microwave), and frequency of going to the bathroom at night (by the light switched on)) [51]. In order to protect consumers' personal privacy from adversaries, the privacy preserving techniques are much needed in the SG environment [52, 53].

13.2.6 Intrusion Detection

Intrusion is an illegal task performed by an intruder (adversary) in the SG network. Depending upon the intruder's capabilities, an attack can be classified as either a passive attack, where eavesdropping of information during communication takes place, or an active attack, where malicious packet injection and packet dropping occur. An intrusion detection system (IDS) is defined as a software application that can detect any dubious activity occurring in the network. Once a dubious smart device is detected by the IDS, quick detection is needed to prevent further damage to the system [54]. Hence, an efficient IDS scheme is also needed for securing the SG environment.

13.2.7 Trusted Computing

Due to the wide range of cyber security threats and severe consequences from cyberattacks, it is necessary for the smart grid to have cyber security protection that matches very closely the cyber security requirements. Therefore, we need a holistic approach to smart grid communication that considers all aspects of smart grid operations. Trusted computing is a component in achieving this goal [55].

13.3 Security Issues in Smart Grid

In this section, we first present the threat model associated with data security in the SG environment. We then discuss various security requirements and possible security attacks in the SG environment.

13.3.1 Threat Model

The well-known Dolev-Yao (DY) threat model [56] is adopted in the SG environment [47]. Based on the DY model, any two nodes in the network communicate over a public channel where the end-point communicating entities, such as users, smart meters, and intelligent sensors are not trusted. An adversary A can then intercept, modify or delete the exchanged messages during transmission. In addition, A can also insert fake messages during the communication.

The current de facto standard model in modeling authenticated key-exchange protocols is Canetti and Krawczyk's adversary model

(CK-adversary model) [57, 58], where A is responsible for delivering messages (as in the DY model), and can also compromise private keys, session keys, and session state. The security of an authenticated key-exchange protocol needs to assure that if some forms of secret information are revealed (e.g., session ephemeral secrets, session keys, or long-term private keys), it should have minimal impact on the security of other secret credentials and session keys of the communicating entities [5]. Thus, an authenticated key-exchange protocol should provide the session key (SK) security under the CK-adversary model.

If the mobile device or smart card of a legitimate registered user is lost or stolen, an adversary can extract all the sensitive credentials stored in it using sophisticated power analysis attacks [59, 60].

Since the monitoring of the installed smart meters and intelligent sensors is not always possible in the SG environment, such devices can be physically captured by an adversary. The adversary can then use the extracted information from the captured devices to launch other attacks (such as impersonation attacks) in the network.

13.3.2 Security Requirements in Smart Grid Environment

Next, we discuss the security requirements in the SG environment [61]:

- *Authentication*: Authentication is a process which gives assurance that the communicating party is the one that it claims to be. It involves authentication of smart meters, intelligent sensors, users, and service providers before permitting access to a restricted resource, or revealing important information.
- *Integrity*: This process deals with the assurance that the data received is exactly what was sent by an authorized party. In other words, the message sent contains no modification, insertion, or deletion. Integrity in the SG environment falls into the following three categories [55]:
  - System integrity: This is a binary property that indicates if the system has a reliable execution environment. With trusted computing (as discussed in Section 13.2.7), it can perform binary attestation (binary attestation in trusted computing provides the ability to reason about the state of a platform using integrity measurements) to check a system's integrity and its administrative capabilities.
  - Process integrity: This heavily depends on the genuineness of a process code. It is crucial not only to detect changes in software but also to assure that the newly developed code is trustworthy.
  - Data integrity: Verifying the genuineness of data depends on whether the data is collected or generated. Collected data is primitive data given to a process and its integrity is application specific.
- *Confidentiality* (*privacy*): Confidentiality or privacy ensures that the data transmitted over the wireless communication channel is protected from unauthorized disclosure of information.

- *Nonrepudiation*: This process provides protection against denial by one of the communicating parties who has participated in all or part of the communication in the SG environment. This can be further classified into two categories:
  - Nonrepudiation, origin: This proves that the message was sent by the specified entity.
  - Nonrepudiation, destination: This proves that the message was received by the specified entity.
- *Authorization*: Authorization ensures that only legitimate devices in the SG environment can supply information to network services.
- *Auditability*: Auditability is the ability to reconstruct the complete history of the system behavior from historical records of all (relevant) actions executed on it [55].
- *Freshness*: This process should assure that the information is fresh and that old messages cannot be replayed by any adversary.
- *Availability*: This process should ensure that only the relevant network services should be made available to authorized entities even under denial-of-service attacks in the SG environment.
- *Third-party protection*: This prevents damage performed by third parties (e.g., the service providers) via the communication systems [55].
- *Forward secrecy*: If any device leaves the SG environment, it must no longer have access to future messages.
- *Backward secrecy*: When a new device is added to the SG environment, it must not know any previously transmitted messages.

13.3.3 Security Attacks in Smart Grid Environment

A smart grid typically faces the following attacks:

- *Eavesdropping*: An eavesdropping attack, also known as a sniffing or snooping attack, occurs when an adversary can intercept the digital communications between two (or more) entities. An eavesdropping attack is a serious attack because it is a prerequisite for other attacks.
- *Traffic analysis*: Traffic analysis intercepts and examines the intercepted messages to know information from patterns in communication.
- *Replay attack*: In a replay attack, an adversary tries to misdirect another authorized party by recording the transmitted messages and later reusing them in an attack.
- *Man-in-the-middle attack*: In a man-in-the-middle attack, an adversary eavesdrops on the transmitted messages and then attempts to modify or delete the contents of the messages delivered to the recipients.
- *Impersonation attack*: In an impersonation attack, an adversary successfully assumes the identity of one of the authorized entities in the system or in a communication protocol.

- *Denial-of-service attack*: A denial-of-service (DoS) attack is one where an adversary takes an action that prevents authorized users from accessing targeted computer systems, devices, or other network resources. A distributed DoS (DDoS) attack occurs in the network when multiple systems flood the resources or bandwidth of a targeted system, typically one or more web servers.
- *Malware attack*: In a malware attack, an adversary executes malicious programs in order to perform various activities such as stealing, deleting, or encrypting sensitive information, altering, or hijacking core computing functions, and also monitoring users' computer activity without their permission. Malware includes worms, computer viruses, spyware, and Trojan horses.
- *Resilience against physical device capture attack*: As we mentioned in the threat model in Section 13.3.1, some devices can be physically captured by an adversary. The adversary can use the extracted information from the captured devices to launch other attacks (such as impersonation attacks) in the network.
- *Resilience against new devices' deployment attacks*: In an access control scheme, several attacks can be launched by an adversary, such as wormhole, Sybil, device replication, and malicious devices deployment attacks.

In a wormhole attack [62], an adversary may tunnel the information between two distant locations with the help of an in-band or out-of-band channel. A wormhole tunnel can be then created by a pair of attacker nodes, which convince two distant devices that they are near to each other. Under this attack, several other attacks, including sniffing, modification, and dropping are possible.

A Sybil attack is possible when an illegal device falsely assumes multiple identities [63, 64]. However, it is not necessary that the claimed identities are from the existing devices' identities.

Using a device replication attack [65], the purpose of an adversary is intentionally to make several replicas of a compromised device, which are then inserted into the network. The attacker can physically capture some devices, extract the credentials from their memory, load such information in some other fake devices, and then deploy them in the network so that those fake devices can connect with the existing devices in the network.

- *Insider attack*: An insider attack is considered as a malicious attack where a privileged-insider user of a network or computer system misuses the authorized system access.
- *Address resolution protocol (ARP) spoofing attack*: This is an attack in which a malicious adversary sends falsified ARP messages over a local area network. This attack leads to linking of an attacker's media access control (MAC) address with the IP address of an authorized computer or server on the network.

- *Database attack*: This refers to several attacks related to the database. For example, in a structured query language (SQL) injection attack an attacker wishing to execute the SQL injection manipulates a standard SQL query in order to exploit nonvalidated input vulnerabilities in a database.
- *Anonymity and untraceability*: With the anonymity property, an adversary does not know the real identity of an entity during the communication. The untraceability property prevents the adversary's actions from being traced back to the adversary. A security protocol in the SG environment should preserve both anonymity and untraceability properties.

13.4 Security Solutions in Smart Grid

In this section, we briefly discuss some security solutions based on the taxonomy presented in Fig. 13.4.

13.4.1 Key Management

Lv et al. [66] designed a remote key generation and distribution scheme for the SG environment. Their scheme uses public key-wrapping in the SG application scenario. Key-wrapping is a technique which helps in designing an adapter for smart meters. The adapter creates a secure channel between the collector device and a smart meter, which is used for secure transmission of the usage data from the smart meter. In this scheme, a smart meter sends a session key request to the collector device connected to the key management server. The server then sends the session key to the smart meter. The scheme is secure under the CK-adversary model (as discussed in the threat model in Section 13.3.1). Since the key-wrapping method has a weaker security assumption and, furthermore, as some computation cost is moved from smart meters to servers, the computation cost for smart meters is reduced in their scheme.

Suhendra et al. [67] proposed a session key management protocol in smart grid. Their scheme has the ability to detect and revoke any compromised device in the SG environment. Since the key distribution center (KDC) stores expiration dates along with the session keys in its key table, this scheme is vulnerable to a stolen verifier attack if an adversary has access to the key table. Benmalek et al. [68] also designed four key management protocols for advanced metering infrastructure (AMI) for secure data communications in the SG environment. Their proposed protocols meet the different requirements of the power utility, such as transmission of data and distribution of electricity requirements. A more in-depth discussion of key-management protocols for the SG environment is given in [69].

13.4.2 User Authentication

To manage demand response during peak hours, user authentication between a vehicle user and a smart meter is essential. To address this important problem,

Wazid et al. [47] designed a three-factor lightweight authentication protocol for the SG environment, called TUAS-RESG. In TUAS-RESG, after successful mutual authentication between a user and a smart meter, they establish a session key for their secure communication in future. The three factors used in TUAS-RESG are the user's mobile device, password and personal biometrics. The local biometric verification at the mobile device is performed by the widely accepted biometric fuzzy extractor technique [70] in TUAS-RESG. TUAS-RESG supports local password and biometric update phase and does not need to contact the trusted authority from changing the password/biometrics of a registered user in the system. In addition, TUAS-RESG also supports dynamic smart meter addition phase after initial deployment of the nodes in the network. Moreover, the authors evaluated TUAS-RESG using the NS2 simulator.

13.4.3 Device Authentication

Wu and Zhou [71] proposed an authentication scheme in which mutual authentication between a smart sensor (device) and a data collector is achieved with the help of a trusted anchor. Their scheme relies on both symmetric key as well as public key (elliptic curve cryptography) techniques. However, their scheme fails to provide perfect forward secrecy, session key security, and strong smart meter credentials' privacy, and it also does not protect against man-in-the-middle attack [47].

Tsai and Lo [31] designed an anonymous authentication scheme for the SG environment in which mutual authentication between a smart meter and a service provider is achieved. Their scheme uses identity-based signature and identity-based encryption mechanisms. Unfortunately, their scheme is vulnerable to ephemeral secret leakage attack, and it also does not provide strong credentials' privacy of the smart meter [5]. To address these drawbacks, Odelu et al. [5] designed a secure authenticated key agreement protocol.

Yan et al. [72] presented an authentication scheme in which authentication is performed between a smart meter and the gateway node of a building area network (BAN). Their scheme is lightweight, as it uses a computationally efficient one-way cryptographic hash function. Although their scheme provides key refreshment and incurs a low computational cost, it does not provide smart meter anonymity and untraceability properties.

13.4.4 Access Control

Jung et al. [73] proposed an access control mechanism in the SG environment. In their mechanism, the data owners specify access policies that identify applications that are allowed to access data related to the data owner. The X.509 certificate used in the scheme is stored in a user smart card along with his/her username and password. Since their scheme is based on public key infrastructure (PKI), it is computationally expensive.

Duan et al. [74] proposed a fata-centric access control framework (DCACF) to support secure access control under a publish/subscribe model. DCACF consists of the following two phases. In the authorization phase, a publisher *P* translates a policy into an access credential for the subscriber *S* with the help of a trusted third party called the broker. In the policy enforcement phase, *P* publishes the encrypted events, and after that, *S* decrypts the subscribed events under the intervention of the connected notification broker. In DCACF, only qualified subscribers can access the encrypted events, whereas unqualified subscribers cannot access the encrypted events even if they collaborate with each other.

Mutsvangwa et al. [75] proposed an access control scheme that allows the authorized users to decrypt data provided that valid attributes are with them. In their scheme, the encryption relies on the public key-based Diffie–Hellman key establishment protocol and hash-based message authentication code. As a result, their scheme is computationally expensive.

13.4.5 Secure Transport Protocol

Existing protocols do not satisfy the scalable secure transport requirements for smart grid sensor data collection [50]. To address this issue, Kim et al. [50] proposed a scalable and secure transport protocol (SSTP) for smart grid data collection. Their experimental evaluation demonstrated that combining transmission control protocol (TCP) with the transport layer security (TLS) protocol has scalability issues for sensor data collection in a large-scale network; in addition, TCP by itself is not appropriate for periodic sensor data collection. In SSTP, this scalability issue has been addressed.

13.4.6 Privacy-Preservation

Vehicle to grid (V2G) network is a significant part of the SG environment with the home area network (HAN), industry area network (IAN), neighborhood area network (NAN), and building area network (BAN) [54, 76]. In a V2G network, electric vehicles communicate with the service providers with the help of aggregator nodes or other networks such as HANs.

Blind signature is one of the promising cryptographic techniques used to design privacy-preserving billing and payment systems for a V2G network. The ID-based partial restrictive blind signature is utilized in the schemes [77, 78], where the unconditional anonymity property is preserved. The advantage of these schemes is that they can protect customers' privacy. The authors of [76] presented a detailed survey on privacy preservation issues for V2G networks in the smart grid environment.

Smart meter reading streams may pose severe privacy threats to the electricity consumers on the power grid environment. To handle the smart meter privacy issues, Hong et al. [51] proposed a privacy model for reading streams and then designed a privacy-preserving streaming mechanism that can efficiently protect the privacy of smart meter readings. In order to protect sensitive energy usage information of consumers, Badra and Zeadally [52] presented a virtual ring architecture that can provide a privacy protection solution using symmetric or asymmetric encryptions of customers' requests belonging to the same group. Their smart grid privacy solution is simple, scalable, cost-effective, and incurs minimal computational processing overheads. In addition, it also protects the privacy of smart meter readings.

13.4.7 Trusted Computing

Smart grid communication needs a holistic approach to enforce end-to-end security, which encompasses all aspects related to smart grid operations. Trusted computing plays a pivotal role in this vision [55]. In [55], the authors discussed a basic trusted computing model that can be used in the SG environment. The main design goal of this model is to provide a minimal, and therefore manageable, and stable security for conventional hardware platforms, embedded systems, servers, and mobile devices.

13.4.8 Intrusion Detection

Beigi-Mohammadi et al. [79] discussed various requirements for intrusion detection systems (IDSs) in NAN as the components of the SG environment. They proposed a distributed IDS that considers the specifications and requirements of NAN. Their IDS can detect wormhole attack, which is considered as a serious attack in NAN.

Ullah and Mahmoud [80] also proposed an IDS framework in the SG environment. Their proposed IDS system can collect and correlate alerts from various sensing devices (e.g., smart meters in the SG infrastructure) which can be deployed in HANs, NANs, and wide area network (WAN) in the SG network.

13.5 Deployment and Implementation of Cyber-Physical Smart Grid Testbeds

In this section, we discuss various cyber security testbed platforms that are being deployed and implemented in the SG environment in order to investigate vulnerabilities of the power critical infrastructure along with the wide-area situational awareness research activities.

The challenges faced in the areas of network device security include firewalls, routers, attack scenarios, encryption, countermeasures, forensic analysis, and intrusion analysis. The network attack scenarios for smart grid applications include several attacks, such as man-in-the-middle attack, DoS attack, and replay attack. Table 13.2 presents a list of cyber-attack testing capabilities of several recent cyber-physical smart grid testbeds [16].

TABLE 13.2 Current Cyber-Physical Smart Grid Testbeds [16]

Testbed	Target Research Area	Area Covered in the Smart Grid Domain	Type of Test Platform
Sandia Lab Virtual Control System Environment (VCSE) [81]	Wide area situational awareness, cyber security	Transmission, operations	Simulator
Virtual Power System Testbed at the University of Illinois at Urbana-Champaign [82]	Cyber security	Transmission, operations	Simulator
Florida International University Smart Grid Testbed [83, 84]	Wide area situational awareness, distributed grid management, network connections, cyber security	Transmissions, operations, customer	Simulator
Intrusion and Defense Testbed at University College Dublin [85]	Cyber security	Transmission, operations	Hybrid
SCADA Security Lab at Mississippi State University [86]	Cyber security, network communications	Transmission, operations	Real-time Simulator
DeterLab at the University of Southern California [87]	Cyber security	—	Hardware
PowerCyber Testbed at Iowa State University [88]	Wide area situational awareness, cyber security	Transmission, operations	Real-time Simulator
Cyber-Physical System Testbed at Texas A&M [89, 90]	Wide area situational awareness, cyber security	Transmission, operations	Real-time Simulator
Cyber-Physical Testbed at Washington State University [91, 92]	Wide area situational awareness, cyber security	Transmission, operations	Real-time Simulator
Network Intrusion Detection System (NIDS) [93] at Sapienza University of Rome & Arizona State University	Cyber security, network communications	Operations	Hybrid
Cybersecurity Testbed for IEC 61850 at Queen's University Belfast [94]	Cyber security, network communications	Operations	Real-time Simulator

Next, we present a brief overview of the existing testbeds shown in Table 13.2 that focuses on security and privacy research in the smart grid environment. In this table, we mainly consider power system operations, such as monitoring, control, protection, analysis, and transmission level operations. Sandia National Laboratories developed the Virtual Control System Environment (VCSE). Supervisory Control and Data Acquisition (SCADA) Security Lab at Mississippi State University and Virginia Tech. designed a testbed that can probe various SCADA vulnerabilities of energy systems [81]. In this testbed, encryption as well as secured data communication channels on the internet protocol (IP)-routed computer networks were investigated. The encryption and authentication are applied to protect all data being transmitted and all transmission links.

The University of Illinois at Urbana-Champaign developed the virtual power system testbed (VPST) [82]. It enables integration with other testbeds to analyze cyber security attacks on a large-scale power grid. This testbed is used to investigate reliability issues that can occur on the national power grid. VPST also provides a remote connection access facility to it.

Energy Systems Research Laboratory (ESRL) [83, 84], developed at Florida International University, is a comprehensive testbed, which facilitates hardware-based testbed setup including a total power capability of up to 136-kW with conventional power generation and storage capabilities, as well as renewable units.

The intrusion and defense testbed [85], designed at University College Dublin, provides a cyber power system, which is composed of a power system simulator and a substation automation platform. In this testbed, cyber security intrusion and anomaly detection can be tested. The SCADA security laboratory [86] was developed at Mississippi State University. It is a cyber-physical testbed that supports the investigation of cyber security vulnerabilities and forensic studies.

DeterLab [87] is a security research and educational platform at the University of Southern California. This testbed provides various tests, such as the detection of buffer overflows, SQL injections, network intrusion detection, computer forensics, ARP spoofing, transmission control protocol (TCP) SYN flooding, which is a type of distributed denial of service (DDoS) attack, man-in-the-middle attack, and worm modeling. The PowerCyber testbed [88], designed at Iowa State University, is another comprehensive smart grid testbed which supports information and communication technology (ICT) and security. The cyber-physical system testbed designed at Texas A&M University [89, 90] supports simulations of various industrial protocols and technologies.

The cyber-physical system testbeds developed at Washington State University [91, 92] provides investigation of various wide-area situational awareness, cyber security and network communications research areas. A network intrusion detection system (NIDS) testbed [93] was developed by at Sapienza University of Rome and Arizona State University, which implements several intrusion detection rules. These rules check abnormality or attack criteria based on a packet sequence as well as the time gap between cycles of packets. Finally, the cyber security testbed for International Electrotechnical Commission (IEC) 61,850 [94] was designed at Queen's University Belfast in the United Kingdom, for focusing on IEC 61850 vulnerabilities. Several cyberattacks, such as DoS, man-in-the-middle, ARP spoofing, and database attacks can be performed using this testbed.

13.6 Conclusion

This chapter has presented and discussed data security issues in the smart grid environment. We classified the security protocols in smart grid into several research areas, namely: key management, secure transport protocol, user and device authentication, access and user access control, trusted computing, intrusion detection system, and privacy-preservation. To review several security protocols, we defined a general threat model and the various security requirements that must be supported to thwart attacks on the Smart Grid environment. We then reviewed some state-of-art security protocols that have been recently proposed for the Smart Grid. We also briefly present several security testbed platforms for the Smart Grid that have been developed over the last few years at various institutions around the world in order to investigate vulnerabilities of the critical power infrastructure along with wide-area situational awareness research.

References

[1] C. Alcaraz, S. Zeadally, Critical control system protection in the 21st century, IEEE Computer 46 (10) (2013) 74–83.

[2] C. Alcaraz, S. Zeadally, Critical infrastructure protection: requirements and challenges for the 21st century, Int. J. Crit. Infrastruct. Prot. 8 (2015) 53–66.

[3] D. Boroyevich, I. Cvetkovic, R. Burgos, D. Dong, Intergrid: a future electronic energy network, IEEE J. Emerg. Sel. Top. Power Electr. 1 (3) (2013) 127–138.

[4] X. Fang, S. Misra, G. Xue, D. Yang, Smart grid-the new and improved power grid: a survey, IEEE Commun. Surv. Tut. 14 (4) (2012) 944–980.

[5] V. Odelu, A.K. Das, M. Wazid, M. Conti, Provably secure authenticated key agreement scheme for smart grid, IEEE Trans. Smart Grid 9 (3) (2018) 1900–1910.

[6] W. Su, H. Eichi, W. Zeng, M.-Y. Chow, A survey on the electrification of transportation in a smart grid environment, IEEE Trans. Ind. Inform. 8 (1) (2012) 1–10.

[7] F. Akhtar, M.H. Rehmani, Energy replenishment using renewable and traditional energy resources for sustainable wireless sensor networks: a review, Renew. Sustain. Energy Rev. 45 (2015) 769–784.

[8] M. Erol-Kantarci, H.T. Mouftah, Energy-efficient information and communication infrastructures in the smart grid: a survey on interactions and open issues, IEEE Commun. Surv. Tut. 17 (1) (2015) 179–197.

[9] A.A. Khan, M.H. Rehmani, M. Reisslein, Cognitive radio for smart grids: survey of architectures, spectrum sensing mechanisms, and networking protocols, IEEE Commun. Surv. Tut. 18 (1) (2016) 860–898.

[10] N. Nezamoddini, S. Mousavian, M. Erol-Kantarci, A risk optimization model for enhanced power grid resilience against physical attacks, Electr. Pow. Syst. Res. 143 (2017) 329–338.
[11] M.H. Rehmani, M.E. Kantarci, A. Rachedi, M. Radenkovic, M. Reisslein, IEEE access special section editorial smart grids: a hub of interdisciplinary research, IEEE Access 3 (2015) 3114–3118.
[12] G.A. Shah, V.C. Gungor, O.B. Akan, A cross-layer QoS-aware communication framework in cognitive radio sensor networks for smart grid applications, IEEE Trans. Ind. Inform. 9 (3) (2013) 1477–1485.
[13] S. Uludag, S. Zeadally, M. Badra, Techniques, Taxonomy, and Challenges of Privacy Protection in the Smart Grid, Springer International Publishing, Cham/London, 2015, pp. 343–390.
[14] R. Yu, W. Zhong, S. Xie, C. Yuen, S. Gjessing, Y. Zhang, Balancing power demand through EV mobility in vehicle-to-grid mobile energy networks, IEEE Trans. Ind. Inform. 12 (1) (2016) 79–90.
[15] G. Locke and P. D. Gallagher. NIST framework and roadmap for smart grid interoperability standards, 2010. Release 1.0, 2010. Avialble at: http://www.nist.gov/sites/default/files/documents/public affairs/releases/smartgrid interoperability final.pdf. Accessed on January 2018.
[16] M.H. Cintuglu, O.A. Mohammed, K. Akkaya, A.S. Uluagac, A survey on smart grid cyber-physical system testbeds, IEEE Commun. Surv. Tutor. 19 (1) (2017) 446–464.
[17] M.H. Albadi, E.F. El-Saadany, A summary of demand response in electricity markets, Electr. Pow. Syst. Res. 78 (11) (2008) 1989–1996.
[18] A. Mohd, E. Ortjohann, A. Schmelter, N. Hamsic, and D. Morton. Challenges in integrating distributed energy storage systems into future smart grid. In IEEE International Symposium on Industrial Electronics, Cambridge, pp. 1627–1632, 2008.
[19] G. Andersson, P. Donalek, R. Farmer, N. Hatziargyriou, I. Kamwa, P. Kundur, N. Martins, J. Paserba, P. Pourbeik, J. Sanchez-Gasca, R. Schulz, A. Stankovic, C. Taylor, V. Vittal, Causes of the 2003 major grid blackouts in North America and Europe, and recommended means to improve system dynamic performance, IEEE Trans. Power Syst. 20 (4) (2005) 1922–1928.
[20] R.R. Mohassel, A. Fung, F. Mohammadi, K. Raahemifar, A survey on advanced metering infrastructure, Int. J. Electr. Power Energy Syst. 63 (2014) 473–484.
[21] S. Mehar, S. Zeadally, G. Remy, S. Senouci, Sustainable transportation management system for a fleet of electric vehicles (STMS), IEEE Trans. Intell. Transp. Syst. 16 (3) (2015) 1401–1414.
[22] R. Das, V. Madani, F. Aminifar, J. McDonald, S.S. Venkata, D. Novosel, A. Bose, M. Shahidehpour, Distribution automation strategies: evolution of technologies and the business case, IEEE Trans. Smart Grid 6 (4) (2015) 2166–2175.
[23] W. Luan, D. Sharp, and S. Lancashire. Smart grid communication network capacity planning for power utilities. In IEEE/PES Transmission & Distribution Conference and Exposition (T&D'10), New Orleans, LA, 1–4, 2010.
[24] Z.M. Fadlullah, M.M. Fouda, N. Kato, A. Takeuchi, N. Iwasaki, Y. Nozaki, Toward intelligent machine-to-machine communications in smart grid, IEEE Commun. Mag. 49 (4) (2011) 60–65.
[25] P. P. Parikh, M. G. Kanabar, and T. S. Sidhu. Opportunities and challenges of wireless communication technologies for smart grid applications. In IEEE PES General Meeting, Providence, RI, 1–7, 2010.
[26] A. Lee and T. Brewer. 2009. Smart grid cyber security strategy and requirements, US Department of Commerce, Draft Interagency Technical Report NISTIR 7628. National Institute of Standards and Technology (NIST): Gaithersburg, MD.

[27] G.N. Ericsson, Cyber Security and power system communication–essential parts of a smart grid infrastructure, IEEE Trans. Power Del. 25 (3) (2010) 1501–1507.

[28] Y. Mo, T.H. Kim, K. Brancik, D. Dickinson, H. Lee, A. Perrig, B. Sinopoli, Cyber–physical security of a smart grid infrastructure, Proc. IEEE 100 (1) (2012) 195–209.

[29] S. Clements and H. Kirkham. Cyber-security considerations for the smart grid. In IEEE PES General Meeting, Providence, RI, 1–5, 2010.

[30] X. Li, X. Liang, R. Lu, X. Shen, X. Lin, H. Zhu, Securing smart grid: cyber attacks, countermeasures, and challenges, IEEE Commun. Mag. 50 (8) (2012) 38–45.

[31] J. Tsai, N. Lo, Secure anonymous key distribution scheme for smart grid, IEEE Trans. Smart Grid 7 (2) (2016) 906–914.

[32] F. Siddiqui, S. Zeadally, C. Alcaraz, and S. Galvao. Smart grid privacy: issues and solutions. In IEEE 21st International Conference on Computer Communications and Networks (ICCCN'12), Munich, 1–5, 2012.

[33] P. McDaniel, S. McLaughlin, Security and privacy challenges in the smart grid, IEEE Secur. Priv. 7 (3) (2009) 75–77.

[34] W. Diffie, M.E. Hellman, New directions in cryptography, IEEE Trans. Inf. Theory 22 (1976) 644–654.

[35] Advanced Encryption Standard (AES) FIPS PUB 197, National Institute of Standards and Technology (NIST), US Department of Commerce 2001. http://csrc.nist.gov/publications/fips/fips197/fips-197.pdf.

[36] C. Blundo, A. De Santis, A. Herzberg, S. Kutten, U. Vaccaro, and M. Yung. Perfectly-secure key distribution for dynamic conferences. In 12th Annual International Cryptology Conference—Advances in Cryptology (CRYPTO'92), vol. 740 Lecture Notes in Computer Science, pp. 471–486, Santa Barbara, CA, Springer, 1993.

[37] H. Chan, A. Perrig, and D. Song. Random key predistribution schemes for sensor networks. In IEEE Symposium on Security and Privacy (S&P'03), Berkeley, CA, pp. 197–213, 2003.

[38] D. Chen, G. Chang, D. Sun, J. Jia, X. Wang, Lightweight key management scheme to enhance the security of internet of things, Int. J. Wirel. Mob. Comput. 5 (2) (2012) 191–198.

[39] A.K. Das, A random key establishment scheme for multi-phase deployment in large-scale distributed sensor networks, Int. J. Inf. Secur. 11 (3) (2012) 189–211.

[40] L. Eschenauer and V. D. Gligor. A key management scheme for distributed sensor networks. In 9th ACM Conference on Computer and Communication Security (CCS'02), Washington, DC, pp. 41–47, 2002.

[41] D. Liu, P. Ning, Improving key pre-distribution with deployment knowledge in static sensor networks, ACM Trans. Sensor Netw. 1 (2) (2005) 204–239.

[42] D. Liu, P. Ning, R. Li, Establishing pairwise keys in distributed sensor networks, ACM Trans. Inf. Syst. Secur. 8 (1) (2005) 41–77.

[43] R. Blom. An optimal class of symmetric key generation systems. In Workshop on the Theory and Application of Cryptographic Techniques—Advances in Cryptology (EUROCRYPT'84), Lecture Notes in Computer Science, vol. 209, pp. 235–238, Paris, Springer-Verlag, 1985.

[44] W. Du, J. Deng, Y. S. Han, and P. K. Varshney. A pairwise key pre-distribution scheme for wireless sensor networks. In ACM Conference on Computer and Communications Security (CCS'03), Washington, DC, pp. 42–51, 2003.

[45] Y. H. Kim, H. Lee, D. H. Lee, and J. Lim. A key management scheme for large scale distributed sensor networks. In Proceedings of the IFIP TC6 11th International Conference (PWC'06), Lecture Notes in Computer Science, Springer-Verlag, Albacete, Spain, vol. 4217, pp. 437–446, 2006.

[46] S. Zhu, S. Setia, S. Jajodia, LEAP+: efficient security mechanisms for large-scale distributed sensor networks, ACM Trans. Sens. Netw. 2 (4) (2006) 500–528.

[47] M. Wazid, A.K. Das, N. Kumar, J.J.P.C. Rodrigues, Secure three-factor user authentication scheme for renewable-energy-based smart grid environment, IEEE Trans. Ind. Inform. 13 (6) (2017) 3144–3153.

[48] F. Jameel, Z. Hamid, F. Jabeen, S. Zeadally, M.A. Javed, A survey of device-to-device communications: research issues and challenges, IEEE Commun. Surv. Tut. 20 (3) (2018) 2133–2168.

[49] Information Technology. X.509: Information technology—Open Systems Interconnection—The Directory: Public-key and attribute certificate frameworks, 2016. https://www.itu.int/rec/T-REC-X.509.

[50] Y. Kim, V. Kolesnikov, H. Kim, and M. Thottan. SSTP: A scalable and secure transport protocol for smart grid data collection. In IEEE International Conference on Smart Grid Communications (SmartGridComm'11), Brussels, Belgium, pp. 161–166, 2011.

[51] Y. Hong, W.M. Liu, L. Wang, Privacy preserving smart meter streaming against information leakage of appliance status, IEEE Trans. Inform. Foren. Sec. 12 (9) (2017) 2227–2241.

[52] M. Badra, S. Zeadally, Design and performance analysis of a virtual ring architecture for smart grid privacy, IEEE Trans. Inform. Foren. Secur. 9 (2) (2014) 321–329.

[53] S. Zeadally, A.-S.K. Pathan, C. Alcaraz, M. Badra, Towards privacy protection in smart grid, Wirel. Pers. Commun. 73 (1) (2013) 23–50.

[54] G. Si, Z. Guan, J. Li, P. Liu, and H. Yao. A comprehensive survey of privacy-preserving in smart grid. In International Conference on Security, Privacy and Anonymity in Computation, Communication and Storage (SpaCCS'16), Zhangjiajie, China, 213–223, 2016.

[55] Y. Yan, Y. Qian, H. Sharif, D. Tipper, A survey on cyber Security for smart grid communications, IEEE Commun. Surv. Tut. 14 (4) (2012) 998–1010.

[56] D. Dolev, A.C. Yao, On the security of public key protocols, IEEE Trans. Inf. Theory 29 (2) (1983) 198–208.

[57] R. Canetti and H. Krawczyk. Analysis of key-exchange protocols and their use for building secure channels. In International Conference on the Theory and Applications of Cryptographic Techniques—Advances in Cryptology (EUROCRYPT'01), Springer, Innsbruck (Tyrol), Austria, pp. 453–474. 2001.

[58] R. Canetti and H. Krawczyk. Universally composable notions of key exchange and secure channels. In International Conference on the Theory and Applications of Cryptographic Techniques–Advances in Cryptology (EUROCRYPT'02), Amsterdam, The Netherlands, pp. 337–351, 2002.

[59] P. Kocher, J. Jaffe, and B. Jun. Differential power analysis. In Proceedings of Advances in Cryptology—CRYPTO'99, Lecture Notes in Computer Science, vol. 1666, pp. 388–397, 1999.

[60] T.S. Messerges, E.A. Dabbish, R.H. Sloan, Examining smart-card security under the threat of power analysis attacks, IEEE Trans. Comput. 51 (5) (2002) 541–552.

[61] W. Stallings, Cryptography and Network Security: Principles and Practice, 5th Edn, Prentice Hall Press, Upper Saddle River, NJ, 2010.

[62] Y. Hu, A. Perrig, and D. B. Johnson. Pachet leashes: a defense against wormhole attacks in wireless networks. In Proceedings of IEEE International Conference on Computer and Communications (INFOCOM'03), San Francisco, CA, vol. 3, 1976–1986, 2003.

[63] J. R. Douceur. The sybil attack. In First International Workshop on Peer-to-Peer Systems (IPTPS'02), Lecture Notes in Computer Science (LNCS), Springer-Verlag vol. 2429, pp. 251–260, Cambridge, MA, 2002.

[64] J. Newsome, E. Shi, D. Song, and A. Perrig. The Sybil attack in sensor networks: analysis and defenses. In Proceedings of third IEEE International Conference on Information Processing in Sensor Networks (IPSN'04), Berkeley, CA, 259–268, 2004.

[65] B. Parno, A. Perrig, and V. Gligor. Distributed detection of node replication attacks in sensor networks. In IEEE Symposium on Security and Privacy (S&P'05), Oakland, CA, pp. 49–63, 2005.

[66] X. Lv, Y. Mu, H. Li, Key management for smart grid based on asymmetric keywrapping, Int. J. Comput. Math. 92 (3) (2015) 498–512.

[67] V. Suhendra, Y. Wu, H. Saputra, Z. Zhao, and F. Bao. Lightweight key management protocols for smart grids. In IEEE International Conference on Internet of Things (iThings) and IEEE Green Computing and Communications (GreenCom) and IEEE Cyber, Physical and Social Computing (CPSCom) and IEEE Smart Data (Smart-Data), Chengdu, China, pp. 345–348, 2016.

[68] M. Benmalek, Y. Challal, A. Derhab, A. Bouabdallah, VerSAMI: Versatile and scalable key management for smart grid AMI systems, Comput. Netw. 132 (2018) 161–179.

[69] M. Badra and S. Zeadally. Key management solutions in the smart grid environment. In 6th Joint IFIP Wireless and Mobile Networking Conference (WMNC), Dubai, United Arab Emirates, 1–7, 2013.

[70] Y. Dodis, L. Reyzin, A. Smith, Fuzzy extractors: how to generate strong keys from biometrics and other noisy data, in: Advances in Cryptology–Eurocrypt 2004, Springer, Interlaken, 2004, pp. 523–540.

[71] D. Wu, C. Zhou, Fault-tolerant and scalable key management for smart grid, IEEE Trans. Smart Grid 2 (2) (2011) 375–381.

[72] L. Yan, Y. Chang, S. Zhang, A lightweight authentication and key agreement scheme for smart grid, Int. J. Distrib. Sens. Netw. 13 (2) (2017) 1–7.

[73] M. Jung, T. Hofer, S. Dobelt, G. Kienesberger, F. Judex, and W. Kastner. Access control for a smart grid SOA. In International Conference for Internet Technology and Secured Transactions, London, pp. 281–287, 2012.

[74] L. Duan, D. Liu, Y. Zhang, S. Chen, R.P. Liu, B. Cheng, J. Chen, Secure data-centric access control for smart grid services based on publish/subscribe systems, ACM Trans. Internet Technol. 16 (4) (2016) 23:1–23:17.

[75] A. Mutsvangwa, B. Nleya, and B. Nleya. Secured access control architecture consideration for smart grids. In IEEE PES Power Africa, Livingstone, Zambia, 228–233, 2016.

[76] W. Han, Y. Xiao, Privacy preservation for V2G networks in smart grid: a survey, Comput. Commun. 91–92 (2016) 17–28.

[77] M.H. Au, J.K. Liu, J. Fang, Z.L. Jiang, W. Susilo, J. Zhou, A new payment system for enhancing location privacy of electric vehicles, IEEE Trans. Veh. Technol. 63 (1) (2014) 3–18.

[78] J. K. Liu, M. H. Au, W. Susilo, and J. Zhou. Enhancing location privacy for electric vehicles (at the right time). In 17th European Research in Computer Security, Pisa, Italy, 397–414, 2012.

[79] N. Beigi-Mohammadi, J. Misic, H. Khazaei, and V. B. Misic. An intrusion detection system for smart grid neighborhood area network. In IEEE International Conference on Communications (ICC'14), Sydney, NSW, 4125–4130, 2014.

[80] I. Ullah and Q. H. Mahmoud. An intrusion detection framework for the smart grid. In IEEE 30th Canadian Conference on Electrical and Computer Engineering (CCECE'17), Windsor, ON, 1–5, 2017.

[81] M.J. McDonald, G.N. Conrad, T.C. Service, R.H. Cassidy, Cyber Effects Analysis Using VCSE, Sandia National Laboratories, Albuquerque, NM, 2008 Technical Report SAND2008-5954.

[82] D. C. Bergman, D. Jin, D. M. Nicol, and T. Yardley. The virtual power system testbed and inter-testbed integration. In Proceedings of the 2nd Conference on Cyber Security Experimentation and Test (CSET'09), Montreal, QC, 5, 2009.

[83] V. Salehi, A. Mohamed, A. Mazloomzadeh, O.A. Mohammed, Laboratory-based smart power system, part I: design and system development, IEEE Trans. Smart Grid 3 (3) (2012) 1394–1404.

[84] V. Salehi, A. Mohamed, A. Mazloomzadeh, O.A. Mohammed, Laboratory-based smart power system, Part II: control, monitoring, and protection, IEEE Trans. Smart Grid 3 (3) (2012) 1405–1417.

[85] J. Hong, S. Wu, A. Stefanov, A. Fshosha, C. Liu, P. Gladyshev, and M. Govindarasu. An intrusion and defense testbed in a cyber-power system environment. In IEEE Power and Energy Society General Meeting, Detroit, MI, 1–5, 2011.

[86] T. Morris, A. Srivastava, B. Reaves, W. Gao, K. Pavurapu, R. Reddi, A control system testbed to validate critical infrastructure protection concepts, Int. J. Crit. Infrastruct. Prot. 4 (2) (2011) 88–103.

[87] J. Mirkovic, T. Benzel, Teaching cyber security with deter lab, IEEE Secur. Priv. 10 (1) (2012) 73–76.

[88] A. Hahn, A. Ashok, S. Sridhar, M. Govindarasu, Cyber-physical Security Testbeds: architecture, application, and evaluation for smart grid, IEEE Trans. Smart Grid 4 (2) (2013) 847–855.

[89] B. Chen, K. L. Butler-Purry, A. Goulart, and D. Kundur. Implementing a real-time cyber-physical system test bed in RTDS and OPNET. In North American Power Symposium (NAPS), Pullman, WA, 1–6, 2014.

[90] B. Chen, N. Pattanaik, A. Goulart, K. L. Butler-Purry, and D. Kundur. Implementing attacks for modbus/TCP protocol in a real-time cyber physical system test bed. In IEEE International Workshop Technical Committee on Communications Quality and Reliability (CQR), Charleston, SC, 1–6, 2015.

[91] S. S. Biswas, F. Shariatzadeh, R. Beckstrom, and A. K. Srivastava. Real time testing and validation of smart grid devices and algorithms. In IEEE Power Energy Society General Meeting, Vancouver, BC, 1–5, 2013.

[92] C.B. Vellaithurai, S.S. Biswas, A.K. Srivastava, Development and application of a real-time test bed for cyber–physical system, IEEE Syst. J. 11 (4) (2017) 2192–2203.

[93] G. Koutsandria, R. Gentz, M. James, A. Scaglione, S. Peisert, and C. McParland. A real-time testbed environment for cyber-physical security on the power grid. In Proceedings of the First ACM Workshop on Cyber-Physical Systems-Security and/or Privacy (CPS-SPC'15), Denver, CO, 67–78, 2015.

[94] Y. Yang, H. T. Jiang, K. McLaughlin, L. Gao, Y. B. Yuan, W. Huang, and S. Sezer. Cybersecurity test-bed for IEC 61850 based smart substations. In IEEE Power Energy Society General Meeting, Denver, CO, 1–5, 2015.

Chapter 14

The Socio-Economic Challenges of Smart Grids

Peter Connor and Oscar Fitch-Roy
University of Exeter, Penryn, United Kingdom

Chapter Outline

14.1 Introduction

Supply and demand for electricity is being reshaped across time and space by shifting social, environmental, and economic priorities, and by technical innovation. On the supply-side, renewable energy generation, often at smaller scales, is rising in nearly all systems, leading to new, more geographically distributed portfolios that are more dependent on temporally variable resources. At the same time, new sources of demand, such as electric mobility and the electrification of heat services through technologies such as heat pumps, is fundamentally changing where, when, and how much electricity is consumed. The traditional approach to electrical grid management, based around centralized generation and unidirectional flows of power are no longer appropriate for managing and balancing the demands of these new technologies, which are more distributed, where power flow is often bidirectional, and where demand in individual homes or other buildings may be higher and more volatile. The emergence of smart grids is an adaptive response by electricity systems to these potentially transformative social and technical factors. The key aim is to deal with the added complexity of the evolved energy system, allowing for balancing

Pathways to a Smarter Power System. https://doi.org/10.1016/B978-0-08-102592-5.00014-4

of more diverse and variable energy generation, and more variable and more volatile demand. It does so by acting as a cheaper alternative to greater volumes of new wire infrastructure, thus minimizing total systemic costs.

The technical adaptations and innovations that enable smart grids are accompanied by profound changes to the socio-economic status-quo, which has remained largely unchanged since the birth of the electricity industry. The business models for commercial electricity generation, transmission, and distribution are changing as new revenue streams, roles, and expectations emerge. This change will require reassessment of the demands on markets and how they function, and what products and services are exchanged. New economics will require re-evaluation of how each system element is regulated and how the wider electricity system is planned. Policymakers, consumers, regulators, and utilities of all kinds must reorient themselves toward a smarter, more dynamic electricity system. The role of each is not fixed; future roles will depend on the starting point and the evolution of its different elements. We can expect changes, and thus also stakeholder roles, to vary from country to country and, at the smaller scale, within countries. Different parts of distribution networks may have different needs, depending on the demands placed on them, with these stemming from local renewable energy potential and deployment, customer choice over new technologies and services, different DNO/DSO operational approaches, and other factors.

One key problem with moving to a smarter grid is the tremendous amount of uncertainty concerning what will be needed for a smarter grid. There are many possible degrees of "smartness" that might become apparent on networks and many possible routes to take in making networks smarter. While the transition to a low-carbon energy system is a major driver underlying the need for enhanced smartness, it remains unclear what specific technologies will be adopted, at what scale, at which locations, and over what timeframe. All these factors will impact the overall economics of both the adopted low-carbon technologies (via learning curves) and of the overall costs of investment to expand and make networks smarter. Infrastructure investment decisions concerning the network will be a major factor impacting decarbonization and the knock-on costs that are passed on to consumers. Further, how these costs are distributed may need to change and decisions as to how they might be allocated in future will impact the welfare of consumers. The different responses of consumers to changing systems, their willingness to accept higher costs and potentially more complex approaches to pricing, and any political influence this might engender, will also be a major determinant in shaping smart adoption.

Consideration of the socio-economic challenges for the smart grid, as with that of other concepts relating to smart grids in this text, is complicated by the potential of the smart grid to include many different and disparate elements. To structure this chapter, we will explore some broad, but pertinent, socio-economic topics. Firstly, in Section 14.2, we ask: "why smart?"; framing the smart grid as a social and economic phenomenon as much as a technical one. Secondly, in Section 14.3, we explore what we see as three of the key socio-economic themes for smart grid development. We look at how public policy in the form of economic regulation of networks and markets is responding to the imperatives for change. As well as regulating for the changing roles of networks, both in terms of investment in infrastructure and possible alternatives, we consider how policy interventions can incentivize "smarter" behavior across the electricity system, from consumers, through markets, to generators. Throughout this chapter, we highlight the crucial and often overlooked role of the consumer-citizen, beyond generation, as adopter or rejecter of new technologies and market instruments from aggregators and other energy service providers, as reducers of demand via programs of demand side management (DSM), and in the role as motivator of policy maker and regulator. Finally, in Section 14.4, we conclude by outlining the most pressing challenges, arguing that in addition to technical innovation, bringing about a smart grid requires unprecedented collaboration between researchers, policymakers, and businesses.

14.2 Why Smart? The Socio-Economic Drivers of Change

This chapter aims to provide an overview of the substantive changes to the historical electricity system, the key challenges these present and how smart grids solve existing problems, and also present new challenges. But before we move on, it is worth briefly outlining the drivers for that change. First, we outline the "four Ds" approach as a useful means of framing and understanding the processes of change in modern energy systems, including the "smartening" of electricity systems. We go on to set out the concrete implications of these imperatives in terms of shifts in the economic foundations of electricity systems.

Change and decision making in energy systems has been considered, for many years, as a series of trade-offs between the three goals of security, social equity, and environmental protection; often seen as mutually incompatible [1]. More recently, however, the pace of change in new technologies and practices, such as smart grids and low-carbon energy technologies, has begun to challenge this "trilemma" framework. Instead, an alternative view, based on four interconnecting, self-reinforcing trends known as the "four Ds," has emerged. Much of the recent change in energy systems has been driven by decarbonization, as a response to the threat of climate change, decentralization of resources and demand, and the emergence of greater digitization of energy systems The latter is conceived as holding huge potential for impacting how consumers interact with energy systems in new ways, as more social and commercial life moves online, yet also has the potential to "exacerbate existing inequalities and risks rather than resolve them" [2]. We can add to the three ds a fourth, with the recognition of greater democratization of the energy system, which allows greater access to energy technology by the wider population, and may also place people at the center of design, planning, and decision making [2]. A possible fifth D—disintermediation—relates to the possibilities of removing intermediaries from the energy supply chain [3].

The commitment in many political spheres to the decarbonization of electricity supply as a response to the threat of climate change has driven change in all areas of the energy system. Decarbonization links rising renewable shares in generation, driven through explicit policy support, greater drives for resource efficiency, and changes in demand use, such as e-mobility and the electrification of heat. The centralized model of electricity generation, transmission, and distribution is challenged simultaneously by the increase in distributed renewable energy capacity and the attendant increase of consumer-owed generation equipment. The physical decentralization of network architecture is mirrored by a greater distribution of decision making within the electricity system. This, itself, is part of a wider recognition that decision making about energy—as a fundamental basis for human prosperity and flourishing, and with socially differentiated costs and benefits—should be, and is perhaps becoming, more participatory or "democratic" [4]. Many of these tendencies are enabled and driven by the rise of digital monitoring, communications, and data analysis technologies and practices, which present new data protection and privacy issues, potentially requiring new or adapted laws and standards [5].

The diverse range of technologies and practices that together constitute the "smart" grid are responses to various aspects of the four Ds outlined above. From a purely economic perspective, however, the smart grid concept can alternatively, to a large degree, be measured in terms of its potential to meet social and environmental goals at lowest cost [6]. While many smart grid deployments are framed as technical solutions to variable, distributed renewable generation, in many cases, more dynamic system operation, better distribution system optimization, and improved demand responsivity can reduce the capital and operational costs of achieving a reliable, resilient electricity system. As recently as 2016, overall progress on smart grid was summed up as slow and far from completion, essentially not yet at the point of international competition, but as "an unbordered community of similar aspirations and shared lessons" [7]. The same authors note the complexity of what is to be achieved, the need for a multiactor approach, and the unpredictability of what a smart grid might look like when it emerges.

Implementing a smart grid, like any socio-technical transition, is as much a social endeavor as it is a technical one. It impacts and requires active change and engagement by all electricity system participants, from utilities and regulators to consumers and citizens. To a large extent, the social aspects of smart grids are seen in terms of "gaining acceptance" from key groups, most notably the users or consumers [8, 9]. Resistance to smart grids on the basis of concerns about privacy and data-security are a particular focus [6, 10]. Economic issues, meanwhile, tend to focus on costs and regulation [11].

One great hope for the emergence of a smarter grid is growth in actions on the customer side of the meter. While DSM and demand side response have been around for some time, many assessments of the future of smart grids feature more advanced forms of demand side action being applied at a greater scale, from commercial applications down to individual households. There are various possible elements, many of which will have implications for redistribution of costs relating to consumption, times of consumption, the role of "prosumers," and the potential for new kinds of energy providers offering new services and tariffs.

The conceptual shift from passive *consumer* to active *prosumer* [12, 13], enabled by digital technology and likely to increasingly require a reshaped regulatory framework, is assumed to be a prerequisite of many conceptions of the smart grid, but achieving full interaction requires an "ideal type" of energy citizen, which may not emerge [4]. Moving to wide-scale consumer buy-in is challenging for smart grids in two ways. Firstly, it reflects the requirement for a wide range of collective decisions needed to reshape the regulatory, financial, economic, and institutional energy context [14]. Secondly it challenges the notion that the individual's role in smart grids, as a technically defined phenomenon, is simply to reject or accept them, rather than playing a full role in defining and legitimizing the objectives and means by which electricity systems evolve.

14.3 Key Socio-Economic Themes for the Smart Grid

While the fundamental functions of electricity systems—generation, transmission, distribution, and supply—may be similar, different territories have evolved different structures governing the responsibilities and routes to earning revenues for the different actors that perform these functions. Key variations include the degree of vertical integration, privately or state-owned monopolies, DNO versus DSO approaches to network ownership and management, and various others [15]. Since national and subnational goals may vary, the body of regulation which governs markets and natural monopolies will also tend to vary, with implications for all stakeholders, for shaping historical development and in the specifics of the framework in which new initiatives must develop. One key commonality is that systems were usually designed based around centralized generation, with unidirectional flows of power and no call to consider the concept of the "prosumer," integration of large volumes of distributed generation, or other needs, such as rewarding innovative approaches to more complex network management. The move to privatization of national power systems saw much re-regulation, but has often failed to deal with these new and emergent issues [15a]. Fundamentally, the historical systems fail to offer enough flexibility, either in approaches to network operation, or in creating the conditions for new services to be offered to power or ancillary markets. Consequently, action by regulators and policymakers is required to enable smart grid systems. In this section, we look first at the changing face of policy and regulation from a smart-grid perspective and secondly we ask what this change means for the engagement of consumers and citizens by decision makers.

14.3.1 Costs, Prices, and Regulation: More of the Same or Innovation in New Directions?

The underlying reshaping of the economics of the electricity business leads to a wide range of implications. Here, we focus on three pertinent policy and regulatory questions: how can changing economics square with the need to sustain existing network infrastructure? How does consumer behavior become more dynamic and what are the implications? And, how can electricity markets adapt?

14.3.1.1 Physical Networks: Who Pays and How?

Despite the changes outlined in this volume, the electricity system of the future most likely requires an—albeit changed—network of wires and other physical infrastructure across which energy and other services can be transported. However, the changing economics inherent in smarter systems reveal difficulties in recouping the revenues needed to sustain adequate networks under traditional models of regulated network charging. The trend toward electricity sector liberalization has had at its core the assignment of systemic costs, such as those of building and maintaining networks, to those actors who incur them. Typically, this means that those who use the network pay regulated "use of system" charges for transmission networks, (TNUoS), distribution networks (DUoS), and balancing (BUoS). TNUoS and DUoS costs are typically passed on to the consumer on a per-unit-of-energy basis, so that those who use more of the energy transported through the networks bear more of the cost of the networks. However, as consumers increasingly generate their own power, the need for them to draw power from the network decreases. Additionally, as more generation connects to distribution rather than transmission grid, the fraction of power needing to move through the transmission grid is reduced.

While this is essentially just a development of the market for the generation and supply elements of the electricity sector, the effect on the network is to reduce the number of paying customers, and the revenues enabled by regulated charging regimes. While there may be some reduced need for network investment stemming from this reduction in usage, it is likely that the upshot will be higher costs per-unit of energy that is accessed via the network. This essentially means that remaining consumers will need to pay more for electricity sourced through distribution and transmission, or that alternative routes to financing grid infrastructure will need to be developed [16–18]. In extremis, it has been suggested that increases in prices faced by a diminishing number of network users creates growing incentives for consumers to reduce or remove their dependence on network supplied power, in a positive feedback loop resulting in the collapse of utility companies, sometimes called a "utility death spiral" [16]. Even if the dreaded "death spiral" does not materialize, network actors and consumer groups will most likely experience real economic consequences [19, 20]. Pollitt [17] highlights various problems that, while severe, are somewhat short of a death spiral. These include the potential to negatively impact on poorer consumers who are unable to access their own generation and who are then impacted by a greater share of network costs.

To understand the challenges posed by shifting network economics, and how updated regulation can mitigate them, an emerging literature considers alternative methods by which network costs can be recouped by investors, as the level of distributed and intermittent generation on a system reaches a higher level [18, 21, 22]. Faerber et al. [18] set out the possible alternative components of charging: fixed-cost, peak pricing, and capacity charging, as well as the potential to move costs to be considered in general taxation; and note the potential for combining these elements. Pollitt highlights the difficulty of balancing support for distributed energy, which creates an economic incentive to make further installations against the possible social disbenefits, noting that the point at which social and economic marginal value becomes negative is generally not simultaneous, and the difficulty of making it so. Essentially, Pollitt [17] argues that support for the physical network closely resembles a fixed systemic cost, which, ultimately, society must bear, potentially through taxation, if there is to be an electricity network. How to do so fairly and effectively remains unsolved.

In the next subsection, we consider one of the key elements of smarter electricity networks: more responsive and flexible consumption. "Dynamic" or real-time pricing is often proposed as a step toward shifting the incentives faced by consumers in order to improve system efficiency. Its effectiveness, however, depends on "shiftability" of domestic level demand, with significant technical, social, and behavioral implications.

14.3.1.2 Dynamic or Real Time Pricing

Many commentators have suggested that dynamic pricing or Real Time Use of System (RTUoS) pricing may be an essential element of the future operation of the Smart Grid [18, 23, 24], with one source suggesting it could enable up to €53bn of savings in the EU alone [23]. Essentially, this would mean that prices available to consumers might vary in real time to match variations in the market price. While many commentators seem to regard the adoption of dynamic pricing as inevitable, there are substantive social, socio-economic, and technical challenges to address before this can happen. These are laid out below.

Realizing the potential for wide-scale adoption of RTUoS is dependent on many factors. First, it requires smart meters that are capable of capturing and transmitting the required high volumes of consumer data and for appropriate data to be fed back to them. While many EU countries are in the process of national programs to roll out smart meters, it does not follow that the meters included in the first wave will be sufficiently "smart" or able to cope with all that is required for useful RTUoS, potentially necessitating a second generation of meters with enhanced capability. Since large amounts of data will be required—data privacy concerns notwithstanding (see Section 14.3.2.2)—a communications system that can handle both the volume of data required, and a data hub that can deal with allocating both incoming and outgoing information in a limited timeframe, will be essential. National regulatory systems will need to allow for new or existing stakeholders to bring new services to market and allow data access to those parties concerned with transactions. It seems likely some of the required changes will have to happen simultaneously to enable a new product or service to come to market.

A willingness by consumers to tolerate large volumes of data about their energy consumption entering the public domain will also be required. Concerns have already come to the fore about this in some territories, since it is fairly easy to reveal quite a lot about domestic activity from a smart meter output of a reading per minute [5] or to figure out when a property is empty (see Section 14.3.2.2 for more on data privacy).

Finally, adoption of an RTUoS will require a willingness on the part of consumers to engage with the new form of market. This is discussed further in Section 14.3.2.1.

Much of the potential for systemic movement, from a demand-led to a supply-led electricity system, is premised on the potential for shifting consumer demand from times of peak demand and, often, with the expectation for the future that this will also be to times of high or excess supply (to avoid constraining wind or solar for example). While this is technically already possible, to some extent, it can be expected that further adoption of various low-carbon technologies will both enable greater potential for shifting and also create more need for it, as part of balancing a greater volume of intermittent generation and of integrating larger and more volatile variations in demand from technologies such as heat pumps and electric vehicles (EV). However, many consumers may have only a limited potential or willingness to shift. Limitations may apply to some industry sectors, for example where operations are 24 h per day, commercial consumers may have little potential for operation outside the standard working day, or other reasons which limit their times of operation. Many domestic consumers have little scope to defer their consumption to other times. A UK study suggested that only 7%–10% of the average domestic consumer's demand could be shifted, depending on demographic factors [25]. The same report does suggest, however, that some consumers were willing to consider shifting their demand patterns as part of a financially incentivized program.

Overall domestic potential for shifting periods of consumption may grow with the expansion of new low-carbon technologies. There have been many predictions that electric vehicle use will increase substantially at the global level [26]. This is likely to raise average domestic electricity consumption,[1] but with the potential for vehicle charging to be matched much more to periods of excess

1. Effectively shifting it from petroleum products in most cases, but not relevant to electrical shifting.

supply and lower prices than typical household demand. Clearly this would be in the interests of EV owners, but several publications have suggested that this could also provide substantial system benefits. Essentially, EVs provide a challenge to network management but also offer a potential solution [27]. Considerable technical work has considered routes to integrating adoption of large volumes of EVs; however, Sovacool et al. [28] point out that much of the focus in EV integration has considered these technical implications with much less focus on consumer response. There is also a substantial literature that posits that electrification offers the most likely route to decarbonizing the heat sector in many locations. While many countries make substantive use of waste heat from generation to displace fossil fuel-based heating, the primary technology for electrification would be various models of heat pump. The technology could also double up to provide cooling where this is also in demand.

14.3.1.3 Smart Grids and Electricity Markets

Smart grids enable new modes of generation and consumption, and in so doing, create new products and sources of revenue. An overriding rationale for smart grids is increased economic efficiency, as energy resources are used more effectively, more reliably, at more appropriate times, and often closer to where they are produced [29]. Together, these factors have the potential to require and enable changes in the way future electricity markets operate and are designed. Among these effects is the creation of real-time consumer pricing, discussed above. Since consumers cannot currently see the prices at the time of consumption, the potential to increase system efficiency is expected to result from consumer decision making about how much electricity to consume and when, on the basis of prices. While consumers can be observed to respond to price fluctuations, the change in consumption is not linear, with larger increases to pricing resulting in marginally smaller changes than more modest increases in price [30]. At the same time, the steps needed to enable effective real-time pricing (set out above) are by no means straightforward and consumers generally remain untroubled by market price signals. In addition to increased demand responsiveness to prices, new services may emerge, necessitating new business models in which products such as balancing, frequency response, demand aggregation, storage services, and so on are contracted across a disintermediated enabling platform, with a range of such platforms already emerging at the local level [31].

14.3.2 Smart Consumers or Energy Citizens?

As discussed above, creating a smart grid requires substantial changes in how people interact with the energy system. When considering smart grids and how to create them, the human element of the electricity system is often reduced to a discussion of "consumer engagement." Numerous studies have considered consumer engagement with various elements of a future smart grid; for example, considering the impacts of smart meter installation [32, 33], incentives for behavior change [8, 34], and use of enhanced consumer goods [35, 36]. Results have been mixed, with some delivering meaningful change while others are limited in scope or show a drop off in engagement over time. However, even where positive, these considerations at an experimental level may not be sufficient in scope to persuade regulators and policy makers to make the necessary systemic changes, or to persuade utilities or other energy sector actors to make the scale of investment necessary for systemic change [37]. Here we discuss various approaches to promoting consumer involvement in smart grid development, as well as one of the key legal and ethical challenges: data privacy.

14.3.2.1 Engaging Consumers

There are multiple images of the role the domestic consumer might play in the smart grid. Goulden et al. [38] sum these up as varying from "dumb" (essentially, a slightly more advanced home interface and some automated services), through much "smarter" options (the consumer as active manager). The "dumb" option allows for much less scope in terms of improved systemic efficiency and economic benefit. The smarter options allow for greater system flexibility and potential costs savings but will be governed by how consumers react to change, incentives, and the wider narrative around their electrical supply, the need for decarbonization, and any new onus it places on them for action. They suggest that the difference will be between the current consumer and the potential of an "energy citizen," acting rationally in their own economic interest in response to price signals to deliver the broader potential for systemic smartness. Encouraging consumers to adopt such a role presents a significant challenge, however, and there is no established roadmap for doing so. Research into the perceptions of consumers by engineers and other energy sector actors warns against the presumption that consumers, suitably informed or incentivized, will spontaneously and "rationally" adopt prescribed practices or technologies. Throndsen [39] notes the tendency of engineers to favor an idealized rational consumer when modeling technology adoption and usage, assigning particular responses to this definition even where contradictory data exists as to actual consumer behavior. In general, to become adopted successfully, new, smarter technology must be open and iterative and meet the evolving needs of real consumers [40].

There are multiple stages to the consumer engagement process, each with its own challenges. First, consumers may not wish to use or pay for smart meters—the key enabling technology. A smart meter roll-out will be costly for whichever utility, network or agency is made responsible for it, with costs eventually being passed to the consumer, which may mean some resistance to the idea. Meters offer the potential for cost savings, initially through providing information that may stimulate the consumer to reduce consumption (though this may be limited) and by making services cheaper, but eventually through enabling new services to be made available down to the household level. However, the political desire to minimize metering costs may lead to limits on the cost per meter, which may constrain the capabilities of initial installations and cause them to be inadequate for evolving needs. Selling one generation seems likely to be easier than selling a second generation, should the first generation prove unequal to any required tasks. Once smart meters are installed, and should additional services come to market, the individual consumer will have to decide whether to adopt more complex tariffs.

There has been a substantial number of studies in different territories that have asked consumers to use smart meters and other smart energy technologies to try to gauge how consumers might respond. The results are mixed, with reports varying substantially from disinterest, to initial engagement with rapid tailing off, through to long-term reduction in consumption [41–43]. However, even where studies tend to indicate that some degree of consumer buy-in is likely, research with UK stakeholders suggests that neither the transmission nor the distribution network companies are likely to make the level of investment required without a high level of certainty that consumers would engage, suggesting a substantial chicken-and-egg problem, which may limit or slow movement toward greater levels of smart solutions involving consumers [24, 44].

One possible solution to the problem of a shortfall in active consumer engagement is the potential for automated services. If such services could be brought to market, then they would take the onus for action from the consumer and pass it to a third-party company. This would still require new actors to bring new technological solutions to the market, and would require relevant institutions to enable market access, as well as for consumers to take the step across from traditional tariff structures. That short journey could still prove a step too far, however, with consumers, by-and-large, unwilling to cede control of their appliances to automated services or the system operator [30, 45]. Geelen et al. note in [46] that the focus has tended to be on technical and financial issues rather than on consumer engagement. Gangale et al. [47] note that information collection from consumers has tended to be the focus for consumer engagement research. Park et al. [8] suggest a need to provide consumers with more information regarding the smart grid and to make energy data more accessible as a route to engendering willingness to take up new approaches, but note the paucity of work in the area to that point. Ellabban and Abu-Rub [9] and Kowalska-Pyzalska [48] provide an overview of the issues around consumer engagement with smart grids, again noting the need for other stakeholders such as policymakers, utilities, and service providers to offer relevant and digestible information as well as incentives for behavior change, and to ensure these are sufficient to overcome the consumer perceived barriers for engagement.

14.3.2.2 Privacy: Security Versus Data Access

Even where consumers do not engage proactively with smart grid technology, increased visibility of their consumption behavior could still have significant implications for the ability of utilities and third-party companies to apply new management approaches. One key conception in most views of the smart grid is that the evolution of the networks will see the generation and capture of potentially vast amounts of new data. Maximizing access to data will potentially allow new services to be brought to the energy market, the application of new technologies, and should allow for more efficient network management. This will occur within the networks and the rights to use the data will belong to the network operator. It will occur within industry, businesses, and households as smart meters are rolled out, and as other new smart technologies become increasingly common on the demand side of the meter. Access to household energy data raises issues relating to personal security and privacy rights concerning personal activities, including rights of assembly and freedom, and rights not to be observed by the state or state actors [5]. Typical conceptions of the smart grid suggest data will have value to energy companies and thus potentially also to the consumer. Conditions for data access will thus have substantive economic implications, but they are also likely to depend on the extent to which consumers will provide access to their data, and what share of the economic benefits they receive in return, if any.

Little work has so far been carried out as to public willingness to provide access to data about their energy use. Privacy relating to smart meter data has arisen as an issue in a number of European countries [49]. This has lead policymakers to err on the side of caution, as can be seen in the UK's decision to restrict distribution companies' access to consumer data by default and limiting access by other companies. Many argue, however, that the greatest potential systemic value from the data can be realized at precisely the distribution level and that such restrictions will limit the scope of possible system benefits. It should be noted that UK energy sector stakeholders expect the limitations to be loosened over time, though it is not clear as to what factors will drive this, or the rate at which this will happen [37]. Other territories may allow for more relaxed provision from the start, with potential to accelerate the move to smarter systems.

Perceptions of privacy, and the axiomatically related issue of security, will be a major determinant of consumer buy-in to smart grids, and seem likely to influence the willingness of consumers to engage with their potentially changing role. Little work has been published to indicate what will constitute a societally acceptable level of access and what might be required in return for the consumer to allow access. This seems likely to be a key variable between and within populations. It should be noted that there are multiple degrees of access—this is not a binary situation. An increased degree of access has the potential to increase the utility of the data, but this must be carefully balanced against the ethical and legal rights of consumers to privacy and autonomy over their personal data. McKenna et al. [50] assess the systemic benefit for variations in data access in relation to the different smart grid applications that might be enabled. There remains substantial potential for assessing appetite for data access among the public (with different national publics reacting differently), the value of different levels of access, and whether this might be monetized, along with development of routes that allow some of the value to go to customers, thus potentially influencing opinion relating to access versus privacy. It seems likely that there will be a continuum of levels of acceptability among the public relating to personal opinions about privacy, which may impact uptake, raising as-yet-unanswered questions about whether provision of personal data could or should become a condition for access to the resultant potential for lower tariffs.

14.4 Conclusion

The scope for smarter grids is strongly influenced by the current regulatory and physical architecture of the sector. A meaningfully smarter grid will require changes to the former to allow for new services to come to market and be monetized, and to allow network actors to develop, deploy, and be rewarded for innovative behaviors. Should this happen, some of the options it may enable include reshaping of the physical infrastructure of the networks, offering alternatives to new wire-based infrastructure in enhancing the flexibility of the network. Traditional approaches will still be used to some extent but, increasingly, the option for a smarter approach will offer cost savings, where the system allows for it. Creating the right incentives for network operators, and the right conditions to allow other actors to take a role in providing new approaches to grid development and management, will determine the scope of future network operation. Finally, consumers will need to be persuaded to buy into the new approach to how they interact with their energy purchasing. This final point leads to what is one of the biggest issues with any consideration of the socio-economics of the smart grid: to what extent will consumers be willing to take an active role?

The role of the consumer will be central to adoption of many of the demand side and market-led initiatives. Building a realistic picture of what consumers will and will not adopt, and in what numbers, is essential to uptake of the products and services with the potential to service smarter functionality of markets and networks. This better understanding of the consumer will need to account for the usefulness of financial incentives and the limits of same. The range of possible modes of consumer engagement is broad, from little change, with consumers remaining as largely passive price-takers at one extreme, through to entrepreneurial adoption of innovative technologies and practices at the other [36]. These, in turn, will be dependent on other factors, such as innovation in technology, regulation, and market provision. Each has the potential to block or deter change. However, even if they come to market, they may be rejected by consumers who favor the status quo, or other options they have already adopted. The body of literature suggests that incentives for behavior change has some impact on uptake, but notes that this can vary considerably between publics and that other factors will also have a role.

One means to driving consumers to match their consumption to supply is through new tariff incentives, replacing the current block tariffs, which tend to dominate in most countries, with tariffs more reflective of market conditions. This dynamic pricing, offering cheaper energy at times of surplus and high pricing at times of scarcity, would essentially place the onus for selecting more or less expensive rates much closer to the consumer, or their agent. Having greater "shiftability" in consumption patterns would potentially allow consumers to access a much lower rate per kWh. This may suit the engaged consumer with scope for flexibility, but others may struggle to move demand appropriately, and pay the price through above average tariffs. This may impact more vulnerable consumers more heavily, for example, those who need to consume energy (for example to heat a space) throughout the day or those unable to plan their energy consumption patterns. The effect may be to push some consumers into—or more deeply into—fuel poverty. Care will be needed to protect vulnerable consumers from being taxed for not taking part in initiatives that are beyond their means. This is also an energy justice issue, since low-carbon technologies, such as domestic solar PV units, EV, and heat pumps have been, or remain, subsidized from public funds to drive innovation, price reduction, and societal decarbonization. Subsidies are met from the public purse, but access is often effectively limited to those with the capital to invest in an initial purchase; costs are then passed on in energy costs or via wider taxation. This is itself a justice issue, but it may be exacerbated if those without access to the new energy technologies are also unable to access preferential RTUoS tariffs. In addition to the uncertainty about appetite and ability to change consumption behaviors, a number of ethical and legal issues surround the importance of data analysis to the realization on the smart-system's benefits. Who owns consumer data, who benefits from its application, and how can consumers' rights to privacy and autonomy be protected, are all questions still to be determined by further multidisciplinary work.

However, despite the challenges laid out in this chapter, there is a general expectation of rising costs in the energy sector as fossil fuel becomes scarcer and is pushed out by decarbonization, infrastructure is updated, and—should environmental externalities become increasingly internalized—as generation becomes more variable and network management becomes more complex. Smart grids offer the possibility of lower overall costs, but within the context of overall price rises, which may impact consumer trust in energy companies and make the consumer less likely to wish to engage.

The move to a smarter grid may send more complex and more rapidly shifting price signals to the consumer, offering rewards to those who are able and willing to shift their demand, and punishing those who cannot or will not, as they require the consumer to bear costs which require knowledge, time, or resources that they cannot afford. This change may be justified by the economics of liberalization, but may be politically unappetizing in some territories. There is a clear need to identify and protect more vulnerable consumers, but resolving the quandary of who should meet which costs will not be a straightforward equation.

The systemic changes required to engender smarter grids will be extensive. They will require concerted action as part of a multistep, long-term plan. The policymakers and regulators who devise this plan will need to give proper consideration to the economics of the new technologies, to overcoming the incumbency of current market actors, to be flexible in both creating conditions for innovation and in checking the impacts on the consumer, and will need to vest forward movement, not just in technology, but in meeting the needs of the consumer.

References

[1] World Energy Council. World Energy Trilemma, Defining measures to Accelerate the energy transition, in: World Energy Council Report, 2016 2016, pp. 1–113.

[2] J. Viitanen, R. Kingston, Smart cities and Green growth: outsourcing democratic and environmental resilience to the global technology sector. Environ Plan A 46 (2014) 803–819, https://doi.org/10.1068/a46242.

[3] C. Copeland, D. Brown, D-Day for UK Energy Policy – BEIS hosts conference on UK energy policy, (2017)http://blogs.sussex.ac.uk/sussexenergygroup/2017/10/26/beis-uk-energy-policy-plan/. Accessed 14 August 2018.

[4] K. Szulecki, Conceptualizing energy democracy. Environ. Polit. 27 (2018) 21–41, https://doi.org/10.1080/09644016.2017.1387294.

[5] R. Herold, C. Hertzog, Data Privacy for the Smart Grid. Auerbach Publications, 2015 https://doi.org/10.1201/b18005.

[6] D. Xenias, C.J. Axon, L. Whitmarsh, P.M. Connor, N. Balta-Ozkan, A. Spence, UK smart grid development: an expert assessment of the benefits, pitfalls and functions. Renew. Energy 81 (2015) 89–102, https://doi.org/10.1016/j.renene.2015.03.016.

[7] M.L. Tuballa, M.L. Abundo, A review of the development of smart grid technologies. Renew. Sust. Energ. Rev. 59 (2016) 710–725, https://doi.org/10.1016/J.RSER.2016.01.011.

[8] C.-K. Park, H.-J. Kim, Y.-S. Kim, A study of factors enhancing smart grid consumer engagement. Energy Policy 72 (2014) 211–218, https://doi.org/10.1016/J.ENPOL.2014.03.017.

[9] O. Ellabban, H. Abu-Rub, Smart grid customers' acceptance and engagement: an overview. Renew. Sust. Energ. Rev. 65 (2016) 1285–1298, https://doi.org/10.1016/J.RSER.2016.06.021.

[10] B.J. Murrill, E.C. Liu, R.M. Thompson II, Smart Meter Data: Privacy and Cybersecurity, Congressional Research Service, Washington D.C., 2012.

[11] S. Bigerna, C.A. Bollino, S. Micheli, Socio-economic acceptability for smart grid development—a comprehensive review. J. Clean. Prod. 131 (2016) 399–409, https://doi.org/10.1016/J.JCLEPRO.2016.05.010.

[12] A. Toffler, The Third Wave, Bantam Books, 1980.

[13] A.B. Lovins, Soft Energy Paths: Toward a Durable Peace, Harper & Row, 1979.

[14] M.J. Burke, J.C. Stephens, Energy democracy: goals and policy instruments for sociotechnical transitions. Energy Res. Soc. Sci. 33 (2017) 35–48, https://doi.org/10.1016/j.erss.2017.09.024.
[15] C. Harris, Electricity Markets: Pricing, Structures and Economics, John Wiley & Sons, 2006.
[15a] C. Mitchell, Neutral regulation—the vital ingredient for a sustainable energy future, Energy Environ. 11 (2000) 377–389.
[16] I. Pérez-Arriaga, A. Bharatkumar, A Framework for Redesigning Distribution Network Use of System Charges Under High Penetration of Distributed Energy Resources: New Principles for New Problems, MIT Center for Energy and Environmental Policy Research, Cambridge, Massachusetts, 2014.
[17] M.G. Pollitt, Electricity network charging for flexibility, in: Cambridge Working Paper Economics 1656, 2016.
[18] L.A. Faerber, N. Balta-Ozkan, P.M. Connor, Innovative network pricing to support the transition to a smart grid in a low-carbon economy. Energy Policy 116 (2018) 210–219, https://doi.org/10.1016/J.ENPOL.2018.02.010.
[19] K.W. Costello, R.C. Hemphill, Electric utilities' "death spiral": hyperbole or reality?. Electr. J. 27 (2014) 7–26, https://doi.org/10.1016/J.TEJ.2014.09.011.
[20] N.D. Laws, B.P. Epps, S.O. Peterson, M.S. Laser, G.K. Wanjiru, On the utility death spiral and the impact of utility rate structures on the adoption of residential solar photovoltaics and energy storage. Appl. Energy 185 (2017) 627–641, https://doi.org/10.1016/J.APENERGY.2016.10.123.
[21] J.D. Jenkins, I.J. Pérez-Arriaga, Improved regulatory approaches for the remuneration of electricity distribution utilities with high penetrations of distributed energy resources, Energy J. 38 (2017) 28–29.
[22] I. Pérez-Arriaga, J.D. Jenkins, C. Batlle, A regulatory framework for an evolving electricity sector: highlights of the MIT utility of the future study. Econ. Energy Environ. Policy 6 (2017) 71–92, https://doi.org/10.5547/2160-5890.6.1.iper.
[23] A. Faruqui, D. Harris, R. Hledik, Unlocking the €53 billion savings from smart meters in the EU: how increasing the adoption of dynamic tariffs could make or break the EU's smart grid investment. Energy Policy 38 (2010) 6222–6231, https://doi.org/10.1016/J.ENPOL.2010.06.010.
[24] N. Balta-Ozkan, T. Watson, P. Connor, C. Axon, L. Whitmarsh, R. Davidson, et al., Scenarios for the Development of Smart Grids in the UK Synthesis Report, (2014).
[25] J. Ward, S. Darcy, The Household Electricity Demand-Side and Participation in the GB Electricity Markets, (2014).
[26] International Energy Agengy, Global EV Outlook 2018, IEA, Paris, 2018.
[27] F. Mwasilu, J.J. Justo, E.-K. Kim, T.D. Do, J.-W. Jung, Electric vehicles and smart grid interaction: a review on vehicle to grid and renewable energy sources integration. Renew. Sust. Energ. Rev. 34 (2014) 501–516, https://doi.org/10.1016/J.RSER.2014.03.031.
[28] B.K. Sovacool, J. Axsen, W. Kempton, The future promise of vehicle-to-grid (V2G) integration: a sociotechnical review and research agenda. Annu. Rev. Environ. Resour. 42 (2017) 377–406, https://doi.org/10.1146/annurev-environ-030117-020220.
[29] G.B. SmartGrid, Smart Grid: A Race Worth Winning? A Report on the Economic Benefits of Smart Grid, SmartGrid GB, London, 2012.
[30] R. Green, R. Webb, Impact on Electricity Markets. Smart Grid Handbook. Wiley, Chichester, 2016. https://doi.org/10.1002/9781118755471.sgd025.
[31] DeX, How The Technology Works – deX, (2018) https://dex.energy/why-join-dex/how-the-technology-works/. Accessed 14 August 2018.
[32] D.J. Hess, Smart meters and public acceptance: comparative analysis and governance implications. Health Risk Soc. 16 (2014) 243–258, https://doi.org/10.1080/13698575.2014.911821.
[33] J. Britton, Smart Meter Data and Public Interest Issues-the Sub-National Perspective Discussion Paper 2, (2016).

[34] S. Owens, L. Driffill, How to change attitudes and behaviours in the context of energy. Energy Policy 36 (2008) 4412–4418, https://doi.org/10.1016/J.ENPOL.2008.09.031.

[35] E. Costanza, J.E. Fischer, J.A. Colley, T. Rodden, S.D. Ramchurn, N.R. Jennings, Doing the laundry with agents: a field trial of a future smart energy system in the home. in: Proceedings of the 32nd annual ACM conference on Human factors in computing systems – CHI 14, 2014, pp. 813–822, https://doi.org/10.1145/2556288.2557167.

[36] Y. Strengers, Smart Energy Technologies in Everyday Life : Smart Utopia? Palgrave Macmillan, Basingstoke, 2013.

[37] P.M. Connor, C.J. Axon, D. Xenias, N. Balta-Ozkan, Sources of risk and uncertainty in UK smart grid deployment: an expert stakeholder analysis. Energy 161 (2018) 1–9, https://doi.org/10.1016/J.ENERGY.2018.07.115.

[38] M. Goulden, B. Bedwell, S. Rennick-Egglestone, T. Rodden, A. Spence, Smart grids, smart users? The role of the user in demand side management. Energy Res. Soc. Sci. 2 (2014) 21–29, https://doi.org/10.1016/J.ERSS.2014.04.008.

[39] W. Throndsen, What do experts talk about when they talk about users? Expectations and imagined users in the smart grid. Energy Efficiency 10 (2017) 283–297, https://doi.org/10.1007/s12053-016-9456-5.

[40] D. Abi Ghanem, S. Mander, Designing consumer engagement with the smart grids of the future: bringing active demand technology to everyday life. Technol. Anal. Strateg. Manag. 26 (2014) 1163–1175, https://doi.org/10.1080/09537325.2014.974531.

[41] H. Allcott, Rethinking real-time electricity pricing. Resour. Energy Econ. 33 (2011) 820–842, https://doi.org/10.1016/J.RESENEECO.2011.06.003.

[42] A. Faruqui, S. Sergici, Household response to dynamic pricing of electricity: a survey of 15 experiments. J. Regul. Econ. 38 (2010) 193–225, https://doi.org/10.1007/s11149-010-9127-y.

[43] A. Faruqui, S. Sergici, A. Sharif, The impact of informational feedback on energy consumption—a survey of the experimental evidence. Energy 35 (2010) 1598–1608, https://doi.org/10.1016/J.ENERGY.2009.07.042.

[44] P.M. Connor, P.E. Baker, D. Xenias, N. Balta-Ozkan, C.J. Axon, L. Cipcigan, Policy and regulation for smart grids in the United Kingdom. Renew. Sust. Energ. Rev. 40 (2014) 269–286, https://doi.org/10.1016/j.rser.2014.07.065.

[45] A. Spence, C. Demski, C. Butler, K. Parkhill, N. Pidgeon, Public perceptions of demand-side management and a smarter energy future. Nat. Clim. Chang. 5 (2015) 550–554, https://doi.org/10.1038/nclimate2610.

[46] D. Geelen, A. Reinders, D. Keyson, Empowering the end-user in smart grids: recommendations for the design of products and services. Energy Policy 61 (2013) 151–161, https://doi.org/10.1016/J.ENPOL.2013.05.107.

[47] F. Gangale, A. Mengolini, I. Onyeji, Consumer engagement: an insight from smart grid projects in Europe. Energy Policy 60 (2013) 621–628, https://doi.org/10.1016/J.ENPOL.2013.05.031.

[48] A. Kowalska-Pyzalska, What makes consumers adopt to innovative energy services in the energy market? A review of incentives and barriers. Renew. Sust. Energ. Rev. 82 (2018) 3570–3581, https://doi.org/10.1016/J.RSER.2017.10.103.

[49] S. Zhou, M.A. Brown, Smart meter deployment in Europe: a comparative case study on the impacts of national policy schemes. J. Clean. Prod. 144 (2017) 22–32, https://doi.org/10.1016/J.JCLEPRO.2016.12.031.

[50] E. McKenna, I. Richardson, M. Thomson, Smart meter data: balancing consumer privacy concerns with legitimate applications. Energy Policy 41 (2012) 807–814, https://doi.org/10.1016/J.ENPOL.2011.11.049.

Index

Note: Page numbers followed by *f* indicate figures, *t* indicate tables, and *b* indicate boxes.

I

K

L

M

N

O

P

V

W

Kolibri 360 GmbH
Ferdinand-Jühlke-Straße 7
99095 Erfurt, Deutschland
produktsicherheit@kolibri360.de